# Fundamentals of Applied Probability and Random Processes

# Fundamentals of Applied Probability and Random Processes

2nd Edition

Oliver C. Ibe
University of Massachusetts, Lowell, Massachusetts

AMSTERDAM • BOSTON • HEIDELBERG • LONDON
NEW YORK • OXFORD • PARIS • SAN DIEGO
SAN FRANCISCO • SINGAPORE • SYDNEY • TOKYO
Academic Press is an imprint of Elsevier

Academic Press is an imprint of Elsevier
525 B Street, Suite 1900, San Diego, CA 92101-4495, USA
225 Wyman Street, Waltham, MA 02451, USA

Second edition 2014


**Library of Congress Cataloging-in-Publication Data**

Ibe, Oliver C. (Oliver Chukwudi), 1947-
Fundamentals of applied probability and random processes / Oliver Ibe. – Second edition.
pages cm
Includes bibliographical references and index.
ISBN 978-0-12-800852-2 (alk. paper)
1. Probabilities. I. Title.
QA273.I24 2014
519.2–dc23

2014005103

**British Library Cataloguing in Publication Data**

A catalogue record for this book is available from the British Library

For information on all Academic Press publications
visit our web site at store.elsevier.com

ISBN: **978-0-12-800852-2**

# Contents

# Acknowledgment

The first edition of this book was well received by many students and professors. It had both Indian and Korean editions and received favorable reviews that include the following: "The book is very clear, with many nice examples and with mathematical proofs whenever necessary. Author did a good job! The book is one of the best ones for self-study." Another comment is the following:

> "This book is written for professional engineers and students who want to do self-study on probability and random processes. I have spent money and time in reading several books on this topic and almost all of them are not for self-study. They lack real world examples and have end-of-chapter problems that are boring with mathematical proofs. In this book the concepts are explained by taking real world problems and describing clearly how to model them. Topics are well-balanced; the depth of material and explanations are very good, the problem sets are intuitive and they make you to think, which makes this book unique and suitable for self-study. Topics which are required for both grad and undergrad courses are covered in this book very well. If you are reading other books and are stuck when applying a concept to a particular problem, you should consider this book. Problem solving is the only way to get familiar with different probability models, and this book fulfills that by taking problems in the real world examples as opposed to some theoretical proofs."

These are encouraging reviews of a book that evolved from a course I teach at the University of Massachusetts, Lowell. I am very grateful to those who wrote these wonderful unsolicited anonymous reviews in Amazon.com. Their observations on the structure of the book are precisely what I had in mind in writing the book.

I want to extend my sincere gratitude to my editor, Paula Callaghan of Elsevier, who was instrumental in the production of this book. I thank her for the effort she made to get the petition for the second edition approved. I also want to thank Jessica Vaughan, the Editorial Project Manager, for her ensuring timely production of the book.

So many students have used the first edition of this book at UMass Lowell and have provided useful information that led to more clarity in the presentation of the material in the book. They are too many to name individually, so I say "thank you" to all of them as a group.

Finally, I would like to thank my wife, Christina, for patiently bearing with me while the book was being revised. I would also like to appreciate the encouragement of our children Chidinma, Ogechi, Amanze and Ugonna. As always, they are a source of joy to me and my wife.

# Preface to the Second Edition

Many systems encountered in science and engineering require an understanding of probability concepts because they possess random variations. These include messages arriving at a switchboard; customers arriving at a restaurant, movie theatre or a bank; component failure in a system; traffic arrival at a junction; and transaction requests arriving at a server.

There are several books on probability and random processes. These books range widely in their coverage and depth. At one extreme are the very rigorous books that present probability from the point of view of measure theory and spend so much time proving exotic theorems. At the other extreme are books that combine probability with statistics without devoting enough time on the applications of probability. In the middle lies a group of books that combine probability and random processes. These books avoid the measure theoretic approach and rather emphasize the axioms upon which the theory is based. This book belongs to this group and is based on the premise that to the engineer, probability is a modeling tool. Therefore, to an engineering student the emphasis on a probability and random processes course should be on the application of probability to the solution of engineering problems. Also, since some of the engineering problems deal with data analysis, the student should also be exposed to some knowledge of statistics. However, it is not necessary for the student to take a separate class on statistics since most of the prerequisites for statistics are covered in a probability course. Thus, this book differs from other books in the sense that it presents two chapters on the essentials of statistics.

The book is designed for juniors and seniors, but can also be used at lower graduate levels. It grew out of the author's fifteen years experience developing and analyzing probabilistic models of systems in the industry as well as teaching an introductory course on probability and random processes for over ten years in two different colleges. The emphasis throughout the book is on the applications of probability, which are demonstrated through several examples that deal with real systems. Since many students learn by "doing," it is suggested that the students solve the exercises at the end of each chapter. Some mathematical knowledge is assumed, especially freshman calculus and algebra.

This second edition of the book differs from the first edition in a few ways. First, the chapters have been slightly rearranged. Specifically, statistics now comes before random processes to enable students understand the basic principles of probability and statistics before studying random processes. Second, Chapter 11 has been split into two chapters: Chapter 8, which deals with descriptive statistics; and Chapter 9, which deals with inferential statistics. Third, the new edition includes more application-oriented examples to enable students to appreciate the application of probability and random processes in science, engineering and management. Finally, after teaching the subject every semester for the past eleven years, I have been able to identify several pain points that hinder student understanding of probability and random processes, and I have introduced several new "smart" methods of solving the problems to help ease the pain.

The book is divided into three parts as follows:

Part 1: Probability and Random Variables, which covers chapters 1 to 7
Part 2: Introduction to Statistics, which covers chapters 8 and 9
Part 3: Basic Random Processes, which covers chapters 10 to 12

A more detailed description of the chapters is as follows. Chapter 1 deals with basic concepts in probability including sample space and events, elementary set theory, conditional probability, independent events, basic combinatorial analysis, and applications of probability.

Chapter 2 discusses random variables including events defined by random variables, discrete random variables, continuous random variables, cumulative distribution function, probability mass function of discrete random variables, and probability distribution function of continuous random variables.

Chapter 3 discusses moments of random variables including the concepts of expectation and variance, higher moments, conditional expectation, and the Chebyshev and Markov inequalities.

Chapter 4 discusses special random variables and their distributions. These include the Bernoulli distribution, binomial distribution, geometric distribution, Pascal distribution, hypergeometric distribution, Poisson distribution, exponential distribution, Erlang distribution, uniform distribution, and normal distribution.

Chapter 5 deals with multiple random variables including the joint cumulative distribution function of bivariate random variables, conditional distributions, covariance, correlation coefficient, functions of multivariate random variables, and multinomial distributions.

Chapter 6 deals with functions of random variables including linear and power functions of one random variable, moments of functions of one random variable, sums of independent random variables, the maximum and minimum of two independent random variables, two functions of two random variables, laws of large numbers, the central limit theorem, and order statistics

Chapter 7 discusses transform methods that are useful in computing moments of random variables. In particular, it discusses the characteristic function, the z-transform of the probability mass functions of discrete random variables and the s-transform of the probability distribution functions of continuous random variables.

Chapter 8 presents an introduction to descriptive statistics and discusses such topics as measures of central tendency, measures of spread, and graphical displays.

Chapter 9 presents an introduction to inferential statistics and discusses such topics as sampling theory, estimation theory, hypothesis testing, and linear regression analysis.

Chapter 10 presents an introduction to random processes. It discusses classification of random processes; characterization of random processes including the autocorrelation function of a random process, autocovariance function, crosscorrelation function and crosscovariance function; stationary random processes; ergodic random processes; and power spectral density.

Chapter 11 discusses linear systems with random inputs. It also discusses the autoregressive moving average process.

Chapter 12 discusses special random processes including the Bernoulli process, Gaussian process, random walk, Poisson process and Markov process.

The author has tried different formats in presenting the different chapters of the book. In one particular semester we were able to go through all the chapters except Chapter 12. However, it was discovered that this put a lot of stress on the students. Thus, in subsequent semesters an attempt was made to cover all the topics in Parts 1 and 2 of the book, and a few selections from Part 3. The instructor can try different formats and adopt the one that works best for him or her.

The beginning of a solved example is indicated by a short line and the end of the solution is also indicated by a short line. This is to separate the continuation of a discussion preceding an example from the example just solved.

# Preface to First Edition

Many systems encountered in science and engineering require an understanding of probability concepts because they possess random variations. These include messages arriving at a switchboard; customers arriving at a restaurant, movie theatre or a bank; component failure in a system; traffic arrival at a junction; and transaction requests arriving at a server.

There are several books on probability and random processes. These books range widely in their coverage and depth. At one extreme are the very rigorous books that present probability from the point of view of measure theory and spend so much time proving exotic theorems. At the other extreme are books that combine probability with statistics without devoting enough time on the applications of probability. In the middle lies a group of books that combine probability and random processes. These books avoid the measure theoretic approach and rather emphasize the axioms upon which the theory is based. This book belongs to this group and is based on the premise that to the engineer, probability is a modeling tool. Therefore, to an engineering student the emphasis on a probability and random processes course should be on the application of probability to the solution of engineering problems. Also, since some of the engineering problems deal with data analysis, the student should also be exposed to some knowledge of statistics. However, it is not necessary for the student to take a separate class on statistics since most of the prerequisites for statistics are covered in a probability course. Thus, this book differs from other books in the sense that it presents a chapter on the essentials of statistics.

The book is designed for juniors and seniors, but can also be used at lower graduate levels. It grew out of the author's fifteen years experience developing and analyzing probabilistic models of systems in the industry as well as teaching an introductory course on probability and random processes for over four years in two different colleges. The emphasis throughout the book is on the applications of probability, which are demonstrated through several examples that deal with real systems. Since many students learn by "doing," it is suggested that the students solve the exercises at the end of each chapter. Some mathematical knowledge is assumed, especially freshman calculus and algebra. The book is divided into three parts as follows:

Part 1: Probability and Random Variables, which covers chapters 1 to 7
Part 2: Basic Random Processes, which covers chapters 8 to 11
Part 3: Introduction to Statistics, which covers chapter 12.

A more detailed description of the chapters is as follows. Chapter 1 deals with basic concepts in probability including sample space and events, elementary set theory, conditional probability, independent events, basic combinatorial analysis, and applications of probability.

Chapter 2 discusses random variables including events defined by random variables, discrete random variables, continuous random variables, cumulative distribution function, probability mass function of discrete random variables, and probability distribution function of continuous random variables.

Chapter 3 deals with moments of random variables including the concepts of expectation and variance, higher moments, conditional expectation, and the Chebyshev and Markov inequalities.

Chapter 4 discusses special random variables and their distributions. These include the Bernoulli distribution, binomial distribution, geometric distribution, Pascal distribution, hypergeometric distribution, Poisson distribution, exponential distribution, Erlang distribution, uniform distribution, and normal distribution.

Chapter 5 deals with multiple random variables including the joint cumulative distribution function of bivariate random variables, conditional distributions, covariance, correlation coefficient, many random variables, and multinomial distribution.

Chapter 6 deals with functions of random variables including linear and power functions of one random variable, moments of functions of one random variable, sums of independent random variables, the maximum and minimum of two independent random variables, two functions of two random variables, laws of large numbers, the central limit theorem, and order statistics

Chapter 7 discusses transform methods that are useful in computing moments of random variables. In particular, it discusses the characteristic function, the z-transform of the probability mass functions of discrete random variables and the s-transform of the probability distribution functions of continuous random variables.

Chapter 8 presents an introduction to random processes. It discusses classification of random processes; characterization of random processes including the autocorrelation function of a random process, autocovariance function, crosscorrelation function and crosscovariance function; stationary random processes; ergodic random processes; and power spectral density.

Chapter 9 discusses linear systems with random inputs.

Chapter 10 discusses such specialized random processes as the Gaussian process, random walk, and Poisson process and Markov process

Chapter 11 presents an introduction to statistics and discusses such topics as sampling theory, estimation theory, hypothesis testing, and linear regression.

The author has tried different formats in presenting the different chapters of the book. In one particular semester we were able to go through all the chapters. However, it was discovered that this put a lot of stress on the students. Thus, in subsequent semesters an attempt was made to cover all the topics in Part 1 of the book, chapters 8 and 9, and a few selections from the other chapters. The instructor can try different formats and adopt the one that works best for him or her.

The symbol Δ is used to indicate the end of the solution to an example. This is to separate the continuation of a discussion preceding an example from the example just solved.

CHAPTER 1

# Basic Probability Concepts

## 1.1 INTRODUCTION

Probability deals with unpredictability and randomness, and probability theory is the branch of mathematics that is concerned with the study of random phenomena. A random phenomenon is one that, under repeated observation, yields different outcomes that are not deterministically predictable. However, these outcomes obey certain conditions of statistical regularity whereby the relative frequency of occurrence of the possible outcomes is approximately predictable. Examples of these random phenomena include the number of electronic mail (e-mail) messages received by all employees of a company in one day, the number of phone calls arriving at the university's switchboard over a given period, the number of components of a system that fail within a given interval, and the number of A's that a student can receive in one academic year.

According to the preceding definition, the fundamental issue in random phenomena is the idea of a repeated experiment with a set of possible outcomes or events. Associated with each of these events is a real number called the probability of the event that is related to the frequency of occurrence of the event in a long sequence of repeated trials of the experiment. In this way it becomes obvious that the probability of an event is a value that lies between zero and one, and the sum of the probabilities of the events for a particular experiment should sum to one.

This chapter begins with events associated with a random experiment. Then it provides different definitions of probability and considers elementary set theory and algebra of sets. Also, it discusses basic concepts in combinatorial analysis that will be used in many of the later chapters. Finally, it discusses how probability is used to compute the reliability of different component configurations in a system.



Fundamentals of Applied Probability and Random Processes. http://dx.doi.org/10.1016/B978-0-12-800852-2.00001-8

## 1.2 SAMPLE SPACE AND EVENTS

The concepts of *experiments* and *events* are very important in the study of probability. In probability, an experiment is any process of trial and observation. An experiment whose outcome is uncertain before it is performed is called a *random* experiment. When we perform a random experiment, the collection of possible elementary outcomes is called the *sample space* of the experiment, which is usually denoted by $\Omega$. We define these outcomes as elementary outcomes because exactly one of the outcomes occurs when the experiment is performed. The elementary outcomes of an experiment are called the *sample points* of the sample space and are denoted by $w_i$, $i = 1, 2, \ldots$. If there are $n$ possible outcomes of an experiment, then the sample space is $\Omega = \{w_1, w_2, \ldots, w_n\}$.

An *event* is the occurrence of either a prescribed outcome or any one of a number of possible outcomes of an experiment. Thus, an event is a subset of the sample space. For example, if we toss a die, any number from 1 to 6 may appear. Therefore, in this experiment the sample space is defined by

$$\Omega = \{1, 2, 3, 4, 5, 6\} \tag{1.1}$$

The event "the outcome of the toss of a die is an even number" is the subset of $\Omega$ and is defined by

$$E = \{2, 4, 6\} \tag{1.2}$$

For a second example, consider a coin tossing experiment in which each toss can result in either a head ($H$) or tail ($T$). If we toss a coin three times and let the triplet $xyz$ denote the outcome "$x$ on first toss, $y$ on second toss and $z$ on third toss," then the sample space of the experiment is

$$\Omega = \{HHH, HHT, HTH, HTT, THH, THT, TTH, TTT\} \tag{1.3}$$

The event "one head and two tails" is the subset of $\Omega$ and is defined by

$$E = \{HTT, THT, TTH\} \tag{1.4}$$

Other examples of events are as follows:

- In a single coin toss experiment with sample space $\Omega = \{H, T\}$, the event $E = \{H\}$ is the event that a head appears on the toss and $E = \{T\}$ is the event that a tail appears on the toss.
- If we toss a coin twice and let $xy$ denote the outcome "$x$ on first toss and $y$ on second toss," where $x$ is head or tail and $y$ is head or tail, then the

sample space is $\Omega=\{HH, HT, TH, TT\}$. The event $E=\{HT, TT\}$ is the event that a tail appears on second toss.

- If we measure the lifetime of an electronic component, such as a chip, the sample space consists of all nonnegative real numbers. That is,

  $$\Omega=\{x|0\leq x<\infty\}$$

  The event that the lifetime is not more than 7 hours is defined as follows:

  $$E=\{x|0\leq x\leq 7\}$$

- If we toss a die twice and let the pair $(x, y)$ denote the outcome "$x$ on first toss and $y$ on second toss," then the sample space is

  $$\Omega=\begin{Bmatrix} (1,1) & (1,2) & (1,3) & (1,4) & (1,5) & (1,6) \\ (2,1) & (2,2) & (2,3) & (2,4) & (2,5) & (2,6) \\ (3,1) & (3,2) & (3,3) & (3,4) & (3,5) & (3,6) \\ (4,1) & (4,2) & (4,3) & (4,4) & (4,5) & (4,6) \\ (5,1) & (5,2) & (5,3) & (5,4) & (5,5) & (5,6) \\ (6,1) & (6,2) & (6,3) & (6,4) & (6,5) & (6,6) \end{Bmatrix} \tag{1.5}$$

  The event that the sum of the two tosses is 8 is denoted by

  $$E=\{(2,6),(3,5),(4,4),(5,3),(6,2)\}$$

For any two events $A$ and $B$ of a sample space $\Omega$, we can define the following new events:

- $A\cup B$ is event that consists of all sample points that are either in $A$ or in $B$ or in both $A$ and $B$. The event $A\cup B$ is called the *union* of events $A$ and $B$.
- $A\cap B$ is event that consists of all sample points that are in both $A$ and $B$. The event $A\cap B$ is called the *intersection* of events $A$ and $B$. Two events are defined to be *mutually exclusive* if their intersection does not contain a sample point; that is, they have no outcomes in common. Events $A_1, A_2, A_3, \ldots$ are defined to be mutually exclusive if no two of them have any outcomes in common and the events collectively have no outcomes in common.
- $A-B=A\backslash B$ is event that consists of all sample points that are in $A$ but not in $B$. The event $A-B$ (also denoted by $A\backslash B$) is called the *difference* of events $A$ and $B$. Note that $A-B$ is different from $B-A$.

The algebra of unions, intersections and differences of events will be discussed in greater detail when we study set theory later in this chapter.

## 1.3 DEFINITIONS OF PROBABILITY

There are several ways to define probability. In this section we consider three definitions; namely the *axiomatic* definition, the *relative-frequency* definition, and the *classical* definition.

### 1.3.1 Axiomatic Definition

Consider a random experiment whose sample space is $\Omega$. For each event $A$ of $\Omega$ we assume that a number $P[A]$, called the *probability* of event $A$, is defined such that the following hold:

1. **Axiom 1**: $0 \leq P[A] \leq 1$, which means that the probability of $A$ is some number between and including 0 and 1.
2. **Axiom 2**: $P[\Omega] = 1$, which states that with probability 1 the outcome will be a sample point in the sample space.
3. **Axiom 3**: For any set of $n$ mutually exclusive events $A_1, A_2, A_3, \ldots, A_n$ defined on the same sample space,

$$P[A_1 \cup A_2 \cup A_3 \cup \ldots \cup A_n] = P[A_1] + P[A_2] + P[A_3] + \cdots + P[A_n] \tag{1.6}$$

That is, for any set of mutually exclusive events defined on the same space, the probability of at least one of these events occurring is the sum of their respective probabilities.

### 1.3.2 Relative-Frequency Definition

Consider a random experiment that is performed $n$ times. If an event $A$ occurs $n_A$ times, then the probability of event $A$, $P[A]$, is defined as follows:

$$P[A] = \lim_{n \to \infty} \left\{ \frac{n_A}{n} \right\} \tag{1.7}$$

The ratio $n_A/n$ is called the *relative frequency* of event $A$. While the relative-frequency definition of probability is intuitively satisfactory for many practical problems, it has a few limitations. One such limitation is the fact that the experiment may not be repeatable, especially when we are dealing with destructive testing of expensive and/or scarce resources. Also, the limit may not exist.

### 1.3.3 Classical Definition

In the classical definition, the probability $P[A]$ of an event $A$ is the ratio of the number of outcomes $N_A$ of an experiment that are favorable to $A$ to the total number $N$ of possible outcomes of the experiment. That is,

$$P[A] = \frac{N_A}{N} \tag{1.8}$$

This probability is determined *a priori* without actually performing the experiment. For example, in a coin toss experiment, there are two possible outcomes: heads or tails. Thus, $N=2$ and if the coin is fair the probability of the event that the toss comes up heads is 1/2.

## EXAMPLE 1.1

Two fair dice are tossed. Find the probability of each of the following events:

a. The sum of the outcomes of the two dice is equal to 7
b. The sum of the outcomes of the two dice is equal to 7 or 11
c. The outcome of second die is greater than the outcome of the first die
d. Both dice come up with even numbers

**Solution:**
We first define the sample space of the experiment. If we let the pair $(x, y)$ denote the outcome "first die comes up $x$ and second die comes up $y$," then the sample space is given by equation (1.5). The total number of sample points is 36. We evaluate the three probabilities using the classical definition method.

a. Let $A_1$ denote the event that the sum of the outcomes of the two dice is equal to seven. Then $A_1=\{(1,6),(2,5),(3,4),(4,3),(5,2),(6,1)\}$. Since the number of sample points in the event is 6, we have that $P[A_1]=6/36=1/6$.
b. Let $B$ denote the event that the sum of the outcomes of the two dice is either seven or eleven, let $A_1$ be as defined in part (a), and let $A_2$ denote the event that the sum of the outcomes of the two dice is eleven. Then, $A_2=\{(6,5),(5,6)\}$ with 2 sample points. Thus, $P[A_2]=2/36=1/18$. Since $B$ is the union of $A_1$ and $A_2$, which are mutually exclusive events, we obtain

$$P[B]=P[A_1 \cup A_2]=P[A_1]+P[A_2]=\frac{1}{6}+\frac{1}{18}=\frac{2}{9}$$

c. Let $C$ denote the event that the outcome of the second die is greater than the outcome of the first die. Then

$$C=\left\{\begin{matrix}(1,2),(1,3),(1,4),(1,5),(1,6),(2,3),(2,4),(2,5), \\ (2,6),(3,4),(3,5),(3,6),(4,5),(4,6),(5,6)\end{matrix}\right\}$$

with 15 sample points. Thus, $P[C]=15/36=5/12$.
d. Let $D$ denote the event that both dice come up with even numbers. Then

$$D=\{(2,2),(2,4),(2,6),(4,2),(4,4),(4,6),(6,2),(6,4),(6,6)\}$$

with 9 sample points. Thus, $P[D]=9/36=1/4$.

Note that the problem can also be solved by considering a two-dimensional display of the state space, as shown in Figure 1.1. The figure shows the different events just defined. The sample points in event D are spread over the entire sample space. Therefore, the event D is not shown in Figure 1.1.

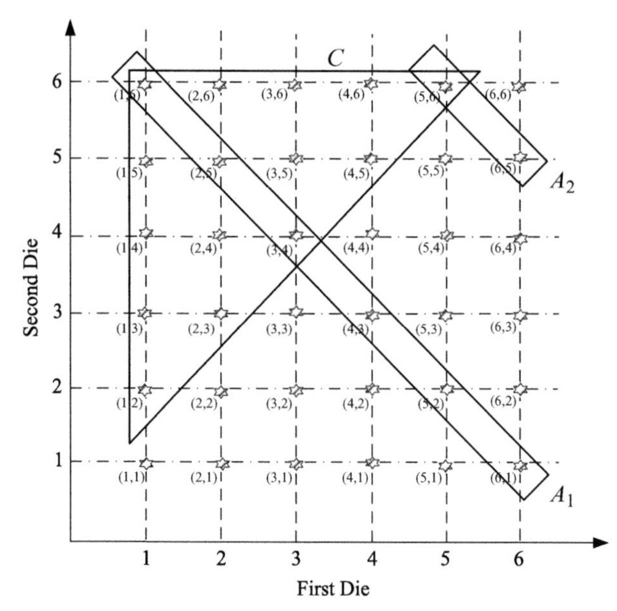


**FIGURE 1.1**
Sample Space for Example 1.1

## 1.4 APPLICATIONS OF PROBABILITY

There are several science and engineering applications of probability. Some of these applications are as follows:

### 1.4.1 Information Theory

Information theory is a branch of mathematics that is concerned with two fundamental questions in communication theory: (a) how we can define and quantify information, and (b) the maximum information that can be sent through a communication channel. It has made fundamental contributions not only in communication systems but also in thermodynamics, computer science and statistical inference. A communication system is modeled by a source of messages, which may be a person or machine producing the message to be transmitted; a channel, which is the medium over which the message is transmitted; and a sink or destination that absorbs the message. While on the channel the message can be corrupted by noise. A model of a communication system is illustrated in Figure 1.2.

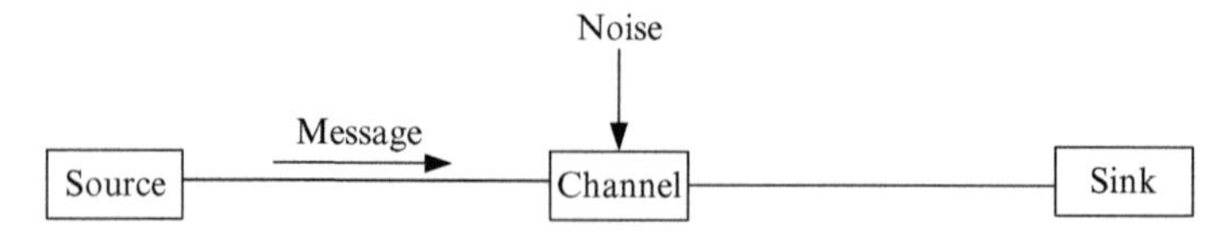


**FIGURE 1.2**
Model of a Communication System

The message generated by the source conveys some information. One of the objectives of information theory is to quantify the information content of messages. This quantitative measure enables communication system designers to provision the channel that can support the message. A good measure of the information content of a message is the probability of occurrence of the message: The higher the probability of occurrence of a message, the less information it conveys; and the smaller the probability of occurrence of a message, the greater its information content. For example, the message "the temperature is 90° F in the northeastern part of the United States in December" has more information content than the message "the temperature is 10° F in the northeastern part of the United States in December" because the second message is more likely to occur than the first.

Thus, information theory uses the probability of occurrence of events to convey information about those events. Specifically, let $P[A]$ denote the probability of occurrence of some event $A$. Then the information content of $A$, $I(A)$, is given by

$$I(A) = \log_2\left(\frac{1}{P[A]}\right) = \log_2(1) - \log_2(P[A]) = 0 - \log_2(P[A]) = -\log_2(P[A])$$

From the preceding equation we observe that the greater the probability that an event will occur, the smaller the information content of the event, as we stated earlier. If event $A$ is certain to occur, then $P[A]=1$ and $I(A)=-\log_2(1)=0$. Similarly, the smaller the probability that an event will occur, the greater is the information content of the event. In particular, if event $A$ is certain not to occur, $P[A]=0$ and $I(A)=-\log_2(0)=\infty$. Thus, when an event that is not expected to occur does actually occur, its information content is infinite.

### 1.4.2 Reliability Engineering

Reliability theory is concerned with the duration of the useful life of components and systems of components. System failure times are unpredictable. Thus, the time until a system fails, which is referred to as the *time to failure* of the system, is usually modeled by a probabilistic function. Reliability applications of probability are considered later in this chapter.

### 1.4.3 Quality Control

Quality control deals with the inspection of finished products to ensure that they meet the desired requirements and specifications. One way to perform the quality control function is to physically test/inspect each product as it comes off the production line. However, this is a very costly way to do it. The practical method is to randomly select a sample of the product from a lot and test each item in the sample. A decision to declare the lot good or defective is thus based on the outcome of the test of the items of the sample. This decision is itself based on a well-designed policy that guarantees that a good lot is rejected with a very small probability and that a bad lot is accepted with a very small probability. A lot is considered good if the parameter that characterizes the quality of the sample has a value that exceeds a predefined threshold value. Similarly the lot is considered to be defective if the parameter that characterizes the quality of the sample has a value that is smaller than the predefined threshold value. For example, one rule for acceptance of a lot can be that the number of defective items in the selected sample be less than some predefined fraction of the sample; otherwise the lot is declared defective.

### 1.4.4 Channel Noise

Noise is an unwanted signal. In Figure 1.1, the message transmitted from the source passes through a channel where it is subject to different kinds of random disturbances that can introduce errors in the message received at the sink. That is, channel noise corrupts messages. Thus, in modeling a communication system, the effect of noise must be taken into consideration. Since channel noise is a random phenomenon, one of the performance issues is the probability that a received message is not corrupted by noise. Thus, probability plays an important role in evaluating the performance of noisy communication channels.

### 1.4.5 System Simulation

Sometimes it is difficult to provide an exact solution of physical problems involving random phenomena. The difficulty arises from the fact that such problems are very complex, which is the case, for example, when a system has unusual properties. One way to deal with these problems is to provide an approximate solution, which attempts to make simplifying assumptions that enable the problem to be solved analytically. Another method is to use computer simulation, which imitates the physical process. Even when an approximate solution is obtained, it is always advisable to use simulation to validate the assumptions.

A simulation model describes the operation of a system in terms of individual events of the individual elements in the system. The model includes the interrelationships among the different elements and allows the effects of the elements' actions on each other to be captured as a dynamic process.

The key to a simulation model is the generation of random numbers that can be used to represent events, such as arrival of customers at a bank, in the system being modeled. Because these events are random in nature, the random numbers are used to drive the probability distributions that characterize them. Thus, knowledge of probability theory is essential for a meaningful simulation analysis.

## 1.5 ELEMENTARY SET THEORY

A set is a collection of objects known as elements. The events that we discussed earlier in this chapter are usually modeled as sets, and the algebra of sets is used to study events. A set can be represented in a number of ways as the following examples illustrate.

Let $A$ denote the set of positive integers between and including 1 and 5. Then,

$$A = \{a | 1 \leq a \leq 5\} = \{1, 2, 3, 4, 5\}$$

Similarly, let $B$ denote the set of positive odd numbers less than 10. Then

$$B = \{1, 3, 5, 7, 9\}$$

If $k$ is an element of the set $E$, we say that $k$ belongs to (or is a member of) $E$ and write $k \in E$. If $k$ is not an element of the set $E$, we say that $k$ does not belong to (or is not a member of) $E$ and write $k \notin E$.

A set $A$ is called a *subset* of set $B$, denoted by $A \subset B$, if every member of $A$ is a member of $B$. Alternatively, we say that the set $B$ contains the set $A$ by writing $B \supset A$.

The set that contains all possible elements is called the *universal set* $\Omega$. The set that contains no elements (or is empty) is called the *null set* Ø (or *empty set*). That is,

$$\emptyset = \{\}$$

### 1.5.1 Set Operations

**Equality**: Two sets $A$ and $B$ are defined to be equal, denoted by $A = B$, if and only if (iff) $A$ is a subset of $B$ and $B$ is a subset of $A$; that is, $A \subset B$ and $B \subset A$.

**Complementation**: Let $A \subset \Omega$. The complement of $A$, denoted by $\overline{A}$, is the set containing all elements of $\Omega$ that are not in $A$. That is,

$$\overline{A} = \{k | k \in \Omega \text{ and } k \notin A\}$$

---

**EXAMPLE 1.2**

Let $\Omega = \{1, 2, 3, 4, 5, 6, 7, 8, 9, 10\}$, $A = \{1, 2, 4, 7\}$ and $B = \{1, 3, 4, 6\}$. Then $\overline{A} = \{3, 5, 6, 8, 9, 10\}$ and $\overline{B} = \{2, 5, 7, 8, 9, 10\}$.

---

**Union**: The union of two sets $A$ and $B$, denoted by $A \cup B$, is the set containing all the elements of either $A$ or $B$ or both $A$ and $B$. That is,

$A \cup B = \{k | k \in A \text{ or } k \in B\}$

In example 1.2, $A \cup B = \{1, 2, 3, 4, 6, 7\}$. Note that if an element is a member of both sets $A$ and $B$, it is listed only once in $A \cup B$.

**Intersection**: The intersection of two sets $A$ and $B$, denoted by $A \cap B$, is the set containing all the elements that are in both $A$ and $B$. That is,

$A \cap B = \{k | k \in A \text{ and } k \in B\}$

In Example 1.2, $A \cap B = \{1, 4\}$.

**Difference**: The difference of two sets $A$ and $B$, denoted by $A - B$ (also $A \backslash B$), is the set containing all elements of $A$ that are not in $B$. That is,

$A - B = \{k | k \in A \text{ and } k \notin B\}$

Note that $A - B \neq B - A$. From Example 1.2 we find that $A - B = \{2, 7\}$ while $B - A = \{3, 6\}$. $A - B$ contains the elements of set $A$ that are not in set $B$ while $B - A$ contains the elements of set $B$ that are not in set $A$.

**Disjoint Sets**: Two sets $A$ and $B$ are called disjoint (or mutually exclusive) sets if they contain no elements in common, which means that $A \cap B = \emptyset$.

### 1.5.2 Number of Subsets of a Set

Let a set $A$ contain $n$ elements labeled $a_1, a_2, \ldots, a_n$. The number of possible subsets of $A$ is $2^n$, which can be obtained as follows for the case of $n = 3$. The eight subsets are given by $\{\overline{a}_1, \overline{a}_2, \overline{a}_3\} = \emptyset$, $\{\overline{a}_1, \overline{a}_2, a_3\}$, $\{\overline{a}_1, a_2, \overline{a}_3\}$, $\{\overline{a}_1, a_2, a_3\}$, $\{a_1, \overline{a}_2, \overline{a}_3\}$, $\{a_1, \overline{a}_2, a_3\}$, $\{a_1, a_2, \overline{a}_3\}$, and $\{a_1, a_2, a_3\} = A$; where $\overline{a}_k$ indicates that the element $a_k$ is not included. By convention, if $a_k$ is not an element of a subset, its complement is not explicitly included in the subset. Thus, the subsets are ø, $\{a_1\}$, $\{a_2\}$, $\{a_3\}$, $\{a_1, a_2\}$, $\{a_1, a_3\}$, $\{a_2, a_3\}$, $\{a_1, a_2, a_3\} = A$. Since the number of subsets includes the null set, the number of subsets that contain at least one element is $2^n - 1$. The result can be extended to the case of $n > 3$.

The set of all subsets of a set $A$ is called the *power set* of $A$ and is denoted by $s(A)$. Thus, for the set $A = \{a_1, a_2, a_3\}$, the power set of $A$ is given by

$s(A) = \{ø, \{a_1\}, \{a_2\}, \{a_3\}, \{a_1, a_2\}, \{a_1, a_3\}, \{a_2, a_3\}, \{a_1, a_2, a_3\}\}$

The number of members of a set $A$ is called the cardinality of $A$ and denoted by $|A|$. Thus, if the cardinality of the set $A$ is $n$, then the cardinality of the power set of $A$ is $|s(A)| = 2^n$.

### 1.5.3 Venn Diagram

The different set operations discussed in the previous section can be graphically represented by the Venn diagram. Figure 1.3 illustrates the complementation,

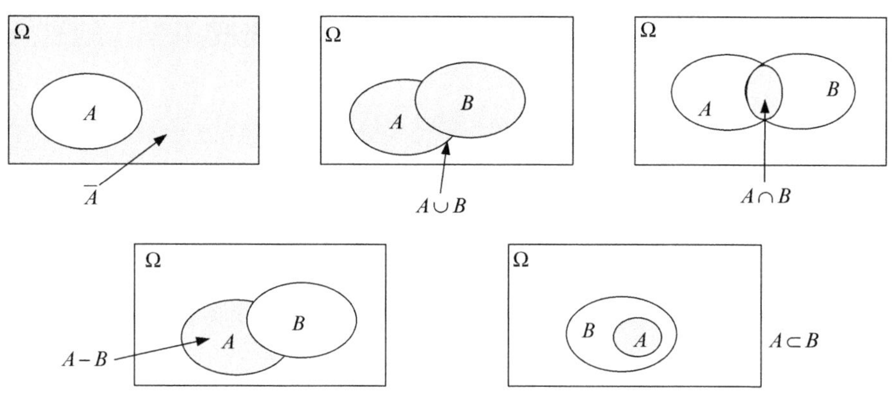


**FIGURE 1.3**
Venn Diagrams of Different Set Operations

union, intersection, and difference operations on two sets $A$ and $B$. The universal set is represented by the set of points inside a rectangle. The sets $A$ and $B$ are represented by the sets of points inside oval objects.

### 1.5.4 Set Identities

The operations of forming unions, intersections and complements of sets obey certain rules similar to the rules of algebra. These rules include the following:

- *Commutative law for unions*: $A \cup B = B \cup A$, which states that the order of the union operation on two sets is immaterial.
- *Commutative law for intersections*: $A \cap B = B \cap A$, which states that the order of the intersection operation on two sets is immaterial.
- *Associative law for unions*: $A \cup (B \cup C) = (A \cup B) \cup C$, which states that in performing the union operation on three sets, we can proceed in two ways: We can first perform the union operation on the first two sets to obtain an intermediate result and then perform the operation on the result and the third set. The same result is obtained if we first perform the operation on the last two sets and then perform the operation on the first set and the result obtained from the operation on the last two sets.
- *Associative law for intersections*: $A \cap (B \cap C) = (A \cap B) \cap C$, which states that in performing the intersection operation on three sets, we can proceed in two ways: We can first perform the intersection operation on the first two sets to obtain an intermediate result and then perform the operation on the result and the third set. The same result is obtained if we first perform the operation on the last two sets and then perform the operation on the first set and the result obtained from the operation on the last two sets.
- *First distributive law*: $A \cap (B \cup C) = (A \cap B) \cup (A \cap C)$, which states that the intersection of a set $A$ and the union of two sets $B$ and $C$ is equal to

the union of the intersection of $A$ and $B$ and the intersection of $A$ and $C$. This law can be extended as follows:

$$A \cap \left( \bigcup_{i=1}^{n} B_i \right) = \bigcup_{i=1}^{n} (A \cap B_i)$$

- *Second distributive law*: $A \cup (B \cap C) = (A \cup B) \cap (A \cup C)$, which states that the union of a set $A$ and the intersection of two sets $B$ and $C$ is equal to the intersection of the union of $A$ and $B$ and the union of $A$ and $C$. The law can also be extended as follows:

$$A \cup \left( \bigcap_{i=1}^{n} B_i \right) = \bigcap_{i=1}^{n} (A \cup B_i) \tag{1.9}$$

- *De Morgan's first law*: $\overline{A \cup B} = \overline{A} \cap \overline{B}$, which states that the complement of the union of two sets is equal to the intersection of the complements of the sets. The law can be extended to include more than two sets as follows:

$$\overline{\bigcup_{i=1}^{n} A_i} = \bigcap_{i=1}^{n} \overline{A_i} \tag{1.10}$$

- *De Morgan's second law*: $\overline{A \cap B} = \overline{A} \cup \overline{B}$, which states that the complement of the intersection of two sets is equal to the union of the complements of the sets. The law can also be extended to include more than two sets as follows:

$$\overline{\bigcap_{i=1}^{n} A_i} = \bigcup_{i=1}^{n} \overline{A_i} \tag{1.11}$$

- Other identities include the following:
  - $A - B = A \cap \overline{B}$, which states that the difference of $A$ and $B$ is equal to the intersection of $A$ and the complement of $B$.
  - $A \cup \Omega = \Omega$, which states that the union of $A$ and the universal set $\Omega$ is equal to $\Omega$.
  - $A \cap \Omega = A$, which states that the intersection of $A$ and the universal set $\Omega$ is equal to $A$.
  - $A \cup \emptyset = A$, which states that the union of $A$ and the null set is equal to $A$.
  - $A \cap \emptyset = \emptyset$, which states that the intersection of $A$ and the null set is equal to the null set.
  - $\overline{\Omega} = \emptyset$, which states that the complement of the universal set is equal to the null set.
  - For any two sets $A$ and $B$, $A = (A \cap B) \cup (A \cap \overline{B})$, which states that the set $A$ is equal to the union of the intersection of $A$ and $B$ and the intersection of $A$ and the complement of $B$.

The way to prove these identities is to show that any point contained in the event on the left side of the equality is also contained in the event on the right side, and vice versa.

### 1.5.5 Duality Principle

The duality principle states that any true result involving sets is also true when we replace unions by intersections, intersections by unions, sets by their complements, and if we reverse the inclusion symbols $\subset$ and $\supset$. For example, if we replace the union in the first distributive law with intersection and intersection with union, we obtain the second distributive law, and vice versa. The same result holds for the two De Morgan's laws.

## 1.6 PROPERTIES OF PROBABILITY

We now combine the results of set identities with those of the axiomatic definition of probability. (See Section 1.3.1.) From these two sections we obtain the following results:

1. $P[\overline{A}] = 1 - P[A]$, which states that the probability of the complement of $A$ is one minus the probability of $A$.
2. $P[\emptyset] = 0$, which states that the impossible (or null) event has probability zero.
3. If $A \subset B$, then $P[A] \leq P[B]$. That is, if $A$ is a subset of $B$, the probability of $A$ is at most the probability of $B$ (or the probability of $A$ cannot exceed the probability of $B$).
4. $P[A] \leq 1$, which means that the probability of event $A$ is at most 1. This follows from the fact that $A \subset \Omega$, and since $P[\Omega] = 1$ the previous result holds.
5. If $A = A_1 \cup A_2 \cup \cdots \cup A_n$, where $A_1, A_2, \ldots, A_n$ are mutually exclusive events, then

   $$P[A] = P[A_1] + P[A_2] + \cdots + P[A_n]$$

6. For any two events $A$ and $B$, $P[A] = P[A \cap B] + P[A \cap \overline{B}]$, which follows from the set identity: $A = (A \cap B) \cup (A \cap \overline{B})$. Since $A \cap B$ and $A \cap \overline{B}$ are mutually exclusive events, the result follows.
7. For any two events $A$ and $B$, $P[A \cup B] = P[A] + P[B] - P[A \cap B]$. This result can be proved by making use of the Venn diagram. Figure 1.4a represents

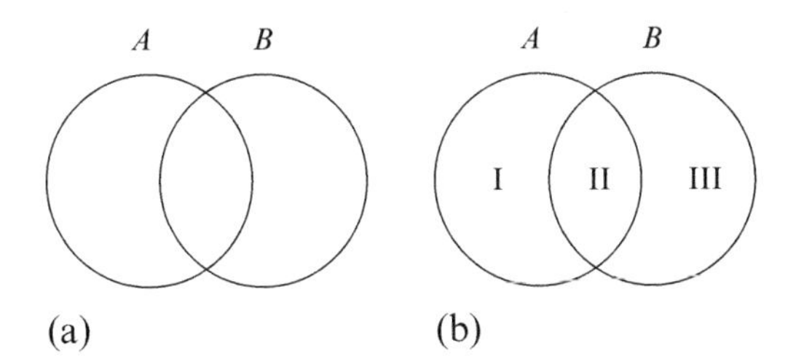


**FIGURE 1.4**
Venn Diagram of $A \cup B$

a Venn diagram in which the left circle represents event $A$ and the right circle represents event $B$. In Figure 1.4b we divide the diagram into three mutually exclusive sections labeled I, II, and III where section I represents all points in $A$ that are not in $B$, section II represents all points in both $A$ and $B$, and section III represents all points in $B$ that are not in $A$. From Figure 1.4b, we observe that:

$$\begin{aligned} A \cup B &= I \cup II \cup III \\ A &= I \cup II \\ B &= II \cup III \end{aligned}$$

Since I, II and III are mutually exclusive, Property 5 implies that

$$\begin{aligned} &P[A \cup B] = P[I] + P[II] + P[III] \\ &P[A] = P[I] + P[II] \Rightarrow P[I] = P[A] - P[II] \\ &P[B] = P[II] + P[III] \Rightarrow P[III] = P[B] - P[II] \end{aligned}$$

Thus, we have that

$$\begin{aligned} P[A \cup B] &= P[A] - P[II] + P[II] + P[B] - P[II] \\ &= P[A] + P[B] - P[II] = P[A] + P[B] - P[A \cap B] \end{aligned} \tag{1.12}$$

8. We can extend Property 7 to the case of three events. If $A_1, A_2$ and $A_3$ are three events in $\Omega$, then 3

$$\begin{aligned} P[A_1 \cup A_2 \cup A_3] = P[A_1] + P[A_2] + P[A_3] - P[A_1 \cap A_2] - P[A_1 \cap A_3] \\ -P[A_2 \cap A_3] + P[A_1 \cap A_2 \cap A_3] \end{aligned} \tag{1.13}$$

This can be further generalized to the case of $n$ arbitrary events in $\Omega$ as follows:

$$\begin{aligned} P[A_1 \cup A_2 \cup \cdots \cup A_n] = \sum_{i=1}^{n} P[A_i] - \sum_{1 \leq i \leq j \leq n} P\left[A_i \cap A_j\right] \\ + \sum_{1 \leq i \leq j \leq k \leq n} P\left[A_i \cap A_j \cap A_k\right] - \cdots \end{aligned} \tag{1.14}$$

That is, to find the probability that at least one of the $n$ events occurs, first add the probability of each event, then subtract the probabilities of all possible two-way intersections, then add the probabilities of all possible three-way intersections, and so on.

## 1.7 CONDITIONAL PROBABILITY

Consider the following experiment. We are interested in the sum of the numbers that appear when two dice are tossed. Suppose we are interested in the event that the sum of the two tosses is 7 and we observe that the first toss is 4. Based on this fact, the six possible and equally likely outcomes of the two tosses are $(4, 1), (4, 2), (4, 3), (4, 4), (4, 5)$ and $(4, 6)$. In the absence of the information that the first toss is 4, there would have been 36 sample points in the sample space. But with the information on the outcome of the first toss, there are now only 6 sample points.

Let $A$ denote the event that the sum of the two dice is 7, and let $B$ denote the event that the first die is 4. The conditional probability of event $A$ given event $B$, denoted by $P[A|B]$, is defined by

$$P[A|B] = \frac{P[A \cap B]}{P[B]} \qquad P[B] \neq 0 \tag{1.15}$$

Thus, for the preceding problem we have that

$$\begin{aligned} P[A|B] &= \frac{P[(4, 3)]}{P[(4, 1)] + P[(4, 2)] + P[(4, 3)] + P[(4, 4)] + P[(4, 5)] + P[(4, 6)]} \\ &= \frac{(1/36)}{(6/36)} = \frac{1}{6} \end{aligned}$$

## EXAMPLE 1.3

A bag contains 8 red balls, 4 green, and 8 yellow balls. A ball is drawn at random from the bag and it is found not to be one of the red balls. What is the probability that it is a green ball?

**Solution:**

Let $G$ denote the event that the selected ball is a green ball and let $\overline{R}$ denote the event that it is not a red ball. Then, $P[G] = 4/20 = 1/5$ since there are 4 green balls out of a total of 20 balls, and $P[\overline{R}] = 12/20 = 3/5$ since there are 12 balls out of 20 that are not red. Now,

$$P[G|\overline{R}] = \frac{P[G \cap \overline{R}]}{P[\overline{R}]}$$

But if the ball is green and not red, it must be green. Thus, we have that $\{G \cap \overline{R}\} = \{G\}$ because $G$ is a subset of $\overline{R}$. Thus,

$$P[G|\overline{R}] = \frac{P[G \cap \overline{R}]}{P[\overline{R}]} = \frac{P[G]}{P[\overline{R}]} = \frac{1/5}{3/5} = \frac{1}{3}$$

## EXAMPLE 1.4

A fair coin was tossed two times. Given that the first toss resulted in heads, what is the probability that both tosses resulted in heads?

**Solution:**

Because the coin is fair, the four sample points of the sample space $\Omega=\{HH, HT, TH, TT\}$ are equally likely. Let $X$ denote the event that both tosses came up heads; that is, $X=\{HH\}$. Let $Y$ denote the event that the first toss came up heads; that is, $Y=\{HH, HT\}$. Because $X$ is a subset of $Y$, the probability that both tosses resulted in heads, given that the first toss resulted in heads, is given by

$$P[X|Y]=\frac{P[X\cap Y]}{P[Y]}=\frac{P[X]}{P[Y]}=\frac{1/4}{2/4}=\frac{1}{2}$$

### 1.7.1 Total Probability and the Bayes' Theorem

A partition of a set $A$ is a set $\{A_1, A_2, \ldots, A_n\}$ with the following properties:

a. $A_i \subseteq A, i=1,2,\cdots,n$, which means that $A$ is a set of subsets.
b. $A_i \cap A_k=\emptyset, i=1,2,\cdots,n; k=1,2,\cdots,n; i\neq k$, which means that the subsets are mutually (or pairwise) disjoint; that is, no two subsets have any element in common.
c. $A_1 \cup A_2 \cup \ldots \cup A_n=A$, which means that the subsets are collectively exhaustive. That is, the subsets together include all possible values of the set $A$.

**Proposition 1.1**

Let $\{A_1, A_2, \ldots, A_n\}$ be a partition of the sample space $\Omega$, and suppose each one of the events $A_1, A_2, \ldots, A_n$ has nonzero probability of occurrence. Let $B$ be any event defined in $\Omega$. Then

$$\begin{aligned}P[B]&=P[B|A_1]P[A_1]+P[B|A_2]P[A_2]+\cdots+P[B|A_n]P[A_n]\\&=\sum_{i=1}^{n}P[B|A_i]P[A_i]\end{aligned}$$

**Proof**

The proof is based on the observation that because $\{A_1, A_2, \ldots, A_n\}$ is a partition of $\Omega$, the set $\{B\cap A_1, B\cap A_2, \ldots, B\cap A_n\}$ is a partition of the event $B$ because if $B$ occurs, then it must occur in conjunction with one of the $A_i$'s. Thus, we can express $B$ as the union of $n$ mutually exclusive events. That is,

$$B=(B\cap A_1)\cup(B\cap A_2)\cup\ldots\cup(B\cap A_n)$$

Since these events are mutually exclusive, we obtain

$$P[B]=P[B\cap A_1]+P[B\cap A_2]+\ldots+P[B\cap A_n]$$

From our definition of conditional probability, $P[B\cap A_i]=P[B|A_i]P[A_i]$, which exists because we assumed in the Proposition that the events $A_1, A_2, \ldots, A_n$ have nonzero probabilities. Substituting the definition of conditional probabilities we obtain the desired result:

$$P[B]=P[B|A_1]P[A_1]+P[B|A_2]P[A_2]+\ldots+P[B|A_n]P[A_n]$$

The above result is defined as the law of *total probability* of event $B$, which will be useful in the remainder of the book.

## EXAMPLE 1.5

A student buys 1000 chips from supplier A, 2000 chips from supplier B, and 3000 chips from supplier C. He tested the chips and found that the probability that a chip is defective depends on the supplier from where it was bought. Specifically, given that a chip came from supplier A, the probability that it is defective is 0.05; given that a chip came from supplier B, the probability that it is defective is 0.10; and given that a chip came from supplier C, the probability that it is defective is 0.10. If the chips from the three suppliers are mixed together and one of them is selected at random, what is the probability that it is defective?

**Solution:**

Let $P[A]$, $P[B]$ and $P[C]$ denote the probability that a randomly selected chip came from supplier A, B and C respectively. Also, let $P[D|A]$ denote the conditional probability that a chip is defective, given that it came from supplier A; let $P[D|B]$ denote the conditional probability that a chip is defective, given that it came from supplier B; and let $P[D|C]$ denote the conditional probability that a chip is defective, given that it came from supplier C. Then the following are true:

$$P[D|A] = 0.05$$

$$P[D|B] = 0.10$$

$$P[D|C] = 0.10$$

$$P[A] = \frac{1000}{1000 + 2000 + 3000} = \frac{1}{6}$$

$$P[B] = \frac{2000}{1000 + 2000 + 3000} = \frac{1}{3}$$

$$P[C] = \frac{3000}{1000 + 2000 + 3000} = \frac{1}{2}$$

Let $P[D]$ denote the unconditional probability that a randomly selected chip is defective. Then, from the law of total probability of $D$ we have that

$$\begin{aligned} P[D] &= P[D|A]P[A] + P[D|B]P[B] + P[D|C]P[C] \\ &= (0.05)(1/6) + (0.10)(1/3) + (0.10)(1/2) \\ &= 0.09167 \end{aligned}$$

We now go back to the general discussion. Suppose event $B$ has occurred but we do not know which of the mutually exclusive and collectively exhaustive events $A_1, A_2, \ldots, A_n$ holds true. The conditional probability that event $A_k$ occurred, given that $B$ occurred, is given by

$$P[A_k|B] = \frac{P[A_k \cap B]}{P[B]} = \frac{P[A_k \cap B]}{\sum_{i=1}^{n} P[B|A_i]P[A_i]}$$

where the second equality follows from the law of total probability of event $B$. Since $P[A_k \cap B] = P[B|A_k]P[A_k]$, the above equation can be rewritten as follows:

$$P[A_k|B] = \frac{P[A_k \cap B]}{P[B]} = \frac{P[B|A_k]P[A_k]}{\sum_{i=1}^{n} P[B|A_i]P[A_i]} \quad (1.16)$$

This result is called the *Bayes' formula* (or *Bayes' rule*).

## EXAMPLE 1.6

In Example 1.5, given that a randomly selected chip is defective, what is the probability that it came from supplier A?

**Solution:**

Using the same notation as in Example 1.5, the probability that the randomly selected chip came from supplier A, given that it is defective, is given by

$$\begin{aligned} P[A|D] &= \frac{P[D \cap A]}{P[D]} = \frac{P[D|A]P[A]}{P[D|A]P[A] + P[D|B]P[B] + P[D|C]P[C]} \\ &= \frac{(0.05)(1/6)}{(0.05)(1/5) + (0.10)(1/3) + (0.10)(1/2)} \\ &= 0.0909 \end{aligned}$$

## EXAMPLE 1.7

(***The Binary Symmetric Channel***) A discrete channel is characterized by an input alphabet $X=\{x_1, x_2, \ldots, x_n\}$; an output alphabet $Y=\{y_1, y_2, \ldots, y_n\}$; and a set of conditional probabilities (called *transition probabilities*), $P_{ij}$, which are defined as follows:

$$P_{ij} = P\left[y_j|x_i\right] = P\left[\text{receiving symbol } y_j, \text{given that symbol } x_i \text{ was transmitted}\right]$$

$i=1,2,\ldots,n; j=1,2,\ldots,m$. The binary channel is a special case of the discrete channel where $n=m=2$. It can be represented as shown in Figure 1.5.

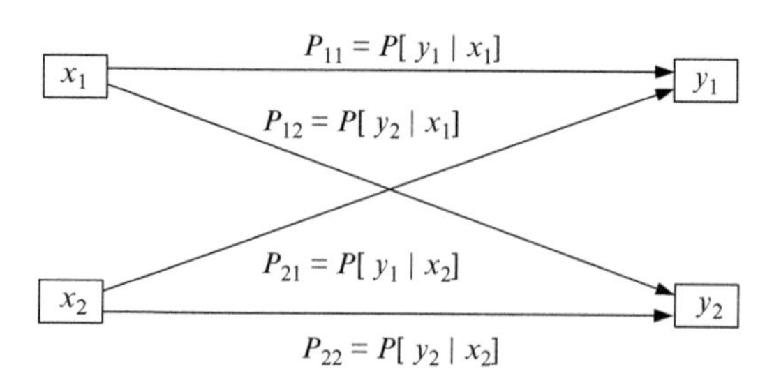


**FIGURE 1.5**
The Binary Channel

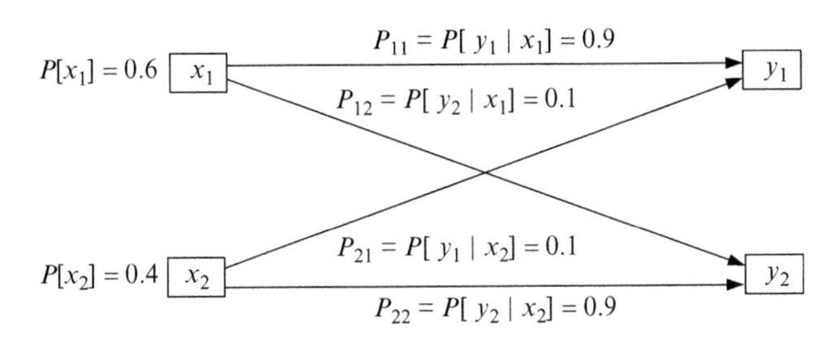


**FIGURE 1.6**
The Binary Symmetric Channel for Example 1.7

In the binary channel, an error occurs if $y_2$ is received when $x_1$ is transmitted or $y_1$ is received when $x_2$ is transmitted. Thus, the probability of error, $P_e$, is given by:

$$\begin{aligned} P_e &= P[x_1 \cap y_2] + P[x_2 \cap y_1] = P[y_2|x_1]P[x_1] + P[y_1|x_2]P[x_2] \\ &= P[x_1]P_{12} + P[x_2]P_{21} \end{aligned} \tag{1.17}$$

If $P_{12} = P_{21}$, we say that the channel is a *binary symmetrical channel* (BSC). Also, if in the BSC $P[x_1] = p$, then $P[x_2] = 1 - p = q$.

Consider the BSC shown in Figure 1.6, with $P[x_1] = 0.6$ and $P[x_2] = 0.4$. Compute the following:

a. The probability that $x_1$ was transmitted, given that $y_2$ was received
b. The probability that $x_2$ was transmitted, given that $y_1$ was received
c. The probability that $x_1$ was transmitted, given that $y_1$ was received
d. The probability that $x_2$ was transmitted, given that $y_2$ was received
e. The unconditional probability of error

**Solution:**

Let $P[y_1]$ denote the probability that $y_1$ was received and $P[y_2]$ the probability that $y_2$ was received. Then

a. The probability that $x_1$ was transmitted, given that $y_2$ was received is given by

$$\begin{aligned} P[x_1|y_2] &= \frac{P[x_1 \cap y_2]}{P[y_2]} = \frac{P[y_2|x_1]P[x_1]}{P[y_2|x_1]P[x_1] + P[y_2|x_2]P[x_2]} \\ &= \frac{(0.1)(0.6)}{(0.1)(0.6) + (0.9)(0.4)} \\ &= 0.143 \end{aligned}$$

b. The probability that $x_2$ was transmitted, given that $y_1$ was received is given by

$$\begin{aligned} P[x_2|y_1] &= \frac{P[x_2 \cap y_1]}{P[y_1]} = \frac{P[y_1|x_2]P[x_2]}{P[y_1|x_1]P[x_1] + P[y_1|x_2]P[x_2]} \\ &= \frac{(0.1)(0.4)}{(0.9)(0.6) + (0.1)(0.4)} \\ &= 0.069 \end{aligned}$$

c. The probability that $x_1$ was transmitted, given that $y_1$ was received is given by

$$P[x_1|y_1]=\frac{P[x_1\cap y_1]}{P[y_1]}=\frac{P[y_1|x_1]P[x_1]}{P[y_1|x_1]P[x_1]+P[y_1|x_2]P[x_2]}$$

$$=\frac{(0.9)(0.6)}{(0.9)(0.6)+(0.1)(0.4)}$$

$$=0.931=1-P[x_2|y_1]$$

d. The probability that $x_2$ was transmitted, given that $y_2$ was received is given by

$$P[x_2|y_2]=\frac{P[x_2\cap y_2]}{P[y_2]}=\frac{P[y_2|x_2]P[x_2]}{P[y_2|x_1]P[x_1]+P[y_2|x_2]P[x_2]}$$

$$=\frac{(0.9)(0.4)}{(0.1)(0.6)+(0.9)(0.4)}$$

$$=0.857=1-P[x_1|y_2]$$

e. The unconditional probability of error is given by

$$P_e=P[x_1]P[y_2|x_1]+P[x_2]P[y_1|x_2]=P[x_1]P_{12}+P[x_2]P_{21}=(0.6)(0.1)+(0.4)(0.1)$$
$$=0.1$$

## EXAMPLE 1.8

The quarterback for a certain football team has a good game with probability 0.6 and a bad game with probability 0.4. When he has a good game, he throws an interception with a probability of 0.2; and when he has a bad game, he throws an interception with a probability of 0.5. Given that he threw an interception in a particular game, what is the probability that he had a good game?

**Solution:**

Let $G$ denote the event that the quarterback has a good game and $B$ the event that he has a bad game. Similarly, let $I$ denote the event that he throws an interception. Then we have that

$$P[G]=0.6$$

$$P[B]=0.4$$

$$P[I|G]=0.2$$

$$P[I|B]=0.5$$

$$P[G|I]=\frac{P[G\cap I]}{P[I]}$$

According to the Bayes' formula, the last equation becomes

$$P[G|I]=\frac{P[G\cap I]}{P[I]}=\frac{P[I|G]P[G]}{P[I|G]P[G]+P[I|B]P[B]}$$

$$=\frac{(0.2)(0.6)}{(0.2)(0.6)+(0.5)(0.4)}=\frac{0.12}{0.32}$$

$$=0.375$$

## EXAMPLE 1.9

Two events $A$ and $B$ are such that $P[A \cap B] = 0.15, P[A \cup B] = 0.65$, and $P[A|B] = 0.5$. Find $P[B|A]$.

**Solution:**

$P[A \cup B] = P[A] + P[B] - P[A \cap B] \Rightarrow 0.65 = P[A] + P[B] - 0.15$. This means that $P[A] + P[B] = 0.65 + 0.15 = 0.80$. Also, $P[A \cap B] = P[A|B]P[B]$. This then means that

$$P[B] = \frac{P[A \cap B]}{P[A|B]} = \frac{0.15}{0.50} = 0.30$$

Thus, $P[A] = 0.80 - 0.30 = 0.50$. Since $P[A \cap B] = P[B|A]P[A]$, we have that

$$P[B|A] = \frac{P[A \cap B]}{P[A]} = \frac{0.15}{0.50} = 0.30$$

## EXAMPLE 1.10

A student went to the post office to send a priority mail to his parents. He gave the postal lady a bill he believed was \$20. However, the postal lady gave him change based on her belief that she received a \$10 bill from the student. The student started to dispute the change. Both the student and the postal lady are honest but may make mistakes. If the postal lady's drawer contains thirty \$20 bills and twenty \$10 bills, and the postal lady correctly identifies bills 90% of the time, what is the probability that the student's claim is valid?

**Solution:**

Let $A$ denote the event that the student gave a \$10 bill and $B$ the event that the student gave a \$20 bill. Let $V$ denote the event that the student's claim is valid. Finally, let $L$ denote the event that the postal lady said that the student gave her a \$10 bill. Since there are 30 \$20 bills and 20 \$10 bills in the drawer, the probability that the money the student gave the postal lady was a \$20 bill is $30/(20+30) = 0.6$, and the probability that it was a \$10 bill is $1 - 0.6 = 0.4$. Thus,

$$\begin{aligned} P[L] &= P[L|A]P[A] + P[L|B]P[B] = (0.90)(0.4) + (0.1)(0.6) \\ &= 0.42 \end{aligned}$$

Therefore, the probability that the student's claim is valid is the probability that he gave a \$20 bill, given that the postal lady said that the student gave her a \$10 bill. Using Bayes' formula we obtain

$$P[V|L] = \frac{P[V \cap L]}{P[L]} = \frac{P[L|V]P[V]}{P[L]} = \frac{(0.10)(0.60)}{0.42} = \frac{1}{7} = 0.1429$$

## EXAMPLE 1.11

An aircraft maintenance company bought an equipment for detecting structural defects in aircrafts. Tests indicate that 95% of the time the equipment detects defects when they actually exist, and 1% of the time it gives a false alarm (that is, it indicates the presence of a structural defect when in fact there is none). If 2% of the aircrafts actually have structural defects, what is the probability that an aircraft actually has a structural defect given that the equipment indicates that it has a structural defect?

**Solution:**

Let $D$ denote the event that an aircraft has a structural defect and $B$ the event that the test indicates that there is a structural defect. Then we are required to find $P[D|B]$. Using Bayes' formula we obtain

$$P[D|B] = \frac{P[D \cap B]}{P[B]} = \frac{P[B|D]P[D]}{P[B|D]P[D] + P[B|\overline{D}]P[\overline{D}]}$$
$$= \frac{(0.95)(0.02)}{(0.95)(0.2) + (0.01)(0.98)} = 0.660$$

Thus, only 66% of the aircrafts that the equipment diagnoses as having structural defects actually have structural defects.

### 1.7.2 Tree Diagram

Conditional probabilities are used to model experiments that take place in stages. The outcomes of such experiments are conveniently represented by a tree diagram. A tree is a connected graph that contains no circuit (or loop). Every two nodes in the tree have a unique path connecting them. Line segments called branches or edges interconnect the nodes. Each branch may split into other branches or it may terminate. When used to model an experiment, the nodes of the tree represent events of the experiment. The number of branches that emanate from a node represents the number of events that can occur, given that the event represented by that node occurs. The node that has no predecessor is called the *root* of the tree, and any node that has no successor or child is called a *leaf* of the tree. The events of interest are usually defined at the leaves by tracing the outcomes of the experiment from the root to each leaf.

The conditional probabilities appear on the branches leading from the node representing an event to the nodes representing the next events of the experiment. A path through the tree corresponds to a possible outcome of the experiment. Thus, the product of all the branch probabilities from the root of the tree to any node is equal to the probability of the event represented by that node.

Consider an experiment that consists of three tosses of a coin. Let $p$ denote the probability of heads in a toss; then $1-p$ is the probability of tails in a toss. Figure 1.7 is the tree diagram for the experiment.

Let $A$ be the event "the first toss came up heads" and let $B$ be the event "the second toss came up tails." Thus, $A = \{HHH, HHT, HTH, HTT\}$ and $B = \{HTH, HTT, TTH, TTT\}$. From Figure 1.7, we have that

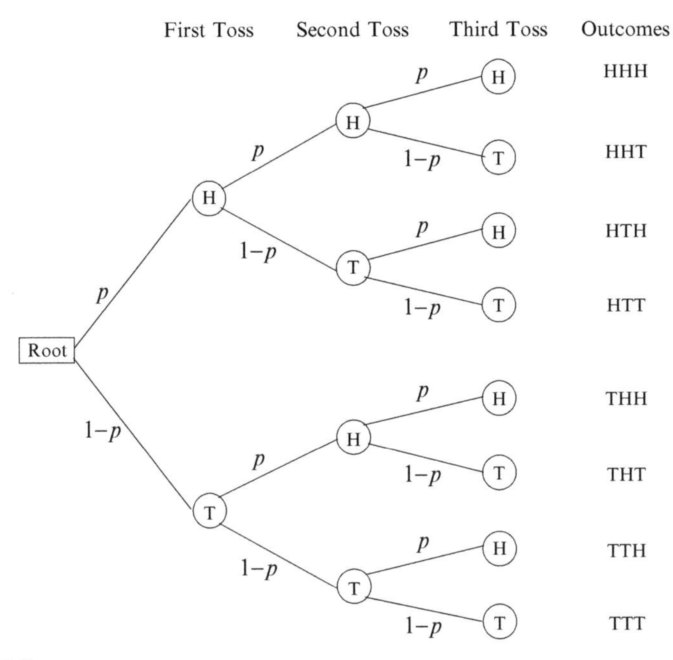


**FIGURE 1.7**
Tree Diagram for Three Tosses of a Coin

$$
\begin{aligned}
P[A] &= P[HHH] + P[HHT] + P[HTH] + P[HTT] \\
&= p^3 + 2p^2(1-p) + p(1-p)^2 \\
&= p
\end{aligned}
$$

$$
\begin{aligned}
P[B] &= P[HTH] + P[HTT] + P[TTH] + P[TTT] \\
&= p^2(1-p) + 2p(1-p)^2 + (1-p)^3 \\
&= 1-p
\end{aligned}
$$

Since $A \cap B = \{HTH, HTT\}$, we have that

$$P[A \cap B] = P[HTH] + P[HTT] = p^2(1-p) + p(1-p)^2 = p(1-p)\{p + 1 - p\} = p(1-p)$$

Now, $A \cup B = \{HHH, HHT, HTH, HTT, TTH, TTT\}$ and we have that

$$
\begin{aligned}
P[A \cup B] &= P[HHH] + P[HHT] + P[HTH] + P[HTT] + P[TTH] + P[TTT] \\
&= p^3 + 2p^2(1-p) + 2p(1-p)^2 + (1-p)^3 \\
&= p^2\{p + 2(1-p)\} + (1-p)^2\{2p + 1 - p\} = 1 - p + p^2 \\
&= 1 - p(1-p)
\end{aligned}
$$

Note that we can obtain the same result as follows:

$$P[A \cup B] = P[A] + P[B] - P[A \cap B] = p + 1 - p - p(1-p)$$
$$= 1 - p(1-p)$$

## EXAMPLE 1.12

A university has twice as many undergraduate students as graduate students. 25% of the graduate students live on campus and 10% of the undergraduate students live on campus.

a. If a student is chosen at random from the student population, what is the probability that the student is an undergraduate student living on campus?
b. If a student living on campus is chosen at random, what is the probability that the student is a graduate student?

**Solution:**

We use the tree diagram to solve the problem. Since there are twice as many undergraduate students as there are graduate students, the proportion of undergraduate students in the population is 2/3, and the proportion of graduate students is 1/3. These as well as the other data are shown as the labels on the branches of the tree in Figure 1.8. In the figure G denotes graduate student, U denotes undergraduate student, ON denotes living on campus, and OFF denotes living off campus.

a. From the figure we see that the probability that a randomly selected student is an undergraduate student living on campus is 0.067. We can also solve the problem directly as follows. We are required to find the probability of choosing an undergraduate student who lives on campus, which is $P[U \rightarrow ON]$, the probability of first going on the U branch and then to the ON branch from there. That is,

$$P[U \rightarrow ON] = \frac{2}{3} \times 0.10 = 0.067$$

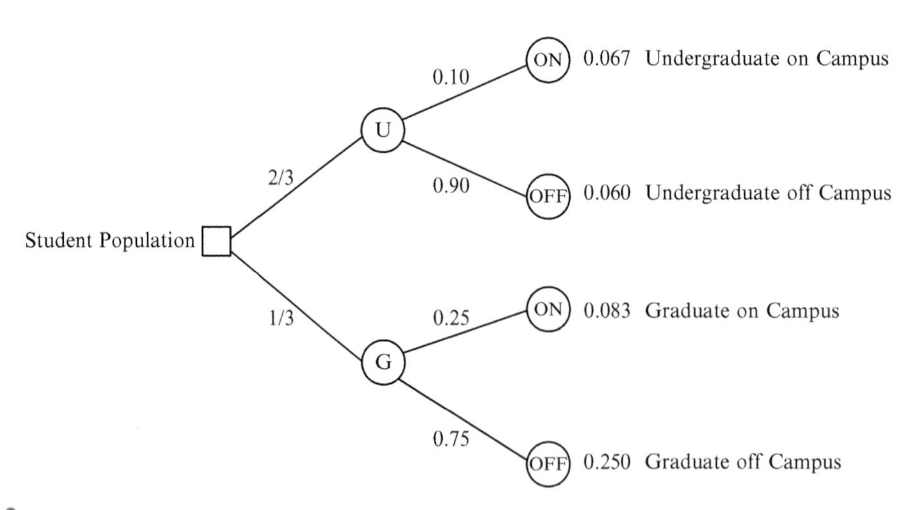


**FIGURE 1.8**
Figure for Example 1.12

b. From the tree, the probability that a student lives on campus is $P[U \rightarrow ON]+P[G \rightarrow ON]=$ $0.067+0.083=0.15$. Thus, the probability that a randomly selected student living on campus is a graduate student is $P[G \rightarrow ON]/\{P[U \rightarrow ON]+P[G \rightarrow ON]\}=0.083/0.15=0.55$. Note that we can also use the Bayes' theorem to solve the problem as follows:

$$P[G|ON]=\frac{P[ON|G]P[G]}{P[ON|U]P[U]+P[ON|G]P[G]}=\frac{(0.25)(1/3)}{(0.25)(1/3)+(0.10)(2/3)}$$
$$=\frac{5}{9}=0.55$$

## EXAMPLE 1.13

A multiple-choice exam consists of 4 choices per question. On 75% of the questions, Pat thinks she knows the answer; and on the other 25% of the questions, she just guesses at random. Unfortunately even when she thinks she knows the answer, Pat is right only 80% of the time.

a. What is the probability that her answer to an arbitrary question is correct?
b. Given that her answer to a question is correct, what is the probability that it was a lucky guess? (This means that it is among the questions whose answers she guessed at random.)

**Solution:**
We can use the tree diagram as follows. There are two branches at the root labeled $K$ for the event "Pat thinks she knows," and $\overline{K}$ for the event "Pat does not know." Under event $K$, she is correct ($C$) with probability 0.80 and not correct ($\overline{C}$) with probability 0.20. Under event $\overline{K}$, she is correct with probability 0.25 because she is equally likely to choose any of the 4 answer; therefore, she is not correct with probability 0.75. The tree diagram is shown in Figure 1.9.

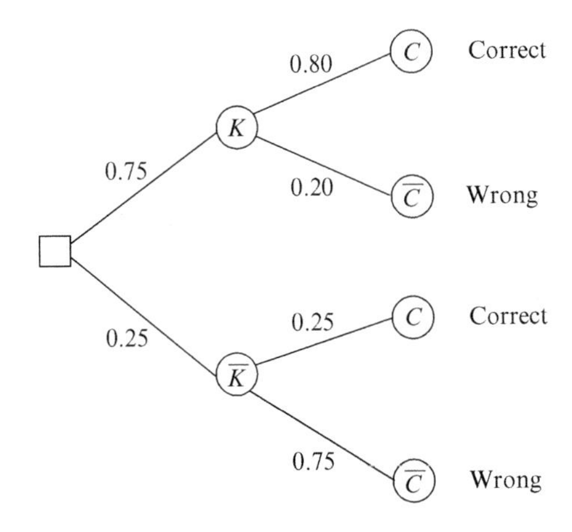


**FIGURE 1.9**
Figure for Example 1.13

a. The probability that Pat's answer to an arbitrary question is correct is given by

$$P[\text{Correct Answer}] = P[K \rightarrow C] + P[\overline{K} \rightarrow C] = (0.75)(0.8) + (0.25)(0.25)$$
$$= 0.6625$$

This can also be obtained by the direct method as follows:

$$P[\text{Correct Answer}] = P[\text{Correct Answer}|K]P[K] + P[\text{Correct Answer}|\overline{K}]P[\overline{K}]$$
$$= (0.80)(0.75) + (0.25)(0.25) = 0.6625$$

b. Given that she gets a question correct, the probability that it was a lucky guess is given by

$$P[\text{Lucky Guess}|\text{Correct Answer}] = P[\overline{K} \rightarrow C]/P[\text{Correct Answer}] = \frac{(0.25)(0.25)}{0.6625}$$
$$= 0.0943$$

We can also use the direct method as follows:

$$P[\text{Lucky Guess}|\text{Correct Answer}] = \frac{P[\text{Correct Answer}|\overline{K}]P[\overline{K}]}{P[\text{Correct Answer}]} = \frac{(0.25)(0.25)}{0.6625}$$
$$= 0.0943$$

## 1.8 INDEPENDENT EVENTS

Two events $A$ and $B$ are defined to be independent if the knowledge that one has occurred does not change or affect the probability that the other will occur. In particular, if events $A$ and $B$ are independent, the conditional probability of event $A$, given event $B$, $P[A|B]$, is equal to the probability of event $A$. That is, events $A$ and $B$ are independent if

$$P[A|B] = P[A] \tag{1.18}$$

Since by definition $P[A \cap B] = P[A|B]P[B]$, an alternative definition of independence of events is that events $A$ and $B$ are independent if

$$P[A \cap B] = P[A]P[B] \tag{1.19}$$

The definition of independence can be extended to multiple events. The $n$ events $A_1, A_2, \ldots, A_n$ are said to be independent if the following conditions are true:

$$P[A_i \cap A_j] = P[A_i]P[A_j]$$
$$P[A_i \cap A_j \cap A_k] = P[A_i]P[A_j]P[A_k]$$
$$\vdots$$
$$P[A_i \cap A_j \cap \cdots \cap A_n] = P[A_i]P[A_j] \cdots P[A_n]$$

This is true for all $1 \leq i < j < k < \cdots \leq n$. That is, these events are pairwise independent, independent in triplets, and so on.

## EXAMPLE 1.14

A red die and a blue die are rolled together. What is the probability that we obtain 4 on the red die and 2 on the blue die?

**Solution:**
Let $R$ denote the event "4 on the red die" and let $B$ denote the event "2 on the blue die." We are, therefore, required to find $P[R \cap B]$. Since the outcome of one die does not affect the outcome of the other die, the events $R$ and $B$ are independent. Thus, since $P[R] = 1/6$ and $P[B] = 1/6$, we have that $P[R \cap B] = P[R]P[B] = 1/36$.

## EXAMPLE 1.15

Two coins are tossed. Let $A$ denote the event "at most one head on the two tosses" and let $B$ denote the event "one head and one tail in both tosses." Are $A$ and $B$ independent events?

**Solution:**
The sample space of the experiment is $\Omega = \{HH, HT, TH, TT\}$. Now, the two events are defined as follows: $A = \{HT, TH, TT\}$ and $B = \{HT, TH\}$. Also, $A \cap B = \{HT, TH\}$. Thus,

$$
\begin{aligned}
P[A] &= \frac{3}{4} \\
P[B] &= \frac{2}{4} = \frac{1}{2} \\
P[A \cap B] &= \frac{2}{4} = \frac{1}{2} \\
P[A]P[B] &= \frac{3}{4} \times \frac{1}{2} = \frac{3}{8}
\end{aligned}
$$

Since $P[A \cap B] \neq P[A]P[B]$, we conclude that events $A$ and $B$ are not independent. Note that $B$ is a subset of $A$, which confirms that they cannot be independent.

**Proposition 1.2**
If $A$ and $B$ are independent events, then so are events $A$ and $\overline{B}$, events $\overline{A}$ and $B$, and events $\overline{A}$ and $\overline{B}$.

**Proof**
Event $A$ can be written as follows: $A = (A \cap B) \cup (A \cap \overline{B})$. Since the events $(A \cap B)$ and $(A \cap \overline{B})$ are mutually exclusive, we may write

$$
\begin{aligned}
P[A] &= P[A \cap B] + P[A \cap \overline{B}] \\
&= P[A]P[B] + P[A \cap \overline{B}]
\end{aligned}
$$

where the last equality follows from the fact that $A$ and $B$ are independent. Thus, we obtain

$$P[A\cap\overline{B}] = P[A] - P[A\cap B] = P[A] - P[A]P[B] = P[A]\{1 - P[B]\} = P[A]P[\overline{B}]$$

which proves that events $A$ and $\overline{B}$ are independent. To prove that events $\overline{A}$ and $B$ are independent, we start with $B = (A\cap B)\cup(\overline{A}\cap B)$. Using the same fact that the two events $A$ and $B$ are mutually exclusive, we establish the independence of $\overline{A}$ and $B$. Finally, to prove that events $\overline{A}$ and $\overline{B}$ are independent, we start with $\overline{A} = (\overline{A}\cap B)\cup(\overline{A}\cap\overline{B})$ and proceed as above using the results already established.

## EXAMPLE 1.16

$A$ and $B$ are two independent events defined in the same sample space. They have the following probabilities: $P[A] = x$ and $P[B] = y$. Find the probabilities of the following events in terms of $x$ and $y$:

a. Neither event $A$ nor event $B$ occurs
b. Event $A$ occurs but event $B$ does not occur
c. Either event $A$ occurs or event $B$ does not occur

**Solution:**
Since events $A$ and $B$ are independent, we have shown earlier in the chapter that events $A$ and $\overline{B}$ are independent, events $\overline{A}$ and $B$ are independent, and events $\overline{A}$ and $\overline{B}$ are also independent.

a. The probability that neither event $A$ nor event $B$ occurs is the probability that event $A$ does not occur and event $B$ does not occur, which is given by

$$P_{\overline{a}\overline{b}} = P[\overline{A}\cap\overline{B}] = P[\overline{A}]P[\overline{B}] = (1-x)(1-y)$$

where the second equality is due to independence of $\overline{A}$ and $\overline{B}$.

b. The probability that event $A$ occurs but event $B$ does not occur is the probability that event $A$ occurs and event $B$ does not occur, which is given by

$$P_{a\overline{b}} = P[A\cap\overline{B}] = P[A]P[\overline{B}] = x(1-y)$$

where the second equality is due to the independence of $A$ and $\overline{B}$.

c. The probability that either event $A$ occurs or event $B$ does not occur is given by

$$\begin{aligned} P[A\cup\overline{B}] &= P[A] + P[\overline{B}] - P[A\cap\overline{B}] \\ &= P[A] + P[\overline{B}] - P[A]P[\overline{B}] = x + (1-y) - x(1-y) \\ &= 1 - y(1-x) \end{aligned}$$

where the second equality is due to the independence of $A$ and $\overline{B}$.

## EXAMPLE 1.17

Jim and Bill are marksmen. Jim can hit a target with a probability of 0.8 while Bill can hit a target with a probability of 0.7. If both fire at a target at the same time, what is the probability that the target is hit at least once?

**Solution:**
Let $J$ denote the event that Jim hits a target, let $\overline{J}$ denote the event that Jim does not hit a target, let B denote the event that Bill hits a target, and let $\overline{B}$ denote the event that Bill does not hit a target. Since the outcome of Bill's shot is not affected by the outcome of Jim's shot and vice versa, the events $J$ and $B$ are independent. Because $B$ and $J$ are independent events, the events $J$ and $\overline{B}$ are independent, and the events $B$ and $\overline{J}$ are independent. Thus, the probability that the target is hit at least once is the probability of the union of its being hit once and its being hit twice. That is, if $p$ is the probability that the target is hit at least once, then

$$\begin{aligned} p &= P[\{J\cap B\}\cup\{J\cap\overline{B}\}\cup\{\overline{J}\cap B\}] = P[J\cap B]+P[J\cap\overline{B}]+P[\overline{J}\cap B] \\ &= P[J]P[B]+P[J]P[\overline{B}]+P[\overline{J}]P[B] = (0.8)(0.7)+(0.8)(0.3)+(0.2)(0.7) \\ &= 0.94 \end{aligned}$$

Note that this is the complement of the probability that neither of them hits the target; that is,

$$p = 1-P[\overline{J}\cap\overline{B}] = 1-P[\overline{J}]P[\overline{B}] = 1-(0.2)(0.3) = 1-0.06 = 0.94$$

## 1.9 COMBINED EXPERIMENTS

Up till now our discussion has been limited to single experiments. Sometimes we are required to form an experiment by combining multiple individual experiments. Consider the case of two experiments in which one experiment has the sample space $\Omega_1$ with $N$ sample points and the other has the sample space $\Omega_2$ with the $M$ sample points. That is,

$$\begin{aligned} \Omega_1 &= \{x_1, x_2, \ldots, x_N\} \\ \Omega_2 &= \{y_1, y_2, \ldots, y_M\} \end{aligned}$$

If we form an experiment that is a combination of these two experiments, the sample space of the combined experiment is called the *combined space* (or the *Cartesian product space*) and is defined by

$$\Omega = \Omega_1 \times \Omega_2 = \left\{x_i, y_j | x_i \in \Omega_1, y_j \in \Omega_2; i = 1, 2, \ldots, N; j = 1, 2, \ldots, M\right\}$$

The combined sample space of an experiment that is a combination of $N$ experiments with sample spaces $\Omega_k, k = 1, 2, \ldots, N$, is given by:

$$\Omega = \Omega_1 \times \Omega_2 \times \cdots \times \Omega_N$$

Note that if $L_k$ is the number of sample points in $\Omega_k, k = 1, 2, \ldots, N$, then the number of sample points in $\Omega$ (also called the *cardinality* of $\Omega$) is given by

$$L = L_1 \times L_2 \times \cdots \times L_N$$

That is, the cardinality of $\Omega$ is the product of the cardinalities of the sample spaces of the different experiments.

### EXAMPLE 1.18

Consider a combined experiment formed from two experiments. The first experiment consists of tossing a coin and the second experiment consists of rolling a die. Let $\Omega_1$ denote the sample space of the first experiment and let $\Omega_2$ denote the sample space of the second experiment. If $\Omega$denotes the sample space of the combined experiment, we obtain the following:

$$\Omega_1 = \{H, T\}$$
$$\Omega_2 = \{1, 2, 3, 4, 5, 6\}$$
$$\Omega = \left\{ \begin{matrix} (H, 1),(H,2),(H,3),(H,4),(H,5),(H,6), \\ (T, 1),(T,2),(T,3),(T,4),(T,5),(T,6) \end{matrix} \right\}$$

As we can see, the number of sample points in $\Omega$ is the product of the number of sample points in the two sample spaces. If we assume that the coin and die are fair, then the sample points in $\Omega$ are equiprobable; that is, each sample point is equally likely to occur. Thus, for example, if we define $X$ to be the event "a head on the coin and an even number of the die," then $X$ and its probability are given by:

$$X = \{(H, 2), (H, 4), (H, 6)\}$$
$$P[X] = \frac{3}{12} = \frac{1}{4}$$

An alternative way to solve the problem is as follows. Let $H$ denote the event that the coin comes up heads, and $E$ the event that the die comes up an even number. Then $X = H \cap E$. Because the events $H$ and $E$ are independent, we obtain

$$P[X] = P[H \cap E] = P[H]P[E] = \frac{1}{2} \times \frac{3}{6} = \frac{1}{4}$$

## 1.10 BASIC COMBINATORIAL ANALYSIS

Combinatorial analysis deals with counting the number of different ways in which an event of interest can occur. Two basic aspects of combinatorial analysis that are used in probability theory are permutation and combination.

### 1.10.1 Permutations

Sometimes we are interested in how the outcomes of an experiment can be arranged; that is, we are interested in the order of the outcomes of an experiment. For example, if the possible outcomes are A, B and C, we can think of six possible arrangements of these outcomes: ABC, ACB, BAC, BCA, CAB, and CBA. Each of these arrangements is called a *permutation*. Thus, there are six permutations of a set of three distinct objects. This number can be derived as follows: There are 3 ways of choosing the first object; after the first object has been chosen, there are two ways of choosing the second object; and after the first two objects have been chosen, there is 1 way to choose the third object. This means that there are $3 \times 2 \times 1 = 6$ permutations.

For a system of $n$ distinct objects we can apply a similar reasoning to obtain the following number of permutations:

$$n \times (n-1) \times (n-2) \times \cdots \times 3 \times 2 \times 1 = n!$$

where $n!$ is read as "$n$ factorial." By convention, $0! = 1$.

Assume that we want to arrange the $n$ objects taking $r$ at a time. The problem now becomes that of finding how many possible sequences of $r$ objects we can get from $n$ objects, where $r \leq n$. This number is denoted by $P(n,r)$ and defined as follows:

$$P(n,r) = \frac{n!}{(n-r)!} = n \times (n-1) \times (n-2) \times \cdots \times (n-r+1), \quad r = 1, 2, \ldots, n \tag{1.20}$$

The number $P(n,r)$ represents the number of permutations (or sequences) of $r$ objects taken at a time from $n$ objects when the arrangement of the objects within a given sequence is important. Note that when $n=r$, we obtain:

$$P(n,n) = \frac{n!}{(n-n)!} = \frac{n!}{0!} = n!$$

## EXAMPLE 1.19

A little girl has six building blocks and is required to select four of them at a time to build a model. If the order of the blocks in each model is important, how many models can she build?

**Solution:**
Since the order of objects is important, this is a permutation problem. Therefore, the number of models is given by:

$$P(6,4) = \frac{6!}{(6-4)!} = \frac{6!}{2!} = \frac{6 \times 5 \times 4 \times 3 \times 2 \times 1}{2 \times 1} = 360$$

Note that if the little girl were to select three blocks at a time, the number of permutations decreases to 120; and if she were to select two blocks at a time, the number of permutations will be 30.

## EXAMPLE 1.20

How many words can be formed from the word SAMPLE? Assume that a formed word does not have to be an actual English word, but it may contain at most as many instances of a letter as there are in the original word (for example, "maa" is not acceptable since "a" does not appear twice in SAMPLE, but "mas" is allowed).

**Solution:**
The words can be single-letter words, two-letter words, three-letter words, four-letter words, five-letter words or six-letter words. Since the letters of the word SAMPLE are all unique, there are $P(6,k)$ ways of forming $k$-letter words, $k=1,2,\ldots,6$. Thus, the number of words that can be formed is

$$\begin{aligned}N &= P(6,1)+P(6,2)+P(6,3)+P(6,4)+P(6,5)+P(6,6)\\ &= 6+30+120+360+720+720=1956\end{aligned}$$

We present the following theorem without proof:

**Theorem**
Given a population of $n$ elements, let $n_1,n_2,\ldots,n_k$ be positive integers such that $n_1+n_2+\cdots+n_k=n$. Then there are

$$N=\frac{n!}{n_1!\times n_2!\times\cdots\times n_k!} \tag{1.21}$$

ways to partition the population into $k$ subgroups of sizes $n_1,n_2,\ldots,n_k$, respectively.

## EXAMPLE 1.21

Five identical red blocks, two identical white blocks and three identical blue blocks are arranged in a row. How many different arrangements are possible?

**Solution:**
In this example, $n_1=5, n_2=2, n_3=3$ and $n=5+2+3=10$. Thus, the number of possible arrangements is given by:

$$N=\frac{10!}{5!\times 2!\times 3!}=2520$$

## EXAMPLE 1.22

How many words can be formed by using all the letters of the word MISSISSIPPI?

**Solution:**
The word contains 11 letters consisting of 1 M, 4 S's, 4 I's, and 2 P's. Thus, the number of words that can be formed is

$$N=\frac{11!}{1!\times 4!\times 4!\times 2!}=34650$$

### 1.10.2 Circular Arrangement

Consider the problem of seating $n$ people in a circle. Assume that the positions are labeled $1,2,\ldots,n$. Then, if after one arrangement everyone moves one place either to the left or to the right, that will also be an arrangement because each person is occupying a new location. However, each person's previous neighbors to the left and right are still his/her neighbors in the new arrangement. This means that such a move does not lead to a new valid arrangement. To solve this problem, one person must remain fixed while the others move.

Thus, the number of people being arranged is $n-1$, which means that the number of possible arrangements is $(n-1)!$ For example, the number of ways that 10 people can be seated in a circle is $(10-1)!=9!=362880$.

### 1.10.3 Applications of Permutations in Probability

Consider a system that contains $n$ distinct objects labeled $a_1, a_2, \ldots, a_n$. Assume that we choose $r$ of these objects in the following manner. We choose the first object, record its type and put it back into the "population." We then choose the second object, record its type and put it back into the population. We continue this process until we have chosen a total of $r$ objects. This gives an "ordered sample" consisting of $r$ of the $n$ objects. The question is to determine the number of distinct ordered samples that can be obtained, where two ordered samples are said to be distinct if they differ in at least one entry in a particular position within the samples. Since the number of ways of choosing an object in each round is $n$, the total number of distinct samples is $n \times n \times \cdots \times n = n^r$.

Assume now that the sampling is done without replacement. That is, after an object has been chosen, it is not put back into the population. Then the next object from the remainder of the population is chosen and not replaced, and so on until all the $r$ objects have been chosen. The total number of possible ways of making this sampling can be obtained by noting that there are $n$ ways to choose the first object, $n-1$ ways to choose the second object, $n-2$ ways to choose the third object, and so on, and finally there are $n-r+1$ ways to choose the $r$th object. Thus, the total number of distinct samples is $n \times (n-1) \times (n-2) \times \cdots \times (n-r+1)$.

## EXAMPLE 1.23

A subway train consists of $n$ cars. The number of passengers waiting to board the train is $k<n$ and each passenger enters a car at random. What is the probability that all the $k$ passengers end up in different cars of the train?

**Solution:**

The probability $p$ of an event of this nature where a constraint or restriction is imposed on the participants is given by

$$p = \frac{\text{Number of Restricted (or Constrained) Arrangements}}{\text{Number of Unrestricted (or Unconstrained) Arrangements}}$$

Now, without any restriction on the occupancy of the cars, each of the $k$ passengers can enter any one of the $n$ cars. Thus, the number of distinct, unrestricted arrangements of the passengers in the cars is $N = n \times n \times \cdots \times n = n^k$.

If the passengers enter the cars in such a way that there is no more than one passenger in a car, then the first passenger can enter any one of the $n$ cars. After the first passenger has entered a car, the second passenger can enter any one of the $n-1$ remaining cars. Similarly, the third passenger can enter any one of the $n-2$ remaining cars, and so on. Finally, the $k$th passenger can enter any one of the $n-k+1$ remaining cars. Thus, the total number of distinct arrangements of passengers when no two passengers can be in the same car is $M = n \times (n-1) \times (n-2) \times \cdots \times (n-k+1)$. Therefore, the probability of this event is

$$p = \frac{M}{N} = \frac{n \times (n-1) \times (n-2) \times \cdots \times (n-k+1)}{n^k}$$

## EXAMPLE 1.24

Ten books are placed in random order on a bookshelf. Find the probability that three given books are placed side by side.

**Solution:**

The number of unrestricted ways of arranging the books is 10!. Consider the three books as a "superbook," which means that there are 8 "books" on the bookshelf. The number of ways of arranging these books is 8!. In all these arrangements, the three books can be arranged among themselves in $3! = 6$ ways. Thus, the total number of arrangements with the three books together is 8!3!, and the required probability $p$ is given by:

$$p = \frac{8!3!}{10!} = \frac{6 \times 8!}{10 \times 9 \times 8!} = \frac{6}{90} = \frac{1}{15}$$

### 1.10.4 Combinations

In permutations, the order of objects within a selection is important; that is, the arrangement of objects within a selection is very important. Thus, the arrangement ABC is different from the arrangement ACB even though they both contain the same three objects. In some problems, the order of objects within a selection is not relevant. For example, consider a student who is required to select four subjects out of six subjects in order to graduate. Here, the order of subjects in not important; all that matters is that the student selects four subjects.

Since the order of the objects within a selection is not important, the number of ways of choosing $r$ objects from $n$ objects will be smaller than when the order is important. The number of ways of selecting $r$ objects at a time from $n$ objects when the order of the objects is not important is called the *combination* of

$r$ objects taken from $n$ distinct objects and denoted by $C(n,r)$. It is defined as follows:

$$C(n,r)=\binom{n}{r}=\frac{P(n,r)}{r!}=\frac{n!}{(n-r)!r!} \tag{1.22}$$

Recall that $r!$ is the number of permutations of $r$ objects taken $r$ at a time. Thus, $C(n,r)$ is equal to the number of permutations of $n$ objects taken $r$ at a time divided by the number of permutations of $r$ objects taken $r$ at a time.

Observe that $C(n,r)=C(n,n-r)$, as can be seen from the above equation. One very useful combinatorial identity is the following:

$$\binom{n+m}{k}=\sum_{i=0}^{k}\binom{n}{i}\binom{m}{k-i} \tag{1.23}$$

This identity can easily be proved by considering how many ways we can select $k$ people from a group of $m$ boys and $n$ girls. In particular, when $m=k=n$ we have that

$$\begin{aligned}\binom{2n}{n}&=\binom{n}{0}\binom{n}{n}+\binom{n}{1}\binom{n}{n-1}+\binom{n}{2}\binom{n}{n-2}+\cdots+\binom{n}{k}\binom{n}{n-k}+\cdots+\binom{n}{n}\binom{n}{0}\\&=\binom{n}{0}^2+\binom{n}{1}^2+\binom{n}{2}^2+\cdots+\binom{n}{k}^2+\cdots+\binom{n}{n}^2\\&=\sum_{k=0}^{n}\binom{n}{k}^2\end{aligned} \tag{1.24}$$

where the second to the last equality follows from the fact that $\binom{n}{k}=\binom{n}{n-k}$.

## EXAMPLE 1.25

Evaluate $\binom{16}{8}$.

**Solution:**

$$\begin{aligned}\binom{16}{8}&=\sum_{k=0}^{8}\binom{8}{k}^2\\&=\binom{8}{0}^2+\binom{8}{1}^2+\binom{8}{2}^2+\binom{8}{3}^2+\binom{8}{4}^2+\binom{8}{5}^2+\binom{8}{6}^2+\binom{8}{7}^2+\binom{8}{8}^2\\&=1^1+8^2+28^2+56^2+70^2+56^2+28^2+8^2+1^1\\&=12{,}870\end{aligned}$$

## EXAMPLE 1.26

A little girl has six building blocks and is required to select four of them at a time to build a model. If the order of the blocks in each model is not important, how many models can she build?

**Solution:**

The number of models is

$$C(6,4) = \frac{6!}{(6-4)!4!} = \frac{6!}{2!4!} = \frac{6 \times 5 \times 4!}{2 \times 1 \times 4!} = \frac{30}{2} = 15$$

Recall that $P(6,4) = 360$. Also $P(6,4)/C(6,4) = 24 = 4!$, which indicates that for each combination, there are 4! arrangements involved.

## EXAMPLE 1.27

Five boys and five girls are getting together for a party.

a. How many couples can be formed, where a couple is defined as a boy and a girl?
b. Suppose one of the boys has two sisters among the five girls, and he would not accept either of them as a partner. How many couples can be formed?
c. Assume that couples can be formed without regard to their sexes. That is, a boy can be paired with either a boy or a girl, and a girl can be paired with either a girl or a boy. How many couples can be formed?

**Solution:**

a. Without any restrictions, the first boy can choose from 5 girls, the second boy can choose from 4 girls, the third boy can choose from 3 girls, the fourth boy can choose from 2 girls, and the fifth boy has only one choice. Thus, the number of couples that can be formed is $5 \times 4 \times 3 \times 2 \times 1 = 5! = 120$.
b. The boy that has two sisters among the girls can only be matched with 3 girls, but each of the other four boys can be matched with any of the girls. The pairing will begin with the boy whose sisters are in the group; he can choose from 3 girls. After him, the first boy can choose from 4 girls, the second boy can choose from 3 girls, the third boy can choose from 2 girls, and the fourth boy has only one choice. Thus, the number of possible couples is given by $3 \times 4 \times 3 \times 2 \times 1 = 3 \times 4! = 72$.
c. Assume that we randomly number the 10 boys and girls (who we will refer to as teenagers or teens) from 1 to 10 and have them paired in an ascending order. The first teen can choose from 9 teens. After the couple is formed, there will be 8 unpaired teens. The next lowest-ordered unpaired teen can choose from 7 teens. After the two couples are formed there will be 6 unpaired teens. The next lowest-ordered unpaired teen can choose from 5 unpaired teens, leaving 4 unpaired teens. The next lowest-ordered unpaired teen can choose from 3 teens, leaving 2 unpaired teens who will be the last couple. Thus, the number of couples is $9 \times 7 \times 5 \times 3 \times 1 = 945$.

### 1.10.5 The Binomial Theorem

The following theorem, which is called the binomial theorem, is presented without proof. The theorem states that

$$(a+b)^n = \sum_{k=0}^{n} \binom{n}{k} a^k b^{n-k} \tag{1.25}$$

This theorem can be used to present a more formal proof of the statement we made earlier in the chapter about the number of subsets of a set with $n$ elements. The number of subsets of size $k$ is $\binom{n}{k}$. Thus, summing this over all possible values of $k$ we obtain the desired result:

$$\sum_{k=0}^{n} \binom{n}{k} = \sum_{k=0}^{n} \binom{n}{k} (1)^k (1)^{n-k} = (1+1)^n = 2^n$$

### 1.10.6 Stirling's Formula

Problems involving permutations and combinations require the calculation of $n!$. Even for a moderately large $n$ the evaluation of $n!$ becomes tedious. An approximate formula for large values of $n$, called the Stirling's formula, is given by

$$n! \sim \sqrt{2\pi n}\left(\frac{n}{e}\right)^n = \sqrt{2\pi n}\, n^n e^{-n} \tag{1.26}$$

where $e = 2.71828\ldots$ is the base of the natural logarithms and the notation $a \sim b$ means that the number on the right is an asymptotic representation of the number on the left. As a check on the accuracy of the formula, by direct computation $10! = 3,628,800 = 3.6288 \times 10^6$ while the value obtained via the Stirling's formula is $3.60 \times 10^6$, which represents an error of 0.79%. In general, the percentage error in the approximation is about $100/12n$, which means that the approximation error becomes smaller as $n$ increases.

## EXAMPLE 1.28

Evaluate 50!

**Solution:**

Using the Stirling's formula we obtain:

$$50! \sim \sqrt{100\pi}(50)^{50} e^{-50} = 10\sqrt{\pi}\left(\frac{50}{2.71828}\right)^{50} = 3.04 \times 10^{64}$$

This compares well with the exact value, which is $50! = 3.04140932 \times 10^{64}$.

## EXAMPLE 1.29

Evaluate 70!

**Solution:**
Using the Stirling's formula we obtain:

$$70! \sim \sqrt{140\pi}(70)^{70}e^{-70} = \sqrt{140\pi}\left(\frac{70}{2.71828}\right)^{70} = 1.20 \times 10^{100}$$

We can obtain the result as follows:

$$N = 70! \sim \sqrt{140\pi}\left(\frac{70}{2.71828}\right)^{70}$$

$$\log N = \frac{1}{2}\log 140 + \frac{1}{2}\log \pi + 70\log 70 - 70\log 2.71828$$

$$= 1.07306 + 0.24857 + 129.15686 - 30.40061 = 100.07788 = 100 + 0.07788$$

$$N = 1.20 \times 10^{100}$$

### 1.10.7 The Fundamental Counting Rule

As we shall see in Chapter 4, combination plays a very important role in the class of random variables that have the binomial distribution as well as those that have the Pascal and hypergeometric distributions. In this section we discuss how it can be applied to the problem of counting the number of selections among items that contain multiple subgroups. To understand these applications we first state the following ***fundamental counting rule***:

> Assume that a number of multiple choices are to be made, which include $m_1$ ways of making the first choice, $m_2$ ways of making the second choice, $m_3$ ways of making the third choice, and so on. If these choices can be made independently, then the total number of possible ways of making these choices is $m_1 \times m_2 \times m_3 \times \cdots$.

## EXAMPLE 1.30

The standard car license plate in a certain state has seven characters that are made up as follows. The first character is one of the digits 1, 2, 3 or 4; the next three characters are letters $(a, b, \ldots, z)$ of which repetition is allowed; and the final three characters are digits $(0, 1, \ldots, 9)$ that also allow repetition.

a. How many license plates are possible?
b. How many of these possible license plates have no repeated characters?

**Solution:**
Let $m_1$ be the number of ways of choosing the first character, $m_2$ the number of ways of choosing the next three characters, and $m_3$ the number of ways of choosing the final three characters. Since

these choices can be made independently, the principle of the fundamental counting rule implies that there are $m_1 \times m_2 \times m_3$ total number of possible ways of making these choices.

a. $m_1 = C(4,1) = 4$; since repetition is allowed, $m_2 = \{C(26,1)\}^3 = 26^3$; and since repetition is allowed, $m_3 = \{C(10,1)\}^3 = 10^3$. Thus, the number of possible license plates is $4 \times 26^3 \times 10^3 = 70,304,000$.

b. When repetition is not allowed, we obtain $m_1 = C(4,1) = 4$. To obtain the new $m_2$ we note that after the first letter has been chosen it cannot be chosen again as the second or third letter, and after the second letter has been chosen it cannot be chosen as the third letter. This means that there are $m_2 = C(26,1) \times C(25,1) \times C(24,1) = 26 \times 25 \times 24$ ways of choosing the next three letters of the plate. Similarly, since repetition is not allowed, the digit chosen as the first character of the license plate cannot appear in the third set of characters. This means that the first digit of the third set of characters will be chosen from 9 digits, the second from 8 digits, and the third from 7 digits. Thus we have that $m_3 = 9 \times 8 \times 7$. Therefore, the number of possible license plates that have no repeated characters is given by

$$M = 4 \times 26 \times 25 \times 24 \times 9 \times 8 \times 7 = 31,449,600$$

## EXAMPLE 1.31

Suppose there are $k$ defective items in a box that contains $m$ items. How many samples of $n$ items of which $j$ items are defective can we get from the box, where $n < m$?

**Solution:**

Since there are two classes of items (defective and non-defective), we can select independently from each group once the number of defective items in the sample has been specified. Thus, since there are $k$ defective items in the box, the total number of ways of selecting $j$ out of the $k$ items at a time is $C(k,j)$, where $0 \le j \le \min(k,n)$. Similarly, since there are $m-k$ non-defective items in the box, the total number of ways of selecting $n-j$ of them at a time is $C(m-k, n-j)$. Since these two choices can be made independently, the total number of ways of choosing $j$ defective items and $n-j$ non-defective items is $C(k,j) \times C(m-k, n-j)$, which is

$$C(k,j) \times C(m-k, n-j) = \binom{k}{j}\binom{m-k}{n-j}$$

## EXAMPLE 1.32

A container has 100 items 5 of which are defective. If we randomly pick samples of 20 items from the container, find the total number of samples with at most one bad item among them.

**Solution:**

Let $A$ be the event that there is no defective item in the selected sample and $B$ the event that there is exactly one defective item in the selected sample. Then event $A$ consists of two subevents: zero defective items and 20 non-defective items. Similarly, event $B$ consists of two subevents: 1 defective item and 19 non-defective items. The number of ways in which event $A$ can occur is

$C(5,0) \times C(95,20) = C(95,20)$. Similarly, the number of ways in which event $B$ can occur is $C(5,1) \times C(95,19) = 5C(95,19)$. Because these two events are mutually exclusive, the total number of samples with at most one defective item is the sum of the two numbers, which is

$$C(95,20) + 5C(95,19) = \frac{95!}{75!20!} + \frac{5 \times 95!}{76!19!} = \frac{176 \times 95!}{76!20!} = 3.9633 \times 10^{20}$$

## EXAMPLE 1.33

A particular department of a small college has 7 faculty members of whom two are full professors, three are associate professors and two are assistant professors. How many committees of three faculty members can be formed if each subgroup (that is, full, associate and assistant professors) must be represented?

**Solution:**
There are $C(2,1) \times C(3,1) \times C(2,1) = 12$ possible committees.

### 1.10.8 Applications of Combinations in Probability

As in the case of permutation, we consider the probability of events that involve combinations. The probability of any such event is the ratio of the number of combinations when restrictions are imposed to the number of combinations when no restrictions are imposed. That is, the probability $p$ of an event is given by

$$p = \frac{\text{Number of Restricted Combinations}}{\text{Number of Unrestricted Combinations}}$$

## EXAMPLE 1.34

A batch of 100 manufactured components is checked by an inspector who examines 10 components selected at random. If none of the 10 components is defective, the inspector accepts the whole batch. Otherwise, the batch is subjected to further inspection. What is the probability that a batch containing 10 defective components will be accepted?

**Solution:**
Let $N$ denote the number of ways of selecting 10 components from a batch of 100 components. Then $N$ is the number of unrestricted combinations and is given by

$$N = C(100,10) = \frac{100!}{90! \times 10!}$$

Let $E$ denote the event "the batch containing 10 defective components is accepted by the inspector." The number of ways that $E$ can occur is the number of ways of selecting 10 components from the 90 non-defective components. This number, $N(E)$, is given by

$$N(E) = C(90, 10) \times C(10, 0) = C(90, 10) = \frac{90!}{80! \times 10!}$$

$N(E)$ is the number of restricted combinations. Because the components are selected at random, the combinations are equiprobable. Thus, the probability of event $E$ is given by

$$P[E] = \frac{N(E)}{N} = \frac{90!}{80! \times 10!} \times \frac{90! \times 10!}{100!} = \frac{90 \times 89 \times \cdots \times 81}{100 \times 99 \times \cdots \times 91} = 0.3305$$

## EXAMPLE 1.35

The Applied Probability professor gave the class a set of 12 review problems and told them that the midterm exam would consist of 6 of the 12 problems selected at random. If Kate memorized the solutions to 8 of the 12 problems but could not solve any of the other 4 problems and thus would fail them if they came in the exam, what is the probability that she got 4 or more problems correct in the exam?

**Solution:**

Kate partitioned the 12 problems into two sets: a set consisting of the 8 problems she memorized and a set consisting of the 4 problems she could not solve. If she got $k$ problems correct in the exam, then the $k$ problems came from the first set and the $6-k$ problems she failed came from the second set, where $k = 0, 1, 2, \ldots, 6$. The number of unrestricted ways of choosing 6 problems from 12 problems is $C(12, 6)$. The number of ways of choosing $k$ problems from the 8 problems that she memorized is $C(8, k)$, and the number of ways of choosing $6-k$ problems from the four she did not memorize is $C(4, 6-k)$, where $6-k \leq 4$ or $2 \leq k \leq 6$. Because the problems have been partitioned, the number of ways in which the 8 problems can be chosen so that Kate could get 4 or more of them correct in the exam is

$$C(8, 4)C(4, 2) + C(8, 5)C(4, 1) + C(8, 6)C(4, 0) = 420 + 224 + 28 = 672$$

This is the number of restricted combinations. Thus, the probability $p$ that she got 4 or more problems correct in the exam is given by

$$p = \frac{C(8, 4)C(4, 2) + C(8, 5)C(4, 1) + C(8, 6)C(4, 0)}{C(12, 6)} = \frac{672}{924} = \frac{8}{11}$$

## 1.11 RELIABILITY APPLICATIONS

As discussed earlier in the chapter, reliability theory is concerned with the duration of the useful life of components and systems of components. That is, it is concerned with determining the probability that a system with possibly many components will be functioning at time $t$. The components of a system can be arranged in two basic configurations: *series* configuration and *parallel* configuration. A real system consists of a mixture of series and parallel components, which can sometimes be reduced to an equivalent system of series configuration or a system of parallel configuration. Figure 1.10 illustrates the two basic configurations between points A and B.

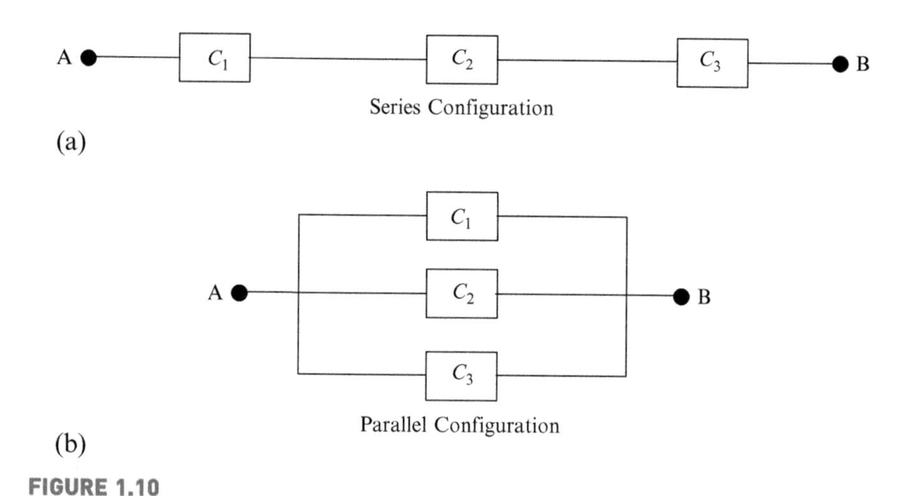


**FIGURE 1.10**
Basic Reliability Models

A system with a series configuration will function iff all its components are functioning while a system with parallel configuration will function iff at least one of the components is functioning. To simplify the discussion, we assume that the different components fail independently.

Consider a system with $n$ components labeled $C_1, C_2, \ldots, C_n$. Let $R_k(t)$ denote the probability that component $C_k$ has not failed in the interval $(0, t]$, where $k = 1, 2, \ldots, n$. That is, $R_k(t)$ is the probability that $C_k$ has not failed up to time $t$ and is called the *reliability function* of $C_k$. For a system of components in series, the system reliability function is given by

$$R(t) = R_1(t) \times R_2(t) \times \ldots \times R_n(t) = \prod_{k=1}^{n} R_k(t) \tag{1.27}$$

This follows from the fact that all components must be operational for the system to be operational. The system has a "single point of failure" because the failure of any one component causes it to fail.

In the case of a system of parallel components, we need at least one path between A and B for the system to be operational. The probability that no such path exists is the probability that all the components have failed. We call this probability the "unreliability function" of the system and it is given by

$$\overline{R}(t) = [1 - R_1(t)] \times [1 - R_2(t)] \times \cdots \times [1 - R_n(t)] \tag{1.28}$$

Thus, the system reliability function is the complement of the unreliability function and is given by

$$R(t) = 1 - \overline{R}(t) = 1 - [1 - R_1(t)] \times [1 - R_2(t)] \times \cdots \times [1 - R_n(t)]$$
$$= 1 - \prod_{k=1}^{n} [1 - R_k(t)] \quad (1.29)$$

We will sometimes denote $R_x(t)$ by $R_x$, omitting the time parameter.

## EXAMPLE 1.36

Find the system reliability function for the system shown in Figure 1.11 in which $C_1$, whose reliability function is $R_1(t)$, and $C_2$, whose reliability function is $R_2(t)$, are in series, and the two are in parallel with $C_3$ whose reliability function is $R_3(t)$.

**Solution:**

We first reduce the series structure into a composite component $C_{1-2}$ whose reliability function is given by $R_{1-2}(t) = R_1(t)R_2(t)$. Thus, we obtain the new structure shown in Figure 1.12.

This means that we obtain two parallel components, and the system reliability function is

$$R(t) = 1 - [1 - R_3(t)][1 - R_{1-2}(t)] = 1 - [1 - R_3(t)][1 - R_1(t)R_2(t)]$$

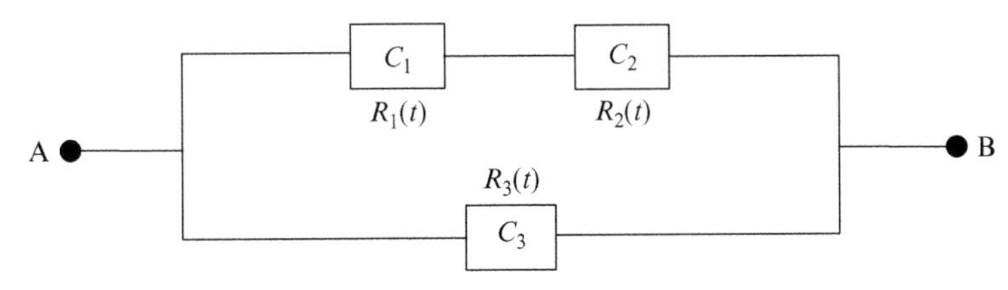


**FIGURE 1.11**
Figure for Example 1.36

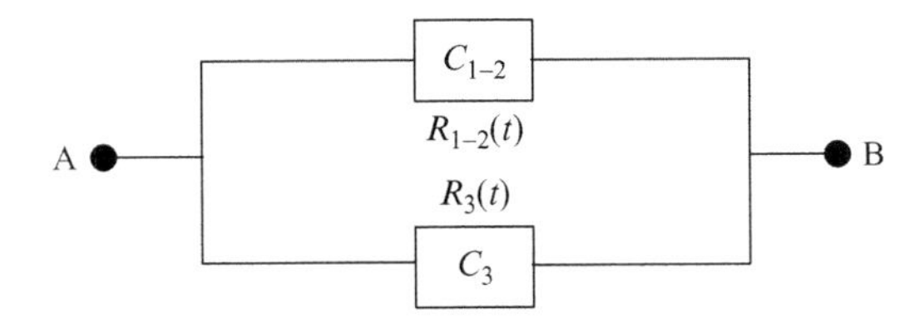


**FIGURE 1.12**
Composite System for Example 1.36

## EXAMPLE 1.37

Consider the network shown in Figure 1.13 that interconnects nodes A and B. The switches $S_1, S_2, S_3$ and $S_4$ have reliability functions $R_1, R_2, R_3$ and $R_4$, respectively, where we have dropped the time function for ease of manipulation. If the switches fail independently, what is the reliability function of the path between A and B?

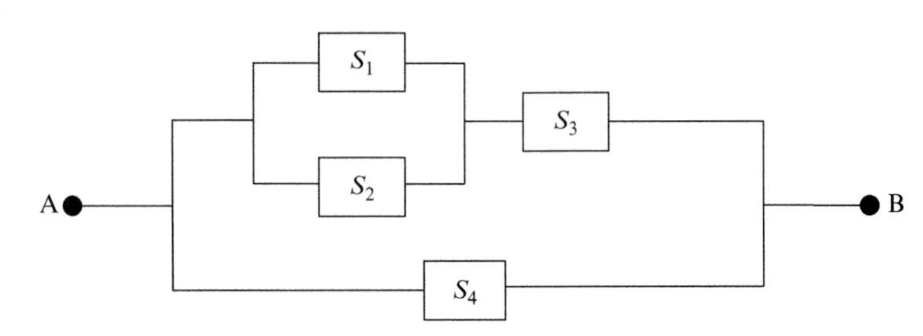


**FIGURE 1.13**
Figure for Example 1.37

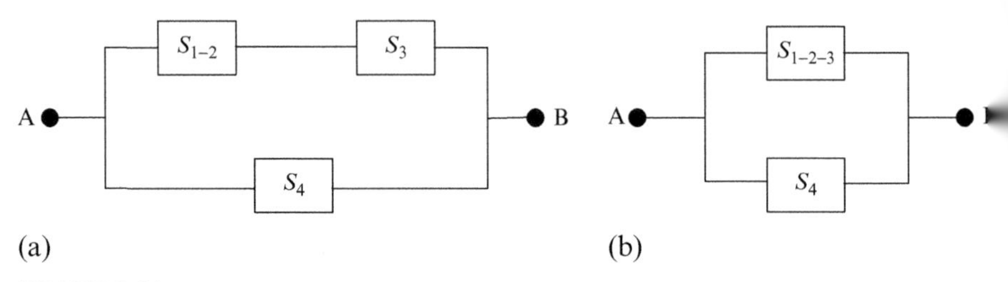


**FIGURE 1.14**
Reduced Forms of Figure 1.13

**Solution:**
We begin by reducing the structure as shown in Figure 1.14 where $S_{1-2}$ is the composite system of $S_1$ and $S_2$, and $S_{1-2-3}$ is the composite system of $S_{1-2}$ and $S_3$.

From Figure 1.14(a), the reliability of $S_{1-2}$ is $R_{1-2}=1-(1-R_1)(1-R_2)$. Similarly, the reliability of $S_{1-2-3}$ is $R_{1-2-3}=R_{1-2}R_3$. Finally, from Figure 1.14(b), the reliability of the path between A and B is given by

$$R_{AB}=1-(1-R_{1-2-3})(1-R_4)$$

where $R_{1-2-3}$ is as defined earlier.

## EXAMPLE 1.38

Find the system reliability function for the system shown in Figure 1.15, which is called a *bridge structure*.

**Solution:**
The system is operational if at least one of the following series arrangements is operational: $C_1C_4$, $C_2C_5$, $C_1C_3C_5$, or $C_2C_3C_4$. Thus, we can replace the system with a system of series-parallel arrangements. However, the different paths will not be independent since they have components in common. To avoid this complication, we use a conditional probability approach. First, we consider the reliability function of the system given that the bridge, $C_3$, is operational. Next we consider the reliability function of the system given that $C_3$ is not operational. Figure 1.16 shows the two cases.

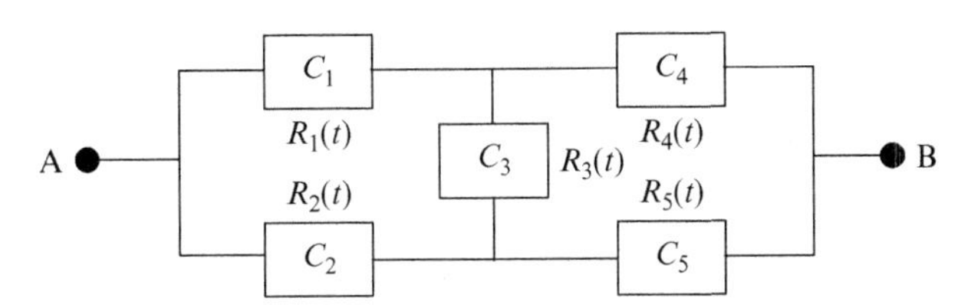


**FIGURE 1.15**
Figure for Example 1.38

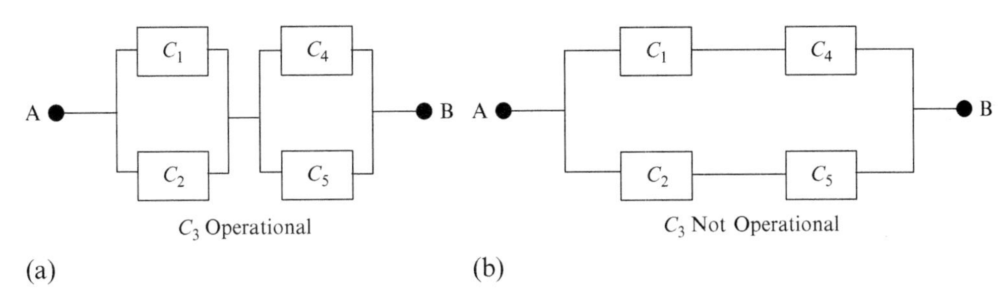


**FIGURE 16**
Decomposing the System into Two Cases

When $C_3$ is operational, the system behaves like a parallel subsystem consisting of $C_1$ and $C_2$, which is in series with another parallel subsystem consisting of $C_4$ and $C_5$, as shown in Figure 1.16(a). Let $X$ denote this event. Thus, if we omit the time function, the reliability of the system under event $X$ becomes

$$R_X = [1-(1-R_1)(1-R_2)][1-(1-R_4)(1-R_5)]$$

When $C_3$ is not operational, no signal flows through that component and the system behaves as shown in Figure 1.16(b). Let $Y$ denote this event. Thus, the reliability of the system under event $Y$ becomes

$$R_Y = 1-(1-R_1R_4)(1-R_2R_5)$$

Now, $P[X]=R_3$ and $P[Y]=1-R_3$. Thus, we use the law of total probability to obtain the system reliability as follows:

$$\begin{aligned} R &= P[X]R_X + {}^X P[Y]R_Y = R_3R_X + {}^X(1-R_3)R_Y \\ &= R_3[1-(1-R_1)(1-R_2)][1-(1-R_4)(1-R_5)] + (1-R_3)\{1-(1-R_1R_4)(1-R_2R_5)\} \\ &= R_1R_4 + R_2R_5 + R_1R_3R_5 + R_2R_3R_4 \\ &\quad -R_1R_2R_3R_4 - R_1R_2R_3R_5 - R_1R_2R_4R_5 - R_1R_3R_4R_5 - R_2R_3R_4R_5 \\ &\quad + 2R_1R_2R_3R_4R_5 \end{aligned}$$

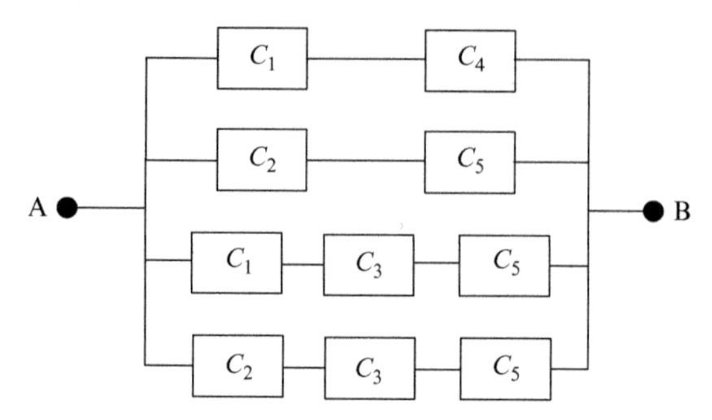


**FIGURE 1.17**
Alternative System Configuration for Example 1.38

The first four positive terms represent the different ways we can pass signals between the input and output. Thus, the equivalent system configuration is as shown in Figure 1.17. The other terms account for the dependencies we mentioned earlier.

## 1.12 CHAPTER SUMMARY

This chapter has developed the basic concepts of probability, random experiments and events. Several examples are solved and applications of probability have been provided in the fields of communications and reliability engineering. Finally, it has introduced the concepts of permutation and combination that will be used in later chapters.

## 1.13 PROBLEMS

### Section 1.2 Sample Space and Events

1.1 A fair die is rolled twice. Find the probability of the following events:
  a. The second number is twice the first
  b. The second number is not greater than the first
  c. At least one number is greater than 3.

1.2 Two distinct dice A and B are rolled. What is the probability of each of the following events?
  a. At least one 4 appears
  b. Just one 4 appears
  c. The sum of the face values is 7
  d. One of the values is 3 and the sum of the two values is 5
  e. One of the values is 3 or the sum of the two values is 5

1.3 Consider an experiment that consists of rolling a die twice.
   a. Plot the sample space $\Omega$ of the experiment
   b. Identify the event A, which is the event that the sum of the two outcomes is equal to 6.
   c. Identify the event B, which is the event that the difference between the two outcomes is equal to 2.

1.4 A 4-sided fair die is rolled twice. What is the probability that the outcome of the first roll is greater than the outcome of the second roll?

1.5 A coin is tossed until the first head appears, and then the experiment is stopped. Define a sample space for the experiment.

1.6 A coin is tossed four times and observed to be either a head or a tail each time. Describe the sample space for the experiment.

1.7 Three friends, Bob, Chuck, and Dan take turns (in that order) throwing a die until the first "six" appears. The person that throws the first six wins the game, and the game ends. Write down a sample space for this game.

## Section 1.3 Definitions of Probability

1.8 A small country has a population of 17 million people of whom 8.4 million are male and 8.6 million are female. If 75% of the male population and 63% of the female population are literate, what percentage of the total population is literate?

1.9 Let $A$ and $B$ be two independent events with $P[A]=0.4$ and $P[A\cup B]=0.7$. What is $P[B]$?

1.10 Consider two events $A$ and $B$ with known probabilities $P[A], P[B]$ and $P[A\cap B]$. Find the expression for the event that exactly one of the two events occurs in terms of $P[A], P[B]$ and $P[A\cap B]$.

1.11 Two events $A$ and $B$ have the following probabilities: $P[A]=1/4$, $P[B|A]=1/2$ and $P[A|B]=1/3$. Compute (a) $P[A\cap B]$, (b) $P[B]$ and (c) $P[A\cup B]$.

1.12 Two events $A$ and $B$ have the following probabilities: $P[A]=0.6$, $P[B]=0.7$ and $P[A\cap B]=p$. Find the range of values that $p$ can take.

1.13 Two events $A$ and $B$ have the following probabilities: $P[A]=0.5$, $P[B]=0.6$ and $P\left[\overline{A}\cap\overline{B}\right]=0.25$. Find the value of $P[A\cap B]$.

1.14 Two events $A$ and $B$ have the following probabilities: $P[A]=0.4$, $P[B]=0.5$ and $P[A\cap B]=0.3$. Calculate the following:
   a. $P[A\cup B]$
   b. $P\left[A\cap\overline{B}\right]$
   c. $P\left[\overline{A}\cup\overline{B}\right]$

1.15 Christie is taking a multiple-choice test in which each question has four possible answers. She knows the answers to 40% of the questions and can narrow the choices down to two answers 40% of the time. If she knows nothing about the remaining 20% of the questions and merely guesses the answers to these questions, what is the probability that she will correctly answer a question chosen at random from the test?

1.16 A box contains 9 red balls, 6 white balls, and 5 blue balls. If 3 balls are drawn successively from the box, determine the following:

a. The probability that they are drawn in the order red, white, and blue if each ball is replaced after it has been drawn.

b. The probability that they are drawn in the order red, white, and blue if each ball is not replaced after it has been drawn.

1.17 Let $A$ be the set of positive even integers, let $B$ be the set of positive integers that are divisible by 3, and let $C$ be the set of positive odd integers. Describe the following events in words:

a. $E_1 = A \cup B$

b. $E_2 = A \cap B$

c. $E_3 = A \cap C$

d. $E_4 = (A \cup B) \cap C$

e. $E_5 = A \cup (B \cap C)$

1.18 A box contains 4 red balls labeled $R_1, R_2, R_3$ and $R_4$; and 3 white balls labeled $W_1, W_2$ and $W_3$. A random experiment consists of drawing a ball from the box. State the outcomes of the following events:

a. $E_1$, the event that the number on the ball (i.e., the subscript of the ball) is even

b. $E_2$, the event that the color of the ball is red and its number is greater than 2

c. $E_3$, the event that the number on the ball is less than 3

d. $E_4 = E_1 \cup E_3$

e. $E_5 = E_1 \cup (E_2 \cap E_3)$

1.19 A box contains 50 computer chips of which 8 are known to be bad. A chip is selected at random and tested.

a. What is the probability that it is bad?

b. If a test on the first chip shows that it is bad, what is the probability that a second chip selected at random will also be bad, assuming the tested chip is not put back into the box?

c. If the first chip tests good, what is the probability that a second chip selected at random will be bad, assuming a tested chip is not put back into the box?

## Section 1.5 Elementary Set Theory

1.20 A set $\Omega$ has four members: $A$, $B$, $C$, and $D$. Determine all possible subsets of $\Omega$.

1.21 For three sets $A$, $B$, and $C$, use the Venn diagram to show the areas corresponding to the sets

a. $(A \cup C) - C$

b. $\overline{B} \cap A$

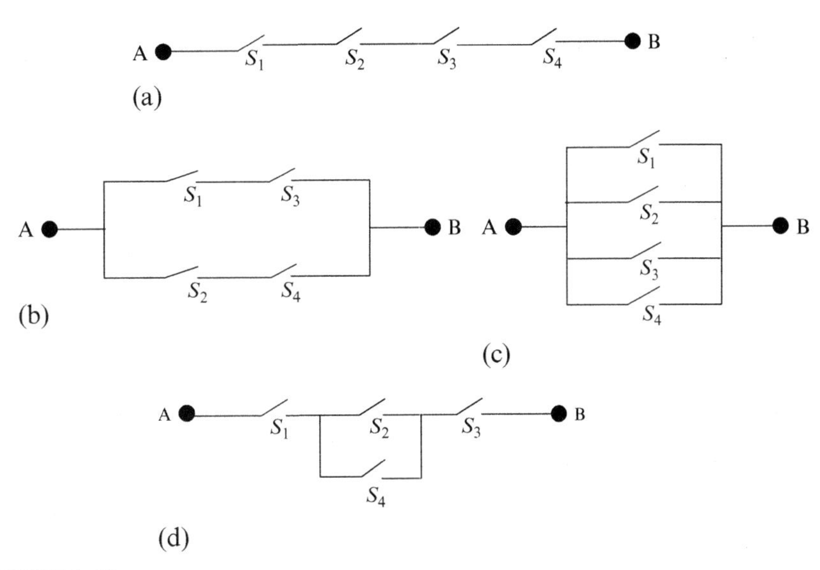


**FIGURE 1.18**
Figure for Problem 1.23

c. $A \cap B \cap C$
d. $(\overline{A \cup B}) \cap C$

1.22 A universal set is given by $\Omega = \{2, 4, 6, 8, 10, 12, 14\}$. If we define two sets $A = \{2, 4, 8\}$ and $B = \{4, 6, 8, 12\}$, determine the following:
a. $\overline{A}$
b. $B - A$
c. $A \cup B$
d. $A \cap B$
e. $\overline{A} \cap B$
f. $(A \cap B) \cup (\overline{A} \cap B)$

1.23 Consider the switching networks shown in Figure 1.18. Let $E_k$ denote the event that switch $S_k$ is closed, $k = 1, 2, 3, 4$. Let $E_{AB}$ denote the event that there is a closed path between nodes $A$ and $B$. Express $E_{AB}$ in terms of the $E_k$ for each network.

1.24 Let $A$, $B$ and $C$ be three events. Write out the expressions for the following events in terms of $A$, $B$ and $C$ using set notation:
a. $A$ occurs but neither $B$ nor $C$ occurs.
b. $A$ and $B$ occur, but not $C$.
c. $A$ or $B$ occurs, but not $C$.
d. Either $A$ occurs and $B$ does not occur, or $B$ occurs and $A$ does not occur.

## Section 1.6 Properties of Probability

1.25 Mark and Lisa registered for Physics 101 class. Mark attends class 65% of the time and Lisa attends class 75% of the time. Their absences are independent. On a given day, what is the probability that

a. at least one of them is in class?

b. exactly one of them is in class?

c. Mark is in class, given that only one of them is in class?

1.26 The probability of rain on a day of the year selected at random is 0.25 in a certain city. The local weather forecast is correct 60% of the time when the forecast is rain and 80% of the time for other forecasts. What is the probability that the forecast on a day selected at random is correct?

1.27 53% of the adults in a certain city are female, and 15% of the adults are unemployed males.

a. What is the probability that an adult chosen at random in this city is an employed male?

b. If the overall unemployment rate in the city is 22%, what is the probability that a randomly selected adult is an employed female?

1.28 A survey of 100 companies shows that 75 of them have installed wireless local area networks (WLANs) on their premises. If three of these companies are chosen at random without replacement, what is the probability that each of the three has installed WLANs?

## Section 1.7 Conditional Probability

1.29 A certain manufacturer produces cars at two factories labeled A and B. 10% of the cars produced at factory A are found to be defective, while 5% of the cars produced at factory B are defective. If factory A produces 100,000 cars per year and factory B produces 50,000 cars per year, compute the following:

a. The probability of purchasing a defective car from the manufacturer

b. If a car purchased from the manufacturer is defective, what is the probability that it came from factory A?

1.30 Kevin rolls two dice and tells you that there is at least one 6. What is the probability that the sum is at least 9?

1.31 Chuck is a fool with probability 0.6, a thief with probability 0.7, and neither with probability 0.25.

a. What is the probability that he is a fool or a thief but not both?

b. What is the conditional probability that he is a thief, given that he is not a fool?

1.32 Studies indicate that the probability that a married man votes is 0.45, the probability that a married woman votes is 0.40, and the probability that a married woman votes given that her husband does is 0.60. Compute the following probabilities:

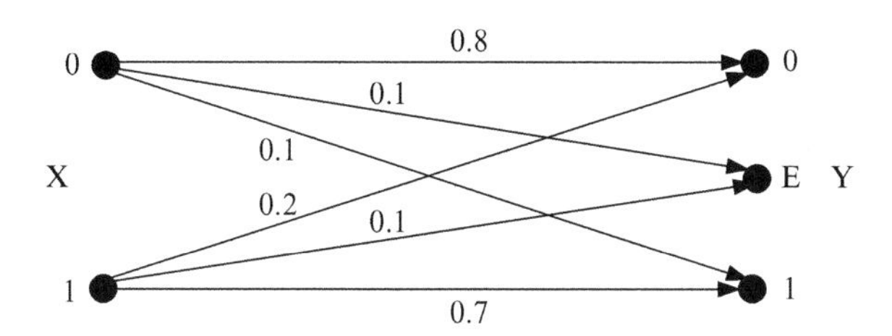


**FIGURE 1.19**
Figure for Problem 1.34

a. Both a man and his wife vote.
b. A man votes given that his wife votes also.

1.33 Tom is planning to pick up a friend at the airport. He has figured out that the plane is late 80% of the time when it rains, but only 30% of the time when it does not rain. If the weather forecast that morning calls for a 40% chance of rain, what is the probability that the plane will be late?

1.34 Consider the communication channel shown in Figure 1.19. The symbols transmitted are 0 and 1. However, three possible symbols can be received: 0, 1, and E. Thus, we define the input symbol set as $X=\{0,1\}$ and the output symbol set as $Y=\{0,1,E\}$. The transition (or conditional) probability $P[Y|X]$ is the probability that $Y$ is received, given that $X$ was transmitted. In particular, $P[0|0]=0.8$, $P[1|0]=0.1$ and $P[E|0]=0.1$. Similarly, $P[0|1]=0.2$, $P[1|1]=0.7$ and $P[E|1]=0.1$. If $P[X=0]=P[X=1]=0.5$, determine the following:
a. $P[Y=0]$, $P[Y=1]$ and $P[Y=E]$
b. If 0 is received, what is the probability that 0 was transmitted?
c. If E is received, what is the probability that 1 was transmitted?
d. If 1 is received, what is the probability that 1 was transmitted?

1.35 A group of students consists of 60% men and 40% women. Among the men, 30% are foreign students, and among the women, 20% are foreign students. A student is randomly selected from the group and found to be a foreign student. What is the probability that the student is a woman?

1.36 Joe frequently gets into trouble at school, and past experience shows that 80% of the time he is guilty of the offense he is accused of. Joe has just gotten into trouble again, and two other students, Chris and Dana, have been called into the principal's office to testify about the incident. Chris is Joe's friend and will definitely tell the truth if Joe is innocent, but he will lie with probability 0.2 if Joe is guilty. Dana does not like Joe and so will definitely tell the truth if Joe is guilty, but will lie with probability 0.3 if Joe is innocent.
a. What is the probability that Chris and Dana give conflicting testimonies?
b. What is the probability that Joe is guilty, given that Chris and Dana give conflicting testimonies?

1.37 Three car brands A, B, and C have all the market share in a certain city. Brand A has 20% of the market share, brand B has 30%, and brand C has 50%. The probability that a brand A car needs a major repair during the first year of purchase is 0.05, the probability that a brand B car needs a major repair during the first year of purchase is 0.10, and the probability that a brand C car needs a major repair during the first year of purchase is 0.15.

a. What is the probability that a randomly selected car in the city needs a major repair during its first year of purchase?

b. If a car in the city needs a major repair during its first year of purchase, what is the probability that it is a brand A car?

## Section 1.8 Independent Events

1.38 If I toss two coins and tell you that at least one is heads, what is probability that the first coin is heads?

1.39 Assume that we roll two dice and define three events $A$, $B$, and $C$, where $A$ is the event that the first die is odd, $B$ is the event that the second die is odd, and $C$ is the event that the sum is odd. Show that these events are pairwise independent but the three are not independent.

1.40 Consider a game that consists of two successive trials. The first trial has outcome A or B, and the second trial has outcome C or D. The probabilities of the four possible outcomes of the game are as follows:

| Outcome | AC | AD | BC | BD |
|---|---|---|---|---|
| Probability | 1/3 | 1/6 | 1/6 | 1/3 |

Determine in a convincing way if A and C are statistically independent.

1.41 Suppose that two events $A$ and $B$ are mutually exclusive and $P[B] > 0$. Under what conditions will $A$ and $B$ be independent?

## Section 1.10 Combinatorial Analysis

1.42 Four married couples bought tickets for eight seats in a row for a football game.

a. In how many different ways can they be seated?

b. In how many ways can they be seated if each couple is to sit together with the husband to the left of his wife?

c. In how many ways can they be seated if each couple is to sit together?

d. In how many ways can they be seated if all the men are to sit together and all the women are to sit together?

1.43 A committee consisting of 3 electrical engineers and 3 mechanical engineers is to be formed from a group of 7 electrical engineers and

5 mechanical engineers. Find the number of ways in which this can be done if

a. any electrical engineer and any mechanical engineer can be included.
b. one particular electrical engineer must be on the committee.
c. two particular mechanical engineers cannot be on the same committee.

1.44 Use Stirling's formula to evaluate 200!

1.45 A committee of 3 members is to be formed consisting of one representative from labor, one from management, and one from the public. If there are 7 possible representatives from labor, 4 from management, and 5 from the public, how many different committees can be formed?

1.46 There are 100 U.S. senators, two from each of the 50 states.

a. If 2 senators are chosen at random, what is the probability that they are from the same state?
b. If 10 senators are randomly chosen to form a committee, what is the probability that they are all from different states?

1.47 A committee of 7 people is to be formed from a pool of 10 men and 12 women.

a. What is the probability that the committee will consist of 3 men and 4 women?
b. What is the probability that the committee will consist of all men?

1.48 Five departments in the college of engineering, which are labeled departments A, B, C, D and E, send three delegates each to the college's convention. A committee of four delegates, selected by lot, is formed. Determine the probability that

a. Department A is not represented on the committee.
b. Department A has exactly one representative on the committee.
c. Neither department A nor department C is represented on the committee.

## Section 1.11 Reliability Applications

1.49 Consider the system shown in Figure 1.20. If the number inside each box indicates the probability that the component will independently fail within the next two years, find the probability that the system fails within two years.

1.50 Consider the structure shown in Figure 1.21. Switches $S_1$ and $S_2$ are in series, and the pair is in parallel with a parallel arrangement of switches $S_3$ and $S_4$ Their reliability functions are $R_1(t), R_2(t), R_3(t)$and $R_4(t)$, respectively. The structure interconnects nodes A and B. What is the reliability function of the composite system in terms of $R_1(t), R_2(t), R_3(t)$ and $R_4(t)$ if the switches fail independently?

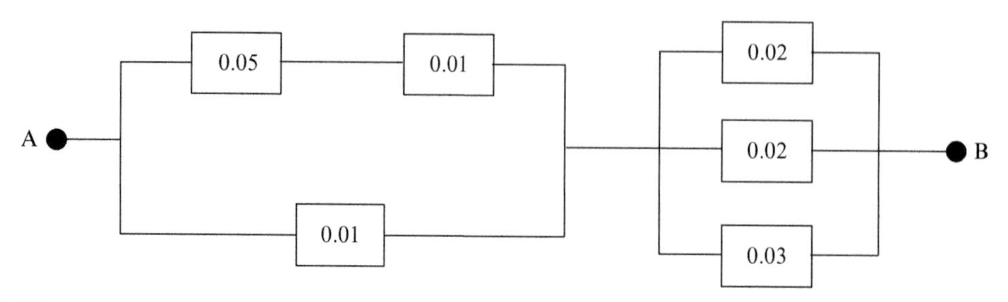


**FIGURE 1.20**
Figure for Problem 1.49

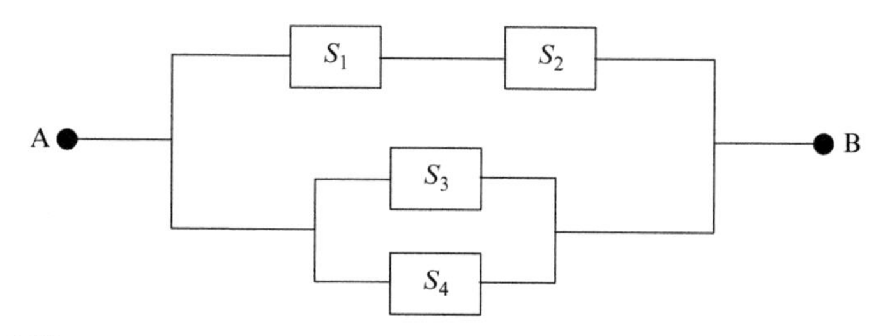


**FIGURE 1.21**
Figure for Problem 1.50

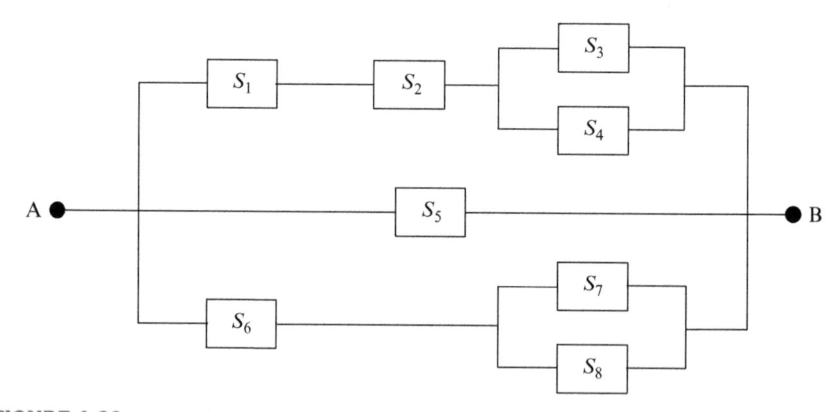


**FIGURE 1.22**
Figure for Problem 1.51

1.51 Consider the network shown in Figure 1.22 that interconnects nodes A and B. The switches labeled $S_1, S_2, \ldots, S_8$ have the reliability functions $R_1(t), R_2(t), \ldots, R_8(t)$, respectively. If the switches fail independently, find the reliability function of the composite system.

1.52 Consider the network shown in Figure 1.23 that interconnects nodes A and B. The switches labeled $S_1, S_2, \ldots, S_8$ have the reliability functions

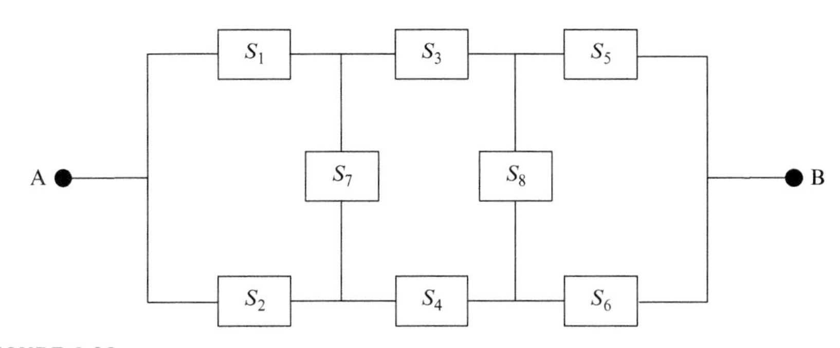


**FIGURE 1.23**
Figure for Problem 1.52

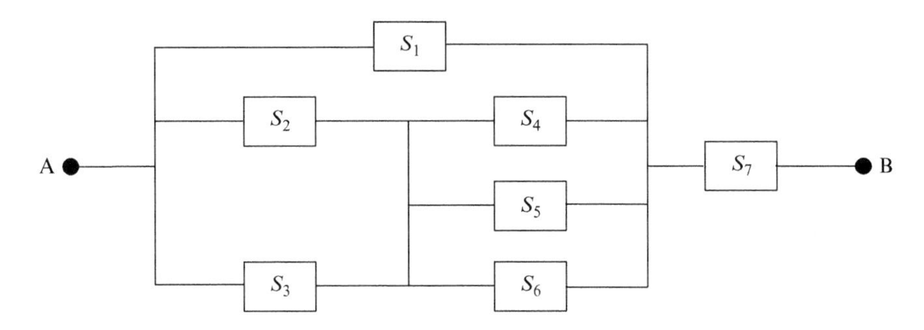


**FIGURE 1.24**
Figure for Problem 1.53

$R_1(t), R_2(t), \ldots, R_8(t)$, respectively. If the switches fail independently, find the reliability function of the composite system.

1.53 Consider the network shown in Figure 1.24 that interconnects nodes A and B. The switches labeled $S_1, S_2, \ldots, S_7$ have the reliability functions $R_1(t), R_2(t), \ldots, R_7(t)$, respectively. If the switches fail independently, find the reliability function of the composite system.

CHAPTER 2

# Random Variables

## 2.1 INTRODUCTION

The concept of a probability space that completely describes the outcome of a random experiment has been developed in Chapter 1. In this chapter we develop the idea of a function defined on the outcome of a random experiment, which is a very high-level definition of a random variable. Thus, the value of a random variable is a random phenomenon and is a numerically valued random phenomenon.

## 2.2 DEFINITION OF A RANDOM VARIABLE

Consider a random experiment with sample space $\Omega$. Let $w$ be a sample point in $\Omega$. We are interested in assigning a real number to each $w \in \Omega$. A random variable, $X(w)$, is a single-valued real function that assigns a real number, called the value of $X(w)$, to each sample point $w \in \Omega$. That is, it is a mapping of the sample space onto the real line.

Generally a random variable is represented by a single letter $X$ instead of the function $X(w)$. Therefore, in the remainder of the book we use $X$ to denote a random variable. The sample space $\Omega$ is called the *domain* of the random variable $X$. Also, the collection of all numbers that are values of $X$ is called the range of the random variable $X$. Figure 2.1 illustrates the concepts of domain and range of $X$.

### EXAMPLE 2.1

A coin-tossing experiment has two sample points: Heads and Tails. Thus, we may define the random variable $X$ associated with the experiment as follows:

$$X(\text{heads}) = 1$$
$$X(\text{tails}) = 0$$

In this case the mapping of the sample space to the real line is as shown in Figure 2.2.



Fundamentals of Applied Probability and Random Processes. http://dx.doi.org/10.1016/B978-0-12-800852-2.00002-X

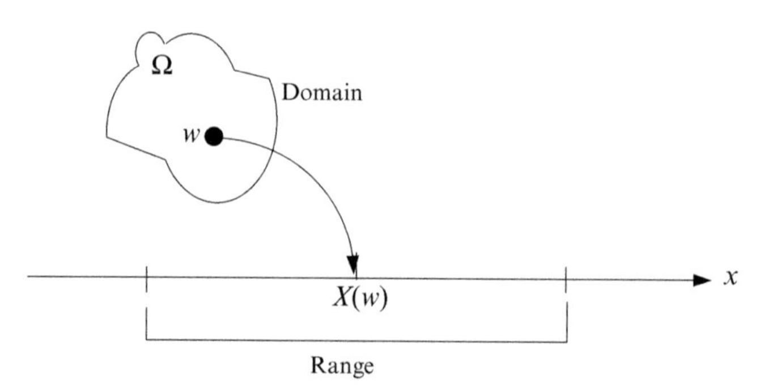


**FIGURE 2.1**
A Random Variable Associated with a Sample Point

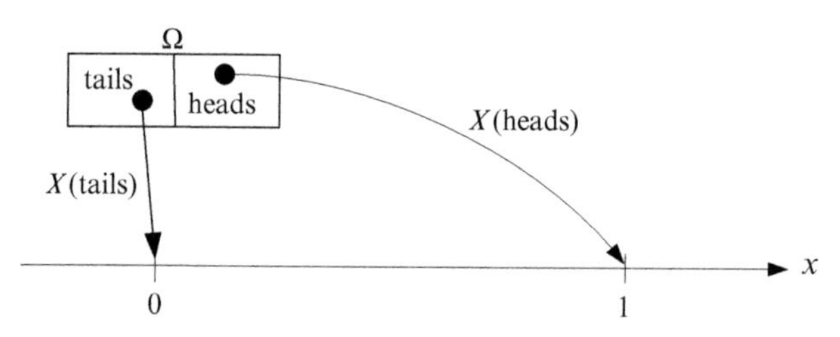


**FIGURE 2.2**
Random Variable Associated with Coin Tossing Experiment

## 2.3 EVENTS DEFINED BY RANDOM VARIABLES

Let $X$ be a random variable and let $x$ be a fixed real value. Let the event $A_x$ define the subset of $\Omega$ that consists of all real sample points to which the random variable $X$ assigns the number $x$. That is,

$$A_x = \{w|X(w) = x\} = [X = x]$$

Since $A_x$ is an event, it will have a probability, which we define as follows:

$$p = P[A_x]$$

We can define other types of events in terms of a random variable. For fixed numbers $x$, $a$, and $b$, we can define the following:

$$[X \leq x] = \{w|X(w) \leq x\}$$
$$[X > x] = \{w|X(w) > x\}$$
$$[a < X < b] = \{w|a < X(w) < b\}$$

These events have probabilities that are denoted by:

- $P[X \leq x]$ is the probability that $X$ takes a value less than or equal to $x$

- $P[X > x]$ is the probability that $X$ takes a value greater than $x$; this is equal to $1 - P[X \leq x]$
- $P[a < X < b]$ is the probability that $X$ takes a value that strictly lies between $a$ and $b$

## EXAMPLE 2.2

Consider an experiment in which a fair coin is tossed twice. The sample space consists of four equally likely sample points:

$$\Omega = \{HH, HT, TH, TT\}$$

Let $X$ denote the random variable that counts the number of heads in each sample point. Thus $X$ has the range $\{0, 1, 2\}$. Figure 2.3 shows how the sample space can be mapped onto the real line. If we consider $[X \leq 1]$, which is the event that the number of heads is at most 1, we obtain:

$$\begin{aligned} [X \leq 1] &= \{TT, TH, HT\} \\ P[X \leq 1] &= P[TT] + \{P[TH] + P[HT]\} \\ &= P[X = 0] + P[X = 1] \\ &= \frac{1}{4} + \frac{1}{2} = \frac{3}{4} \end{aligned}$$

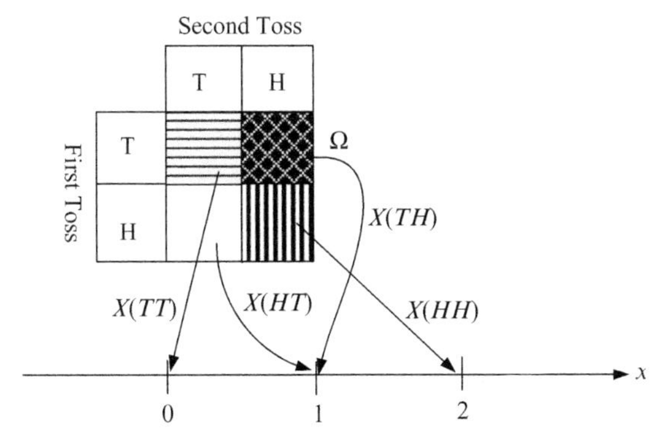


**FIGURE 2.3**
Definition of Random Variable for Example 2.2

## 2.4 DISTRIBUTION FUNCTIONS

Let $X$ be a random variable and let $x$ be a number. As stated earlier, we can define the event $[X \leq x] = \{x | X(w) \leq x\}$. The distribution function (or the cumulative distribution function (CDF)) of $X$ is defined by

$$F_X(x) = P[X \leq x] \qquad -\infty < x < \infty \tag{2.1}$$

That is, $F_X(x)$ denotes the probability that the random variable $X$ takes on a value that is less than or equal to $x$. Some properties of $F_X(x)$ include:

1. $F_X(x)$ is a non-decreasing function, which means that if $x_1 < x_2$, then $F_X(x_1) \leq F_X(x_2)$. Thus, $F_X(x)$ can increase or stay level, but it cannot go down, as $x$ increases.
2. $0 \leq F_X(x) \leq 1$
3. $F_X(\infty) = 1$
4. $F_X(-\infty) = 0$
5. $P[a < X \leq b] = F_X(b) - F_X(a)$
6. $P[X > a] = 1 - P[X \leq a] = 1 - F_X(a)$

## EXAMPLE 2.3

The CDF of the random variable $X$ is given by

$$F_X(x) = \begin{cases} 0 & x < 0 \\ x + \frac{1}{2} & 0 \leq x < \frac{1}{2} \\ 1 & x \geq \frac{1}{2} \end{cases}$$

a. Draw the graph of the CDF
b. Compute $P[X > \frac{1}{4}]$

Solution:

a. The graph of the CDF is as shown in Figure 2.4.

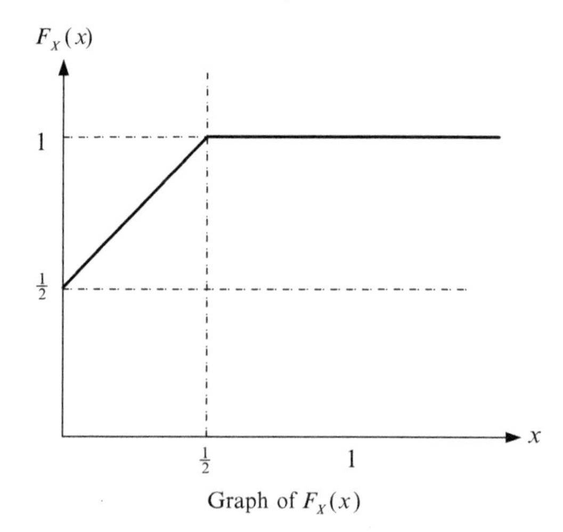


**FIGURE 2.4**
Graph of $F_X(x)$

b. The probability that $X$ is greater than $\frac{1}{4}$ is given by

$$P\left[X > \frac{1}{4}\right] = 1 - P\left[X \leq \frac{1}{4}\right] = 1 - F_X\left(\frac{1}{4}\right) = 1 - \left(\frac{1}{4} + \frac{1}{2}\right)$$
$$= \frac{1}{4}$$

## 2.5 DISCRETE RANDOM VARIABLES

A discrete random variable is a random variable that can take on at most a countable number of possible values. The number can be finite or infinite; that is, the random variable can have a countably finite number of values or a countably infinite number of values. For a discrete random variable $X$, the *probability mass function* (PMF), $p_X(x)$, is defined as follows:

$$p_X(x) = P[X = x] \tag{2.2}$$

where $\sum_{x=-\infty}^{\infty} p_X(x) = 1$. The PMF is nonzero for at most a countable number of values of $x$. In particular, if we assume that $X$ can only assume one of the values $x_1, x_2, \ldots, x_n$, then

$$\begin{array}{ll} p_X(x_i) \geq 0 & i = 1,2,\ldots,n \\ p_X(x_i) = 0 & \text{otherwise} \end{array}$$

The CDF of $X$ can be expressed in terms of $p_X(x)$ as follows:

$$F_X(x) = \sum_{k \leq x} p_X(k) \tag{2.3}$$

The CDF of a discrete random variable is a series of step functions. That is, if $X$ takes on values at $x_1, x_2, x_3, \ldots$, where $x_1 < x_2 < x_3 < \cdots$, then the value of $F_X(x)$ is constant in the interval between $x_{i-1}$ and $x_i$ and then takes a jump of size $p_X(x_i)$ at $x_i, i = 1, 2, 3, \ldots$. Thus, in this case, $F_X(x)$ represents the sum of all the probability masses we have encountered as we move from $-\infty$ to $x$.

### EXAMPLE 2.4

Assume that $X$ has the PMF given by

$$p_X(x) = \begin{cases} \frac{1}{4} & x = 0 \\ \frac{1}{2} & x = 1 \\ \frac{1}{4} & x = 2 \\ 0 & \text{otherwise} \end{cases}$$

The PMF of $X$ is given in Figure 2.5(a), and its CDF is given by

$$F_X(x) = \begin{cases} 0 & x < 0 \\ \frac{1}{4} & 0 \leq x < 1 \\ \frac{3}{4} & 1 \leq x < 2 \\ 1 & x \geq 2 \end{cases}$$

Thus, the graph of the CDF of $X$ is as shown in Figure 2.5(b).

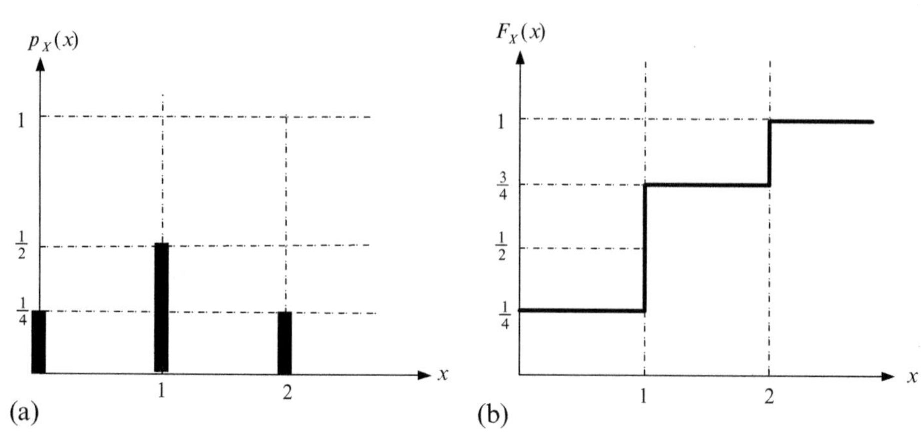


**FIGURE 2.5**
Graph of $F_X(x)$ for Example 2.4

## EXAMPLE 2.5

Let the random variable $X$ denote the number of heads in three tosses of a fair coin. (a) What is the PMF of $X$? (b) Sketch the CDF of $X$.

Solution:

a. The sample space of the experiment is

$$\Omega = \{HHH, HHT, HTH, HTT, THH, THT, TTH, TTT\}$$

The different events defined by the random variable $X$ are as follows:

$$[X=0] = \{TTT\}$$
$$[X=1] = \{HTT, THT, TTH\}$$
$$[X=2] = \{HHT, HTH, THH\}$$
$$[X=3] = \{HHH\}$$

Since the eight sample points in $\Omega$ are equally likely, the PMF of $X$ is as follows:

$$p_X(x) = \begin{cases} \frac{1}{8} & x=0 \\ \frac{3}{8} & x=1 \\ \frac{3}{8} & x=2 \\ \frac{1}{8} & x=3 \\ 0 & \text{otherwise} \end{cases}$$

The PMF is graphically illustrated in Figure 2.6(a).

b. The CDF of $X$ is given by

$$F_X(x) = \begin{cases} 0 & x < 0 \\ \frac{1}{8} & 0 \le x < 1 \\ \frac{1}{2} & 1 \le x < 2 \\ \frac{7}{8} & 2 \le x < 3 \\ 1 & x \ge 3 \end{cases}$$

The graph of $F_X(x)$ is shown in Figure 2.6(b).

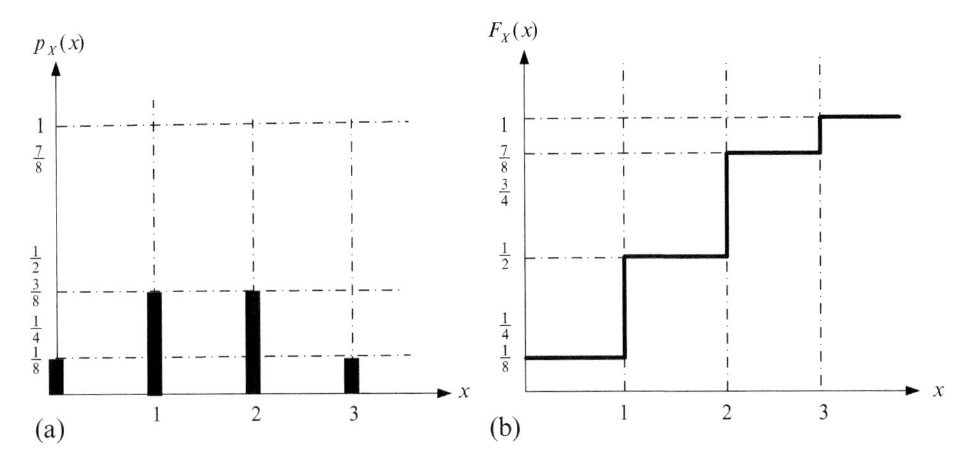


**FIGURE 2.6**
Graphs of $p_X(x)$ and $F_X(x)$ for Example 2.5

## EXAMPLE 2.6

Let the random variable $X$ denote the sum obtained in rolling a pair of fair dice. Determine the PMF of $X$.

Solution:

Let the pair $(a, b)$ denote the outcomes of the roll, where $a$ is the outcome of one die and $b$ is the outcome of the other. Thus, the sum of the outcomes is $X = a + b$. The different events defined by the random variable $X$ are as follows:

$$\begin{aligned}
[X=2] &= \{(1,1)\} \\
[X=3] &= \{(1,2),(2,1)\} \\
[X=4] &= \{(1,3),(2,2),(3,1)\} \\
[X=5] &= \{(1,4),(2,3),(3,2),(4,1)\} \\
[X=6] &= \{(1,5),(2,4),(3,3),(4,2),(5,1)\} \\
[X=7] &= \{(1,6),(2,5),(3,4),(4,3),(5,2),(6,1)\} \\
[X=8] &= \{(2,6),(3,5),(4,4),(5,3),(6,2)\} \\
[X=9] &= \{(3,6),(4,5),(5,4),(6,3)\} \\
[X=10] &= \{(4,6),(5,5),(6,4)\} \\
[X=11] &= \{(5,6),(6,5)\} \\
[X=12] &= \{(6,6)\}
\end{aligned}$$

Since there are 36 equally likely sample points in the sample space, the PMF of $X$ is given by:

$$p_X(x)=\begin{cases}\frac{1}{36} & x=2\\ \frac{2}{36} & x=3\\ \frac{3}{36} & x=4\\ \frac{4}{36} & x=5\\ \frac{5}{36} & x=6\\ \frac{6}{36} & x=7\\ \frac{5}{36} & x=8\\ \frac{4}{36} & x=9\\ \frac{3}{36} & x=10\\ \frac{2}{36} & x=11\\ \frac{1}{36} & x=12\\ 0 & \text{otherwise}\end{cases}$$

## EXAMPLE 2.7

The PMF of the number of components $K$ of a system that fail is defined by

$$p_K(k)=\begin{cases}\binom{4}{k}(0.2)^k(0.8)^{4-k} & k=0,1,\ldots,4\\ 0 & \text{otherwise}\end{cases}$$

a. What is the CDF of $K$?
b. What is the probability that less than 2 components of the system fail?

**Solution:**

a. The CDF of $K$ is given by

$$F_K(k)=P[K\leq k]=\sum_{m\leq k}p_K(m)=\sum_{m=0}^{k}p_K(m)=\sum_{m=0}^{k}\frac{4!}{(4-m)!m!}(0.2)^m(0.8)^{4-m}$$

$$=\begin{cases}0 & k<0\\ (0.8)^4 & 0\leq k<1\\ (0.8)^4+4(0.2)(0.8)^3 & 1\leq k<2\\ (0.8)^4+4(0.2)(0.8)^3+6(0.2)^2(0.8)^2 & 2\leq k<3\\ (0.8)^4+4(0.2)(0.8)^3+6(0.2)^2(0.8)^2+4(0.2)^3(0.8) & 3\leq k<4\\ (0.8)^4+4(0.2)(0.8)^3+6(0.2)^2(0.8)^2+4(0.2)^3(0.8)+(0.2)^4 & k\geq 4\end{cases}$$

$$=\begin{cases}0 & k<0\\ 0.4096 & 0\leq k<1\\ 0.8192 & 1\leq k<2\\ 0.9728 & 2\leq k<3\\ 0.9984 & 3\leq k<4\\ 1.0 & k\geq 4\end{cases}$$

b. The probability that less than 2 components of the system fail is the probability that either no component fails or one component fails, which is given by

$$P[K<2]=P[\{K=0\}\cup\{K=1\}]=P[K=0]+P[K=1]=F_K(1)=0.8192$$

where the second equality is due to the fact that the two events are mutually exclusive.

## EXAMPLE 2.8

The PMF of the number $N$ of customers that arrive at a local library within one hour interval is defined by

$$p_N(n)=\begin{cases}\frac{5^n}{n!}e^{-5} & n=0,1,\ldots \\ 0 & \text{otherwise}\end{cases}$$

What is the probability that at most two customers arrive at the library within one hour?

Solution:
The probability that at most two customers arrive at the library within one hour is the probability that 0 or 1 or 2 customers arrive at the library within one hour, which is

$$\begin{aligned}P[N\leq 2]&=P[\{N=0\}\cup\{N=1\}\cup\{N=2\}]=P[N=0]+P[N=1]+P[N=2]\\&=p_N(0)+p_N(1)+p_N(2)=e^{-5}\left\{1+5+\frac{25}{2}\right\}=18.5e^{-5}\\&=0.1246\end{aligned}$$

where the second equality on the first line is due to the fact that the three events are mutually exclusive.

### 2.5.1 Obtaining the PMF from the CDF

So far we have shown how to obtain the CDF from the PMF; namely, for a discrete random variable $X$ with PMF $p_X(x)$, the CDF is given by

$$F_X(x)=\sum_{k\leq x}p_X(k)$$

Sometimes we are given the CDF of a discrete random variable and are required to obtain its PMF. From Figures 2.5 and 2.6 we observe that the CDF of a discrete random variable has the staircase plot with jumps at those values of the random variable where the PMF has a nonzero value. The size of a jump at a value of a random variable is equal to the value of the PMF at the value.

Thus, given the plot of the CDF of a discrete random variable, we can obtain the PMF of the random variable by noting that the random variable only takes on values that have nonzero probability at those points where jumps occur. The probability that the random variable takes on any other value than where the jumps occur is zero. More importantly, the probability that the random variable takes a value where a jump occurs is equal to the size of the jump.

## EXAMPLE 2.9

The plot of the CDF of a discrete random variable $X$ is shown in Figure 2.7. Find the PMF of $X$.

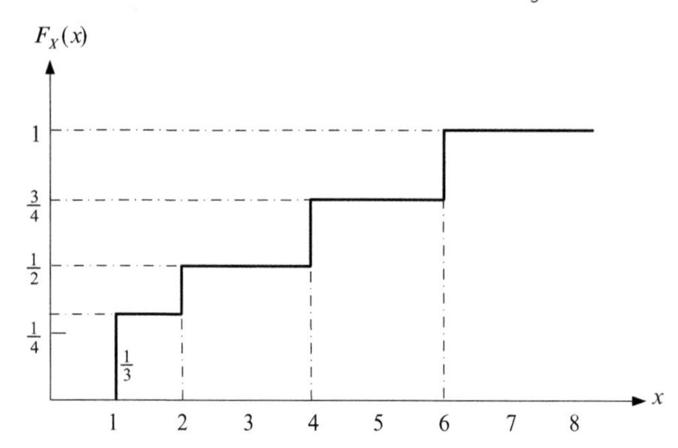


**FIGURE 2.7**
Graph of $F_X(x)$ for Example 2.9

**Solution**

The random variable takes on values with nonzero probability at $X=1, X=2, X=4$ and $X=6$. The size of the jump at $X=1$ is $\frac{1}{3}$, the size of the jump at $X=2$ is $\frac{1}{2}-\frac{1}{3}=\frac{1}{6}$, the size of the jump at $X=4$ is $\frac{3}{4}-\frac{1}{2}=\frac{1}{4}$, and the size of the jump at $X=6$ is $1-\frac{3}{4}=\frac{1}{4}$. Thus, the PMF of $X$ is given by

$$p_X(x)=\begin{cases}\frac{1}{3} & x=1\\ \frac{1}{6} & x=2\\ \frac{1}{4} & x=4\\ \frac{1}{4} & x=6\\ 0 & \text{otherwise}\end{cases}$$

## EXAMPLE 2.10

Find the PMF of a discrete random variable $X$ whose CDF is given by:

$$F_X(x)=\begin{cases}0 & x<0\\ \frac{1}{6} & 0\leq x<2\\ \frac{1}{2} & 2\leq x<4\\ \frac{5}{8} & 4\leq x<6\\ 1 & x\geq 6\end{cases}$$

Solution:
In this example, we do not need to plot the CDF. We observe that it changes values at $X=0, X=2$, $X=4$ and $X=6$, which means that these are the values of the random variable that have nonzero probabilities. The next task after isolating these values with nonzero probabilities is to determine their probabilities. The first value is $p_X(0)$, which is $\frac{1}{6}$. At $X=2$ the size of the jump is $\frac{1}{2}-\frac{1}{6}=\frac{1}{3}=p_X(2)$. Similarly, at $X=4$ the size of the jump is $\frac{5}{8}-\frac{1}{2}=\frac{1}{8}=p_X(4)$. Finally, at $X=6$ the size of the jump is $1-\frac{5}{8}=\frac{3}{8}=p_X(6)$. Therefore, the PMF of $X$ is given by

$$p_X(x)=\begin{cases}\frac{1}{6} & x=0\\ \frac{1}{3} & x=2\\ \frac{1}{8} & x=4\\ \frac{3}{8} & x=6\\ 0 & \text{otherwise}\end{cases}$$

## 2.6 CONTINUOUS RANDOM VARIABLES

Discrete random variables have a set of possible values that are either finite or countably infinite. However, there exists another group of random variables that can assume an uncountable set of possible values. Such random variables are called continuous random variables. Thus, we define a random variable $X$ to be a continuous random variable if there exists a nonnegative function $f_X(x)$, defined for all real $x \in (-\infty, \infty)$, having the property that for any set $A$ of real numbers,

$$P[X \in A]=\int_A f_X(x)dx \tag{2.4}$$

The function $f_X(x)$ is called the *probability density function* (PDF) of the random variable $X$ and is defined by

$$f_X(x)=\frac{dF_X(x)}{dx} \tag{2.5}$$

The properties of $f_X(x)$ are as follows:

1. $f_X(x) \geq 0$; that is, it is a nonnegative function, as stated earlier.
2. Since $X$ must assume some value, $\int_{-\infty}^{\infty} f_X(x)dx=1$
3. $P[a \leq X \leq b]=\int_a^b f_X(x)dx$, which means that $P[X=a]=\int_a^a f_X(x)dx=0$.
   Thus, the probability that a continuous random variable will assume any fixed value is zero.

4. $P[X < a] = P[X \le a] = F_X(a) = \int_{-\infty}^{a} f_X(x)dx$

From property (3) we observe that

$$P[x \le X \le x + dx] = f_X(x)dx \tag{2.6}$$

## EXAMPLE 2.11

Assume that $X$ is a continuous random variable with the following PDF:

$$f_X(x)dx = \begin{cases} A(2x - x^2) & 0 < x < 2 \\ 0 & \text{otherwise} \end{cases}$$

a. What is the value of $A$?
b. Find $P[X > 1]$.

Solution:

a. Since $f_X(x)$ is a PDF, we have that

$$\int_{-\infty}^{\infty} f_X(x)dx = \int_{-\infty}^{0} 0dx + \int_{0}^{2} A(2x - x^2)dx + \int_{2}^{\infty} 0dx = \int_{0}^{2} A(2x - x^2)dx = 1$$

Thus, we obtain

$$A\left[x^2 - \frac{x^3}{3}\right]_{x=0}^{2} = 1 = A\left(4 - \frac{8}{3}\right) = \frac{4A}{3} \Rightarrow A = \frac{3}{4}$$

b. Therefore,

$$P[X > 1] = \int_{1}^{\infty} f_X(x)dx = \frac{3}{4}\int_{1}^{2}(2x - x^2)dx = \frac{3}{4}\left[x^2 - \frac{x^3}{3}\right]_{1}^{2} = \frac{3}{4}\left[\left(4 - \frac{8}{3}\right) - \left(1 - \frac{1}{3}\right)\right] = \frac{1}{2}$$

## EXAMPLE 2.12

Is the following function a legitimate PDF?

$$g(x) = \begin{cases} \dfrac{x^2}{9} & 0 \le x \le 3 \\ 0 & \text{otherwise} \end{cases}$$

Solution:

For $g(x)$ to be a legitimate PDF, we need to check to see if $\int_{-\infty}^{\infty} g(x)dx = 1$. Thus,

$$\int_{-\infty}^{\infty} g(x)dx = \int_{-0}^{3} \frac{x^2}{9}dx = \left[\frac{x^3}{27}\right]_{0}^{3} = 1$$

Since $g(x) \ge 0$, we conclude that $g(x)$ is a legitimate PDF.

## EXAMPLE 2.13

Consider the function

$$h(x)=\begin{cases} c & a \le x \le b \\ 0 & \text{otherwise} \end{cases}$$

a. For what value of $c$ is $h(x)$ a legitimate PDF?
b. Find the CDF of the random variable $X$ with the above PDF.

Solution:

a. For $h(x)$ to be a legitimate PDF, we must have that $\int_{-\infty}^{\infty} h(x)dx = 1$. That is,

$$\int_a^b c dx = 1 = [cx]_a^b = c(b-a)$$

This implies that $c = \frac{1}{b-a}$.

b. The CDF is given by

$$H(x)=\int_{-\infty}^{x} h(u)du = \int_a^x \frac{du}{b-a} = \begin{cases} 0 & x < a \\ \dfrac{x-a}{b-a} & a \le x < b \\ 1 & x \ge b \end{cases}$$

## EXAMPLE 2.14

Consider the function

$$f(x)=\begin{cases} 2x & 0 \le x \le b \\ 0 & \text{otherwise} \end{cases}$$

a. For what value of $b$ is $f(x)$ a legitimate PDF?
b. Find the CDF of the random variable $X$ with the above PDF.

Solution:

a. For $f(x)$ to be a valid PDF in the specified range, we must have that $\int_0^b f(x)dx = 1$. That is,

$$\int_0^b 2x dx = \left[x^2\right]_0^b = b^2 = 1$$

Thus, $b = 1$.

b. The CDF of $X$ is given by

$$F(x)=\int_{-\infty}^{x} f(u)du = \int_0^x 2u du = \begin{cases} 0 & x < 0 \\ x^2 & 0 \le x < 1 \\ 1 & x \ge 1 \end{cases}$$

## EXAMPLE 2.15

The lifetime in hours of a component is represented by the random variable $X$ whose PDF is given by $f_X(x) = 0.1e^{-0.1x}, \ x \geq 0$.

a. What is the CDF of $X$ ?
b. What is the probability that the component will last at least 5 hours?

Solution:

a. The CDF of $X$ is given by

$$F_X(x) = \int_{-\infty}^{x} f_X(u)du = \int_{0}^{x} 0.1e^{-0.1u}du = \left[-e^{-0.1u}\right]_0^x = \begin{cases} 0 & x < 0 \\ 1 - e^{-0.1x} & x \geq 0 \end{cases}$$

b. Since $X$ is a continuous random variable, the probability that the component will last at least 5 hours is given by $P[X \geq 5] = P[X > 5] = 1 - P[X \leq 5] = 1 - F_X(5)$. Thus, we have that

$$P[X \geq 5] = 1 - F_X(5) = 1 - \left\{1 - e^{-0.1(5)}\right\} = e^{-0.5} = 0.6065.$$

## EXAMPLE 2.16

The PDF of the time $T$ it takes a bank teller to serve a customer is defined by

$$f_T(t) = \begin{cases} \frac{1}{6} & 2 \leq t \leq 8 \\ 0 & \text{otherwise} \end{cases}$$

a. What is the CDF of $T$?
b. What is the probability that a customer is served in less than 5 minutes?

Solution:

a. The CDF of $T$ is given by

$$F_T(t) = P[T \leq t] = \int_{-\infty}^{t} f_T(u)du = \int_{2}^{t} \frac{du}{6} = \left[\frac{u}{6}\right]_2^t$$
$$= \begin{cases} 0 & t < 2 \\ \frac{t-2}{6} & 2 \leq t < 8 \\ 1 & t \geq 8 \end{cases}$$

b. Since $T$ is a continuous random variable, the probability that a customer is served in less than 5 minutes is given by

$$P[T < 5] = P[T \leq 5] = F_T(5) = \frac{5-2}{6} = 0.5$$

## EXAMPLE 2.17

The CDF of the time $T$ it takes a bank teller to serve a customer is defined by

$$F_X(x) = \begin{cases} 0 & t < 2 \\ A(x-2) & 2 \le x < 6 \\ 1 & x \ge 6 \end{cases}$$

a. What is the value of $A$?
b. With the above value of $A$, what is $P[X > 4]$?
c. With the above value of $A$, what is $P[3 \le X \le 5]$?

Solution:

a. To find $A$, we know that $F_X(6) = 1$. Thus, from the definition of the CDF we have that

$$F_X(6) = A(6-2) = 4A = 1 \Rightarrow A = \frac{1}{4}$$

b. The probability that $X$ is greater than 4 is given by

$$P[X > 4] = 1 - P[X \le 4] = 1 - F_X(4) = 1 - \frac{1}{4}(4-2) = \frac{1}{2}$$

c. The probability that $X$ lies between 3 and 5 is given by

$$P[3 \le X \le 5] = F_X(5) - F_X(3) = \frac{1}{4}\{(5-2) - (3-2)\} = \frac{1}{2}$$

Note that we can also solve the problem by first finding the PDF of $X$ as follows:

$$f_X(x) = \frac{d}{dx}F_X(x) = \begin{cases} A & 2 \le x \le 6 \\ 0 & \text{otherwise} \end{cases}$$

Then the remainder of the problem is solved using the PDF, as follows:

$$\int_{-\infty}^{\infty} f(x)dx = \int_{2}^{6} Adx = 1 = A[x]_2^6 = 4A \Rightarrow A = \frac{1}{4}$$

$$P[X > 4] = \int_{4}^{6} Adx = \frac{1}{4}[x]_4^6 = \frac{1}{2}$$

$$P[3 \le X \le 5] = \int_{3}^{5} Adx = \frac{1}{4}[x]_3^5 = \frac{1}{2}$$

## EXAMPLE 2.18

The CDF of the random variable $Y$ is defined by

$$F_Y(y) = \begin{cases} 0 & y < 0 \\ K\{1 - e^{-2y}\} & y \ge 0 \end{cases}$$

a. What is the value of $K$?
b. With the above value of $K$, what is $F_Y(3)$?
c. With the above value of $K$, what is $P[2<Y<\infty]$?

Solution:

a. To find $K$, we know that $F_Y(\infty)=1$. Thus, from the definition of the CDF we have that

$$F_Y(\infty)=K\{1-e^{-\infty}\}=K\{1-0\}=1 \Rightarrow K=1$$

b. $F_Y(3)=K\{1-e^{-6}\}=1-e^{-6}=0.9975$
c. $P[2<Y<\infty]=F_X(\infty)-F_X(2)=1-F_X(2)=1-(1-e^{-4})=e^{-4}=0.0183.$

As in Example 2.17, the problem can also be solved by first obtaining the PDF of $Y$ and then integrating over the appropriate intervals as follows:

$$f_Y(y)=\frac{d}{dy}F_Y(y)=2Ke^{-2y},\ y\geq 0$$

$$\int_0^\infty f_Y(y)dy=1=K\left[-e^{-2y}\right]_0^\infty \Rightarrow K=1$$

$$F_Y(3)=\int_0^3 f_Y(y)dy=\left[-e^{-2y}\right]_0^3=1-e^{-6}$$

$$P[2<Y<\infty]=\int_2^\infty f_Y(y)dy=\left[-e^{-2y}\right]_2^\infty=e^{-4}$$

## 2.7 CHAPTER SUMMARY

This chapter developed the concept of functions defined on the outcomes of random phenomena. These functions, which are called random variables, can be classified into two types: discrete random variables that have a set of possible values that are either finite or countably infinite, and continuous random variables that can assume an uncountable set of possible values.

Associated with both types of random variables is the concept of cumulative distribution function, $F_X(x)$, that denotes the probability that a random variable $X$ takes on a value that is less than or equal to $x$. The probability mass function, $p_X(x)$, has been defined for discrete random variables and is the probability that the random variable $X$ has a value $x$. Similarly, the probability density function (PDF), $f_X(x)$, is a nonnegative function associated with a continuous random variable such that integrating the PDF between two distinct values of the random variable gives the probability that the random variable takes a value that lies between these two values. Thus, the area under the curve defined by the PDF is the probability that the random variable lies between the values limiting the area. Because of this, the probability that a continuous random variable takes on a particular value is zero since the area associated with a point is zero. Understanding the concept of random variables is key to understanding the rest of this book.

## 2.8 PROBLEMS

### Section 2.4 Distribution Functions

2.1 Bob claims that he can model his experimental study of a process by the following CDF:

$$F_X(x) = \begin{cases} 0 & -\infty < x \leq 1 \\ B\left[1 - e^{-(x-1)}\right] & 1 < x < \infty \end{cases}$$

a. For what value of $B$ is the function a valid CDF?
b. With the above value of $B$, what is $F_X(3)$?
c. With the above value of $B$, what is $P[2 < X < \infty]$?
d. With the above value of $B$, what is $P[1 < X \leq 3]$?

2.2 The CDF of a random variable $X$ is given by

$$F_X(x) = \begin{cases} 0 & x < 0 \\ 3x^2 - 2x^3 & 0 \leq x < 1 \\ 1 & x \geq 1 \end{cases}$$

What is the PDF of $X$?

2.3 A random variable $X$ has the CDF

$$F_X(x) = \begin{cases} 0 & x < 0 \\ 1 - e^{-x^2/2\sigma^2} & x \geq 0 \end{cases}$$

where $\sigma$ is a positive constant.
a. Find $P[\sigma \leq X \leq 2\sigma]$
b. Find $P[X > 3\sigma]$.

2.4 The CDF of a random variable $T$ is given by

$$F_T(t) = \begin{cases} 0 & t < 0 \\ t^2 & 0 \leq t < 1 \\ 1 & t \geq 1 \end{cases}$$

a. What is the PDF of $T$?
b. What is $P[T > 0.5]$?
c. What is $P[0.5 < T \leq 0.75]$?

2.5 The CDF of a continuous random variable $X$ is given by

$$F_X(x) = \begin{cases} 0 & x < -\frac{\pi}{2} \\ k\{1 + \sin(x)\} & -\frac{\pi}{2} \leq x < \frac{\pi}{2} \\ 1 & x \geq \frac{\pi}{2} \end{cases}$$

a. Find the value of $k$.
b. Find the PDF of $X$.

2.6 The CDF of a random variable $X$ is given by

$$F_X(x) = \begin{cases} 0 & x < 2 \\ 1 - \dfrac{4}{x^2} & x \geq 2 \end{cases}$$

a. Find $P[X < 3]$
b. Find $P[4 < X < 5]$.

2.7 The CDF of a discrete random variable $K$ is given by

$$F_K(k) = \begin{cases} 0.0 & k < -1 \\ 0.2 & -1 \leq k < 0 \\ 0.7 & 0 \leq k < 1 \\ 1.0 & k \geq 1 \end{cases}$$

a. Draw the graph of $F_K(k)$
b. Find $p_K(k)$, the PMF of $K$.

2.8 The random variable $N$ has the CDF

$$F_N(n) = \begin{cases} 0.0 & n < -2 \\ 0.3 & -2 \leq n < 0 \\ 0.5 & 0 \leq n < 2 \\ 0.8 & 2 \leq n < 4 \\ 1.0 & n \geq 4 \end{cases}$$

a. Draw the graph of $F_N(n)$
b. Find $p_N(n)$ the PMF of $N$.
c. Draw the graph of $p_N(n)$.

2.9 The CDF of a discrete random variable $Y$ is given by

$$F_Y(y) = \begin{cases} 0.0 & y < 2 \\ 0.3 & 2 \leq y < 4 \\ 0.8 & 4 \leq y < 6 \\ 1.0 & y \geq 6 \end{cases}$$

a. What is $P[3 < Y < 4]$?
b. What is $P[3 < Y \leq 4]$?

2.10 Determine the PMF of the random variable $Y$ if its CDF is given by

$$F_Y(y) = \begin{cases} 0.0 & y < 0 \\ 0.50 & 0 \leq y < 2 \\ 0.75 & 2 \leq y < 3 \\ 0.90 & 3 \leq y < 5 \\ 1.0 & y \geq 5 \end{cases}$$

2.11 The CDF of a discrete random variable $X$ is given as follows:

$$F_X(x) = \begin{cases} 0 & x < 0 \\ \frac{1}{4} & 0 \leq x < 1 \\ \frac{1}{2} & 1 \leq x < 3 \\ \frac{5}{8} & 3 \leq x < 4 \\ 1 & x \geq 4 \end{cases}$$

a. Determine $p_X(x)$, the PMF of $X$, and draw its graph.
b. Determine the values of (i) $P[X < 2]$ and (ii) $P[0 \leq X < 4]$.

## Section 2.5 Discrete Random Variables

2.12 Let the random variable $K$ denote the number of heads in 4 flips of a fair coin.
a. Plot the graph of $p_K(k)$.
b. What is $P[K \geq 3]$?
c. What is $P[2 \leq K \leq 4]$?

2.13 Ken was watching some people play poker, and he wanted to model the PMF of the random variable $N$ that denotes the number of plays up to and including the play in which his friend Joe won a game. He conjectured that if $p$ is the probability that Joe wins any game and the games are independent, then the PMF of $N$ is given by

$$p_N(n) = p(1-p)^{n-1} \qquad n = 1, 2, \ldots$$

a. Show that $p_N(n)$ is a proper PMF.
b. Find the CDF of $N$.

2.14 The discrete random variable $K$ has the following PMF:

$$p_K(k) = \begin{cases} b & k = 0 \\ 2b & k = 1 \\ 3b & k = 2 \\ 0 & \text{otherwise} \end{cases}$$

a. What is the value of $b$?
b. Determine the values of (i) $P[K < 2]$, (ii) $P[K \leq 2]$ and (iii) $P[0 < K < 2]$.
c. Determine the CDF of $K$.

2.15 A student got a summer job at a bank, and his assignment was to model the number of customers who arrive at the bank. The student observed that the number of customers $K$ that arrive over this period had the PMF

$$p_K(k) = \begin{cases} \frac{\lambda^k}{k!} e^{-\lambda} & k = 0, 1, 2, \ldots; \lambda > 0 \\ 0 & \text{otherwise} \end{cases}$$

a. Show that $p_K(k)$ is a proper PMF.
b. What is $P[K > 1]$?
c. What is $P[2 \leq K \leq 4]$?

2.16 Let $X$ be the random variable that denotes the number of times we roll a fair die until the first time the number 5 appears. Find the probability that $X = k$, if the outcomes of the rolls are independent.

2.17 The PMF of a random variable $X$ is given by $p_X(x) = b\lambda^x / x!, x = 0, 1, 2, \ldots,$ where $\lambda > 0$. Find the values of (a) the parameter $b$, (b) $P[X = 1]$ and (c) $P[X > 3]$.

2.18 A random variable $K$ has the PMF

$$p_K(k) = \binom{5}{k} (0.1)^k (0.9)^{5-k} \qquad k = 0, 1, \ldots, 5$$

Obtain the values of the following:
a. $P[K = 1]$
b. $P[K \geq 1]$

2.19 A biased four-sided die has faces labeled 1, 2, 3, and 4. Let the random variable $X$ denote the outcome of a roll of the die. Extensive testing of the die shows that the PMF of $X$ is given by

$$p_X(x) = \begin{cases} 0.4 & x = 1 \\ 0.2 & x = 2 \\ 0.3 & x = 3 \\ 0.1 & x = 4 \end{cases}$$

a. Find the CDF of $X$.
b. What is the probability that a number less than 3 appears on a roll of the die?
c. What is the probability of obtaining a number whose value is at least 3 on a roll of the die?

2.20 The number $N$ of calls arriving at a switchboard during a period of one hour has the PMF

$$p_N(n) = \frac{10^n}{n!} e^{-10} \qquad n = 0, 1, 2, \ldots$$

a. What is the probability that at least 2 calls arrive within one hour?
b. What is the probability that at most 3 calls arrive within one hour?
c. What is the probability that the number of calls that arrive within one hour is greater than 3 but less than or equal to 6?

2.21 Assume that the random variable $K$ denotes the number of successes in $n$ trials of an experiment. The probability of success in any trial of the experiment is 0.6, and the PMF of $K$ is given by

$$p_K(k) = \binom{n}{k}(0.6)^k(0.4)^{n-k} \qquad k=0,1,\ldots,n;\ n=1,2,\ldots$$

a. What is the probability of at least 1 success in 5 trials of the experiment?
b. What is the probability of at most 1 success in 5 trials of the experiment?
c. What is the probability that the number of successes is greater than 1 but less than 4 in 5 trails of the experiment?

2.22 Prove that the function $p(x)$ is a legitimate PMF of a discrete random variable $X$, where $p(x)$ is defined by

$$p(x) = \begin{cases} \frac{2}{3}\left(\frac{1}{3}\right)^x & x=0,1,2,\ldots \\ 0 & \text{otherwise} \end{cases}$$

## Section 2.6 Continuous Random Variables

2.23 Consider the following function.

$$g(x) = \begin{cases} a(1-x^2) & -1<x<1 \\ 0 & \text{otherwise} \end{cases}$$

a. Determine the value of $a$ that makes $g(x)$ a valid probability density function.
b. If $X$ is the random variable with this PDF, determine the value of $P[0<K<0.5]$.

2.24 The PDF of a continuous random variable $X$ is defined as follows for $\lambda>0$,

$$f_X(x) = \begin{cases} bxe^{-\lambda x} & 0\le x<\infty \\ 0 & \text{otherwise} \end{cases}$$

a. What is the value of $b$?
b. What is the CDF of $X$?
c. What is $P[0\le X\le 1/\lambda]$?

2.25 Find the PDF of the continuous random variable $X$ with the following CDF:

$$F_X(x) = \begin{cases} 0 & x<0 \\ 2x^2-x^3 & 0\le x<1 \\ 1 & x\ge 1 \end{cases}$$

2.26 A random variable $X$ has the following PDF, where $K>0$:

$$f_X(x)=\begin{cases}0 & x<1\\ K(x-1) & 1\leq x<2\\ K(3-x) & 2\leq x<3\\ 0 & x\geq 3\end{cases}$$

a. What is the value of $K$?
b. Sketch $f_X(x)$
c. What is the CDF of $X$?
d. What is $P[1\leq X\leq 2]$?

2.27 A random variable $X$ has the CDF

$$F_X(x)=\begin{cases}0 & x<-1\\ A(1+x) & -1\leq x<1\\ 1 & x\geq 1\end{cases}$$

a. What is the value of $A$?
b. With the above value of $A$, what is $P\left[X>\frac{1}{4}\right]$?
c. With the above value of $A$, what is $P[-0.5\leq X\leq 0.5]$?

2.28 The lifetime $X$ of a system in weeks is given by the following PDF:

$$f_X(x)=\begin{cases}0.25e^{-0.25x} & x\geq 0\\ 0 & \text{otherwise}\end{cases}$$

a. What is the probability that the system will not fail within two weeks?
b. Given that the system has not failed by the end of the fourth week, what is the probability that it will fail between the fourth and sixth weeks?

2.29 A military radar is set at a remote site with no repair facility. The time $T$ until the radar fails in years has the PDF given by

$$f_T(t)=\begin{cases}0.2e^{-0.2t} & t\geq 0\\ 0 & \text{otherwise}\end{cases}$$

What is the probability that the radar lasts for at least four years?

2.30 The PDF of a random variable $X$ is given by

$$f_X(x)=\begin{cases}\dfrac{A}{x^2} & x>10\\ 0 & \text{otherwise}\end{cases}$$

a. What is the value of $A$?
b. With the above value of $A$, determine the CDF of $X$.
c. With the above value of $A$, what is $P[X>20]$?

2.31 Assume that $X$ is a continuous random variable with the following PDF:

$$f_X(x) = \begin{cases} A(3x^2 - x^3) & 0 \leq x < 3 \\ 0 & \text{otherwise} \end{cases}$$

a. What is the value of $A$?
b. Find $P[1 < X < 2]$.

2.32 A random variable $X$ has the PDF

$$f_X(x) = \begin{cases} k(1 - x^4) & -1 \leq x \leq 1 \\ 0 & \text{otherwise} \end{cases}$$

a. Find the value of $k$.
b. Find the CDF of $X$.
c. Find $P\left[X < \frac{1}{2}\right]$.

2.33 The PDF of a random variable $X$ is given by

$$f_X(x) = \begin{cases} x & 0 < x < 1 \\ 2 - x & 1 \leq x \leq 2 \\ 0 & \text{otherwise} \end{cases}$$

a. Find the CDF of $X$.
b. Find $P[0.2 < X < 0.8]$.
c. Find $P[0.6 < X < 1.2]$.

2.34 The mileage (in thousands of miles) that car owners get with a certain type of tire is a random variable $X$ with PDF

$$f_X(x) = \begin{cases} Ae^{-x/20} & x \geq 0 \\ 0 & \text{otherwise} \end{cases}$$

a. Find the value of $A$.
b. Find the CDF of $X$.
c. What is $P[X < 10]$?
d. What is $P[16 < X < 24]$?

2.35 Assume that $X$ is a continuous random variable with the following PDF:

$$f_X(x) = \begin{cases} 0 & x < 0.5 \\ ke^{-2(x-0.5)} & x \geq 0.5 \end{cases}$$

a. Find the value of $k$.
b. Find the CDF of $X$.
c. What is $P[X < 1.5]$?
d. What is $P[1.2 < X < 2.4]$?

CHAPTER 3

# Moments of Random Variables

## 3.1 INTRODUCTION

Given the set of data $X_1, X_2, \ldots, X_N$, we know that the arithmetic average (or arithmetic mean) is given by

$$\overline{X} = \frac{X_1 + X_2 + \cdots + X_N}{N}$$

When the above numbers occur with different frequencies, we usually assign weights $w_1, w_2, \ldots, w_N$ to them and the so-called weighted arithmetic mean becomes

$$\overline{X} = \frac{w_1 X_1 + w_2 X_2 + \cdots + w_N X_N}{w_1 + w_2 + \cdots + w_N} = \frac{w_1}{w} X_1 + \frac{w_2}{w} X_2 + \cdots + \frac{w_N}{w} X_N \tag{3.1}$$

where $w = w_1 + w_2 + \cdots + w_N$. The average is a value that is representative or typical of a set of data and tends to lie centrally within a set of data that are arranged according to their magnitudes. Thus, it is usually called a measure of central tendency. Observe in equation (3.1) that the relative weight $\frac{w_k}{w}$ is the probability of occurrence of the data point $X_k$ in the set of observations.

The term *expectation* is used for the process of averaging when a random variable is involved. It is a number used to locate the "center" of the distribution of a random variable. In many situations we are primarily interested in the central tendency of a random variable, and as will be seen later in this chapter, the expectation (or mean or average) of a random variable can be likened to the weighted arithmetic average defined above.

Another measure of central tendency of a random variable is its variance, which measures the degree to which a random variable is spread out relative to the average value. This chapter deals with how the expectation (or mean or average) and variance of a random variable can be computed.



Fundamentals of Applied Probability and Random Processes. http://dx.doi.org/10.1016/B978-0-12-800852-2.00003-1

## 3.2 EXPECTATION

If $X$ is a random variable, then the *expectation* (or *expected value* or *mean*) of $X$, denoted by $E[X]$ or $\overline{X}$, is defined by

$$E[X] = \overline{X} = \begin{cases} \sum_k x_k p_X(x_k) & X \text{ discrete} \\ \int_{-\infty}^{\infty} x f_X(x) dx & X \text{ continuous} \end{cases} \tag{3.2}$$

Thus, for a discrete random variable $X$, the expected value is the weighted average of the possible values that $X$ can take, where each value is weighted by the probability that $X$ takes that value.

### EXAMPLE 3.1

Find the expected value of the random variable $X$ whose PDF is defined by

$$f_X(x) = \begin{cases} \frac{1}{b-a} & a \leq x \leq b \\ 0 & \text{otherwise} \end{cases}$$

**Solution:**

The PDF of $X$ is as shown in Figure 3.1. We have that

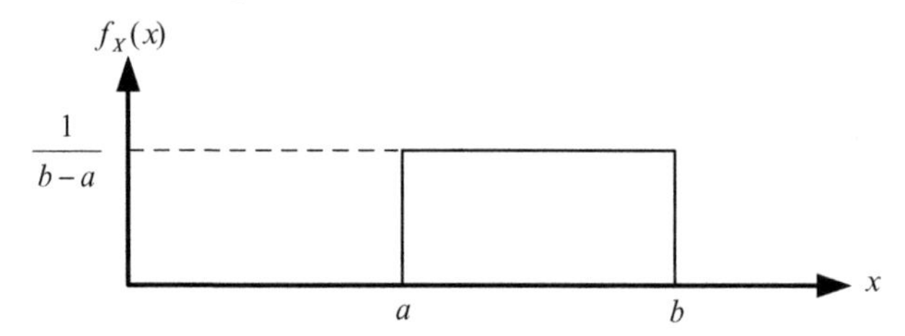


**FIGURE 3.1**
PDF of $X$ in Example 3.1

$$E[X] = \int_{-\infty}^{\infty} x f_X(x) dx = \int_a^b \frac{x}{b-a} dx = \left[\frac{x^2}{2(b-a)}\right]_a^b = \frac{b^2 - a^2}{2(b-a)} = \frac{(b+a)(b-a)}{2(b-a)}$$
$$= \frac{b+a}{2}$$

### EXAMPLE 3.2

Find the expected value of the discrete random variable $X$ with the following PMF:

$$p_X(x) = \begin{cases} \frac{1}{3} & x = 0 \\ \frac{2}{3} & x = 2 \\ 0 & \text{otherwise} \end{cases}$$

**Solution:**

$$E[X]=\sum_{x=-\infty}^{\infty} x p_X(x)=0p_X(0)+2p_X(2)=0\left(\frac{1}{3}\right)+2\left(\frac{2}{3}\right)=\frac{4}{3}$$

## EXAMPLE 3.3

Find the expected value of the random variable $K$ with the following PMF:

$$p(k)=\frac{\lambda^k}{k!}e^{-\lambda} \qquad k=0,1,2,\ldots$$

**Solution:**
$E[K]$ is given by

$$E[K]=\sum_{k=0}^{\infty} k p_K(k)=\sum_{k=0}^{\infty} k\left(\frac{\lambda^k}{k!}e^{-\lambda}\right)=\sum_{k=1}^{\infty}\frac{\lambda^k}{(k-1)!}e^{-\lambda}=\lambda e^{-\lambda}\sum_{k=1}^{\infty}\frac{\lambda^{k-1}}{(k-1)!}=\lambda e^{-\lambda}\sum_{m=0}^{\infty}\frac{\lambda^m}{m!}$$

Since $\sum_{m=0}^{\infty}\frac{a^m}{m!}=e^a$, we have that

$$E[K]=\lambda e^{-\lambda}\sum_{m=0}^{\infty}\frac{\lambda^m}{m!}=\lambda e^{-\lambda}e^{\lambda}=\lambda$$

## EXAMPLE 3.4

Find the expected value of the random variable $X$ whose PDF is given by

$$f_X(x)=\begin{cases}\lambda e^{-\lambda x} & x\geq 0\\ 0 & x<0\end{cases}$$

**Solution:**
The expected value of $X$ is given by

$$E[X]=\int_{-\infty}^{\infty} x f_X(x)dx=\int_0^{\infty} x\lambda e^{-\lambda x}dx$$

Let $u=x$ and $dv=\lambda e^{-\lambda x}dx$. This means that $du=dx$ and $v=-e^{-\lambda x}$. Thus, integrating by parts, we obtain

$$E[X]=[uv]_0^{\infty}-\int_0^{\infty} v\,du=\left[-xe^{-\lambda x}\right]_0^{\infty}+\int_0^{\infty} e^{-\lambda x}dx=0-\left[\frac{e^{-\lambda x}}{\lambda}\right]_0^{\infty}$$
$$=\frac{1}{\lambda}$$

where we have assumed that the function $xe^{-\lambda x}$ goes to zero faster than it goes to infinity as $x\rightarrow\infty$.

## 3.3 EXPECTATION OF NONNEGATIVE RANDOM VARIABLES

Some random variables assume only nonnegative values. For example, the time $X$ until a component fails cannot be negative. In Chapter 1 we defined the reliability function $R(t)$ of a component as the probability that the component has not failed by time $t$. Thus, if the PDF of $X$ is $f_X(x)$ and the CDF is $F_X(x)$, we can define the reliability function of the component by $R_X(t)$, which is related to the CDF and PDF as follows:

$$R_X(t) = P[X > t] = 1 - P[X \leq t] = 1 - F_X(t) \tag{3.3a}$$

$$F_X(t) = 1 - R_X(t) \tag{3.3b}$$

**Proposition 3.1:**
For a nonnegative random variable $X$ with CDF $F_X(x)$, the expected value is given by

$$E[X] = \begin{cases} \int_0^{\infty} P[X > x]dx = \int_0^{\infty} \{1 - F_X(x)\}dx & X \text{ continuous} \\ \sum_{x=0}^{\infty} P[X > x] = \sum_{x=0}^{\infty} \{1 - F_X(x)\} & X \text{ discrete} \end{cases}$$

**Proof:**
We first prove the case for a continuous random variable. Since

$$P[X > x] = \int_x^{\infty} f_X(u)du$$

we have that

$$\int_0^{\infty} P[X > x]dx = \int_0^{\infty} \int_x^{\infty} f_X(u)dudx$$

The region of integration $\{(x, u) | 0 \leq x < \infty; x \leq u < \infty\}$ is as shown in Figure 3.2.

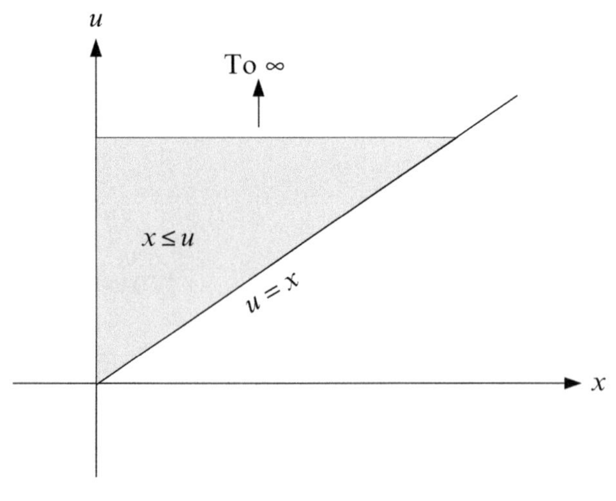


**FIGURE 3.2**
Region of Integration

From the figure we observe that the region of integration can be transformed into $\{(x,u)|0 \leq x \leq u; 0 \leq u < \infty\}$, which gives

$$\int_0^{\infty} P[X > x]dx = \int_0^{\infty}\int_x^{\infty} f_X(u)du dx = \int_{u=0}^{\infty}\left\{\int_{x=0}^{u} dx\right\} f_X(u)du = \int_{u=0}^{\infty} u f_X(u)du$$
$$= E[X]$$

This proves the proposition for the case of a continuous random variable. For a discrete random variable $X$ that assumes only nonnegative values, we have that

$$\begin{aligned}
E[X] &= \sum_{x=0}^{\infty} x p_X(x) = \sum_{x=1}^{\infty} x p_X(x) = 1p_X(1) + 2p_X(2) + 3p_X(3) + 4p_X(4) + \cdots \\
&= p_X(1) \\
&\quad + p_X(2) + p_X(2) \\
&\quad + p_X(3) + p_X(3) + p_X(3) \\
&\quad + p_X(4) + p_X(4) + p_X(4) + p_X(4) \\
&\quad + \cdots \\
&= \sum_{x=1}^{\infty} p_X(x) + \sum_{x=2}^{\infty} p_X(x) + \sum_{x=3}^{\infty} p_X(x) + \cdots \\
&= \{1 - F_X(0)\} + \{1 - F_X(1)\} + \{1 - F_X(2)\} + \cdots \\
&= \sum_{x=0}^{\infty}\{1 - F_X(x)\} = \sum_{x=0}^{\infty} P[X > x]
\end{aligned}$$

## EXAMPLE 3.5

Use the above method to find the expected value of the random variable $X$ whose PDF is given in Example 3.4.

**Solution:**

From Example 3.4, the PDF of $X$ is given by

$$f_X(x) = \begin{cases} \lambda e^{-\lambda x} & x \geq 0 \\ 0 & x < 0 \end{cases}$$

Thus, the CDF of $X$ is given by

$$F_X(x) = \int_0^x f_X(u)du = \int_0^x \lambda e^{-\lambda u} du = 1 - e^{-\lambda x}$$

Since $X$ is a nonnegative random variable, its expected value is given by

$$E[X] = \int_0^{\infty} \{1 - F_X(x)\}dx = \int_0^{\infty} e^{-\lambda x} dx = \frac{1}{\lambda}$$

which is the same result that we obtained in Example 3.4.

## 3.4 MOMENTS OF RANDOM VARIABLES AND THE VARIANCE

The $n$th moment of the random variable $X$, denoted by $E[X^n]=\overline{X^n}$, is defined by

$$E[X^n]=\overline{X^n}=\begin{cases}\sum_k x_k^n p_X(x_k) & X \text{ discrete}\\ \int_{-\infty}^{\infty} x^n f_X(x)dx & X \text{ continuous}\end{cases} \tag{3.4}$$

for $n=1,2,3,\ldots$. The first moment (i.e., $n=1$), $E[X]$, is the expected value of $X$.

We can also define the *central moments* (or *moments about the mean*) of a random variable. These are the moments of the difference between a random variable and its expected value. The $n$th central moment is defined by

$$E\left[(X-\overline{X})^n\right]=\overline{(X-\overline{X})^n}=\begin{cases}\sum_k (x_k-\overline{X})^n p_X(x_k) & X \text{ discrete}\\ \int_{-\infty}^{\infty} (x-\overline{X})^n f_X(x)dx & X \text{ continuous}\end{cases} \tag{3.5}$$

The central moment for the case of $n=2$ is very important and carries a special name, the *variance*, which is usually denoted by $\sigma_X^2$. Thus,

$$\sigma_X^2=E\left[(X-\overline{X})^2\right]=\begin{cases}\sum_k (x_k-\overline{X})^2 p_X(x_k) & X \text{ discrete}\\ \int_{-\infty}^{\infty} (x-\overline{X})^2 f_X(x)dx & X \text{ continuous}\end{cases} \tag{3.6}$$

Before we can simplify the expression for the variance, we first state and prove the following two propositions.

---

**Proposition 3.2:**
Let $X$ be a random variable with the PDF $f_X(x)$ and mean $E[X]$. Let $a$ and $b$ be constants. Then, if $Y$ is the random variable defined by $Y=aX+b$, the expected value of $Y$ is given by $E[Y]=aE[X]+b$.

**Proof:**
Since $Y$ is a function of $X$, its expected value is given by

$$\begin{aligned}E[Y]=E[aX+b]&=\int_{-\infty}^{\infty}(ax+b)f_X(x)dx=\int_{-\infty}^{\infty}axf_X(x)dx+\int_{-\infty}^{\infty}bf_X(x)dx\\ &=a\int_{-\infty}^{\infty}xf_X(x)dx+b\int_{-\infty}^{\infty}f_X(x)dx\\ &=aE[X]+b\end{aligned}$$

where the first term of the last line follows from the definition of expectation, and the second term follows from the fact that the integration is 1.

---

The topic "Functions of Random Variables" is discussed in detail in Chapter 6.

**Proposition 3.3:**
Let $X$ be a random variable with the PDF $f_X(x)$ and mean $E[X]$. Let $g_1(x)$ and $g_2(x)$ be two functions of the random variable $X$, and let $g_3(x)$ be defined by $g_3(x) = g_1(x) + g_2(x)$. The expected value of $g_3(x)$ is given by

$$E[g_3(x)] = E[g_1(x)] + E[g_2(x)]$$

**Proof:**
Since $g_3(x)$ is a function of $X$, its expected value is given by

$$\begin{aligned} E[g_3(x)] &= \int_{-\infty}^{\infty} g_3(x) f_X(x) dx = \int_{-\infty}^{\infty} \{g_1(x) + g_2(x)\} f_X(x) dx \\ &= \int_{-\infty}^{\infty} g_1(x) f_X(x) dx + \int_{-\infty}^{\infty} g_2(x) f_X(x) dx \\ &= E[g_1(x)] + E[g_2(x)] \end{aligned}$$

Using these two propositions, and noting that $\overline{X}$ is a constant, we obtain the variance of $X$ as follows:

$$\begin{aligned} \sigma_X^2 &= E\left[(X - \overline{X})^2\right] = E[X^2 - 2X\overline{X} + (\overline{X})^2] = E[X^2] - 2\overline{X}E[X] + (\overline{X})^2 \\ &= E[X^2] - 2(\overline{X})^2 + (\overline{X})^2 = E[X^2] - (\overline{X})^2 \\ &= E[X^2] - \{E[X]\}^2 \end{aligned} \tag{3.7}$$

The square root of the variance is called the *standard deviation*, $\sigma_X$. The variance is a measure of the "spread" of a PDF or PMF about the mean. If a random variable has a concentrated PDF or PMF, it will have a small variance. Similarly, if it has a widely spread PDF or PMF, it will have a large variance. For example, consider the random variables $X_1, X_2$ and $X_3$ with the following PDFs:

$$f_{X_1}(x) = \begin{cases} \frac{1}{4} & 0 \le x \le 4 \\ 0 & \text{otherwise} \end{cases}$$

$$f_{X_2}(x) = \begin{cases} \frac{1}{3} & 0.5 \le x \le 3.5 \\ 0 & \text{otherwise} \end{cases}$$

$$f_{X_3}(x) = \begin{cases} \frac{1}{2} & 1 \le x \le 3 \\ 0 & \text{otherwise} \end{cases}$$

We can plot these PDFs as shown in Figure 3.3.

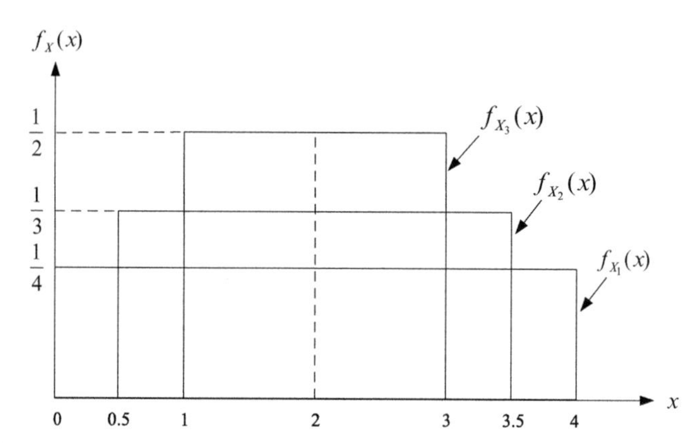


**FIGURE 3.3**
PDFs of $X_1, X_2$ and $X_3$

From Example 3.1, we find the mean value of this class of distributions is $\frac{b+a}{2}$, where $b$ is the upper limit and $a$ is the lower limit of the random variable. Thus, $E[X_1]=E[X_2]=E[X_3]=2$. However, their spreads about the mean value are different. Specifically, $X_1$ has the largest spread about the mean, while $X_3$ has the smallest spread about the mean. In other words, values of $X_3$ tend to cluster around the mean value the most, while the values of $X_1$ tend to cluster around the mean value the least. In terms of variance, we therefore say that $X_1$ has the largest variance, while $X_3$ has the smallest variance.

## EXAMPLE 3.6

Let $X$ be a continuous random variable with the PDF

$$f_X(x)=\begin{cases}\frac{1}{b-a} & a \leq x \leq b \\ 0 & \text{otherwise}\end{cases}$$

Find the variance of $X$.

**Solution:**
From Example 3.1, the expected value of $X$ is given by

$$E[X]=\frac{b+a}{2}$$

The second moment of $X$ is given by

$$\begin{aligned}E\left[X^2\right] &= \int_a^b x^2 f_X(x)dx = \int_a^b \frac{x^2}{b-a}dx = \left[\frac{x^3}{3(b-a)}\right]_a^b = \frac{b^3-a^3}{3(b-a)} \\ &= \frac{(b-a)\left(b^2+ab+a^2\right)}{3(b-a)} = \frac{b^2+ab+a^2}{3}\end{aligned}$$

To find the variance of $X$, we use the formula

$$\sigma_X^2 = E\left[X^2\right] - (E[X])^2 = \frac{b^2+ab+a^2}{3} - \left(\frac{b+a}{2}\right)^2$$
$$= \frac{b^2+ab+a^2}{3} - \frac{b^2+2ab+a^2}{4} = \frac{b^2-2ab+a^2}{12}$$
$$= \frac{(b-a)^2}{12}$$

This result reinforces the statement we made about the PDFs displayed in Figure 3.3. Applying the preceding result to the figure we have that

$$\sigma_{X_1}^2 = \frac{(4-0)^2}{12} = \frac{16}{12} = 1.33$$
$$\sigma_{X_2}^2 = \frac{(3.5-0.5)^2}{12} = \frac{9}{12} = 0.75$$
$$\sigma_{X_3}^2 = \frac{(3-1)^2}{12} = \frac{4}{12} = 0.33$$

Thus, while $E[X_1] = E[X_2] = E[X_3] = 2$, the variances are different with $X_1$ having the largest variance and $X_3$ having the smallest variance and hence the smallest spread about the mean.

## EXAMPLE 3.7

Find the variance of the random variable $K$ with the following PMF:

$$p(k) = \frac{\lambda^k}{k!}e^{-\lambda} \qquad k = 0, 1, 2, \ldots$$

(This is the continuation of Example 3.3.)

**Solution:**
From Example 3.3, we know that $E[K] = \lambda$. The second moment of $K$ is given by

$$E\left[K^2\right] = \sum_{k=0}^{\infty} k^2 p_K(k) = \sum_{k=0}^{\infty} k^2 \frac{\lambda^k}{k!} e^{-\lambda} = \lambda e^{-\lambda} \sum_{k=1}^{\infty} \frac{k\lambda^{k-1}}{(k-1)!}$$

Now, we apply the following trick:

$$\sum_{k=1}^{\infty} \frac{\lambda^k}{(k-1)!} = \lambda \sum_{k=1}^{\infty} \frac{\lambda^{k-1}}{(k-1)!} = \lambda \sum_{m=0}^{\infty} \frac{\lambda^m}{m!} = \lambda e^{\lambda}$$

But

$$\frac{d}{d\lambda}\left[\sum_{k=1}^{\infty} \frac{\lambda^k}{(k-1)!}\right] = \sum_{k=1}^{\infty} \frac{d}{d\lambda}\left[\frac{\lambda^k}{(k-1)!}\right] = \sum_{k=1}^{\infty} \frac{k\lambda^{k-1}}{(k-1)!}$$

Therefore,

$$\sum_{k=1}^{\infty} \frac{k\lambda^{k-1}}{(k-1)!} = \frac{d}{d\lambda}\{\lambda e^{\lambda}\} = (1+\lambda)e^{\lambda}$$

Thus, the second moment of $K$ is given by

$$E\left[K^2\right] = \lambda e^{-\lambda}\sum_{k=1}^{\infty}\frac{k\lambda^{k-1}}{(k-1)!} = \lambda e^{-\lambda}(1+\lambda)e^{\lambda} = \lambda + \lambda^2$$

Finally, the variance of $K$ is given by

$$\sigma_K^2 = E\left[K^2\right] - (E[K])^2 = \lambda + \lambda^2 - \lambda^2 = \lambda$$

It can be seen that the expected value and variance of $K$ have identical values.

## EXAMPLE 3.8

A test engineer discovered that the CDF of the lifetime of an equipment in years is given by

$$F_X(x) = \begin{cases} 0 & x < 0 \\ 1 - e^{-x/5} & 0 \le x < \infty \end{cases}$$

a. What is the expected lifetime of the equipment?
b. What is the variance of the lifetime of the equipment?

**Solution:**

From the definition of its CDF, we can see that $X$ is a random variable that takes only nonnegative values. Thus, we proceed as follows:

a. The expected lifetime of the equipment is given by

$$E[X] = \int_0^{\infty} P[X > x]dx = \int_0^{\infty}[1 - F_X(x)]dx = \int_0^{\infty} e^{-x/5}dx = 5$$

b. To find the variance we first evaluate the PDF:

$$f_X(x) = \frac{d}{dx}F_X(x) = \frac{1}{5}e^{-x/5} \qquad x \ge 0$$

Thus, the second moment of $X$ is given by

$$E\left[X^2\right] = \int_0^{\infty} x^2 f_X(x)dx = \frac{1}{5}\int_0^{\infty} x^2 e^{-x/5}dx$$

Let $u = x^2 \Rightarrow du = 2xdx$, and let $dv = e^{-x/5}dx \Rightarrow v = -5e^{-x/5}$. Thus,

$$E\left[X^2\right] = \left[-\frac{5x^2e^{-x/5}}{5}\right]_0^{\infty} + \frac{10}{5}\int_0^{\infty} xe^{-x/5}dx = 0 + 2\int_0^{\infty} xe^{-x/5}dx = 2\int_0^{\infty} xe^{-x/5}dx$$

Let $u = x \Rightarrow du = dx$, and let $dv = e^{-x/5}dx \Rightarrow v = -5e^{-x/5}$. Then we have that

$$E\left[X^2\right] = 2\left[-5xe^{-x/5}\right]_0^{\infty} + 10\int_0^{\infty} e^{-x/5}dx = 0 + 10\left[-5xe^{-x/5}\right]_0^{\infty} = 50$$

Finally, the variance of $X$ is given by

$$\sigma_X^2 = E\left[X^2\right] - (E[X])^2 = 50 - 25 = 25$$

## EXAMPLE 3.9

A shopping cart contains ten books whose weights are as follows: There are four books with a weight of 1.8 lbs each, one book with a weight of 2 lbs, two books with a weight of 2.5 lbs each, and three books with a weight of 3.2 lbs each.

a. What is the mean weight of the books?
b. What is the variance of the weights of the books?

**Solution:**
The total number of books is 10. The fractions of books in each weight category are as follows:

- The fraction of books with weight 1.8 lbs is 4/10 = 0.4
- The fraction of books with weight 2.0 lbs is 1/10 = 0.1
- The fraction of books with weight 2.5 lbs is 2/10 = 0.2
- The fraction of books with weight 3.2 lbs is 3/10 = 0.3

Let $Y$ be a random variable that denotes the weights of books. Since these fractions are essentially the probabilities of occurrence of these weights, we have that

$$E[Y] = (0.4 \times 1.8) + (0.1 \times 2.0) + (0.2 \times 2.5) + (0.3 \times 3.2) = 2.38$$

$$\begin{aligned}\sigma_Y^2 &= \sum_{k=1}^{4} (y_k - E[Y])^2 p_Y(y_k)\\ &= \left\{(1.18 - 2.38)^2 \times 0.4\right\} + \left\{(2.0 - 2.38)^2 \times 0.1\right\} + \left\{(2.5 - 2.38)^2 \times 0.2\right\}\\ &\quad + \left\{(3.2 - 2.38)^2 \times 0.3\right\} = 0.3536\end{aligned}$$

Note that we can also obtain $\sigma_Y^2$ by first computing $E[Y^2]$ as follows:

$$E\left[Y^2\right] = \left(0.4 \times 1.8^2\right) + \left(0.1 \times 2.0^2\right) + \left(0.2 \times 2.5^2\right) + \left(0.3 \times 3.2^2\right) = 6.018$$

$$\sigma_Y^2 = E\left[Y^2\right] - (E[Y])^2 = 6.018 - 2.38^2 = 0.3536$$

## EXAMPLE 3.10

Find the variance of the random variable $X$ whose PDF is given by

$$f_X(x) = \begin{cases} \lambda e^{-\lambda x} & x \geq 0 \\ 0 & x < 0 \end{cases}$$

(This is the continuation of Example 3.4.)

**Solution:**
From Example 3.4, the expected value of $X$ is given by $E[X] = 1/\lambda$. The second moment of $X$ is given by

$$E\left[X^2\right] = \int_0^{\infty} x^2 f_X(x)dx = \int_0^{\infty} x^2 \lambda e^{-\lambda x} dx$$

Let $u = x^2 \Rightarrow du = 2xdx$, and let $dv = \lambda e^{-\lambda x}dx \Rightarrow v = -e^{-\lambda x}$. Thus, integrating by parts we obtain

$$E[X^2] = [-x^2 e^{-\lambda x}]_0^{\infty} + 2\int_0^{\infty} xe^{-\lambda x}dx = 0 + 2\int_0^{\infty} xe^{-\lambda x}dx = 2\int_0^{\infty} xe^{-\lambda x}dx$$

Let $u = x \Rightarrow du = dx$, and let $dv = e^{-\lambda x}dx \Rightarrow v = -e^{-\lambda x}/\lambda$. Then we have that

$$E[X^2] = 2\left[-\frac{xe^{-\lambda x}}{\lambda}\right]_0^{\infty} + \frac{2}{\lambda}\int_0^{\infty} e^{-\lambda x}dx = 0 + \frac{2}{\lambda^2}[-e^{-\lambda x}]_0^{\infty} = \frac{2}{\lambda^2}$$

Note also that

$$\frac{2}{\lambda}\int_0^{\infty} e^{-\lambda x}dx = \frac{2}{\lambda^2}\int_0^{\infty} \lambda e^{-\lambda x}dx = \frac{2}{\lambda^2}$$

where the last equality follows from the fact that the integrand is a PDF and as a result the value of the integration is one. Thus, the variance of $X$ is given by

$$\sigma_X^2 = E[X^2] - (E[X])^2 = \frac{2}{\lambda^2} - \frac{1}{\lambda^2} = \frac{1}{\lambda^2}$$

---

## EXAMPLE 3.11

Find the variance of the discrete random variable $X$ with the following PMF:

$$p_X(x) = \begin{cases} \frac{1}{3} & x = 0 \\ \frac{2}{3} & x = 2 \\ 0 & \text{otherwise} \end{cases}$$

(This is the continuation of Example 3.2.)

**Solution:**
From Example 3.2, we know that $E[X] = 4/3$. Thus, the second moment and variance of $X$ are given by

$$E[X^2] = \sum_{x=-\infty}^{\infty} x^2 p_X(x) = 0^2 p_X(0) + 2^2 p_X(2) = 0^2\left(\frac{1}{3}\right) + 2^2\left(\frac{2}{3}\right) = \frac{8}{3}$$

$$\sigma_X^2 = E[X^2] - (E[X])^2 = \frac{8}{3} - \left(\frac{4}{3}\right)^2 = \frac{8}{3} - \frac{16}{9} = \frac{8}{9}$$

---

## EXAMPLE 3.12

A student doing a summer internship in a company was asked to model the lifetime of a certain equipment that the company makes. After a series of tests the student proposed that the lifetime of the equipment can be modeled by a random variable $X$ that has the PDF

$$f(x) = \begin{cases} \frac{xe^{-x/10}}{100} & x \geq 0 \\ 0 & \text{otherwise} \end{cases}$$

a. Show that $f(x)$ is a valid PDF.
b. What is the probability that the lifetime of the equipment exceeds 20?
c. What is the expected value of $X$?

**Solution:**

a. For $f(x)$ to be a valid PDF, we must have that $\int_{-\infty}^{\infty} f(x)dx = 1$. Now,

$$\int_{-\infty}^{\infty} f(x)dx = \int_0^{\infty} \frac{xe^{-x/10}}{100} dx$$

Let $u = x \Rightarrow du = dx$, and let $dv = e^{-x/10}dx \Rightarrow v = -10e^{-x/10}$. Thus,

$$\int_{-\infty}^{\infty} f(x)dx = \int_0^{\infty} \frac{xe^{-x/10}}{100} dx = \frac{1}{100}\left\{\left[-10xe^{-x/10}\right]_0^{\infty} + 10\int_0^{\infty} e^{-x/10}dx\right\}$$
$$= \frac{1}{100}\left\{0 - \left[100e^{-x/10}\right]_0^{\infty}\right\} = \frac{100}{100} = 1$$

This proves that the function is a valid PDF.

b. Using the results obtained in part (a), the probability that the lifetime of the equipment exceeds 20 is given by

$$P[X > 20] = \int_{20}^{\infty} f(x)dx = \frac{1}{100}\left\{\left[-10xe^{-x/10}\right]_{20}^{\infty} + 10\int_{20}^{\infty} e^{-x/10}dx\right\}$$
$$= \frac{1}{100}\left\{200e^{-2} - \left[100e^{-x/10}\right]_{20}^{\infty}\right\} = 2e^{-2} + e^{-2} = 3e^{-2}$$
$$= 0.4060$$

c. The expected value of $X$ is given by

$$E[X] = \int_{-\infty}^{\infty} xf(x)dx = \int_0^{\infty} \frac{x^2e^{-x/10}}{100} dx$$

Let $u = x^2 \Rightarrow du = 2xdx$, and let $dv = e^{-x/10}dx \Rightarrow v = -10e^{-x/10}$. Thus,

$$E[X] = \left[-\frac{10x^2e^{-x/10}}{100}\right]_0^{\infty} + 20\int_0^{\infty} \frac{xe^{-x/10}}{100} dx = 0 + 20 = 20$$

where the second equality follows from the fact that the integrand is $f(x)$, which has been shown to be a valid PDF and thus integrates to one from zero to infinity.

## EXAMPLE 3.13

A high school student has a personal goal of scoring at least 1500 in the scholastic aptitude test (SAT). He plans to keep taking the test until he achieves this goal; time is not an issue for him. The probability that he scores 1500 or higher in any one SAT test is $p$. His performance in each SAT test is independent of his performance in other SAT tests. His friend who has just completed a course on probability informed him that if $K$ is the number of times he has to take the test before getting a score of 1500 or higher, then the PMF of $K$ is given by

$$p_K(k) = p(1-p)^{k-1} \qquad k = 1, 2, \ldots$$

a. Prove that the function proposed by his friend is a true PMF.
b. Given the above PMF, what is the probability that he takes the test more than 5 times?
c. What is the expected number of times he takes the test to achieve his goal?

**Solution:**

a. To prove that the friend's proposed function is a true PMF, we need to show that it sums to one; that is,

$$\sum_{k=1}^{\infty} p_K(k) = \sum_{k=1}^{\infty} p(1-p)^{k-1} = p\sum_{m=0}^{\infty}(1-p)^m = \frac{p}{1-(1-p)} = 1$$

Thus, the proposed function is a true PMF.

b. The probability that he takes the test more than 5 times is

$$P[K>5] = \sum_{k=6}^{\infty} p(1-p)^{k-1} = p(1-p)^5 \sum_{k=6}^{\infty}(1-p)^{k-6}$$

Let $m = k-6$. Since $(1-p) < 1$, we obtain

$$P[K>5] = p(1-p)^5 \sum_{m=0}^{\infty}(1-p)^m = \frac{p(1-p)^5}{1-(1-p)} = (1-p)^5$$

This is precisely the probability that he did not make the grade in the first 5 attempts.

c. The expected number of times he takes the test is given by

$$E[K] = \sum_{k=1}^{\infty} k p_K(k) = \sum_{k=1}^{\infty} kp(1-p)^{k-1} = p\sum_{k=1}^{\infty} k(1-p)^{k-1}$$

Let

$$G = \sum_{k=0}^{\infty}(1-p)^k = \frac{1}{p}$$

$$\frac{dG}{dp} = \frac{d}{dp}\sum_{k=0}^{\infty}(1-p)^k = \sum_{k=0}^{\infty}\frac{d}{dp}(1-p)^k = -\sum_{k=0}^{\infty}k(1-p)^{k-1} = -\sum_{k=1}^{\infty}k(1-p)^{k-1}$$

Thus,

$$E[K] = p\sum_{k=1}^{\infty}k(1-p)^{k-1} = -p\frac{dG}{dp} = -p\frac{d}{dp}\left(\frac{1}{p}\right) = -p\left(-\frac{1}{p^2}\right) = \frac{1}{p}$$

This shows that the smaller the value of $p$, the more times on the average that he has to take the test. For example, when $p = 1/5$, he will take the test an average of 5 times; and when $p = 1/10$, he will on the average take the test 10 times.

## 3.5 CONDITIONAL EXPECTATIONS

Sometimes we are interested in computing the mean of a subset of a population that shares some property. For example, we may be interested in the mean grade of those students who have passed an exam or the average age of professors who have doctoral degrees.

The conditional expectation of $X$ given that an event $A$ has occurred is given by

$$E[X|A] = \begin{cases} \sum_k x_k p_{X|A}(x_k|A) & X \text{ discrete} \\ \int_{-\infty}^{\infty} x f_{X|A}(x|A)\,dx & X \text{ continuous} \end{cases} \tag{3.8}$$

where the conditional PMF $p_{X|A}(x|A)$ and the conditional PDF $f_{X|A}(x|A)$ are defined as follows:

$$p_{X|A}(x|A) = \frac{p_X(x)}{P[A]} \quad P[A] > 0$$

$$f_{X|A}(x|A) = \frac{f_X(x)}{P[A]} \quad P[A] > 0$$

where $x \in A$, and $P[A]$ is the probability that event $A$ occurs.

### EXAMPLE 3.14

Let $A$ be the event: $A = \{X \leq a\}, -\infty < a < \infty$. What is $E[X|A]$?

**Solution:**

We first obtain the conditional PDF as follows:

$$f_{X|A}(x|X \leq a) = \begin{cases} \dfrac{f_X(x)}{P[X \leq a]} = \dfrac{f_X(x)}{F_X(a)} & x \leq a \\ 0 & x > a \end{cases}$$

Thus,

$$E[X|A] = \frac{\int_{-\infty}^{a} x f_X(x) dx}{F_X(a)}$$

### EXAMPLE 3.15

With reference to Example 3.14, assume that the random variable $X$ has the PDF

$$f_X(x) = \begin{cases} \dfrac{1}{20} & 40 \leq x \leq 60 \\ 0 & \text{otherwise} \end{cases}$$

Find the conditional probability when $a=55$.

**Solution:**

$$F_X(55)=\int_{40}^{55} f_X(x)dx=\left[\frac{x}{20}\right]_{40}^{55}=\frac{55-40}{20}=\frac{3}{4}$$

Thus, $E[X|X\leq 55]$ is given by

$$E[X|X\leq 55]=\frac{\int_{40}^{55} xf_X(x)dx}{F_X(55)}=\frac{\int_{40}^{55}\left(\frac{x}{20}\right)dx}{F_X(55)}=\frac{4}{3}\left[\frac{x^2}{40}\right]_{40}^{55}$$
$$=\frac{4}{3}\left(\frac{3025-1600}{40}\right)=\frac{4}{3}\times\frac{1425}{40}=47.5$$

This problem has an intuitive meaning. Recall that in Example 3.1 the expected value of a random variable that is uniformly distributed between $a$ and $b$ is $(b+a)/2$. In this example, given that $X\leq 55$, the expected value is $(55+40)/2=47.5$.

## 3.6 THE MARKOV INEQUALITY

The Markov inequality applies to random variables that take only nonnegative values. If $X$ is a random variable that takes only nonnegative values, then for any value $a>0$,

$$P[X\geq a]\leq\frac{E[X]}{a} \tag{3.9}$$

This inequality can be proved as follows.

$$E[X]=\int_0^{\infty} xf_X(x)dx=\int_0^{a} xf_X(x)dx+\int_a^{\infty} xf_X(x)dx\geq\int_a^{\infty} xf_X(x)dx$$
$$\geq\int_a^{\infty} af_X(x)dx=aP[X\geq a]$$

### EXAMPLE 3.16

Consider a random variable $X$ that takes values between 0 and 6. If the expected value of $X$ is 3.5, use the Markov inequality to obtain an upper bound for $P[X\geq 5]$.

**Solution:**

$$P[X\geq 5]\leq\frac{3.5}{5}=0.7.$$

The Markov inequality can be uninformative sometimes. For example, in Example 3.16,

$$P[X \geq 2] \leq \frac{3.5}{2} = 1.75$$

which provides no useful information because probability cannot exceed 1.

## 3.7 THE CHEBYSHEV INEQUALITY

The Chebyshev inequality is a statement that places a bound on the probability that an experimental value of a random variable $X$ with finite mean $E[X] = \mu_X$ and variance $\sigma_X^2$ will differ from the mean by more than a fixed positive number $a$. The statement says that the bound is directly proportional to the variance and inversely proportional to $a^2$. That is,

$$P[|X - E[X]| \geq a] \leq \frac{\sigma_X^2}{a^2}, \quad a > 0 \tag{3.10}$$

This is a loose bound that can be obtained as follows.

$$\begin{aligned}
\sigma_X^2 &= \int_{-\infty}^{\infty} (x-\mu_X)^2 f_X(x)dx \\
&= \int_{-\infty}^{-(\mu_X - a)} (x-\mu_X)^2 f_X(x)dx + \int_{-(\mu_X - a)}^{(\mu_X - a)} (x-\mu_X)^2 f_X(x)dx \\
&\quad + \int_{\mu_X - a}^{\infty} (x-\mu_X)^2 f_X(x)dx
\end{aligned}$$

If we omit the middle integral, we obtain

$$\begin{aligned}
\sigma_X^2 &\geq \int_{-\infty}^{-(\mu_X - a)} (x-\mu_X)^2 f_X(x)dx + \int_{\mu_X - a}^{\infty} (x-\mu_X)^2 f_X(x)dx \\
&= \int_{|x-\mu_X| \geq a} (x-\mu_X)^2 f_X(x)dx \geq \int_{|x-\mu_X| \geq a} a^2 f_X(x)dx \\
&= a^2 \int_{|x-\mu_X| \geq a} f_X(x)dx \\
&= a^2 P[|X - \mu_X| \geq a]
\end{aligned}$$

which gives the Chebyshev inequality.

### EXAMPLE 3.17

A random variable $X$ has a mean 4 and variance 2. Use the Chebyshev inequality to obtain an upper bound for $P[|X-4| \geq 3]$.

**Solution:**

From the Chebyshev inequality,

$$P[|X-4| \geq 3] \leq \frac{\sigma_X^2}{3^2} = \frac{2}{9}$$

It must be emphasized that the Chebyshev inequality is only an approximation that is sometimes uninformative, as in the Markov inequality. For example, suppose we are required to use the data in Example 3.17 to obtain $P[|X-4|\geq 1]$. In this case we would get

$$P[|X-4|\geq 1]\leq \frac{\sigma_X^2}{1^2}=\frac{2}{1}=2$$

which provides no useful information.

**Comments:**
The Chebyshev inequality tends to be more powerful than the Markov inequality, which means that it provides a more accurate bound than the Markov inequality, because in addition to the mean of a random variable, it also uses information on the variance of the random variable.

## 3.8 CHAPTER SUMMARY

This chapter developed methods of evaluating the two most commonly used measures of central tendency of random variables, namely, the expectation (or mean or average) and the variance. Two limit theorems associated with these measures are also considered; these are the Markov inequality and the Chebyshev inequality. Several examples are solved to demonstrate how these measures can be computed for both discrete random variables and continuous random variables.

## 3.9 PROBLEMS

### Section 3.2 Expected Values

3.1 Find the mean and variance of the random variable whose triangular PDF is given in Figure 3.4.

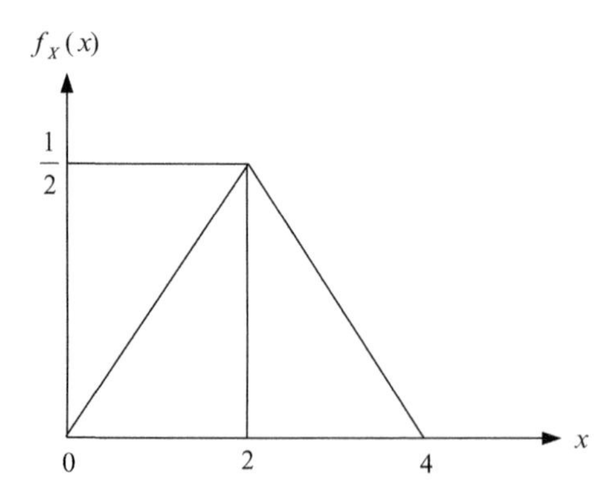


**FIGURE 3.4**
PDF of $X$ for Problem 3.1

3.2 An insurance company has 1000 policies of men of age 50. The company's estimate is that the probability that a man of age 50 dies within one year is 0.02. What is the expected number of claims that the company can expect from the beneficiaries of these men within one year?

3.3 A class has 20 students whose heights are as follows: There are 4 students with a height of 5.5 feet, there are 5 students with a height of 5.8 feet, there are 3 students with a height of 6.0 feet, there are 5 students with a height of 6.2 feet, and there are 3 students with a height of 6.5 feet. If a student is randomly selected from the class, what is his expected height?

3.4 The time it takes a machine to perform an operation depends on the state it enters when the start button is pushed. There are three known states: *fast, moderate,* and *slow*. When it is in the fast state, it takes 2 minutes to perform the operation. When it is in the moderate state, it takes 4 minutes to perform the operation. When it is in the slow state, it takes 7 minutes to perform the operation. Studies indicate that the machine goes into the fast state when the start button is pushed 60% of the time; it goes into the moderate state 25% of the time; and it goes into the slow state 15% of the time. What is the expected time it takes the machine to perform the operation?

3.5 A student has three methods of solving a problem. It takes him 1 hour to solve the problem using method A. It takes him 45 minutes to solve the problem using method B. It takes him 30 minutes to solve the problem using method C. His friends discovered that the student uses method A 10% of the time, method B 40% of the time, and method C 50% of the time. What is the expected time it takes the student to solve a problem?

3.6 Consider the following game that involves tossing a fair die. If the outcome of a toss is an even number, you win \$2. If the outcome is 1 or 3, you lose \$1. If the outcome is 5, you lose \$3. What are the expected winnings?

3.7 Three vans are used to carry 45 students in the applied probability class to participate in a state competition on random phenomena. The first van carried 12 students, the second van carried 15 students, and the third van carried 18 students. Upon their arriving at the place where the competition was taking place, one student was randomly selected from the entire group to receive a gift certificate for a free lunch at a local restaurant. What is the expected number of students in the van that carried the selected student?

3.8 Find the expected value of a discrete random variable $N$ whose PMF is given by

$$p_N(n) = p(1-p)^{n-1} \qquad n = 1, 2, \ldots$$

3.9 Find the expected value of a discrete random variable $K$ whose PMF is given by

$$p_K(k) = \frac{5^k e^{-5}}{k!} \qquad k = 0, 1, 2, \ldots$$

3.10 Find the expected value of a continuous random variable $X$ whose PDF is given by

$$f_X(x) = 2e^{-2x} \qquad x \geq 0$$

3.11 Assume that a random variable $X$ takes on discrete values $x_1, x_2, \ldots, x_n$ with probabilities $p_1, p_2, \ldots, p_n$ respectively. That is, $P[X = x_i] = p_i$, $i = 1, 2, \ldots, n$. The "entropy" of $X$, which is also defined as the amount of information provided by an observation of $X$, is defined by

$$H(X) = \sum_{i=1}^{n} p_i \log\left(\frac{1}{p_i}\right) = -\sum_{i=1}^{n} p_i \log(p_i)$$

When the logarithm is taken base 2, the unit of entropy is bits. Let $X$ represent the outcome of a single roll of a fair die. What is the entropy of $X$ in bits?

## Section 3.4 Moments of Random Variables and the Variance

3.12 A random variable $X$ assumes two values 4 and 7 with probabilities $p$ and $q$, respectively, where $q = 1 - p$. Determine the mean and standard deviation of $X$.

3.13 Find the mean and variance of the discrete random variable $X$ with the following PMF:

$$p_X(x) = \begin{cases} \frac{2}{5} & x = 3 \\ \frac{3}{5} & x = 6 \end{cases}$$

3.14 Let $N$ be a random variable with the following CDF:

$$F_N(n) = \begin{cases} 0 & n < 1 \\ 0.2 & 1 \leq n < 2 \\ 0.5 & 2 \leq n < 3 \\ 0.8 & 3 \leq n < 4 \\ 1 & n \geq 4 \end{cases}$$

a. What is the PMF of $N$?
b. What is the expected value of $N$?
c. What is the variance of $N$?

3.15 Let $X$ be the random variable that denotes the outcome of tossing a fair die once.
a. What is the PMF of $X$?
b. What is the expected value of $X$?
c. What is the variance of $X$?

3.16 Suppose the random variable $X$ has the PDF $f_X(x)=ax^3,\ 0<x<1$.
a. What is the value of $a$?
b. What is the expected value of $X$?
c. What is the variance of $X$?
d. What is the value of $m$ so that $P[X\leq m]=0.5$?

3.17 A random variable $X$ has the CDF

$$F_X(x)=\begin{cases}0 & x<1\\ 0.5(x-1) & 1\leq x<3\\ 1 & x\geq 3\end{cases}$$

a. What is the PDF of $X$?
b. What is the expected value of $X$?
c. What is the variance of $X$?

3.18 A random variable $X$ has the PDF $f_X(x)=x^2/9,\ 0\leq x\leq 3$. Find the mean, variance, and third moment of $X$.

3.19 Assume the random variable $X$ has the PDF $f_X(x)=\lambda e^{-\lambda x},\ x\geq 0$. Find the third moment of $X$, $E[X^3]$.

3.20 Suppose $X$ is a random variable with PDF $f_X(x)$, mean $E[X]$, and variance $\sigma_X^2$. If we define the random variable $Y=X^2$, determine the mean and variance of $Y$ in terms of the mean, variance, and other higher moments of $X$.

3.21 The PDF of a random variable $X$ is given by $f_X(x)=4x(9-x^2)/81$, $0\leq x\leq 3$. Find the mean, variance, and third moment of $X$.

## Section 3.5 Conditional Expectations

3.22 If the PDF of a random variable $X$ is given by $f_X(x)=4x(9-x^2)/81$, $0\leq x\leq 3$, find the conditional expected value of $X$, given that $X\leq 2$.

3.23 The PDF of a continuous random variable $X$ is given by $f_X(x)=2e^{-2x}$, $x\geq 0$. Find the conditional expected value of $X$, given that $X\leq 3$.

3.24 The PDF of a random variable $X$ is given by

$$f_X(x)=\begin{cases}0.1 & 30\leq x\leq 40\\ 0 & \text{otherwise}\end{cases}$$

Find the conditional expected value of $X$, given that $X\leq 35$.

3.25 A fair die is rolled once. Let $N$ denote the outcome of the experiment. Find the expected value of $N$, given that the outcome is an even number.

3.26 The life of a lightbulb in months is denoted by a random variable $X$ with the PDF $f_X(x)=0.5e^{-0.5x}, \ x \geq 0$. Find the conditional expected value of $X$, given that $X \leq 1.5$.

### Sections 3.6 and 3.7 Markov and Chebyshev Inequalities

3.27 A random variable $X$ has the PDF $f_X(x)=2e^{-2x}, \ x \geq 0$. Obtain an upper bound for $P[X \geq 1]$ using the Markov inequality.

3.28 A random variable $X$ has the PDF $f_X(x)=2e^{-2x}, \ x \geq 0$. Obtain an upper bound for $P[|X-E[X]| \geq 1]$.

3.29 A random variable $X$ has a mean 4 and variance 2. Use the Chebyshev inequality to obtain an upper bound for $P[|X-4| \geq 2]$.

3.30 A random variable $X$ has the PDF

$$f_X(x)=\begin{cases} \frac{1}{3} & 1 \leq x \leq 4 \\ 0 & \text{otherwise} \end{cases}$$

Use the Chebyshev inequality to estimate $P[|X-2.5| \geq 2]$.

CHAPTER 4

# Special Probability Distributions

## 4.1 INTRODUCTION

Chapters 2 and 3 deal with the general properties of random variables. Random variables with special probability distributions are encountered in different fields of science and engineering. The goal of this chapter is to describe some of these distributions, including their expected values and variances. These include the Bernoulli distribution, binomial distribution, geometric distribution, Pascal distribution, hypergeometric distribution, Poisson distribution, exponential distribution, Erlang distribution, uniform distribution, and normal distribution. Some of these random variables are discrete random variables and some are continuous random variables. Figure 4.1 shows the relationships among these random variables.

## 4.2 THE BERNOULLI TRIAL AND BERNOULLI DISTRIBUTION

A Bernoulli trial is an experiment that results in two outcomes: *success* and *failure*. One example of a Bernoulli trial is the coin tossing experiment, which results in heads or tails. In a Bernoulli trial we define the probability of success and probability of failure as follows:

$$P[\text{success}] = p \qquad 0 \le p \le 1$$
$$P[\text{failure}] = 1 - p$$

Let us associate the events of the Bernoulli trial with a random variable $X$ such that when the outcome of the trial is a success we define $X=1$ and when the outcome is a failure we define $X=0$. The random variable $X$ is called a Bernoulli random variable and its PMF is given by

$$p_X(x) = \begin{cases} 1-p & x=0 \\ p & x=1 \end{cases} \tag{4.1}$$

An alternative way to define the PMF of $X$ is as follows:

$$p_X(x) = p^x(1-p)^{1-x} \qquad x=0,1 \tag{4.2}$$



Fundamentals of Applied Probability and Random Processes. http://dx.doi.org/10.1016/B978-0-12-800852-2.00004-3

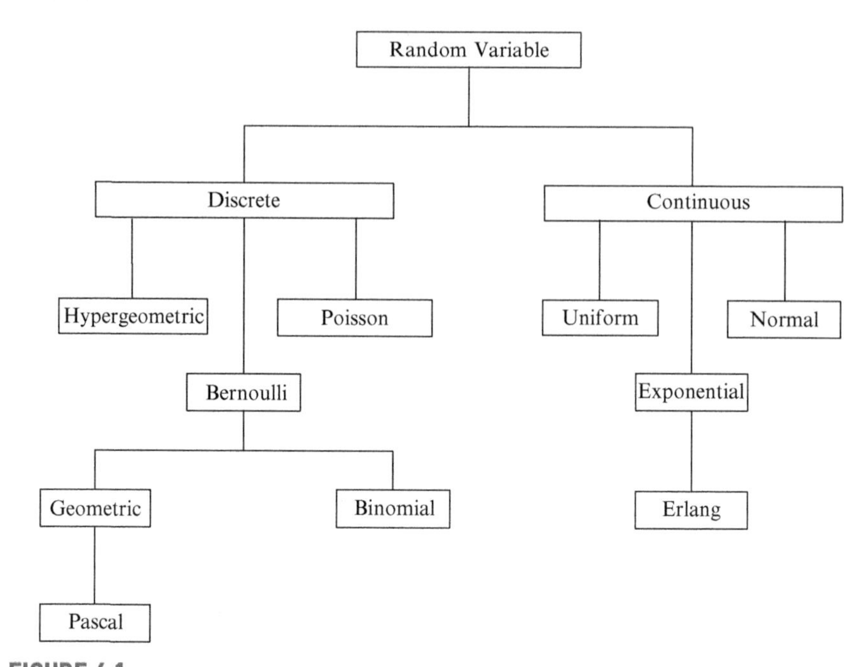


**FIGURE 4.1**
Relationships of the Random Variables

The PMF of $X$ can be plotted as shown in Figure 4.2. The CDF is given by

$$F_X(x) = \begin{cases} 0 & x < 0 \\ 1-p & 0 \leq x < 1 \\ 1 & x \geq 1 \end{cases} \tag{4.3}$$

The expected value of $X$ is given by

$$E[X] = 0(1-p) + 1(p) = p \tag{4.4}$$

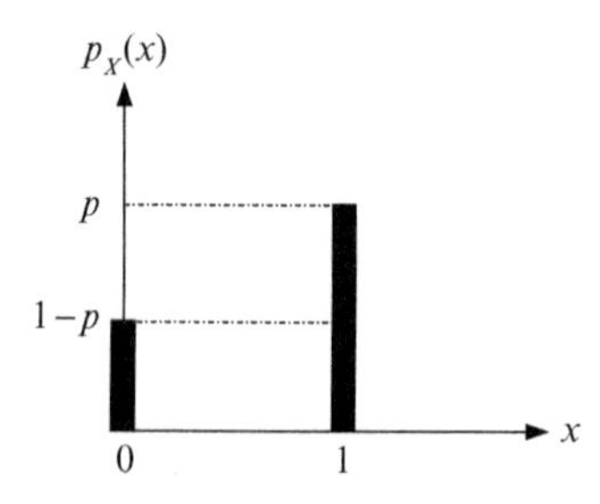


**FIGURE 4.2**
PMF of the Bernoulli Random Variable

Similarly, the second moment of $X$ is given by

$$E[X^2] = 0^2(1-p) + 1^2(p) = p \tag{4.5}$$

Thus, the variance of $X$ is given by

$$\sigma_X^2 = E[X^2] - \{E[X]\}^2 = p - p^2 = p(1-p) \tag{4.6}$$

## 4.3 BINOMIAL DISTRIBUTION

Suppose we conduct $n$ independent Bernoulli trials and we represent the number of successes in those $n$ trials by random variable $X(n)$. Then $X(n)$ is defined as a binomial random variable with parameters $(n, p)$. The PMF of a random variable $X(n)$ with parameters $(n, p)$ is given by

$$p_{X(n)}(x) = \binom{n}{x} p^x (1-p)^{n-x} \quad x = 0, 1, \ldots, n \tag{4.7}$$

Sometimes we use the notation $X(n) \sim B(n, p)$ to define the binomial distribution. The binomial coefficient, $\binom{n}{x}$, represents the number of ways of arranging $x$ successes and $n-x$ failures. The shape of the PMF of $X(n)$ depends on the parameters $n$ and $p$. However, in the case when $p = 0.5$, the graph is symmetrical about the mean value, which we will shortly show is equal to $np$. Thus, for this class of parameters, the PMF can be graphically represented as shown in Figure 4.3.

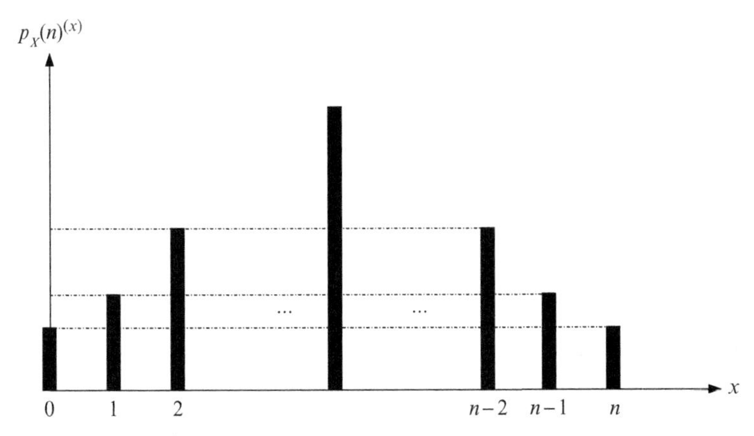


**FIGURE 4.3**
PMF of Binomial Random Variable with $p = 0.5$

Note that

$$\sum_{x=0}^{n} p_{X(n)}(x) = \sum_{x=0}^{n} \binom{n}{x} p^x (1-p)^{n-x} = [p + (1-p)]^n = 1^n = 1$$

which follows from the binomial theorem that we stated in Chapter 1.

The mean of $X(n)$ can be obtained as follows:

$$\begin{aligned} E[X(n)] &= \sum_{x=0}^{n} x p_{X(n)}(x) = \sum_{x=0}^{n} x \binom{n}{x} p^x (1-p)^{n-x} \\ &= \sum_{x=0}^{n} x \frac{n!}{(n-x)!x!} p^x (1-p)^{n-x} = \sum_{x=1}^{n} \frac{n!}{(n-x)!(x-1)!} p^x (1-p)^{n-x} \\ &= \sum_{x=1}^{n} \frac{n(n-1)!}{(n-x)!(x-1)!} p^x (1-p)^{n-x} = np \sum_{x=1}^{n} \frac{(n-1)!}{(n-x)!(x-1)!} p^{x-1} (1-p)^{n-x} \end{aligned}$$

Let $m = x-1$, which means that when $x=1, m=0$; and when $x=n, m=n-1$. Thus, we obtain

$$\begin{aligned} E[X(n)] &= np \sum_{m=0}^{n-1} \frac{(n-1)!}{(n-m-1)!m!} p^m (1-p)^{n-m-1} = np \sum_{m=0}^{n-1} \binom{n-1}{m} p^m (1-p)^{n-m-1} \\ &= np[p + (1-p)]^{n-1} = np(1)^{n-1} \\ &= np \end{aligned} \tag{4.8}$$

To obtain the variance of $X(n)$, we first obtain the second moment as follows:

$$\begin{aligned} E[X(n)\{X(n)-1\}] &= \sum_{x=0}^{n} x(x-1) p_{X(n)}(x) = \sum_{x=0}^{n} x(x-1) \binom{n}{x} p^x (1-p)^{n-x} \\ &= \sum_{x=0}^{n} \frac{x(x-1)n!}{(n-x)!x!} p^x (1-p)^{n-x} \\ &= p^2 \sum_{x=2}^{n} \frac{x(x-1)n(n-1)(n-2)!}{(n-x)!x(x-1)(x-2)!} p^{x-2} (1-p)^{n-x} \\ &= n(n-1)p^2 \sum_{x=2}^{n} \frac{(n-2)!}{(n-x)!(x-2)!} p^{x-2} (1-p)^{n-x} \end{aligned}$$

Let $m = x-2$, which means that when $x=2, m=0$; and when $x=n, m=n-2$. Thus, we obtain

$$\begin{aligned} E[X(n)\{X(n)-1\}] &= n(n-1)p^2 \sum_{m=0}^{n-2} \frac{(n-2)!}{(n-m-2)!m!} p^m (1-p)^{n-m-2} \\ &= n(n-1)p^2 \sum_{m=0}^{n-2} \binom{n-2}{m} p^m (1-p)^{n-m-2} \\ &= n(n-1)p^2 [p + (1-p)]^{n-2} \\ &= n(n-1)p^2 \end{aligned}$$

But $E[X(n)\{X(n)-1\}]=E[X^2(n)]-E[X(n)]$. Thus,

$$E[X^2(n)]=E[X(n)\{X(n)-1\}]+E[X(n)]=n(n-1)p^2+np$$

From this we obtain the variance of $X(n)$ as follows:

$$\begin{aligned}\sigma^2_{X(n)}&=E[X^2(n)]-(E[X(n)])^2=n(n-1)p^2+np-n^2p^2\\&=np(1-p)\end{aligned} \tag{4.9}$$

The CDF of $X(n)$ is given by:

$$F_{X(n)}(x)=P[X(n)\leq x]=\sum_{k=0}^{x}\binom{n}{k}p^k(1-p)^{n-k} \quad x=0,1,\ldots,n \tag{4.10}$$

## EXAMPLE 4.1

Four fair coins are tossed. If the outcomes are assumed to be independent, find the PMF of the number of heads obtained.

**Solution:**

Let $X(n)$ denote the number of heads that appear in the four tosses. Then $X(n)$ is a binomial random variable with $n=4$ and $p=1/2$. Thus, the PMF of $X(n)$ is

$$p_{X(4)}(x)=\binom{4}{x}\left(\frac{1}{2}\right)^x\left(\frac{1}{2}\right)^{4-x}=\binom{4}{x}\left(\frac{1}{2}\right)^4$$

$$p_{X(4)}(0)=\binom{4}{0}\left(\frac{1}{2}\right)^4=\frac{1}{16}$$

$$p_{X(4)}(1)=\binom{4}{1}\left(\frac{1}{2}\right)^4=\frac{4}{16}$$

$$p_{X(4)}(2)=\binom{4}{2}\left(\frac{1}{2}\right)^4=\frac{6}{16}$$

$$p_{X(4)}(0)=\binom{4}{3}\left(\frac{1}{2}\right)^4=\frac{4}{16}$$

$$p_{X(4)}(0)=\binom{4}{4}\left(\frac{1}{2}\right)^4=\frac{1}{16}$$

## EXAMPLE 4.2

100 balls are tossed into 50 boxes. What is the expected number of balls in the tenth box?

**Solution:**

If we think of the balls tossed as Bernoulli trials in which a success is defined as getting a ball in the tenth box, then $p=1/50$. Let $X$ denote the number of balls that go into the tenth box.

Then $X$ is a binomial random variable with parameters $(100, 1/50)$; that is, $X \sim B(100, 1/50)$. Thus,

$$E[X] = np = 100 \times \frac{1}{50} = 2$$

## EXAMPLE 4.3

A coin is tossed 10 times. Given that there are 6 heads in the 10 tosses, what is the expected number of heads in the first 5 tosses?

**Solution:**

Since there are 6 heads and hence 4 tails in the 10 tosses, we can liken the situation to a bag that contains 6 heads and 4 tails. Thus, the probability of a head is $p = 6/10 = 0.6$. If $Y$ denotes the number of heads in the first 5 tosses of the coin, then $Y \sim B(5, 0.6)$; that is, $Y$ is a binomial random variable with parameters $(5, 0.6)$ and

$$E[Y] = np = 5 \times 0.6 = 3$$

## 4.4 GEOMETRIC DISTRIBUTION

The geometric random variable is used to describe the number of Bernoulli trials until the first success occurs. A modified version, which is used to describe the number of Bernoulli trials until the first failure occurs, is discussed later.

Let $X$ be a random variable that denotes the number of Bernoulli trials up to and including the trial that results in the first success. If the first success occurs on the $x$th trial, then we know that the first $x-1$ trials resulted in failures. Since the trials are independent, the PMF of a geometric random variable, $X$, is given by

$$p_X(x) = p(1-p)^{x-1} \quad x = 1, 2, \ldots \tag{4.11}$$

The PMF of $X$ is illustrated in Figure 4.4, which is an exponentially decreasing function as $x$ increases.

## EXAMPLE 4.4

A bag contains 6 blue balls and 4 red balls. Balls are randomly drawn from the bag, one at a time, until a red ball is obtained. If we assume that each drawn ball is replaced before the next one is drawn, what is the probability that exactly 5 balls are drawn?

**Solution:**

Let $X$ denote the number of balls needed to select a red ball. Since balls are replaced after they are drawn from the bag, the probability that a particular ball is drawn from the bag in any trial is constant. Thus, each time a red ball is drawn we have a success with probability $p = 4/(4+6) = 0.4$, and

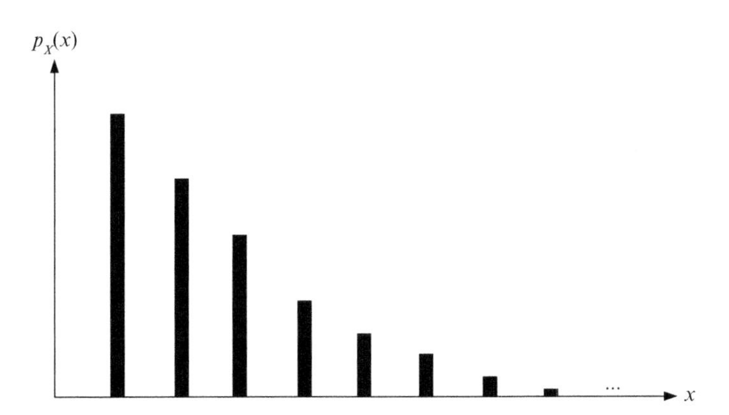


**FIGURE 4.4**
PMF of the Geometric Random Variable

when a blue ball is drawn we have a failure with probability $1-p=0.6$. If exactly 5 balls are drawn before the experiment is stopped, then we drew four blue balls consecutively (or we had four consecutive failures) and the fifth ball is a red ball, which is a success. Therefore, the desired probability is given by

$$P[X=5]=p_X(5)=p(1-p)^{5-1}=p(1-p)^4=(0.4)(0.6)^4=0.05184$$

The expected value of $X$ is given by

$$E[X]=\sum_{x=1}^{\infty} xp_X(x)=\sum_{x=1}^{\infty} xp(1-p)^{x-1}=p\sum_{x=1}^{\infty} x(1-p)^{x-1}$$

But we know that

$$\sum_{x=0}^{\infty}(1-p)^x=\frac{1}{1-(1-p)}=\frac{1}{p}$$

Also,

$$\frac{d}{dp}\sum_{x=0}^{\infty}(1-p)^x=\sum_{x=0}^{\infty}\frac{d}{dp}(1-p)^x=-\sum_{x=0}^{\infty}x(1-p)^{x-1}=-\sum_{x=1}^{\infty}x(1-p)^{x-1}$$

Thus,

$$\begin{aligned}E[X]&=p\sum_{x=1}^{\infty}x(1-p)^{x-1}=-p\frac{d}{dp}\sum_{x=0}^{\infty}(1-p)^x\\&=-p\frac{d}{dp}\left(\frac{1}{p}\right)=-p\left(-\frac{1}{p^2}\right)\\&=\frac{1}{p}\end{aligned} \tag{4.12}$$

We can use the same technique to show that the second moment is given by

$$E[X^2] = \sum_{x=1}^{\infty} x^2 p_X(x) = \frac{2-p}{p^2}$$

Thus, the variance is given by

$$\sigma_X^2 = E[X^2] - (E[X])^2 = \frac{2-p}{p^2} - \frac{1}{p^2} = \frac{1-p}{p^2} \tag{4.13}$$

## EXAMPLE 4.5

Balls are tossed at random into 50 boxes. Find the expected number of tosses required to get the first ball in the fourth box.

**Solution:**

Let $Y$ denote the number of tosses until the first ball goes into the fourth box. Then $Y$ has a geometric distribution with $p = 1/50$. The expected value of Y is

$$E[X] = \frac{1}{p} = 50$$

The probability that the number of trials until the first success is greater than $n$ is given by

$$\begin{aligned} P[X > n] &= \sum_{x=n+1}^{\infty} p_X(x) = \sum_{x=n+1}^{\infty} p(1-p)^{x-1} = p(1-p)^n \sum_{x=n+1}^{\infty} (1-p)^{x-n-1} \\ &= p(1-p)^n \sum_{m=0}^{\infty} (1-p)^m = p(1-p)^n \left\{ \frac{1}{1-(1-p)} \right\} = \frac{p(1-p)^n}{p} \\ &= (1-p)^n \end{aligned} \tag{4.14}$$

Observe that this is the probability that the first $n$ trials of the experiment resulted in failures.

## EXAMPLE 4.6

The probability that a missile hits a target is $p$. If missiles are fired independently at a target until it is hit, what is the probability that it takes more than 3 missiles to get the target?

**Solution:**

Since $p$ is the probability of success, the problem merely states that the number of attempts before a success is greater than 3. That is, there have so far been 3 consecutive failures. Since these attempts are independent, the required answer is $(1-p)^3$.

### 4.4.1 CDF of the Geometric Distribution

We use the result in (4.14) to compute the CDF of $X$ as follows:

$$\begin{aligned} F_X(x) &= P[X \leq x] = \sum_{k=1}^{x} p(1-p)^{k-1} = 1 - P[X > x] \\ &= 1 - (1-p)^x \end{aligned} \tag{4.15}$$

Note that we can obtain the same result directly as follows:

$$\begin{aligned} F_X(x) &= \sum_{k=1}^{x} p(1-p)^{k-1} = p\sum_{m=0}^{x-1}(1-p)^m = p\left\{\frac{1-(1-p)^x}{1-(1-p)}\right\} = p\left\{\frac{1-(1-p)^x}{p}\right\} \\ &= 1-(1-p)^x \end{aligned}$$

Since $(1-p)^x$ is the probability that no success has occurred by the end of the $x$th trial, the result indicates that $F_X(x)$ is the probability that at least one success has occurred in $x$ trials of the experiment.

### 4.4.2 Modified Geometric Distribution

As stated earlier, the geometric random variable can also be used to describe the number of trials until the first failure in a series of Bernoulli trials. If we denote this random variable by $Y$, then $Y$ is defined to be a modified geometric distribution and its PMF is given by

$$p_Y(y) = (1-p)p^{y-1} \quad y = 1, 2, \ldots \tag{4.16}$$

By interchanging $p$ and $1-p$ in the results for the "traditional" geometric distribution we obtain the results for the expected value and variance of $Y$ as follows:

$$E[Y] = \frac{1}{1-p} \tag{4.17a}$$

$$\sigma_Y^2 = \frac{p}{(1-p)^2} \tag{4.17b}$$

Finally, the CDF of $Y$ is given by

$$F_Y(y) = P[Y \leq y] = \sum_{k=1}^{y} (1-p)p^{k-1} = 1 - P[Y > y] = 1 - p^y \tag{4.18}$$

---

**EXAMPLE 4.7**

A couple is planning to have children in a community where the probability of having a boy is $p$.

a. If they keep trying until they have a girl and then stop, what is the expected number of children that they will have?

b. Suppose they plan to have a minimum of 4 children. However, if the first 4 are all boys, they will continue until they get a girl and then stop. If there is at least one girl among the first 4 children, they will stop at 4. What is the expected number of children that they will have?

**Solution:**

The births are Bernoulli trials where the probability of success is $1-p$, the probability of having a girl.

a. Thus, the number of trials until the couple has a girl has the modified geometric distribution with a mean of $1/(1-p)$.

b. The number of children $Y$ that the couple will have depends on whether the first 4 are all boys or not. If the first 4 are all boys, the expected number of attempts until they have a girl after the four boys are born is $1/(1-p)$. Thus, the expected number of children, given that the first four are all boys, is $4+1/(1-p)$. If at least one of the first four is a girl, then the expected number of children is 4. Since the probability that the first four are all boys is $p^4$, the probability that there is at least one girl among the four children is $1-p^4$. Thus, the expected number of children that the couple will have is

$$E[Y]=p^4\left\{4+\frac{1}{1-p}\right\}+(1-p^4)(4)=4+\frac{p^4}{1-p}$$

### 4.4.3 "Forgetfulness" Property of the Geometric Distribution

Let $X$ be a random variable that denotes the number of Bernoulli trials until the first success. Suppose we have observed a fixed number $n$ of these trials and they are all failures. We would like to know the number $K$ of additional trials until the first success. To do this we know that the problem tells us that $K=X-n$. This is illustrated in Figure 4.5.

Thus, we are required to compute the conditional probability mass function of $K$, given that $X>n$. We proceed as follows:

$$\begin{aligned} p_{K|X>n}(k|X>n) &= P[K=k|X>n]=P[X-n=k|X>n] \\ &= P[X=n+k|X>n]=\frac{P[\{X=n+k\}\cap\{X>n\}]}{P[X>n]} \end{aligned}$$

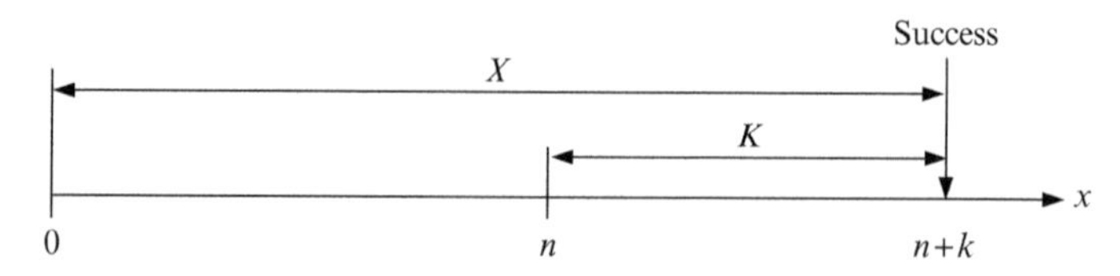


**FIGURE 4.5**

Illustration of Additional Trials Until Success

Because $\{X=n+k\}\subset\{X>n\}$, we have that $\{X=n+k\}\cap\{X>n\}=\{X=n+k\}$. Thus,

$$\begin{aligned} p_{K|X>n}(k|X>n) &= \frac{P[X=n+k]}{P[X>n]} = \frac{p(1-p)^{n+k-1}}{(1-p)^n} \\ &= p(1-p)^{k-1} = p_X(k) \quad k=1, 2, \ldots \end{aligned} \tag{4.19}$$

where the second equality follows from our earlier result that $P[X>n]=(1-p)^n$.

The above result shows that the conditional probability that the number of trials remaining until the first success, given that no success occurred in the first $n$ trials, has the same PMF as $X$ had originally. This property is called the *forgetfulness* or *memorylessness property* of the geometric distribution. It means that the distribution "forgets" the past by not remembering how long the sequence has lasted if no success has already occurred. Thus, each time we want to know the number of trials until the first success, given that the first success has not yet occurred, the process "starts" from the beginning.

## EXAMPLE 4.8

Tony is tossing balls randomly into 50 boxes and his goal is to stop when he gets the first ball into the eighth box. Given that he has tossed 20 balls without getting a ball into the eighth box, what is the expected number of additional tosses he needs to get a ball into the eighth box?

**Solution:**

With respect to the eighth box, each toss is a Bernoulli trial with probability of success $p=1/50$. Let $X$ be the random variable that denotes the number of tosses required to get a ball into the eighth box. Then $X$ has a geometric distribution. Let $K$ denote the number of additional tosses required to get a ball into the eighth box, given that no ball is in the box after 20 tosses. Then, because of the forgetfulness property of the geometric distribution, $K$ has the same distribution as $X$. Thus

$$E[K] = \frac{1}{p} = 50$$

## 4.5 PASCAL DISTRIBUTION

The Pascal random variable is an extension of the geometric random variable. The $k$th-order Pascal (or Pascal-$k$) random variable, $X_k$, describes the number of trials until the $k$th success, which is why it is sometimes called the "$k$th-order interarrival time for a Bernoulli process." The Pascal distribution is also called the *negative binomial distribution*.

Let the $k$th success in a Bernoulli sequence of trials occur at the $n$th trial. This means that $k-1$ successes occurred during the past $n-1$ trials. From our knowledge of the binomial distribution, we know that the probability of this event is

$$p_{X(n-1)}(k-1) = \binom{n-1}{k-1} p^{k-1}(1-p)^{n-k}$$

where $X(n-1)$ is the binomial random variable with parameters $(n-1, p)$. The next trial independently results in a success with probability $p$. This is illustrated in Figure 4.6.

Thus, if we define the Bernoulli random variable as $X_0$ whose PMF we defined earlier as $p_{X_0}(x) = p^x(1-p)^{1-x}$, where $x=0$ or 1, the PMF of the $k$th-order Pascal random variable, $X_k$, is obtained by knowing that $X_k$ is derived from two non-overlapping processes: a $B(n-1, p)$ process that results in $k-1$ successes, and a Bernoulli process that results in a success. Thus, we have that

$$\begin{aligned} p_{X_k}(n) &= P[\{X(n-1)=k-1\} \cap \{X_0=1\}] = P[X(n-1)=k-1]P[X_0=1] \\ &= p_{X(n-1)}(k-1)p_{X_0}(1) = \binom{n-1}{k-1} p^{k-1}(1-p)^{n-k}p \\ &= \binom{n-1}{k-1} p^k(1-p)^{n-k} \qquad k=1,2,\ldots; n=k, k+1,\ldots \end{aligned} \tag{4.20}$$

Since $p_{X_k}(n)$ is a true PMF, we must have that

$$\sum_{n=k}^{\infty} p_{X_k}(n) = \sum_{n=k}^{\infty} \binom{n-1}{k-1} p^k(1-p)^{n-k} = p^k \sum_{n=k}^{\infty} \binom{n-1}{k-1}(1-p)^{n-k} = 1$$

Thus,

$$\sum_{n=k}^{\infty} \binom{n-1}{k-1}(1-p)^{n-k} = \frac{1}{p^k} \tag{4.20a}$$

The expected value and variance of $X_k$ are obtained as follows:

$$E[X_k] = \sum_{n=k}^{\infty} n p_{X_k}(n) = \sum_{n=k}^{\infty} n \binom{n-1}{k-1} p^k(1-p)^{n-k} = p^k \sum_{n=k}^{\infty} n \binom{n-1}{k-1}(1-p)^{n-k} \tag{*}$$

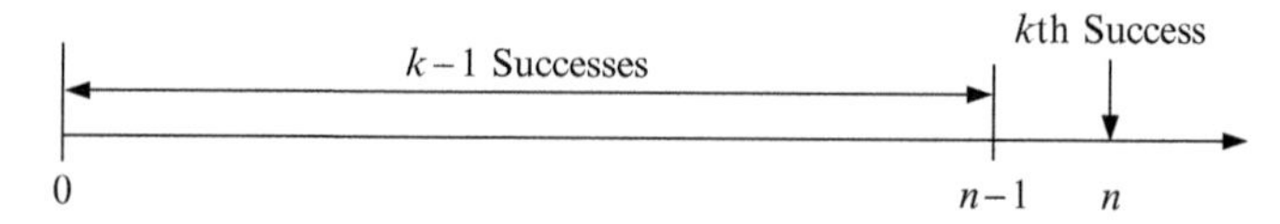


**FIGURE 4.6**
Illustration of the Pascal Random Variable

Define $q = 1 - p$. Then we obtain

$$E[X_k] = (1-q)^k \sum_{n=k}^{\infty} n \binom{n-1}{k-1} q^{n-k}$$

Consider the following Maclaurin's series expansion of the negative binomial:

$$g(q) = (1-q)^{-k} = \sum_{m=0}^{\infty} \binom{k+m-1}{k-1} q^m = \sum_{n=k}^{\infty} \binom{n-1}{k-1} q^{n-k}$$

Taking the derivative of $g(q)$, we obtain

$$\begin{aligned}
\frac{d}{dq} g(q) &= \frac{k}{(1-q)^{k+1}} = \sum_{n=k}^{\infty} (n-k) \binom{n-1}{k-1} q^{n-k-1} \\
&= \frac{1}{q} \left\{ \sum_{n=k}^{\infty} n \binom{n-1}{k-1} q^{n-k} - k \sum_{n=k}^{\infty} \binom{n-1}{k-1} q^{n-k} \right\}
\end{aligned}$$

Thus, combining this result with (*) we obtain

$$\frac{k}{(1-q)^{k+1}} = \frac{1}{q} \left\{ \frac{E[X_k] - k}{(1-q)^k} \right\}$$

which gives

$$E[X_k] = \frac{k}{1-q} = \frac{k}{p} \tag{4.21}$$

The second moment of $X_k$ is given by

$$\begin{aligned}
E\left[X_k^2\right] &= \sum_{n=k}^{\infty} n^2 p_{X_k}(n) = \sum_{n=k}^{\infty} n^2 \binom{n-1}{k-1} p^k (1-p)^{n-k} \\
&= p^k \sum_{n=k}^{\infty} n^2 \binom{n-1}{k-1} (1-p)^{n-k}
\end{aligned}$$

The second derivative of $g(q)$ is given by

$$\begin{aligned}
\frac{d^2}{dq^2} g(q) &= \frac{k(k+1)}{(1-q)^{k+2}} = \sum_{n=k}^{\infty} (n-k)(n-k-1) \binom{n-1}{k-1} q^{n-k-2} \\
&= \frac{1}{q^2} \sum_{n-k}^{\infty} (n^2 - 2nk - n + k^2 + k) \binom{n-1}{k-1} q^{n-k} \\
&= \frac{1}{q^2} \left\{ \frac{E\left[X_k^2\right] - (2k+1)E[X_k] + k(k+1)}{(1-q)^k} \right\}
\end{aligned}$$

This gives

$$E[X_k^2] = \frac{k^2 + kq}{(1-q)^2} = \frac{k^2 + k(1-p)}{p^2}$$

Thus, the variance of $X_k$ is given by

$$\sigma_{X_k}^2 = E[X_k^2] - (E[X_k^2])^2 = \frac{k(1-p)}{p^2} \tag{4.22}$$

The CDF of $X_k$ is given by

$$F(x) = P[X_k \leq x] = \sum_{n=k}^{x} \binom{n-1}{k-1} p^k (1-p)^{n-k} \tag{4.23}$$

## EXAMPLE 4.9

Consider independent Bernoulli trials in which each trial results in a success with probability $p$. What is the probability that there are $m$ successes before $r$ failures?

**Solution:**

This is an example of the Pascal distribution. Since there are $m$ successes and $r$ failures, the total number of trials is $m+r$. In order for $m$ successes to occur before $r$ failures the $m$th success must occur no later than the $(m+r-1)$th trial. This is so because if the $m$th success occurs before or at the $(m+r-1)$th trial, it would have occurred before the $r$th failure. Thus, as long as the $m$th success occurs any time from the $m$th trial up to the $(m+r-1)$th trial, the prescribed condition would have been met. Therefore, the desired probability is given by

$$\gamma = \sum_{n=m}^{m+r-1} \binom{n-1}{m-1} p^m (1-p)^{n-m}$$

## EXAMPLE 4.10

A student's summer job is to call the university alumni for support for the university's scholarship fund. Studies indicate that the probability that each of the student's calls is answered is 1/3. What is the probability that the second call to be answered on one particular day is the student's sixth call?

**Solution:**

This is an example of the second-order Pascal distribution with $p = 1/3$. Thus, the desired probability is given by

$$p_{X_2}(6) = \binom{6-1}{2-1}\left(\frac{1}{3}\right)^2\left(\frac{2}{3}\right)^4 = \binom{5}{1}\left(\frac{1}{9}\right)\left(\frac{16}{81}\right) = \frac{80}{729} = 0.1097$$

## EXAMPLE 4.11

The members of a Girl Scout troop are selling cookies from house to house in a suburban town. The probability that they sell a set of cookies (that is, one or more packs of cookies) at any house they visit is 0.4.

a. If they visited eight houses in one evening, what is the probability that they sold cookies in exactly five of these houses?
b. If they visited eight houses in one evening, what is the expected number of sets of cookies that they sold?
c. What is the probability that they sold their third set of cookies in the sixth house they visited?

**Solution:**
Let $X$ denote the number of sets of cookies that the troop sold.

a. If they visited eight houses, then $X$ is a binomial random variable; that is, $X \sim B(8, 0.4)$. Thus, $X = X(8)$ and its PMF is given by

$$p_{X(8)}(x) = \binom{8}{x}(0.4)^x(0.6)^{8-x} \qquad x = 0, 1, 2, \ldots, 8$$

Therefore, the probability that they sold exactly five sets of cookies is the probability of five successes in the eight attempts, which is

$$p_{X(8)}(5) = \binom{8}{5}(0.4)^5(0.6)^3 = \frac{8!}{5!3!}(0.4)^5(0.6)^3 = 0.1239$$

b. The expected number of sets of cookies they sold after visiting eight houses is

$$E[X(8)] = 8p = 8 \times 0.4 = 3.2$$

c. Let $X_k$ denote the number of houses they visited up to and including the house where they sold their third set of cookies. Then $X_k$ is a third-order Pascal random variable; that is, $X_k = X_3$, and

$$P[X_3 = 6] = p_{X_3}(6) = \binom{6-1}{3-1}(0.4)^3(0.6)^3 = \binom{5}{2}(0.24)^3$$
$$= \frac{5!}{3!2!}(0.24)^3 = 0.1382$$

### 4.5.1 Distinction Between Binomial and Pascal Distributions

From (4.7), the PMF of the binomial random variable $X(n)$ is given by

$$p_{X(n)}(x) = \binom{n}{x}p^x(1-p)^{n-x} \qquad x = 0, 1, \ldots, n$$

Similarly, from (4.20), the PMF of the Pascal-$k$ random variable $X_k$ is given by

$$p_{X_k}(n) = \binom{n-1}{k-1}p^k(1-p)^{n-k} \qquad k = 1, 2, \ldots; n = k, k+1, \ldots$$

Structurally, the two PMFs look alike; the difference is primarily in the combinatorial term and their range of values. However, these random variables address different problems. The binomial random variable assumes that a fixed number of trials of an experiment have been completed before it asks for the number of successes in those trials. The Pascal random variable deals with a process that has a prescribed termination point; the experiment stops once the condition is met. This means that the Pascal random variable deals with an experiment that can continue forever and as a result the number of times that the experiment is performed can be unbounded. On the other hand the binomial random variable deals with an experiment that has a finite number of trials. Also, the Pascal random variable deals with ordinal numbers like the third, sixth, etc; while the binomial random variable deals with cardinal numbers like three, six, etc.

## 4.6 HYPERGEOMETRIC DISTRIBUTION

Suppose we have $N$ objects of which $N_1 < N$ are of type A and $N - N_1$ are of type B. If we draw a random sample of $n$ objects from this population and note the configuration of the sample and replace the objects after each trial (that is, we sample with replacement), the number of type A objects included in the sample is binomially distributed with parameters $(N, p)$, where $p = N_1/N$. This follows from the fact that the probability of selecting an item from a particular group remains constant from trial to trial. That is, the outcomes are independent and the PMF of $X$, which is the random variable that denotes the number of type A objects in the sample of size $n$, is given by:

$$p_X(x) = \binom{n}{x}\left(\frac{N_1}{N}\right)^x\left(\frac{N-N_1}{N}\right)^{n-x} \qquad x = 0, 1, 2, \ldots, n$$

However, if we do not replace the objects after drawing each sample, the outcomes are no longer independent and the binomial distribution no longer applies. Suppose the sample of size $n$ is drawn without replacement. Let $K_n$ be the random variable that denotes the number of type A objects in the sample of $n$ objects drawn at a time. The random variable $K_n$ is said to have a *hypergeometric* PMF that depends on $n$, $N_1$ and $N$, and this PMF is defined as follows:

$$p_{K_n}(k) = \frac{\binom{N_1}{k}\binom{N-N_1}{n-k}}{\binom{N}{n}} \qquad k = 0, 1, 2, \ldots, \min(n, N_1) \tag{4.24}$$

This PMF is obtained from the fundamental counting rule defined in Chapter 1. Specifically, since there are two types of objects, we can select from each type independently once the number of objects to be selected from each type has been specified. Thus, given that we need $k$ objects from type A and hence

$n-k$ objects from type B, the selection process in each group becomes independent of that of the other group. With this in mind, the $k$ objects of type A in the sample can be selected in $\binom{N_1}{k}$ ways. Similarly, the $n-k$ objects of type B in the sample can be selected in $\binom{N-N_1}{n-k}$ ways. The total number of possible samples of size $n$ that are of the specified configuration is the product of the two numbers. There are $\binom{N}{n}$ possible samples of size $n$ that can be drawn from the population, which is essentially the number of unrestricted ways of choosing $n$ objects from $N$ objects. If the sample is drawn randomly, then each sample has the same probability of being selected, which accounts for the above PMF.

The mean and variance of $K_n$ are given by

$$E[K_n] = \frac{nN_1}{N} = np \tag{4.25a}$$

$$\sigma_{K_n}^2 = \frac{nN_1(N-N_1)(N-n)}{N^2(N-1)} = \frac{np(1-p)(N-n)}{N-1} \tag{4.25b}$$

where $p = N_1/N$. Note that in the limit as $N \to \infty$ (or $N$ becomes large compared to $n$), these equations become identical to those of the binomial distribution; that is,

$$E[K_n] = np$$
$$\sigma_{K_n}^2 = np(1-p)$$

The hypergeometric distribution is widely used in quality control, as the following examples illustrate.

## EXAMPLE 4.12

Suppose there are $M_1 < M$ defective items in a box that contains $M$ items. What is the probability that a random sample of $n$ items from the box contains $k$ defective items?

**Solution:**

Since the sample was selected randomly, all samples of size $n$ are equally likely. Therefore, the total number of possible samples with unrestricted composition is

$$C(M, n) = \binom{M}{n}$$

Since there are $M_1$ defective items in the box, the total number of ways of selecting $k$ out of the items at a time is $C(M_1, k) = \binom{M_1}{k}$. Similarly, since there are $M - M_1$ non-defective items in the box, the total number of ways of selecting $n-k$ of them at a time is $C(M-M_1, n-k)=\binom{M-M_1}{n-k}$. Since these two selections can be made independently, the total number of ways of choosing $k$ defective items and $n-k$ non-defective items is $C(M_1,k)\times C(M-M_1,n-k)$. Therefore, the probability that the random sample contains $k$ defective and $n-k$ non-defective items is given by the ratio of $C(M_1,k)\times C(M-M_1,n-k)$ to the total number of ways of selecting samples of size $n$, $C(M,n)$. Thus, we obtain the result as

$$p=\frac{C(M_1,k)\times C(M-M_1,n-k)}{C(M,n)}=\frac{\binom{M_1}{k}\binom{M-M_1}{n-k}}{\binom{M}{n}}$$

## EXAMPLE 4.13

A container has 100 items 5 of which the worker who packed the container knows are defective. A merchant wants to buy the container without knowing the above information. However, he will randomly pick 20 items from the container and will accept the container as good if there is at most one bad item in the selected sample. What is the probability that the merchant will declare the container to be good?

**Solution:**

Let $A$ be the event that there is no defective item in the selected sample and $B$ the event that there is exactly one defective item in the selected sample. Let $q$ denote the probability that the merchant declares the container to be good. Then event $A$ consists of two subevents: zero defective items and 20 non-defective items. Similarly, event $B$ consists of two subevents: 1 defective item and 19 non-defective items. Since events $A$ and $B$ are mutually exclusive we obtain the following results:

$$P[A]=\frac{C(5,0)C(95,20)}{C(100,20)}=0.3193$$

$$P[B]=\frac{C(5,1)C(95,19)}{C(100,20)}=0.4201$$

$$q=P[A\cup B]=P[A]+P[B]=0.7394$$

## EXAMPLE 4.14

We repeat Example 4.13 with a different sample size. Instead of testing the quality of the container with a sample size of 20 the merchant decides to test it with a sample size of 50. As before, he will accept the container as good if there is at most one bad item in the selected sample. What is the probability that the merchant will declare the container to be good?

**Solution:**

Let $A$, $B$ and $q$ be as previously defined in Example 4.13. Then event $A$ consists of two subevents: zero defective items and 50 non-defective items. Similarly, event $B$ consists of two subevents: 1

defective item and 49 non-defective items. Since events $A$ and $B$ are mutually exclusive we obtain the following results:

$$P[A] = \frac{C(5,0)C(95,50)}{C(100,50)} = 0.0281$$

$$P[B] = \frac{C(5,1)C(95,49)}{C(100,50)} = 0.1530$$

$$q = P[A \cup B] = P[A] + P[B] = 0.1811$$

The above examples illustrate the impact of the sample size on decisions that can be made. With a bigger sample size, the merchant is more likely to make a better decision because there is a greater probability that more of the defective items will be included in the sample. For example, if the decision was based on the result of a random sample of 10 items, we would get $q = 0.9231$. Similarly, if it was based on the result of a random sample of 40 items, we would get $q = 0.3316$, while with a sample of 60 items, we would get $q = 0.0816$. Note that while a bigger sample size has the tendency to reveal the fact that the container has a few bad items, it also involves more testing. Thus, to get better information we must be prepared to do more testing, which is a basic rule in quality control.

## EXAMPLE 4.15

A certain library has a collection of 10 books on probability theory. Six of these books were written by American authors and four were written by foreign authors.

a. If I randomly select one of these books, what is the probability that it was written by an American author?
b. If I select five of these books at random, what is the probability that two of them were written by American authors and three of them were written by foreign authors?

**Solution:**

a. There are $\binom{6}{1} = 6$ ways to choose a book written by an American author and $\binom{10}{1} = 10$ ways to choose a book at random. Therefore, the probability that a book chosen at random was written by an American author is $p = 6/10 = 0.6$.
b. This is an example of the hypergeometric distribution. Thus, the probability that of the five of these books selected at random, two of them were written by American authors and three of them were written by foreign authors is given by

$$p = \frac{\binom{6}{2}\binom{4}{3}}{\binom{10}{5}} = \frac{15 \times 4}{252} = 0.2381$$

## 4.7 POISSON DISTRIBUTION

A discrete random variable $K$ is called a Poisson random variable with parameter $\lambda$, where $\lambda > 0$, if its PMF is given by

$$p_K(k) = \frac{\lambda^k}{k!}e^{-\lambda} \quad k = 0, 1, 2, \ldots \tag{4.26}$$

The CDF of $K$ is given by

$$F_K(k) = P[K \leq k] = \sum_{n=0}^{k} \frac{\lambda^n}{n!}e^{-\lambda} \tag{4.27}$$

The expected value of $K$ is given by

$$\begin{aligned} E[K] &= \sum_{k=0}^{\infty} k p_K(k) = \sum_{n=0}^{\infty} k \frac{\lambda^k}{k!}e^{-\lambda} = \sum_{k=1}^{\infty} \frac{\lambda^k}{(k-1)!}e^{-\lambda} = \lambda e^{-\lambda} \sum_{k=1}^{\infty} \frac{\lambda^{k-1}}{(k-1)!} \\ &= \lambda e^{-\lambda} \sum_{m=0}^{\infty} \frac{\lambda^m}{m!} = \lambda e^{-\lambda} e^{\lambda} \\ &= \lambda \end{aligned} \tag{4.28}$$

The second moment of $K$ is given by

$$E\left[K^2\right] = \sum_{k=0}^{\infty} k^2 p_K(k) = \sum_{n=0}^{\infty} k^2 \frac{\lambda^k}{k!}e^{-\lambda} = \lambda e^{-\lambda} \sum_{k=1}^{\infty} \frac{k\lambda^{k-1}}{(k-1)!}$$

But

$$\sum_{k=1}^{\infty} \frac{\lambda^k}{(k-1)!} = \lambda \sum_{k=1}^{\infty} \frac{\lambda^{k-1}}{(k-1)!} = \lambda \sum_{m=0}^{\infty} \frac{\lambda^m}{m!} = \lambda e^{\lambda}$$

$$\frac{d}{d\lambda} \sum_{k=1}^{\infty} \frac{\lambda^k}{(k-1)!} = \sum_{k=1}^{\infty} \frac{d}{d\lambda}\left\{\frac{\lambda^k}{(k-1)!}\right\} = \sum_{k=1}^{\infty} \frac{k\lambda^{k-1}}{(k-1)!} = \frac{d}{d\lambda}\{\lambda e^{\lambda}\} = e^{\lambda}(1+\lambda)$$

Thus, the second moment is given by

$$E\left[K^2\right] = \lambda e^{-\lambda} \sum_{k=1}^{\infty} \frac{k\lambda^{k-1}}{(k-1)!} = \lambda e^{-\lambda} e^{\lambda}(1+\lambda) = \lambda^2 + \lambda$$

The variance of $K$ is given by

$$\sigma_K^2 = E\left[K^2\right] - (E[K])^2 = \lambda^2 + \lambda - \lambda^2 = \lambda \tag{4.29}$$

This means that the mean and the variance of the Poisson random variable are equal. The Poisson distribution has many applications in science and engineering. For example, the number of telephone calls arriving at a switchboard during various intervals of time and the number of customers arriving at a bank during various intervals of time are usually modeled by Poisson random variables.

---

## EXAMPLE 4.16

The number of messages that arrive at a switchboard per hour is a Poisson random variable with a mean of 6. Find the probability for each of the following events:

a. Exactly two messages arrive within one hour
b. No message arrives in one hour
c. At least three messages arrive within one hour.

**Solution:**
Let $K$ be the random variable that denotes the number of messages arriving at the switchboard within a one-hour interval. The PMF of $K$ is given by

$$p_K(k) = \frac{6^k}{k!}e^{-6} \quad k = 0, 1, 2, \ldots$$

a. The probability that exactly two messages arrive within one hour is

$$p_K(2) = \frac{6^2}{2!}e^{-6} = 18e^{-6} = 0.0446$$

b. The probability that no message arrives within one hour is

$$p_K(0) = \frac{6^0}{0!}e^{-6} = e^{-6} = 0.0024$$

c. The probability that at least three messages arrive within one hour is

$$\begin{aligned} P[K \geq 3] &= 1 - P[K < 3] = 1 - \{p_K(0) + p_K(1) + p_K(2)\} \\ &= 1 - e^{-6}\left\{\frac{6^0}{0!} + \frac{6^1}{1!} + \frac{6^2}{2!}\right\} = 1 - e^{-6}\{1 + 6 + 18\} = 1 - 25e^{-6} \\ &= 0.9380 \end{aligned}$$

---

### 4.7.1 Poisson Approximation of the Binomial Distribution

Let $X$ be a binomial random variable with parameter $(n, p)$ and PMF

$$p_X(x) = \binom{n}{x}p^x(1-p)^{n-x} \qquad x = 0, 1, 2, \ldots, n$$

Since the PMF involves evaluating $n!$, which can become very large even for moderate values of $n$, we would like to develop an approximate method for the case of large values of $n$. We know that $E[X] = np$. Let $n$ and $p$ change in such a way that $np = \lambda \Rightarrow p = \lambda/n$. Therefore, substituting for $p$ in the PMF we obtain

$$
\begin{aligned}
p_X(x) &= \binom{n}{x}\left(\frac{\lambda}{n}\right)^x\left(1-\frac{\lambda}{n}\right)^{n-x} = \frac{n(n-1)(n-2)(n-3)\cdots(n-x+1)}{x!n^x}\lambda^x\left(1-\frac{\lambda}{n}\right)^{n-x} \\
&= \frac{n^x\left(1-\frac{1}{n}\right)\left(1-\frac{2}{n}\right)\left(1-\frac{3}{n}\right)\cdots\left(1-\frac{x-1}{n}\right)}{x!n^x}\lambda^x\left(1-\frac{\lambda}{n}\right)^{n-x} \\
&\approx \frac{\left(1-\frac{1}{n}\right)\left(1-\frac{2}{n}\right)\left(1-\frac{3}{n}\right)\cdots\left(1-\frac{x-1}{n}\right)}{x!}\lambda^x\left(1-\frac{\lambda}{n}\right)^{n} \quad \text{if } x \text{ is small relative to } n
\end{aligned}
$$

We know that in the limit as $n$ becomes very large

$$\lim_{n\to\infty}\left(1-\frac{a}{n}\right)=1 \quad a<n$$

Also, since by definition

$$\lim_{n\to\infty}\left(1+\frac{a}{n}\right)^n=e^a$$

we conclude that in the limit as $n\to\infty$, if $\lambda$ remains fixed, which means that $p\to 0$, then we obtain

$$\lim_{n\to\infty}p_X(x)=\frac{\lambda^x}{x!}e^{-\lambda}$$

which is the Poisson distribution.

## 4.8 EXPONENTIAL DISTRIBUTION

A continuous random variable $X$ is defined to be an exponential random variable (or $X$ has an exponential distribution) if for some parameter $\lambda>0$, its PDF is given by

$$f_X(x)=\begin{cases}\lambda e^{-\lambda x} & x\geq 0\\ 0 & x<0\end{cases} \tag{4.30}$$

The CDF of X is given by

$$
\begin{aligned}
F_X(x) &= P[X\leq x]=\int_0^x f_X(y)dy=\int_0^x \lambda e^{-\lambda y}dy=\left[-e^{-\lambda y}\right]_0^x \\
&= 1-e^{-\lambda x}
\end{aligned} \tag{4.31}
$$

The expected value of $X$ is given by

$$E[X]=\int_0^\infty xf_X(x)dx=\int_0^\infty x\lambda e^{-\lambda x}dx$$

Let $u=x$ and $dv=\lambda e^{-\lambda x}dx$, which means that $du=dx$ and $v=-e^{-\lambda x}$. Thus, integrating by parts we obtain

$$E[X]=\int_0^{\infty} x\lambda e^{-\lambda x}dx=\left[-xe^{-\lambda x}\right]_0^{\infty}+\int_0^{\infty} e^{-\lambda x}dx=0+\left[\frac{1}{\lambda}e^{-\lambda x}\right]_0^{\infty} \\ =\frac{1}{\lambda} \quad (4.32)$$

where we have assumed that as $x$ goes to infinity $xe^{-\lambda x}$goes to zero faster than it goes to infinity. Note also that because $X$ is a nonnegative random variable we have that

$$E[X]=\int_0^{\infty}\{1-F(x)\}dx=\int_0^{\infty}\{1-\left[1-e^{-\lambda x}\right]\}dx=\int_0^{\infty} e^{-\lambda x}dx=\frac{1}{\lambda}$$

By repeated use of integration by parts it can be shown that the $n$th moment of $X$ is given by

$$E[X^n]=\int_0^{\infty} x^n f_X(x)dx=\frac{n!}{\lambda^n} \quad n=1, 2, \ldots \quad (4.33)$$

Thus, the variance of $X$ is given by

$$\sigma_X^2=E\left[X^2\right]-(E[X])^2=\frac{2}{\lambda^2}-\frac{1}{\lambda^2}=\frac{1}{\lambda^2} \quad (4.34)$$

This means that the mean and the standard deviation (i.e., the square root of the variance) of an exponentially distributed random variable are equal.

## EXAMPLE 4.17

Assume that the length of phone calls made at a particular telephone booth is exponentially distributed with a mean of 3 minutes. If you arrive at the telephone booth just as Chris was about to make a call, find the following:

a. The probability that you will wait more than 5 minutes before Chris is done with the call
b. The probability that Chris' call will last between 2 minutes and 6 minutes

**Solution:**

Let $X$ be a random variable that denotes the length of calls made at the telephone booth. Since the mean length of calls is $1/\lambda=3$, we have that the PDF of $X$ is given by

$$f_X(x)=\lambda e^{-\lambda x}=\frac{1}{3}e^{-x/3}$$

a. The probability that you will wait more than 5 minutes is the probability that $X$ is greater than 5 minutes, which is given by

$$P[X>5]=\int_5^{\infty}\frac{1}{3}e^{-x/3}dx=\left[-e^{-x/3}\right]_5^{\infty}=e^{-5/3}=0.1889$$

Note that we can also obtain the result as follows:

$$P[X>5]=1-P[X\leq 5]=1-F_X(5)=1-\left\{1-e^{-5/3}\right\}=e^{-5/3}=0.1889$$

b. The probability that the call lasts between 2 and 6 minutes is given by:

$$P[2 \le X \le 6] = \int_2^6 \frac{1}{3} e^{-x/3} dx = \left[-e^{-x/3}\right]_2^6 = e^{-2/3} - e^{-2} = 0.3781$$

The same result can be obtained as follows:

$$P[2 \le X \le 6] = F_X(6) - F_X(2) = e^{-2/3} - e^{-2} = 0.3781$$

## 4.8.1 "Forgetfulness" Property of the Exponential Distribution

The exponential random variable is used extensively in reliability engineering to model the lifetimes of systems. Suppose the life $X$ of an equipment is exponentially distributed with a mean of $1/\lambda$. Assume that the equipment has not failed by time $t$. We are interested in the conditional PDF of $X$, given that the equipment has not failed by time $t$. We start by finding the probability that $X \le t+s$, given that $X > t$ for some nonnegative additional time $s$. This is illustrated in Figure 4.7.

From the figure we have that

$$\begin{aligned} P[X \le t+s | X > t] &= \frac{P[\{X \le t+s\} \cap \{X > t\}]}{P[X > t]} = \frac{P[t < X \le t+s]}{P[X > t]} \\ &= \frac{F_X(t+s) - F_X(t)}{1 - F_X(t)} = \frac{\{1 - e^{-\lambda(t+s)}\} - \{1 - e^{-\lambda t}\}}{e^{-\lambda t}} \\ &= \frac{e^{-\lambda t} - e^{-\lambda(t+s)}}{e^{-\lambda t}} = 1 - e^{-\lambda s} = F_X(s) \\ &= P[X \le s] \end{aligned}$$

This indicates that the process only remembers the present and not the past. Because $x = t+s$, we have that $s = x - t$ and the above result becomes

$$\begin{aligned} P[X \le t+s | X > t] &= 1 - e^{-\lambda(x-t)} = F_{X|X>t}(x|X > t) \\ f_{X|X>t}(x|X > t) &= \frac{d}{dx} F_{X|X>t}(x|X > t) \\ &= \lambda e^{-\lambda(x-t)} \end{aligned} \tag{4.35}$$

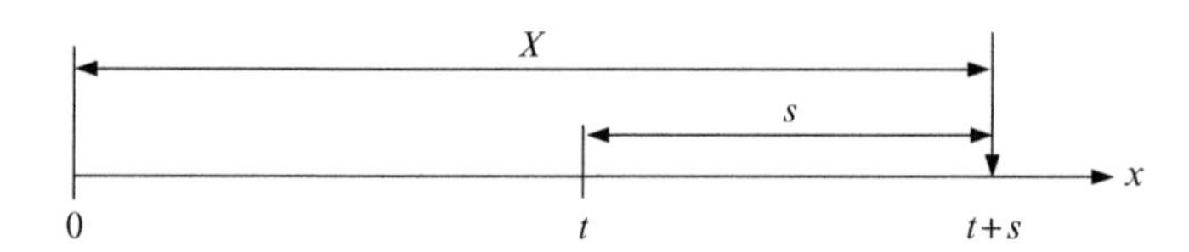


**FIGURE 4.7**
Illustration of Residual Life

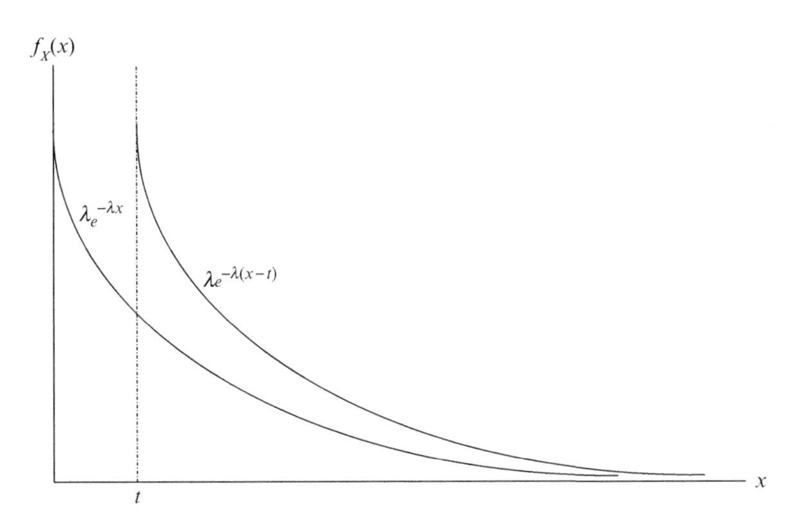


**FIGURE 4.8**
PDF of the Exponential Random Variable

Thus, the conditional PDF is a shifted version of the original PDF, as shown in Figure 4.8. Similar to the geometric distribution, this is referred to as the *forgetfulness* or *memorylessness* property of the exponential distribution. This means that given that the equipment has not failed by time $t$, the residual life of the equipment has a PDF that is identical to that of the life of the equipment before $t$.

## EXAMPLE 4.18

Assume that in Example 4.17 Chris, who is using the phone at the telephone booth had already talked for 2 minutes before you arrived. According to the forgetfulness property of the exponential distribution, the mean time until Chris is done with the call is still 3 minutes. The random variable forgets the length of time the call had lasted before you arrived.

### 4.8.2 Relationship between the Exponential and Poisson Distributions

Let $\lambda$ denote the mean (or average) number of Poisson arrivals per unit time, say per second. Then the average number of arrivals in $t$ seconds is $\lambda t$. If $K$ denotes this Poisson random variable, the PMF of the number of arrivals during an interval of $t$ seconds is

$$p_K(k, t) = \frac{(\lambda t)^k}{k!} e^{-\lambda t} \quad k = 0, 1, 2, \ldots; \quad t \geq 0$$

The probability that there is no arrival in an interval of $t$ seconds is $p_K(0, t) = e^{-\lambda t}$. Thus, the probability that there is at least one arrival in an interval of $t$ seconds is $1-e^{-\lambda t}$. But for an exponential random variable $Y$ with parameter $\lambda$, the probability that an event occurs no later than time $t$ is given by:

$$P[Y \leq t] = F_Y(t) = 1 - e^{-\lambda t}$$

Thus, an exponential distribution $Y$ with parameter $\lambda$ describes the intervals between occurrence of events defined by a Poisson random variable $K$ with mean $\lambda t$. If we define $\lambda$ as the average number of Poisson arrivals per unit time, it becomes very clear why the mean of $Y$ is $1/\lambda$ time units. Thus, if intervals between events in a given process can be modeled by an exponential random variable, then the number of those events that occur during a specified time interval can be modeled by a Poisson random variable. Similarly, if the number of events that occur within a specified time interval can be modeled by a Poisson random variable, then the interval between successive events can be modeled by an exponential random variable. As the saying goes, "Poisson implies exponential and exponential implies Poisson."

## 4.9 ERLANG DISTRIBUTION

The Erlang distribution is a generalization of the exponential distribution. While the exponential random variable describes the time between adjacent events, the Erlang random variable of order $k$ describes the time interval between any event and the $k$th following event. Thus, the Erlang random variable is to the exponential random variable as the Pascal random variable is to the geometric random variable.

A random variable $X_k$ is referred to as a $k$th-order Erlang (or Erlang-$k$) random variable with parameter $\lambda$ if its PDF is given by

$$f_{X_k}(x) = \begin{cases} \dfrac{\lambda^k x^{k-1} e^{-\lambda x}}{(k-1)!} & k = 1, 2, 3, \ldots; x \geq 0 \\ 0 & x < 0 \end{cases} \tag{4.36}$$

This PDF can be obtained by considering Figure 4.9, which shows the $k$th event arriving between the time $x$ and $x+dx$. If the $k$th event arrives within this interval, then $k-1$ events occurred/arrived by time $x$. The probability of this event is the probability that a Poisson random variable with parameter $\lambda x$ has a value of

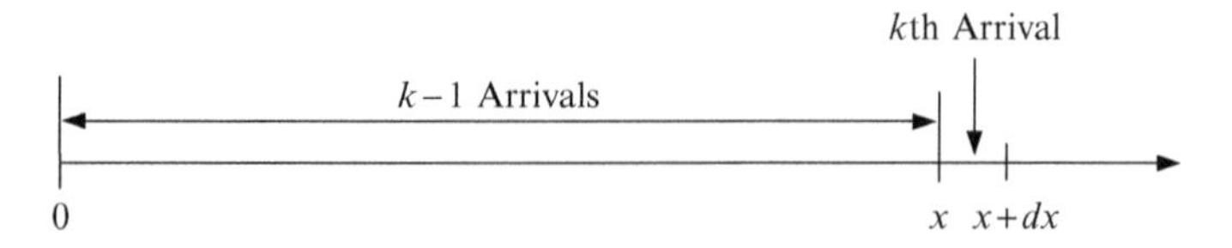


**FIGURE 4.9**
Illustration of the Erlang Random Variable

$k-1$, which is $(\lambda x)^{k-1}e^{-\lambda x}/(k-1)!$ According to the Poisson process, the probability that exactly one event occurs between $x$ and $x+dx$ is $\lambda dx$. Since the two intervals are disjoint, the events occurring in them are independent. Thus, we have that

$$f_{X_k}(x)dx = P[x \le X_k \le x+dx] = \left\{\frac{(\lambda x)^{k-1}}{(k-1)!}e^{-\lambda x}\right\} \times \{\lambda dx\} = \frac{\lambda^k x^{k-1}}{(k-1)!}e^{-\lambda x}dx$$

From this we obtain

$$f_{X_k}(x) = \frac{\lambda^k x^{k-1}}{(k-1)!}e^{-\lambda x} \qquad x \ge 0$$

The CDF of $X_k$ is obtained through repeated application of integration by parts as

$$F_{X_k}(x) = P[X_k \le x] = \int_0^x f_{X_k}(x)dx = 1 - \sum_{m=0}^{k-1}\frac{(\lambda x)^m}{m!}e^{-\lambda x} \qquad x \ge 0 \tag{4.37}$$

A plot of the PDF of $X_k$ for the case when $k=2$ is shown in Figure 4.10.

We will use two methods to obtain the moments of $X_k$: the direct method and the "smart" method.

a. **Direct Method**

The expected value of $X_k$ is given by

$$E[X_k] = \int_0^\infty x f_{X_k}(x)dx = \int_0^\infty x\frac{\lambda^k x^{k-1}}{(k-1)!}e^{-\lambda x}dx = \frac{1}{(k-1)!}\int_0^\infty (\lambda x)^k e^{-\lambda x}dx$$

Let $u=\lambda x$, which means that $dx = du/\lambda$. Thus, the preceding result becomes

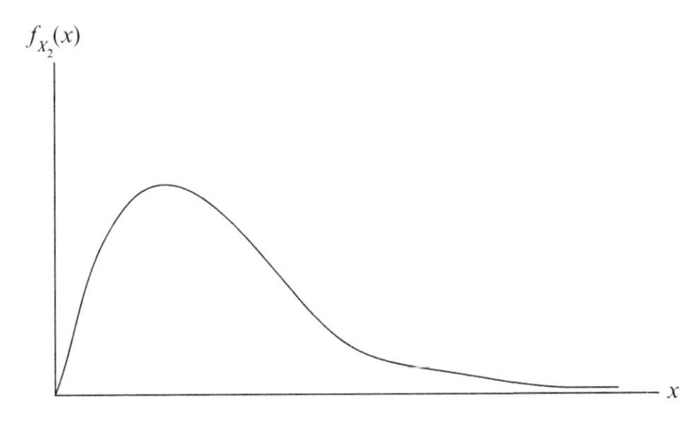


**FIGURE 4.10**
PDF of the Erlang-2 Random Variable

$$E[X_k]=\frac{1}{(k-1)!}\int_0^{\infty}u^k e^{-u}\frac{du}{\lambda}=\frac{1}{\lambda(k-1)!}\int_0^{\infty}u^k e^{-u}du$$

The integral $\int_0^{\infty}u^k e^{-u}du$ is called the gamma function of $k+1$, $\Gamma(k+1)$, which satisfies the condition

$$\Gamma(k+1)=\int_0^{\infty}u^k e^{-u}du=k\Gamma(k)$$

When $k$ is an integer, the gamma function is given by

$$\Gamma(k+1)=k! \quad k=0, 1, 2, \ldots$$

Thus, the expected value of $X_k$ becomes

$$\begin{aligned}E[X_k]&=\frac{1}{\lambda(k-1)!}\int_0^{\infty}u^k e^{-u}du=\frac{\Gamma(k+1)}{\lambda(k-1)!}=\frac{k!}{\lambda(k-1)!}=\frac{k(k-1)!}{\lambda(k-1)!}\\&=\frac{k}{\lambda}\end{aligned} \tag{4.38}$$

Another direct method of computing the expected value of $X_k$ is by recognizing that it is a nonnegative random variable. Therefore,

$$E[X_k]=\int_0^{\infty}\{1-F_{X_k}(x)\}dx=\int_0^{\infty}\left\{\sum_{m=0}^{k-1}\frac{(\lambda x)^m}{m!}e^{-\lambda x}\right\}dx=\sum_{m=0}^{k-1}\frac{1}{m!}\int_0^{\infty}(\lambda x)^m e^{-\lambda x}dx$$

Let $u=\lambda x$, which means that $dx=du/\lambda$. Thus, we obtain

$$\begin{aligned}E[X_k]&=\sum_{m=0}^{k-1}\frac{1}{m!}\int_0^{\infty}(\lambda x)^m e^{-\lambda x}dx=\sum_{m=0}^{k-1}\frac{1}{m!}\left(\frac{1}{\lambda}\right)\int_0^{\infty}u^m e^{-u}dx=\sum_{m=0}^{k-1}\frac{1}{m!}\left(\frac{1}{\lambda}\right)m!\\&=\sum_{m=0}^{k-1}\left(\frac{1}{\lambda}\right)=\frac{k}{\lambda}\end{aligned}$$

Note that this is $k$ times the expected value of the underlying exponential distribution. Similarly, the second moment of $X_k$ is given by

$$E\left[X_k^2\right]=\int_0^{\infty}x^2 f_{X_k}(x)dx=\int_0^{\infty}x^2\frac{\lambda^k x^{k-1}}{(k-1)!}e^{-\lambda x}dx=\frac{1}{\lambda(k-1)!}\int_0^{\infty}(\lambda x)^{k+1}e^{-\lambda x}dx$$

If we let $u=\lambda x$, which implies that $dx=du/\lambda$, we obtain

$$\begin{aligned}E\left[X_k^2\right]&=\frac{1}{\lambda^2(k-1)!}\int_0^{\infty}u^{k+1}e^{-u}du=\frac{\Gamma(k+2)}{\lambda^2(k-1)!}=\frac{(k+1)!}{\lambda^2(k-1)!}=\frac{(k+1)k(k-1)!}{\lambda^2(k-1)!}\\&=\frac{k(k+1)}{\lambda^2}\end{aligned} \tag{4.39}$$

Therefore, the variance of $X_k$ is given by

$$\sigma_{X_k}^2=E\left[X_k^2\right]-(E[X_k])^2=\frac{k(k+1)}{\lambda^2}-\frac{k^2}{\lambda^2}=\frac{k}{\lambda^2} \tag{4.40}$$

which is k times the variance of the underlying exponential distribution.

b. **Smart Method**

A smart method that can be used to obtain the moments of $X_k$ is by noting from equation (4.33) that the *n*th moment of the exponential random variable $X$ is given by

$$E[X^n]=\int_0^\infty x^n f_X(x)dx=\int_0^\infty x^n \lambda e^{-\lambda x}dx=\frac{n!}{\lambda^n} \quad n=1,2,\ldots$$

Thus we have that

$$E[X_k]=\int_0^\infty x\frac{\lambda^k x^{k-1}}{(k-1)!}e^{-\lambda x}dx=\frac{1}{(k-1)!}\int_0^\infty \lambda^k x^k e^{-\lambda x}dx=\frac{\lambda^{k-1}}{(k-1)!}\int_0^\infty \lambda x^k e^{-\lambda x}dx$$

$$=\frac{\lambda^{k-1}}{(k-1)!}E\left[X^k\right]=\frac{\lambda^{k-1}}{(k-1)!}\times\frac{k!}{\lambda^k}=\frac{\lambda^{k-1}}{(k-1)!}\times\frac{k(k-1)!}{\lambda^k}=\frac{k}{\lambda}$$

$$E\left[X_k^2\right]=\int_0^\infty x^2\frac{\lambda^k x^{k-1}}{(k-1)!}e^{-\lambda x}dx=\frac{1}{(k-1)!}\int_0^\infty \lambda^k x^{k+1} e^{-\lambda x}dx=\frac{\lambda^{k-1}}{(k-1)!}\int_0^\infty \lambda x^{k+1} e^{-\lambda x}dx$$

$$=\frac{\lambda^{k-1}}{(k-1)!}E\left[X^{k+1}\right]=\frac{\lambda^{k-1}}{(k-1)!}\times\frac{(k+1)!}{\lambda^{k+1}}=\frac{\lambda^{k-1}}{(k-1)!}\times\frac{(k+1)k(k-1)!}{\lambda^{k+1}}=\frac{k(k+1)}{\lambda^2}$$

## EXAMPLE 4.19

On a day a particular professor has office hours, the times between student visits to his office have been found to be exponentially distributed with a mean of 10 minutes. What is the probability that the time between the arrival of the second student and the arrival of the sixth student is greater than 20 minutes?

**Solution:**

Relative to the second student, the sixth student is the fourth arrival. Thus, the random variable that describes the time between the two arrivals is a 4th-order Erlang (or Erlang-4) random variable. That is, we want the probability that $X_4 > 20$ minutes. Since the CDF of $X_4$ is

$$F_{X_4}(x)=P[X_4\le x]=1-\sum_{m=0}^{3}\frac{(\lambda x)^m}{m!}e^{-\lambda x}$$

where $1/\lambda=10$, we have that

$$P[X_4>20]=1-P[X_4\le 20]=\sum_{m=0}^{3}\frac{2^m}{m!}e^{-2}=e^{-2}+\frac{2}{1}e^{-2}+\frac{4}{2}e^{-2}+\frac{8}{6}e^{-2}=\frac{19}{3}e^{-2}$$

$$=0.8571$$

Note that the result is essentially the probability that less than 4 students arrive during an interval of 20 minutes.

## EXAMPLE 4.20

The lengths of phone calls at a certain phone booth are exponentially distributed with a mean of 4 minutes. I arrived at the booth while Tom was using the phone, and I was told that he had already spent 2 minutes on the call before I arrived.

a. What is the average time I will wait until he ends his call?
b. What is the probability that Tom's call will last between 3 minutes and 6 minutes after my arrival?
c. Assume that I am the first in line at the booth to use the phone after Tom, and by the time he finished his call more than 5 people were waiting behind me to use the phone. What is the probability that the time between the instant I start using the phone and the time the 4th person behind me starts her call is greater than 15 minutes?

**Solution:**

Let $X$ denote the lengths of calls made at the phone booth. Then the PDF of $X$ is

$$f_X(x) = \begin{cases} \lambda e^{-\lambda x} & x \geq 0 \\ 0 & x < 0 \end{cases}$$

where $\lambda = 1/4$.

a. Because of the forgetfulness property of the exponential distribution, the average time I wait until Tom's call ends is the same as the mean length of a call, which is 4 minutes.
b. Due to the forgetfulness property of the exponential distribution, Tom's call "started from scratch" when I arrived. Therefore, the probability that it lasts between 3 minutes and 6 minutes after my arrival is the probability that an arbitrary call lasts between 3 minutes and 6 minutes, which is

$$\begin{aligned} P[3 < X < 6] &= F_X(6) - F_X(3) = \{1 - e^{-6\lambda}\} - \{1 - e^{-3\lambda}\} = e^{-3\lambda} - e^{-6\lambda} \\ &= e^{-3/4} - e^{-6/4} = 0.2492 \end{aligned}$$

c. Let $Y_k$ denote the time that elapses from the instant I commence my call until the end of the call initiated by the third person behind me, which is the instant the 4th person behind me can initiate her call. Then $Y_k$ is a 4th-order Erlang random variable $Y_4$ because it is the sum of the duration of my call and the durations of the calls of the three people behind me. Thus, the PDF of $Y_4$ is

$$f_{Y_4}(x) = \begin{cases} \dfrac{\lambda^4 y^3 e^{-\lambda y}}{3!} & y \geq 0 \\ 0 & \text{otherwise} \end{cases}$$

where $\lambda = 1/4$. The CDF of $Y_4$ is given by

$$F_{Y_4}(y) = P[Y_4 \leq y] = 1 - \sum_{k=0}^{3} \frac{(\lambda y)^k}{k!} e^{-\lambda y}$$

Therefore,

$$\begin{aligned} P[Y_4 > 15] &= 1 - F_{Y_4}(15) = \sum_{k=0}^{3} \frac{(15\lambda)^k}{k!} e^{-15\lambda} = \sum_{k=0}^{3} \frac{(3.75)^k}{k!} e^{-3.75} \\ &= e^{-3.75} \left\{ 1 + 3.75 + \frac{(3.75)^2}{2} + \frac{(3.75)^3}{6} \right\} = 0.4838 \end{aligned}$$

## 4.10 UNIFORM DISTRIBUTION

A continuous random variable $X$ is said to have a uniform distribution over the interval $[a, b]$ if its PDF is given by

$$f_X(x) = \begin{cases} \dfrac{1}{b-a} & a \le x \le b \\ 0 & \text{otherwise} \end{cases} \tag{4.41}$$

The plot of the PDF is shown in Figure 4.11.

The CDF of $X$ is given by

$$F_X(x) = PX \le x] = \begin{cases} 0 & x < a \\ \dfrac{x-a}{b-a} & a \le x < b \\ 1 & x \ge b \end{cases} \tag{4.42}$$

The expected value of $X$ is given by

$$\begin{aligned} E[X] &= \int_{-\infty}^{\infty} x f_X(x)dx = \int_a^b \frac{x}{b-a}dx = \left[\frac{x^2}{2(b-a)}\right]_a^b = \frac{b^2-a^2}{2(b-a)} \\ &= \frac{(b+a)(b-a)}{2(b-a)} = \frac{b+a}{2} \end{aligned} \tag{4.43}$$

The second moment of $X$ is given by

$$\begin{aligned} E[X^2] &= \int_{-\infty}^{\infty} x^2 f_X(x)dx = \int_a^b \frac{x^2}{b-a}dx = \left[\frac{x^3}{3(b-a)}\right]_a^b = \frac{b^3-a^3}{3(b-a)} \\ &= \frac{(b-a)(b^2+ab+a^2)}{3(b-a)} = \frac{b^2+ab+a^2}{3} \end{aligned} \tag{4.44}$$

Thus, the variance of $X$ is given by

$$\begin{aligned} \sigma_X^2 &= E[X^2] - (E[X])^2 = \frac{b^2+ab+a^2}{3} - \left(\frac{b^2+2ab+a^2}{4}\right) \\ &= \frac{b^2-2ab+a^2}{12} = \frac{(b-a)^2}{12} \end{aligned} \tag{4.45}$$

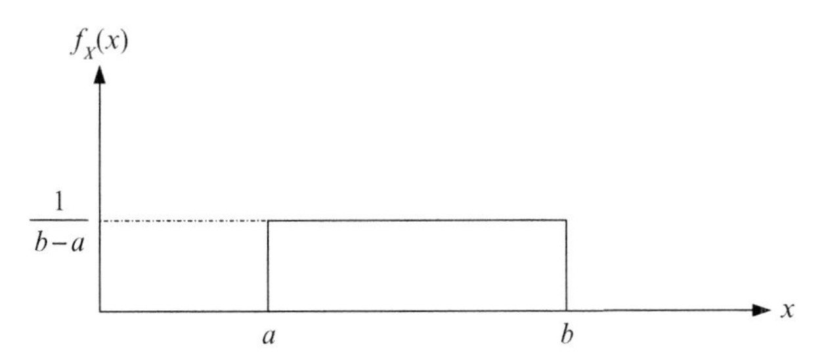


**FIGURE 4.11**
PDF of the Uniform Random Variable

## EXAMPLE 4.21

The time that Joe, the teaching assistant, takes to grade a paper is uniformly distributed between 5 minutes and 10 minutes. Find the mean and variance of the time he takes to grade a paper.

**Solution:**

Let $X$ be a random variable that denotes the time it takes Joe to grade a paper. Since $X$ is uniformly distributed we find that the mean and variance are as follows:

$$E[X]=\frac{10+5}{2}=7.5$$

$$\sigma_X^2=\frac{(10-5)^2}{12}=\frac{25}{12}$$

### 4.10.1 The Discrete Uniform Distribution

A discrete random variable $K$ is said to have a uniform distribution in the range $k=a, a+1, a+2, \ldots, a+N-1$ if it has the PMF

$$p_K(k)=\begin{cases}\frac{1}{N} & k=a,\ a+1,\ a+2,\ \ldots,\ a+N-1\\ 0 & \text{otherwise}\end{cases} \tag{4.46}$$

The mean of $K$ is given by

$$\begin{aligned}E[K]&=\sum_{k=a}^{a+N-1} kp_K(k)=\sum_{k=a}^{a+N-1}\frac{k}{N}=\frac{1}{N}\left\{\sum_{k=1}^{a+N-1}k-\sum_{k=1}^{a-1}k\right\}\\ &=\frac{1}{N}\left\{\frac{(a+N-1)(a+N)}{2}-\frac{a(a-1)}{2}\right\}=\frac{1}{2}\{N+2a-1\}\\ &=\frac{(a+N-1)+a}{2}\end{aligned} \tag{4.47}$$

where the second line follows from the fact that

$$\sum_{k=1}^{n}k=\frac{n(n+1)}{2}$$

Thus, the expected value is the arithmetic average of the lowest and highest values of the random variable as in the case of the continuous uniform random variable. The second moment is given by

$$E[K^2]=\sum_{k=a}^{a+N-1}k^2p_K(k)=\sum_{k=a}^{a+N-1}\frac{k^2}{N}=\frac{1}{N}\left\{\sum_{k=1}^{a+N-1}k^2-\sum_{k=1}^{a-1}k^2\right\}$$

Using the fact that

$$\sum_{k=1}^{n} k^2 = \frac{n(n+1)(2n+1)}{6}$$

we obtain

$$E[K^2] = \frac{2N^2 + 6aN + 6a^2 - 6a - 3N + 1}{6}$$

Thus, the variance is given by

$$\sigma_K^2 = \frac{N^2 - 1}{12} \tag{4.48}$$

## EXAMPLE 4.22

Let $X$ be the random variable that denotes the outcome of the roll of a fair die. We know that the PMF of $X$ is given by

$$p_X(x) = \frac{1}{6} \qquad x = 1, 2, \ldots, 6$$

We can compute the mean and variance of $X$ either directly or via the above formulas. First, by the direct method,

$$E[X] = \frac{1+2+3+4+5+6}{6} = \frac{21}{6} = \frac{7}{2}$$

$$E[X^2] = \frac{1^2+2^2+3^2+4^2+5^2+6^2}{6} = \frac{91}{6}$$

$$\sigma_X^2 = E[X^2] - (E[X])^2 = \frac{91}{6} - \frac{49}{4} = \frac{35}{12}$$

Next, we apply the formula by noting that $N = 6$ and $a = 1$. Then

$$E[X] = \frac{(a+N-1)+a}{2} = \frac{7}{2}$$

$$\sigma_K^2 = \frac{N^2-1}{12} = \frac{35}{12}$$

Thus, the two methods yield the same results.

## 4.11 NORMAL DISTRIBUTION

A continuous random variable $X$ is defined to be a normal random variable with parameters $\mu_X$ and $\sigma_X^2$ if its PDF is given by

$$f_X(x) = \frac{1}{\sqrt{2\pi\sigma_X^2}} e^{-(x-\mu_X)^2/2\sigma_X^2} \qquad -\infty < x < \infty \tag{4.49}$$

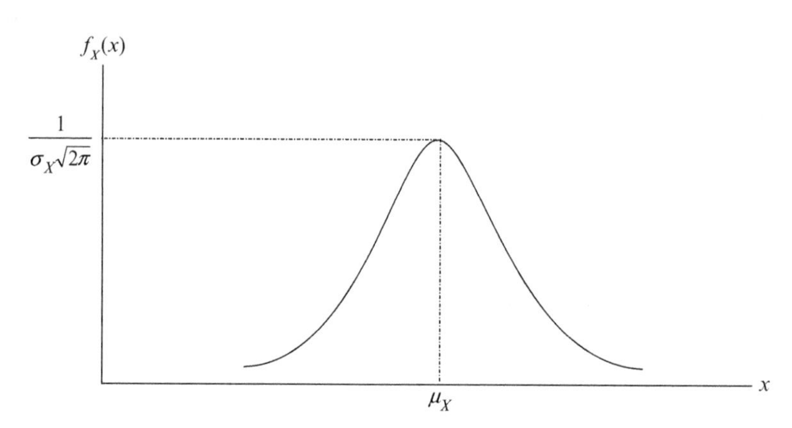


**FIGURE 4.12**
PDF of the Normal Random Variable

The PDF is a bell-shaped curve that is symmetric about $\mu_X$, which is the mean of X. The parameter $\sigma_X^2$ is the variance. Figure 4.12 illustrates the shape of the PDF. Since the variance (or more precisely, the standard deviation) is a measure of the spread around the mean, the larger the variance, the lower the peak of the curve and the more spread out it will be.

The normal random variable $X$ with parameters $\mu_X$ and $\sigma_X^2$ is usually designated $N(\mu_X, \sigma_X^2)$. Thus, we use the notation $X \sim N(\mu_X, \sigma_X^2)$ to indicate that $X$ is a normally distributed random variable with parameters $\mu_X$ and $\sigma_X^2$. The special case of zero mean and unit variance (i.e., $\mu_X = 0$ and $\sigma_X^2 = 1$) is designated $X \sim N(0, 1)$ and is called the *standard normal random variable*.

The CDF of $X$ is given by

$$F_X(x) = P[X \leq x] = \frac{1}{\sigma_X\sqrt{2\pi}}\int_{-\infty}^{x} e^{-(u-\mu_X)^2/2\sigma_X^2}du$$

Let $y = (u - \mu_X)/\sigma_X \Rightarrow du = \sigma_X dy$. Thus, the CDF of $X$ becomes

$$F_X(x) = P[X \leq x] = \frac{1}{\sqrt{2\pi}}\int_{-\infty}^{(x-\mu_X)/\sigma_X} e^{-y^2/2}dy$$

Thus, with the above transformation, $X$ becomes a standard normal random variable. The above integral cannot be evaluated in closed form. It is usually evaluated numerically through the function $\Phi(x)$, which is defined as follows:

$$\Phi(x) = \frac{1}{\sqrt{2\pi}}\int_{-\infty}^{x} e^{-y^2/2}dy \quad (4.50)$$

Thus, the CDF of $X$ is given by

$$F_X(x) = \frac{1}{\sqrt{2\pi}} \int_{-\infty}^{(x-\mu_X)/\sigma_X} e^{-y^2/2} dy = \Phi\left(\frac{x-\mu_X}{\sigma_X}\right) \tag{4.51}$$

The values of $\Phi(x)$ are usually given for nonnegative values of $x$. For negative values of $x$, $\Phi(x)$ can be obtained from the following relationship:

$$\Phi(-x) = 1 - \Phi(x) \tag{4.52}$$

Values of $\Phi(x)$ are given in Table 1 in Appendix 1.

## EXAMPLE 4.23

If $X \sim N(3, 9)$, which means that $X$ is a normal random variable with $\mu_X = 3$ and $\sigma_X^2 = 9$, find the probability that $X$ lies between 2 and 5.

**Solution:**

We are required to evaluate $P[2 < X < 5]$, which is given by

$$\begin{aligned} P[2<X<5] &= F_X(5) - F_X(2) = \Phi\left(\frac{5-3}{3}\right) - \Phi\left(\frac{2-3}{3}\right) = \Phi\left(\frac{2}{3}\right) - \Phi\left(-\frac{1}{3}\right) \\ &= \Phi(0.67) - \Phi(-0.33) = \Phi(0.67) - \{1 - \Phi(0.33)\} \\ &= \Phi(0.67) + \Phi(0.33) - 1 \end{aligned}$$

From Table 1 of Appendix 1, $\Phi(0.67) = 0.7486$ and $\Phi(0.33) = 0.6293$. Thus,

$$P[2<X<5] = 0.7486 + 0.6293 - 1 = 0.3779$$

Note that because the argument of the $\Phi$-function is only defined in the table for two decimal places, 2/3 can also be approximated by 0.66, which means that we could also get $\Phi(2/3) = \Phi(0.66) = 0.7454$, and the desired result would be 0.3747.

## EXAMPLE 4.24

The weights in pounds of parcels arriving at a package delivery company's warehouse can be modeled by an $N(5, 16)$ normal random variable, $X$.

a. What is the probability that a randomly selected parcel weighs between 1 and 10 pounds?
b. What is the probability that a randomly selected parcel weighs more than 9 pounds?

**Solution:**

Since $X$ is an $N(5, 16)$ normal random variable, we have that its mean $\mu_X = 5$ and its variance $\sigma_X^2 = 16$. Therefore, its standard deviation is $\sigma_X = 4$.

a. The probability that a randomly selected parcel weighs between 1 and 10 pounds is

$$P[1 < X < 10] = F_X(10) - F_X(1) = \Phi\left(\frac{10-5}{4}\right) - \Phi\left(\frac{1-5}{4}\right) = \Phi\left(\frac{5}{4}\right) - \Phi\left(-\frac{4}{4}\right)$$
$$= \Phi(1.25) - \Phi(-1.00) = \Phi(1.25) - \{1 - \Phi(1.00)\}$$
$$= \Phi(1.25) + \Phi(1.00) - 1 = 0.8944 + 0.8413 - 1$$
$$= 0.7357$$

b. The probability that a randomly selected parcel weighs more than 9 pounds is

$$P[X > 9] = 1 - P[X \leq 9] = 1 - F_X(9) = 1 - \Phi\left(\frac{9-5}{4}\right) = 1 - \Phi(1)$$
$$= 1 - 0.8413 = 0.1587$$

### 4.11.1 Normal Approximation of the Binomial Distribution

Let $X$ be a binomial random variable with parameter $(n, p)$. Thus, the PMF is given by

$$p_X(x) = \binom{n}{x} p^x (1-p)^{n-x} \qquad x = 0, 1, 2, \ldots, n$$

When $n$ is large with neither $p$ nor $q = 1 - p$ close to zero, the binomial distribution can be approximated by the standard normal random variable with the normalized score as

$$Z = \frac{X - \mu_X}{\sigma_X} = \frac{X - np}{\sqrt{np(1-p)}}$$

Thus,

$$P[a \leq X \leq b] = P\left[\frac{a - np}{\sqrt{np(1-p)}} \leq Z \leq \frac{b - np}{\sqrt{np(1-p)}}\right]$$

The approximation is very good when both $np$ and $n(1-p)$ are greater than 5.

## EXAMPLE 4.25

A coin is tossed 10 times. Find the probability of getting between 4 and 7 heads inclusive using (a) the binomial distribution and (b) the normal approximation to the binomial distribution.

**Solution:**

Let $X$ be the random variable that denotes the number of heads in 10 tosses of the coin. Then we are required to find $P[4 \leq X \leq 7]$.

a. Using the binomial distribution,

$$P[4 \leq X \leq 7] = \sum_{x=4}^{7} p_X(x) = \sum_{x=4}^{7} \binom{10}{x} \left(\frac{1}{2}\right)^x \left(\frac{1}{2}\right)^{10-x} = \left(\frac{1}{2}\right)^{10} \sum_{x=4}^{7} \binom{10}{x}$$

$$= \frac{\frac{10!}{4!6!} + \frac{10!}{5!5!} + \frac{10!}{6!4!} + \frac{10!}{7!3!}}{1024} = \frac{792}{1024} = 0.7734$$

b. Using the normal approximation to the binomial distribution, we obtain the result

$$np = 5$$
$$np(1-p) = 2.5$$
$$P[4 \leq X \leq 7] = P\left[\frac{4-5}{\sqrt{2.5}} \leq Z \leq \frac{7-5}{\sqrt{2.5}}\right] = P[-0.63 \leq Z \leq 1.26]$$
$$= \Phi(1.26) - \Phi(-0.63) = \Phi(1.26) - \{1 - \Phi(0.63)\}$$
$$= \Phi(1.26) + \Phi(0.63) - 1 = 0.8962 + 0.7357 - 1$$
$$= 0.6319$$

Observe that the result obtained by the approximation is much smaller than the result obtained by the direct method. This is because we have not converted the integers 4 and 7 into appropriate non-integral values that the continuous normal random variable can take. In statistical studies, it is a common practice to convert the integers 4 and 7 to the values 3.5 and 7.5, respectively, before using them for the preceding computation. Therefore, with this in mind we proceed as follows:

$$P[4 \leq X \leq 7] = P\left[\frac{3.5-5}{\sqrt{2.5}} \leq Z \leq \frac{7.5-5}{\sqrt{2.5}}\right] = P[-0.95 \leq Z \leq 1.58]$$
$$= \Phi(1.58) - \Phi(-0.95) = \Phi(1.58) - \{1 - \Phi(0.95)\}$$
$$= \Phi(1.58) + \Phi(0.95) - 1 = 0.9429 + 0.8289 - 1$$
$$= 0.7718$$

Thus, we see that the result gives a better approximation than when the conversion is not made.

### 4.11.2 The Error Function

Recall that the normal distribution is given by

$$f_X(x) = \frac{1}{\sqrt{2\pi\sigma_X^2}} e^{-(x-\mu_X)^2/2\sigma_X^2} \qquad -\infty < x < \infty$$

As stated earlier, the CDF of $X$ is given by

$$F_X(x) = P[X \leq x] = \frac{1}{\sigma_X\sqrt{2\pi}} \int_{-\infty}^{x} e^{-(u-\mu_X)^2/2\sigma_X^2} du$$

Another transformation of the random variable $X$ is as follows. Let $y = (u - \mu_X)/\sigma_X\sqrt{2} \Rightarrow du = \sqrt{2}\sigma_X dy$. Then the transformed random variable is an $N(0,\frac{1}{2})$ normal random variable. The CDF of $X$ becomes

$$F_X(x) = \frac{1}{\sqrt{\pi}} \int_{-\infty}^{(x-\mu_X)/\sigma_X\sqrt{2}} e^{-y^2} dy$$

Suppose we are interested in obtaining the probability that $X$ takes on values in the range $-V \leq X \leq V$, where $V$ is a constant parameter. Thus, we obtain the following:

$$P[-V \leq X \leq V] = \frac{1}{\sqrt{\pi}} \int_{-(V-\mu_X)/\sigma_X\sqrt{2}}^{(V-\mu_X)/\sigma_X\sqrt{2}} e^{-y^2} dy = \frac{2}{\sqrt{\pi}} \int_{0}^{(V-\mu_X)/\sigma_X\sqrt{2}} e^{-y^2} dy$$

where the last equality is due to the symmetrical nature of the distribution. In the mathematical literature a quantity called the *error function* is defined as follows:

$$\operatorname{erf}(x) = \frac{2}{\sqrt{\pi}} \int_0^x e^{-y^2} dy \tag{4.53}$$

Values of the error function are usually provided in mathematical tables. The *complementary error function* is defined as

$$\operatorname{erfc}(x) = 1 - \operatorname{erf}(x) = \frac{2}{\sqrt{\pi}} \int_x^{\infty} e^{-y^2} dy \tag{4.54}$$

Thus, the probability that $X$ is greater than $V$ is given by

$$P[X > V] = \frac{1}{\sqrt{\pi}} \int_{(V-\mu_X)/\sigma_X\sqrt{2}}^{\infty} e^{-y^2} dy = \frac{1}{2} \operatorname{erfc}\left(\frac{V-\mu_X}{\sigma_X\sqrt{2}}\right)$$

The error function is related to $\Phi(x)$ as follows:

$$\Phi(x) = \frac{1}{2}\left[1 + \operatorname{erf}\left(\frac{x}{\sqrt{2}}\right)\right] \tag{4.55}$$

Alternatively, given the $\Phi(x)$ table, we can obtain the error function by

$$\operatorname{erf}(x) = 2\Phi\left(x\sqrt{2}\right) - 1 \tag{4.56}$$

### 4.11.3 The *Q*-Function

Another function that is closely related to $\Phi(x)$ and that is commonly used in electrical engineering is the Q-function, which is defined as follows:

$$Q(x) = \frac{1}{\sqrt{2\pi}} \int_x^{\infty} e^{-y^2/2} dy \tag{4.57}$$

The Q-function is essentially the tail probability of the standard normal distribution. That is, $Q(x) = P[X > x]$, where $X \sim N(0,1)$. Thus,

$$F_X(x) = 1 - Q\left(\frac{x-\mu_X}{\sigma_X}\right)$$

which means that $Q(x)$ is related to $\Phi(x)$ by

$$Q(x) = 1 - \Phi(x) = \Phi(-x) \tag{4.58}$$

It also has the property that

$$Q(-x) = 1 - Q(x) \tag{4.59}$$

Similarly, the $Q$-function is related to the error function as follows:

$$Q(x) = \frac{1}{2}\left[1 - \operatorname{erf}\left(\frac{x}{\sqrt{2}}\right)\right] \tag{4.60}$$

## 4.12 THE HAZARD FUNCTION

Let $X$ be a random variable that represents the time until an equipment fails. Let $f_X(x)$ and $F_X(x)$ be the PDF and CDF of $X$, respectively. The hazard function of $X$, denoted by $h_X(x)$, is defined by

$$h_X(x) = \frac{f_X(x)}{1 - F_X(x)} \qquad F_X(x) \neq 1 \tag{4.61}$$

That is, $h_X(x)$ is the instantaneous rate at which a component will fail given that it has already survived a length of time $x$, and $h_X(x)dx$ is the conditional probability that the equipment will fail between $x$ and $x+dx$, given that it has survived a time greater than $x$. Alternatively, based on our knowledge of conditional probability, we can consider $h_X(x)$ to be the conditional PDF of $X$, given that $X > x$.

Recall that in Chapter 1 we defined the reliability function $R(t)$ of a component as the probability that the component has not failed by time $t$. Thus, we can relate the reliability function $R_X(x)$ of $X$ to its CDF as follows:

$$R_X(x) = P[X > x] = 1 - P[X \leq x] = 1 - F_X(x) \tag{4.62}$$

$R_X(x)$ is sometimes called the *survival function*. Thus, the hazard function can be defined in terms of the reliability function as follows:

$$h_X(x) = \frac{f_X(x)}{R_X(x)} \tag{4.63}$$

We now show that by specifying the hazard function, we uniquely specify the reliability function and, hence, the CDF of a random variable. Since $F_X(x) = 1 - R_X(x)$, we have that

$$f_X(x) = \frac{d}{dx}F_X(x) = \frac{d}{dx}[1 - R_X(x)] = -\frac{d}{dx}R_X(x)$$

Thus,

$$h_X(x) = \frac{f_X(x)}{R_X(x)} = \frac{\frac{d}{dx}F_X(x)}{R_X(x)} = \frac{-\frac{d}{dx}R_X(x)}{R_X(x)} = -\frac{d}{dx}\ln R_X(x)$$

Integrating both sides from 0 to $x$ we obtain

$$[\ln R_X(u)]_0^x = -\int_0^x h_X(u)du$$

Since $R_X(0) = 1$, we have that

$$\begin{aligned} \ln R_X(x) &= -\int_0^x h_X(u)du \\ R_X(x) &= 1 - F_X(x) = \exp\left\{-\int_0^x h_X(u)du\right\} \end{aligned} \tag{4.64}$$

## EXAMPLE 4.26

The time until a component fails is exponentially distributed with a mean of 200 hours. Given that the component has not failed after operating for 150 hours, calculate the hazard function.

**Solution:**

Let $T$ denote the time until the component fails. Then the PDF and CDF of $T$ are given by

$$\begin{aligned} f_T(t) &= \lambda e^{-\lambda t} \qquad \lambda = 1/200,\ t \geq 0 \\ F_T(t) &= 1 - e^{-\lambda t} \end{aligned}$$

Thus, the hazard function is given by

$$h_T(t) = \frac{f_T(t)}{1 - F_T(t)} = \frac{\lambda e^{-\lambda t}}{e^{-\lambda t}} = \lambda$$

Since $\lambda$ is the failure rate, we see that the hazard function in this case is a constant that is equal to the failure rate.

## EXAMPLE 4.27

Determine the hazard function of a component whose time to failure $X$ is the so-called Weibull random variable with parameters $\lambda$ and $\rho$, and whose PDF and CDF are given by

$$\begin{aligned} f_X(x) &= \lambda\rho x^{\rho-1}e^{-\lambda x^\rho} \qquad x \geq 0;\ \lambda,\ \rho \geq 0 \\ F_X(x) &= 1 - e^{-\lambda x^\rho} \end{aligned}$$

**Solution:**

The Weibull distribution is widely used in reliability modeling. When $\rho = 1$, we obtain the exponential distribution; and when $\rho = 2$, we obtain the Rayleigh distribution that is popularly used to model different types of interference in communication systems. The hazard function of the Weibull distribution is given by

$$h_X(x) = \frac{f_X(x)}{1 - F_X(x)} = \frac{\lambda p x^{p-1} e^{-\lambda x^p}}{e^{-\lambda x^p}} = \lambda p x^{p-1}$$

## EXAMPLE 4.28

The hazard function of a certain random variable $Y$ is given by

$h_Y(y) = 0.5y,\ y \geq 0$

What is the PDF of $Y$?

**Solution:**
The reliability function of $Y$ is given by

$$R_Y(y) = \exp\left\{-\int_0^y h_Y(u)du\right\} = \exp\left\{-\int_0^y 0.5u du\right\} = e^{-0.25y^2} = 1 - F_Y(y)$$

Therefore, the CDF of $Y$ is given by

$$F_Y(y) = 1 - R_Y(y) = 1 - e^{-0.25y^2}$$

Finally, the PDF of $Y$ is given by

$$f_Y(y) = \frac{d}{dy} F_Y(y) = 0.5ye^{-0.25y^2} \qquad y \geq 0$$

## 4.13 TRUNCATED PROBABILITY DISTRIBUTIONS

In some applications, the range of values that a random variable can take is limited by physical constraints. For example, if the random variable $X$ denotes the number of people in a car on a highway, then $X$ cannot be equal to zero; unless the car is driving itself, which is not currently possible! A probability distribution for a random variable $X$ is defined to be truncated if some set of values in the range of $X$ are excluded.

Suppose the random variable $X$ has the PDF $f_X(x)$ and CDF $F_X(x)$. Assume that we wish to know how $X$ is distributed, given that $a < X \leq b$. Let the random variable $Y$ be defined as follows: $Y = X|a < X \leq b$. Then the PDF of $Y$ can be obtained from the relationship

$$f_Y(y) = \theta f_X(y) \qquad a < y \leq b$$

where $\theta$ is a constant of proportionality that can be obtained from the fact that for $f_Y(y)$ to be a true PDF then

$$\int_{-\infty}^{\infty} f_Y(y)dy = 1 = \theta \int_a^b f_X(y)dy = \theta[F_X(b) - F_X(a)]$$

From this we obtain

$$\theta = \frac{1}{F_X(b) - F_X(a)}$$

Thus, we have that

$$f_Y(y) = \frac{f_X(y)}{F_X(b) - F_X(a)} \qquad a < y \leq b \tag{4.65}$$

The expected value of $Y$ is given by

$$E[Y] = \int_a^b y f_Y(y)dy = \frac{\int_a^b y f_X(y)dy}{F_X(b) - F_X(a)} \tag{4.66}$$

If truncation is done by removing the lower portion of the distribution we obtain

$$\begin{aligned} f_X(x|X > b) &= \frac{f_X(x)}{1 - F_X(b)} \qquad x > b \\ E[X|X > b] &= \frac{\int_b^{\infty} x f_X(x)dx}{1 - F_X(b)} \end{aligned} \tag{4.67}$$

Similarly, if truncation is done by removing the upper portion of the distribution, we obtain

$$\begin{aligned} f_X(x|X \leq a) &= \frac{f_X(x)}{F_X(a)} \qquad x \leq a \\ E[X|X \leq a] &= \frac{\int_{-\infty}^{a} x f_X(x)dx}{F_X(a)} \end{aligned} \tag{4.68}$$

In the remainder of this section we consider four different truncated probability distributions, namely

a. Truncated binomial distribution
b. Truncated geometric distribution
c. Truncated Poisson distribution
d. Truncated normal distribution

These distributions are frequently encountered in statistical and econometric studies.

### 4.13.1 Truncated Binomial Distribution

A random variable $X$ is said to have a truncated binomial distribution if its PMF is a binomial distribution and the values of the random variable are strictly greater than 0. Thus, the PMF is given by

$$p_X(x) = \frac{\binom{n}{x} p^x (1-p)^{n-x}}{1-(1-p)^n} \qquad x = 1, 2, \ldots, n \tag{4.69}$$

The expected value and variance of $X$ are given by

$$\begin{aligned} E[X] &= \frac{np}{1-(1-p)^n} \\ \text{Var}(X) &= \frac{\sigma_X^2 + n^2p^2}{1-(1-p)^n} - \frac{n^2p^2}{[1-(1-p)^n]^2} \end{aligned} \tag{4.70}$$

where $\sigma_X^2$ is the variance of the untruncated binomial random variable; that is, $\sigma_X^2 = np(1-p)$.

### 4.13.2 Truncated Geometric Distribution

A random variable $K$ is said to have a truncated geometric distribution if its PMF is given by

$$p_K(k) = \frac{p(1-p)^{k-1}}{1-(1-p)^n} \qquad k = 1, 2, \ldots, n \tag{4.71}$$

Thus, $K$ cannot take on values that are greater than $n$. The expected value of $K$ is

$$E[K] = \sum_{k=1}^{n} k p_K(k) = \frac{p}{1-(1-p)^n} \sum_{k=1}^{n} k(1-p)^{k-1} = \frac{1-(np+1)(1-p)^n}{p[1-(1-p)^n]}$$

If we require that $K$ be limited to values that lie between $a$ and $n$ for $a \geq 1$, then the PMF is given by

$$p_K(k) = \frac{p(1-p)^{k-1}}{(1-p)^{a-1} - (1-p)^n} \qquad k = a, a+1, \ldots, n; \; a \geq 1 \tag{4.72}$$

### 4.13.3 Truncated Poisson Distribution

A random variable $Y$ is defined to be a truncated version of a standard Poisson random variable $X$ if its PMF is given by

$$p_Y(y) = P[X = y | y > 0] = \frac{p_X(y)}{1-p_X(0)} = \frac{\lambda^y e^{-\lambda}}{y!(1-e^{-\lambda})} \qquad y = 1, 2, \ldots \tag{4.73}$$

Thus, $Y$ takes on values that are greater than 0. The expected value and variance of $Y$ are given by

$$
\begin{aligned}
E[Y] &= \frac{\lambda}{1-e^{-\lambda}} \\
\sigma_Y^2 &= \frac{\lambda}{(1-e^{-\lambda})^2}
\end{aligned}
\tag{4.74}
$$

### 4.13.4 Truncated Normal Distribution

Suppose $X \sim N(\mu_X, \sigma_X^2)$; that is, $X$ has a normal distribution with mean $\mu_X$ and variance $\sigma_X^2$. Let be the random variable $Y = X|a < X < b$, where $-\infty < a < b < \infty$. Then $Y$ has a truncated normal distribution, and its PDF, $f_Y(y)$, is given by

$$
f_Y(y) = \frac{\frac{1}{\sigma_X}\varphi\left(\frac{y-\mu_X}{\sigma_X}\right)}{\Phi\left(\frac{b-\mu_X}{\sigma_X}\right) - \Phi\left(\frac{a-\mu_X}{\sigma_X}\right)} \qquad a \le y \le b \tag{4.75}
$$

where $\varphi(v) = \frac{1}{\sqrt{2\pi}}\exp\left(-\frac{1}{2}v^2\right)$ is the PDF of the standard normal distribution and $\Phi(\cdot)$ is its CDF. It can be shown that the mean of $Y$ is given by

$$
E[Y] = \mu_X + \sigma_X \left\{ \frac{\varphi\left(\frac{a-\mu_X}{\sigma_X}\right) - \varphi\left(\frac{b-\mu_X}{\sigma_X}\right)}{\Phi\left(\frac{b-\mu_X}{\sigma_X}\right) - \Phi\left(\frac{a-\mu_X}{\sigma_X}\right)} \right\} \tag{4.76}
$$

## 4.14 CHAPTER SUMMARY

This chapter introduced some of the many classes of random variables. The *Bernoulli random variable* is used to model experiments that have only two possible outcomes, which are referred to as success and failure. Assume that we perform the experiment $n$ times and then stop. The random variable that is used to denote the number of successes that occurred in those $n$ Bernoulli trials is the *binomial random variable*. If the goal is to keep performing the experiment until a success occurs, then the random variable that denotes the number of Bernoulli trials until success occurs is called the *geometric random variable*. Sometimes we are not interested in the first success but in the $k$th success. The random variable that denotes the number of Bernoulli trial up to and including that trial in which the $k$th success occurs is called the $k$th-order *Pascal random variable*.

One popular area of application of probability is quality control. Some of the items coming off a production line are good and some are bad. If we know beforehand the fraction of the items in a production batch that are good, we may want to know the probability that the sample contains a specified number of bad items. The random variable that is used to denote this number is the *hypergeometric random variable*.

The *Poisson random variable* is used to count the number of arrivals over a given interval. It is a popularly used model for such events as the number of customers arriving at a restaurant, the number of messages that arrive at the switchboard, and the number of equipment failures over a given interval.

All of the above random variables are discrete random variables. Continuous random variables include the *exponential random variable* that is used to denote the length of time between occurrences of an event. Associated with the exponential random variable is the *Erlang-k random variable*, which is used to denote the length of the interval from the beginning of the observation of the occurrence of an event until the point in time when the $k$th occurrence of that event takes place.

Two other continuous random variables are covered in this chapter. One is the *uniformly distributed random variable*, which is used to denote events that are equally likely to occur at any time within a given interval. The other is the *normal random variable* that is used to denote events that have a high probability of occurrence around the mean value and a smaller probability of occurrence the farther away from the mean value we move.

Finally, we considered some random variables that exclude some set of values in the range of a parent random variable. These random variables are defined to have *truncated probability distributions*. They occur regularly in statistical studies and include the *truncated binomial distribution, truncated geometric distribution, truncated Poisson distribution* and the *truncated normal distribution*.

Table 4.1 is a summary of the PMFs, CDFs, means, and variances of the different discrete random variables. Similarly, Table 4.2 is a summary of the PDFs, CDFs, means, and variances of the different continuous random variables.

## 4.15 PROBLEMS

### Section 4.3 Binomial Distribution

4.1 Suppose four dice are tossed. What is the probability that at most one 6 appears?

4.2 An equipment consists of 9 components, each of which will independently fail with a probability of $p$. If the equipment is able to function properly when at least 6 of the components are operational, what is the probability that it is functioning properly?

4.3 A fair coin is tossed three times. Let the random variable $X$ denote the number of heads that turn up. Determine the mean and variance of $X$.

**Table 4.1** Summary of Discrete Random Variables

| Random Variable | PMF | CDF | Mean | Variance |
| --- | --- | --- | --- | --- |
| Bernoulli | $p_X(x)=\begin{cases}1-p & x=0\\ 0 & x=1\end{cases}$ | $F_X(x)=\begin{cases}0 & x<0\\ 1-p & 0\le x<1\\ 1 & x\ge 1\end{cases}$ | $p$ | $p(1-p)$ |
| Binomial | $p_{X(n)}(x)=\binom{n}{x}p^x(1-p)^{n-x}$ <br> $x=0,1,2,\ldots,n$ | $F_{X(n)}(x)=\begin{cases}0 & x<0\\ \sum_{k=0}^{x}\binom{n}{k}p^k(1-p)^{n-k} & 0\le x<n\\ 1 & x\ge n\end{cases}$ | $np$ | $np(1-p)$ |
| Geometric | $p_X(x)=p(1-p)^{x-1}$ <br> $x=1,2,\ldots$ | $F_X(x)=\begin{cases}1-(1-p)^x & x\ge 1\\ 0 & \text{otherwise}\end{cases}$ | $\frac{1}{p}$ | $\frac{1-p}{p^2}$ |
| Pascal-$k$ | $p_{X_k}(n)=\binom{n-1}{k-1}p^k(1-p)^{n-k}$ <br> $k=1,2,\ldots;\ n=k,k+1,\ldots$ | $F_{X_k}(x)=\begin{cases}0 & x<k\\ \sum_{n=k}^{x}\binom{n-1}{k-1}p^k(1-p)^{n-k} & 0\le x<n\\ 1 & x\ge n\end{cases}$ | $\frac{k}{p}$ | $\frac{k(1-p)}{p^2}$ |
| Hypergeometric | $p_{K_n}(k)=\frac{\binom{N_1}{k}\binom{N-N_1}{n-k}}{\binom{N}{k}}$ <br> $k=0,1,\ldots,\min(n,N_1)$ | $F_{K_n}(k)=\sum_{m=0}^{k}\left\{\frac{\binom{N_1}{m}\binom{N-N_1}{n-m}}{\binom{N}{k}}\right\}$ <br> $k=0,1,\ldots,\min(n,N_1)$ <br> $F_{K_n}(k)=1\ \ k\ge\min(n,N_1)$ | $np$, where $p=\frac{N_1}{N}$ | $\frac{np(1-p)(N-n)}{N-1}$ |
| Poisson | $p_K(k)=\frac{\lambda^k}{k!}e^{-\lambda}$ <br> $k=0,1,\ldots$ | $F_K(k)=\sum_{r=0}^{k}\frac{\lambda^r}{r!}e^{-\lambda}\ \ k\ge 0$ | $\lambda$ | $\lambda$ |

**Table 4.2** Summary of Continuous Random Variables

| Random Variable | PDF | CDF | Mean | Variance |
|---|---|---|---|---|
| Exponential | $f_X(x)=\lambda e^{-\lambda x}$<br>$x \geq 0$ | $F_X(x)=\begin{cases}1-e^{-\lambda x} & x \geq 0 \\ 0 & \text{otherwise}\end{cases}$ | $\frac{1}{\lambda}$ | $\frac{1}{\lambda^2}$ |
| Erlang-$k$ | $f_{X_k}(x)=\frac{\lambda^k x^{k-1}}{(k-1)!}e^{-\lambda x}$<br>$k=1, 2, \ldots; x \geq 0$ | $F_{X_k}(x)=1-\sum_{r=0}^{k-1}\frac{(\lambda x)^r}{r!}e^{-\lambda x}$<br>$x \geq 0$ | $\frac{k}{\lambda}$ | $\frac{k}{\lambda^2}$ |
| Uniform | $f_X(x)=\begin{cases}\frac{1}{b-a} & a \leq x \leq b \\ 0 & \text{otherwise}\end{cases}$ | $F_X(x)=\begin{cases}0 & x<a \\ \frac{x-a}{b-a} & a \leq x < b \\ 1 & x \geq b\end{cases}$ | $\frac{b+a}{2}$ | $\frac{(b-a)^2}{12}$ |
| Normal | $f_X(x)=\frac{e^{-(x-\mu_X)^2/2\sigma_X^2}}{\sqrt{2\pi\sigma_X^2}}$<br>$-\infty < x < \infty$ | $F_X(x)=\Phi\left(\frac{x-\mu_X}{\sigma_X}\right)$ | $\mu_X$ | $\sigma_X^2$ |

4.4 A certain student is known to be late to the Signals and Systems class 30% of the time. If the class meets four times a week, find
   a. the probability that the student is late for at least three classes in a given week.
   b. the probability that the student will not be late at all during a given week.

4.5 A multiple-choice exam has 6 problems, each of which has 3 possible answers. What is the probability that John will get 4 or more correct answers by just guessing?

4.6 A block of 100 bits is to be transmitted over a binary channel with a probability of bit error of $p=0.001$. What is the probability that 3 or more bits are received in error?

4.7 An office has 4 phone lines. Each is busy about 10% of the time. Assume that the phone lines act independently.
   a. What is the probability that all 4 phones are busy?
   b. What is the probability that 3 of the phones are busy?

4.8 The laptops made by the XYZ Corporation have a probability of 0.10 of being defective as they come out of the assembly line. The ABC company has purchased 8 of these laptops for office use.
   a. What is the PMF of $K$, the number of defective laptops out of the 8 that the ABC company purchased?
   b. What is the probability that at most one laptop is defective out of the 8?
   c. What is the probability that exactly 1 laptop is defective out of the 8?
   d. What is the expected number of defective laptops out of the 8?

4.9 On the average, 25% of the products manufactured by a certain company are found to be defective. If we select four of these products at random and denote the number of the four products that are defective by the random variable $X$, determine the mean and variance of $X$.

4.10 Five fair coins are tossed. Assuming that the outcomes are independent, find the PMF of the number of heads obtained in the experiment.

4.11 A company makes gadgets that it sells in packages of eight. It has been found that the probability that a gadget made by the company is defective is 0.1 independently of other gadgets. If the company offers a money-back guarantee for any package that contains more than one defective gadget, what is probability that the person that bought a given package will be refunded?

4.12 At least 10 of 12 people in a jury are required to find a person guilty before the person can be convicted. Assume that each juror acts independently of other jurors and each juror has a probability of 0.7 of finding a person guilty. What is the probability that a person is convicted?

4.13 A radar system has a probability of 0.1 of detecting a certain target during a single scan. Find the probability that the target will be detected

a. at least two times in four consecutive scans.
b. at least once in 20 consecutive scans.

4.14 A machine makes errors in a certain operation with probability $p$. There are two types of errors: type A error and type B error. The fraction of errors of type A is $a$, and the fraction of errors of type B is $1-a$.

a. What is the probability of $k$ errors in $n$ operations?
b. What is the probability of $k_A$ type A errors in $n$ operations?
c. What is the probability of $k_B$ type B errors in $n$ operations?
d. What is the probability of $k_A$ type A errors and $k_B$ type B errors in $n$ operations?

4.15 Studies indicate that 40% of marriages end in divorce, where it is assumed that divorces are independent of each other. Out of 10 married couples, determine the following probabilities:

a. That only the Arthurs and the Martins will stay married
b. That exactly 2 of the 10 couples will stay married.

4.16 A car has five traffic lights on its route. Independently of other traffic lights, each traffic light turns red as the car approaches the light (and thus forces the car to stop at the light) with a probability of 0.4.

a. Let $K$ be a random variable that denotes the number of lights at which the car stops. What is the PMF of $K$?
b. What is the probability that the car stops at exactly two lights?
c. What is the probability that the car stops at more than two lights?
d. What is the expected value of $K$?

4.17 In a class of 18 boys and 12 girls, boys have a probability of 1/3 of knowing the answer to a typical question that the teacher asks and girls have a probability of 1/2 of knowing the answer to the question. Assume that each student acts independently. Let $K$ be a random variable that denotes the number of students who know the answer to a question that the teacher asks in class. Determine the following:

a. The PMF of $K$
b. The mean of $K$
c. The variance of $K$.

4.18 A bag contains 2 red balls and 6 green balls. A ball is randomly selected from the bag, its color is noted and the ball is put back into the bag, which is then thoroughly mixed. Determine the probability that in 10 such selections a red ball is selected exactly 4 times using

a. the binomial distribution
b. the Poisson approximation to the binomial distribution

4.19 Ten balls are randomly tossed into 5 boxes labeled $B_1, B_2, \ldots, B_5$. Determine the following probabilities:

a. Each box gets 2 balls
b. Box $B_3$ is empty
c. Box $B_2$ has 6 balls.

## Section 4.4 Geometric Distribution

4.20 A fair die is rolled repeatedly until a 6 appears.

a. What is the probability that the experiment stops at the fourth roll?
b. Given that the experiment stops at the third roll, what is the probability the sum of all the three rolls is at least 12?

4.21 A certain door can be opened by exactly one of six keys. If you try the keys one after another, what is the expected number of keys you will have to try before the door is opened?

4.22 A box contains $R$ red balls and $B$ blue balls. An experiment is conducted with the balls as follows. A ball is randomly selected from the box, its color is noted and the ball is put back into the box. The process is repeated until a blue ball is selected.

a. What is the probability that the experiment stops after exactly $n$ trials?
b. What is the probability that the experiment requires at least $k$ trials before it stops?

4.23 Twenty percent of the population of a particular city wear glasses. If you randomly stop people from that city, determine the following probabilities:

a. It takes exactly 10 tries to get a person who wears glasses
b. It takes at least 10 tries to get a person who wears glasses.

4.24 A student is planning to take the scholastic aptitude test (SAT) exam to gain admission to a top college. She hopes to keep taking the exam until she gets a score of at least 1350 and then she will stop. Her score in any of the exams is uniformly distributed between 1100 and 1500, and her score in one exam is independent of her score in any other exam.

a. What is the probability that she reaches her goal of scoring at least 1350 points in any exam?
b. What is the PMF of the number of times she will take the exam before reaching her goal?
c. What is the expected number of times she will take the exam?

## Section 4.5 Pascal Distribution

4.25 Sam is fond of taking long-distance trips. During each trip his car has a tire failure in each 100-mile stretch with a probability of 0.05. He recently embarked on an 800-mile trip and took two spare tires with him on the trip.

a. What is the probability that the first change of tire occurred 300 miles from his starting point?
b. What is the probability that his second change of tire occurred 500 miles from his starting point?
c. What is the probability that he completed the trip without having to change tires?

4.26 Six applicants for a job were all found to be qualified by the company. The company then ranked these applicants in a priority order. Three positions are open. Past experience has shown that 20% of applicants who are offered this kind of position by the company do not accept the offer. What is the probability that the sixth-ranked applicant will be offered one of the positions?

4.27 Twenty percent of the population of a particular city wear glasses. If you randomly stop people from that city, determine the following probabilities:

a. It takes exactly 10 tries to get the third person who wears glasses
b. It takes at least 10 tries to get the third person who wears glasses.

4.28 The probability of getting a head in a single toss of a biased coin is $q$. In an experiment that consists of repeated tosses of the coin, what is the probability that the 18th head occurs on the 30th toss?

4.29 Pete makes house calls to give away free books to families with children. He gives away books to those families that open the door for him when he rings their doorbell and have children living at home. He gives exactly one book to a qualified family. Studies indicate that the probability that the door is opened when Pete rings the doorbell is 0.75, and the probability that a family has children living at home is 0.5. If the events "door opened" and "family has children" are independent, determine the following:

a. The probability that Pete gives away his first book at the third house he visits

b. The probability that he gives away his second book to the fifth family he visits
c. The conditional probability that he gives away the fifth book to the eleventh family he visits, given that he has given away exactly four books to the first eight families he visited.
d. Given that he did not give away the second book at the second house, what is the probability that he will give it out at the fifth house?

4.30 The Carter family owns a bookstore and their son, who just completed an introductory course in probability at a local college, has determined that the probability that anyone who comes to the store actually buys a book is 0.3. If the family gives a coupon to the local ice cream place to every customer who buys a book from the store, what is the probability that on a particular day the third coupon was given to the eighth customer?

4.31 A telemarketer is paid $1 for each sale she makes. The probability that any call she makes results in a sale is 0.6.
a. What is the probability that she earned her third dollar on the sixth call she made?
b. If she made 6 calls per hour, what is the probability that she earned $8 in two hours?

## Section 4.6 Hypergeometric Distribution

4.32 A list contains the names of 4 girls and 6 boys. If 5 students are randomly selected from the list, what is the probability that those selected will consist of 2 girls and 3 boys?

4.33 There are 50 states in the United States, and each state is represented by 2 senators in the U.S. Senate. A group of 20 U.S. senators is chosen randomly from the 100 senators in the U.S. Senate to visit a troubled part of the world.
a. What is the probability that the two Massachusetts senators are among those chosen?
b. What is the probability that neither of the two Massachusetts senators is among those selected?

4.34 A professor provides 12 review problems for an exam and tells the students that the actual exam will consist of 6 problems chosen randomly from the 12 review problems. Alex decides to memorize the solutions to 8 of the 12 problems. If Alex cannot solve any of the other 4 problems that he did not memorize, what is the probability that he is able to solve 4 or more problems correctly in the exam?

4.35 A class has 18 boys and 12 girls. If the teacher randomly selects a group of 15 students to represent the class in a competition, determine the following:
a. The probability that 8 members of the group are girls
b. The expected number of boys in the group.

4.36 A drawer contains 10 left gloves and 12 right gloves. If you randomly pull out a set of 4 gloves, what is the probability that the set consists of 2 right gloves and 2 left gloves?

## Section 4.7 Poisson Distribution

4.37 The number of cars that arrive at a gas station is a Poisson random variable with a mean of 50 cars per hour. The station has only one attendant, and each car requires exactly one minute to fill up. If we define a waiting line as the condition in which two or more cars are found at the same time at the gas station, what is the probability that a waiting line will occur at the station?

4.38 The number of traffic tickets that a certain traffic officer gives out on any day has been shown to have a Poisson distribution with a mean of 7.

a. What is the probability that on one particular day the officer gave out no ticket?

b. What is the probability that she gives out fewer than 4 tickets in one day?

4.39 A Geiger counter counts the particles emitted by radioactive material. If the number of particles emitted per second by a particular radioactive material has a Poisson distribution with a mean of 10 particles, determine the following:

a. The probability of at most 3 particles in one second

b. The probability of more than 1 particle in one second.

4.40 The number of cars that arrive at a drive-in window of a certain bank over a 20-minute period is a Poisson random variable with a mean of four cars. What is the probability that more than three cars will arrive during any 20-minute period?

4.41 The number of phone calls that arrive at a secretary's desk has a Poisson distribution with a mean of 4 per hour.

a. What is the probability that no phone calls arrive in a given hour?

b. What is the probability that more than 2 calls arrive within a given hour?

4.42 The number of typing mistakes that Ann makes on a given page has a Poisson distribution with a mean of 3 mistakes.

a. What is the probability that she makes exactly 7 mistakes on a given page?

b. What is the probability that she makes fewer than 4 mistakes on a given page?

c. What is the probability that Ann makes no mistake on a given page?

## Section 4.8 Exponential Distribution

4.43 The PDF of a certain random variable $T$ is given by

$$f_T(t) = ke^{-4t} \quad t \geq 0$$

a. What is the value of $k$?
b. What is the expected value of $T$?
c. Find $P[T<1]$.

4.44 The lifetime $X$ of a system in weeks is given by the following PDF:

$$f_X(x) = 0.25e^{-0.25x} \qquad x \geq 0$$

a. What is the expected value of $X$?
b. What is the CDF of $X$?
c. What is the variance of $X$?
d. What is the probability that the system will not fail within two weeks?
e. Given that the system has not failed by the end of the fourth week, what is the probability that it will fail between the fourth and sixth weeks?

4.45 The time $T$ in hours between bus arrivals at a bus station in downtown Lowell is a random variable with the following PDF:

$$f_T(t) = 2e^{-2t} \qquad t \geq 0$$

a. What is the expected value of $T$?
b. What is the variance of $T$?
c. What is $P[T>1]$?

4.46 The PDF of the times between successive bus arrivals at a suburban bus stop is given by

$$f_T(t) = 0.1e^{-0.1t} \qquad t \geq 0$$

where $T$ is in minutes. A turtle that requires exactly 15 minutes to cross the street starts crossing the street at the bus station immediately after a bus has left the station. What is the probability that the turtle will not be on the road when the next bus arrives?

4.47 The PDF of the times between successive bus arrivals at a suburban bus stop is given by

$$f_T(t) = 0.2e^{-0.2t} \qquad t \geq 0$$

where $T$ is in minutes. An ant that requires exactly 10 minutes to cross the street starts crossing the street at the bus station immediately after a bus has left the station. Given that no bus has arrived in the past 8 minutes since the ant started its journey across the street, determine the following:

a. The probability that the ant will completely cross the road before the next bus arrives.
b. The expected time until the next bus arrives.

4.48 The times between telephone calls that arrive at a switchboard are exponentially distributed with a mean of 30 minutes. Given that a call has just arrived, what is the probability that it takes at least 2 hours before the next call arrives?

4.49 The durations of calls to a radio talk show are known to be exponentially distributed with a mean of 3 minutes.

a. What is the probability that a call will last less than 2 minutes?
b. What is the probability that a call will last longer than 4 minutes?
c. Given that a call has already lasted 4 minutes, what is the probability that it will last at least another 4 minutes?
d. Given that a call has already lasted 4 minutes, what is the expected remaining time until it ends?

4.50 The life of a particular brand of batteries is exponentially distributed with a mean of 4 weeks. You just replaced the battery in your gadget with the particular brand.

a. What is the probability that the battery life exceeds 2 weeks?
b. Given that the battery has lasted 6 weeks, what is the probability that it will last at least another 5 weeks?

4.51 The PDF of the times $T$ in weeks between employee strikes at a certain company is given by

$$f_T(t) = 0.02e^{-0.02t} \qquad t \geq 0$$

a. What is the expected time between strikes at the company?
b. Find $P[T \leq t | T < 40]$ for all $t$.
c. Find $P[40 < T < 60]$.

4.52 Find the PDF of the random variable $X$ whose hazard function is given by $h_X(x) = 0.05$.

## Section 4.9 Erlang Distribution

4.53 A communication channel fades in a random manner. The duration $X$ of each fade is exponentially distributed with a mean of $1/\lambda$. The duration $T$ of the interval between the end of one fade and the beginning of the next fade is an Erlang random variable with PDF

$$f_T(t) = \frac{\mu^4 t^3 e^{-\mu t}}{3!} \qquad t \geq 0$$

If we observe the channel at a randomly selected instant, what is the probability that it is in the fade state?

4.54 The random variable $X$, which denotes the interval between two consecutive events, has the PDF

$$f_X(x) = 4x^2 e^{-2x} \qquad x \geq 0$$

If we assume that intervals between events are independent, determine the following:

a. The expected value of $X$.
b. The expected value of the interval between the 11th and 13th events
c. The probability that $X \leq 6$.

4.55 The students in the electrical and computer engineering department of a certain college arrive at the departmental lounge to view the video of the lecture for a particular course according to a Poisson process with a rate of 5 students per hour. The person in charge of the operating the VCR will not turn the machine on until there are at least 5 students in the lounge.

a. Given that there is currently no student in the lounge, what is the expected waiting time until the VCR is turned on?
b. Given that there is currently no student in the lounge, what is the probability that the VCR is not turned on within one hour from now?

## Section 4.10 Uniform Distribution

4.56 Jack is the only employee of an auto repair shop that specializes in installing mufflers. The time $T$ minutes that it takes Jack to install a muffler has the PDF

$$f_T(t) = \begin{cases} 0.05 & 10 \leq t \leq 30 \\ 0 & \text{otherwise} \end{cases}$$

a. What is the expected time it takes Jack to install a muffler?
b. What is the variance of the time it takes Jack to install a muffler?

4.57 A random variable $X$ is uniformly distributed between 0 and 10. Find the probability that $X$ lies between the standard deviation $\sigma_X$ and the mean $E[X]$.

4.58 A random variable $X$ is uniformly distributed between 3 and 15. Find the following parameters:

a. The expected value of $X$
b. The variance of $X$
c. The probability that $X$ lies between 5 and 10
d. The probability that $X$ is less than 6.

4.59 Starting at 7 am, buses arrive at a particular bus stop in a college campus at intervals of 15 minutes (that is, at 7 am, 7:15, 7:30, etc.). Joe is a frequent passenger on this route, and the time he arrives each morning to catch a bus is known to be uniformly distributed between 7 am and 7:30 am.

a. What is the probability that Joe waits less than 5 minutes for a bus?
b. What is the probability that he waits more than 10 minutes for a bus?

4.60 The time it takes a bank teller to serve a customer is uniformly distributed between 2 and 6 minutes. A customer has just stepped up to the window, and you are next in line.

a. What is the expected time you will wait before it is your turn to be served?
b. What is the probability that you wait less than 1 minute before being served?
c. What is the probability that you wait between 3 and 5 minutes before being served?

## Section 4.11 Normal Distribution

4.61 The mean weight of 200 students in a certain college is 140 lbs, and the standard deviation is 10 lbs. If we assume that the weights are normally distributed, evaluate the following:
a. The expected number of students that weigh between 110 and 145 lbs
b. The expected number of students that weigh less than 120 lbs
c. The expected number of students that weigh more than 170 lbs.

4.62 The weights of parcels that are dropped off at a local shipping center can be represented by a random variable $X$ that is normally distributed with mean $\mu_X = 70$ and standard deviation $\sigma_X = 10$. Determine the following:
a. $P[X > 50]$
b. $P[60 < X]$
c. $P[60 < X < 90]$.

4.63 Find the probability of getting between 4 and 8 heads in 12 tosses of a fair coin by using
a. the binomial distribution
b. the normal approximation to the binomial distribution

4.64 The test scores $X$ of a certain subject were found to be approximately normally distributed with mean $\mu_X$ and standard deviation $\sigma_X$. So the professor decided to assign grades according to Table 4.3. What fraction of the class gets A, B, C, D, and F?

4.65 The random variable $X$ is a normal random variable with zero mean and standard deviation $\sigma_X$. Compute the probability $P[|X| \leq 2\sigma_X]$.

4.66 The annual rainfall in inches in a certain region has a normal distribution with a mean of 40 and variance of 16. What is the probability that the rainfall in a given year is between 30 and 48 inches?

**Table 4.3** Table for Grading Score

| Test Score | Letter Grade |
|---|---|
| $X < \mu_X - 2\sigma_X$ | F |
| $\mu_X - 2\sigma_X \leq X < \mu_X - \sigma_X$ | D |
| $\mu_X - \sigma_X \leq X < \mu_X$ | C |
| $\mu_X \leq X < \mu_X + \sigma_X$ | B |
| $X \geq \mu_X + \sigma_X$ | A |

CHAPTER 5

# Multiple Random Variables

## 5.1 INTRODUCTION

We have so far been concerned with the properties of a single random variable defined on a given sample space. Sometimes we encounter problems that deal with two or more random variables defined on the same sample space. In this chapter we consider these multivariate systems. We first consider bivariate random variables and later consider systems with more than two random variables.

## 5.2 JOINT CDFS OF BIVARIATE RANDOM VARIABLES

Consider two random variables $X$ and $Y$ defined on the same sample space. For example, $X$ can denote the grade of a student and $Y$ can denote the height of the same student. The joint cumulative distribution function (joint CDF) of $X$ and $Y$ is given by

$$F_{XY}(x,y) = P[\{X \leq x\} \cap \{Y \leq y\}] = P[X \leq x, Y \leq y] \tag{5.1}$$

The pair $(X,Y)$ is referred to as a *bivariate* random variable. If we define $F_X(x) = P[X \leq x]$ as the *marginal* CDF of $X$ and $F_Y(y) = P[Y \leq y]$ as the *marginal* CDF of $Y$, then we define the random variables $X$ and $Y$ to be independent if

$$F_{XY}(x,y) = F_X(x)F_Y(y)$$

for every value of $x$ and $y$. We consider only bivariate random variables in which both $X$ and $Y$ are discrete random variables or both are continuous random variables. We will not consider the case where one random variable is discrete and the other is continuous.

### 5.2.1 Properties of the Joint CDF

As a probability function $F_{XY}(x,y)$ has certain properties, which include the following:

a. Since $F_{XY}(x,y)$ is a probability measure, $0 \leq F_{XY}(x,y) \leq 1$ for $-\infty < x < \infty$ and $-\infty < y < \infty$.



Fundamentals of Applied Probability and Random Processes. http://dx.doi.org/10.1016/B978-0-12-800852-2.00005-5

b. If $x_1 \leq x_2$ and $y_1 \leq y_2$, then $F_{XY}(x_1, y_1) \leq F_{XY}(x_2, y_1) \leq F_{XY}(x_2, y_2)$. Similarly, $F_{XY}(x_1, y_1) \leq F_{XY}(x_1, y_2) \leq F_{XY}(x_2, y_2)$. This follows from the fact that $F_{XY}(x, y)$ is a non-decreasing function of $x$ and $y$.
c. $\lim_{\substack{x \to \infty \\ y \to \infty}} F_{XY}(x, y) = F_{XY}(\infty, \infty) = 1$
d. $\lim_{x \to -\infty} F_{XY}(x, y) = F_{XY}(-\infty, y) = 0$
e. $\lim_{y \to -\infty} F_{XY}(x, y) = F_{XY}(x, -\infty) = 0$
f. $P[x_1 < X \leq x_2, Y \leq y] = F_{XY}(x_2, y) - F_{XY}(x_1, y)$
g. $P[X \leq x, y_1 < Y \leq y_2] = F_{XY}(x, y_2) - F_{XY}(x, y_1)$
h. If $x_1 \leq x_2$ and $y_1 \leq y_2$, then

$$P[x_1 < X \leq x_2, y_1 < Y \leq y_2] = F_{XY}(x_2, y_2) - F_{XY}(x_1, y_2) - F_{XY}(x_2, y_1) + F_{XY}(x_1, y_1) \geq 0$$

Also, the marginal CDFs are obtained as follows:

$$F_X(x) = F_{XY}(x, \infty) \tag{5.2a}$$

$$F_Y(y) = F_{XY}(\infty, y) \tag{5.2b}$$

From the above properties we can answer questions about $X$ and $Y$.

---

## EXAMPLE 5.1

Given two random variables $X$ and $Y$ with the joint CDF $F_{XY}(x, y)$ and marginal CDFs $F_X(x)$ and $F_Y(y)$, respectively, compute the joint probability that $X$ is greater than $a$ and $Y$ is greater than $b$.

Solution:
We can obtain the desired probability as follows. From the De Morgan's second law, we know that $\overline{A \cap B} = \overline{A} \cup \overline{B}$. Thus,

$$\begin{aligned} P[X > a, Y > b] &= P[\{X > a\} \cap \{Y > b\}] = 1 - P\left[\overline{\{X > a\} \cap \{Y > b\}}\right] \\ &= 1 - P\left[\overline{\{X > a\}} \cup \overline{\{Y > b\}}\right] = 1 - P[\{X \leq a\} \cup \{Y \leq b\}] \\ &= 1 - \{P[X \leq a] + P[Y \leq b] - P[X \leq a, Y \leq b]\} \\ &= 1 - F_X(a) - F_Y(b) + F_{XY}(a, b) \end{aligned}$$

---

## 5.3 DISCRETE BIVARIATE RANDOM VARIABLES

When both $X$ and $Y$ are discrete random variables, we define their joint PMF as follows:

$$p_{XY}(x, y) = P[X = x, Y = y] \tag{5.3}$$

The properties of the joint PMF include the following:

a. As a probability measure, the PMF can neither be negative nor exceed unity, which means that $0 \leq p_{XY}(x, y) \leq 1$.

b. $\sum_x \sum_y p_{XY}(x, y) = 1$

c. $\sum_{x \le a} \sum_{y \le b} p_{XY}(x, y) = F_{XY}(a, b)$

The marginal PMFs are obtained as follows:

$$p_X(x) = \sum_y p_{XY}(x, y) = P[X = x] \tag{5.4a}$$

$$p_Y(y) = \sum_x p_{XY}(x, y) = P[Y = y] \tag{5.4b}$$

If $X$ and $Y$ are independent random variables, then

$$p_{XY}(x, y) = p_X(x)p_Y(y)$$

## EXAMPLE 5.2

The joint PMF of two random variables $X$ and $Y$ is given by

$$p_{XY}(x, y) = \begin{cases} k(2x+y) & x = 1,2;\ y = 1, 2, 3 \\ 0 & \text{otherwise} \end{cases}$$

where $k$ is a constant.

a. What is the value of $k$?
b. Find the marginal PMFs of $X$ and $Y$.
c. Are $X$ and $Y$ independent?

**Solution:**

a. To evaluate $k$, we remember that

$$\sum_x \sum_y p_{XY}(x, y) = 1 = \sum_{x=1}^{2} \sum_{y=1}^{3} k(2x+y)$$

Thus,

$$\sum_{x=1}^{2} \sum_{y=1}^{3} k(2x+y) = k\sum_{x=1}^{2}\{(2x+1)+(2x+2)+(2x+3)\}$$

$$= k\sum_{x=1}^{2}\{6x+6\} = k\{(6+6)+(12+6)\} = 30k = 1$$

This gives $k = 1/30$.

b. The marginal PMFs are

$$p_X(x) = \sum_y p_{XY}(x, y) = \frac{1}{30}\sum_{y=1}^{3}(2x+y) = \frac{1}{30}\{(2x+1)+(2x+2)+(2x+3)\} = \frac{1}{30}(6x+6)$$

$$= \frac{1}{5}(x+1) \quad x = 1,2$$

$$p_Y(y) = \sum_x p_{XY}(x, y) = \frac{1}{30}\sum_{x=1}^{2}(2x+y) = \frac{1}{30}\{(2+y)+(4+y)\} = \frac{1}{30}(2y+6)$$

$$= \frac{1}{15}(y+3) \quad y = 1,2,3$$

c. Since $p_X(x)p_Y(y) = \frac{1}{75}(x+1)(y+3) \neq p_{XY}(x, y)$, we conclude that $X$ and $Y$ are not independent.

## EXAMPLE 5.3

The number of emergency calls $X$ to a police station of a certain town has a Poisson distribution with mean $\lambda$. The probability that any one of these calls is about robbery is $p$. What is the PMF of $Y$, the number of calls about robbery?

Solution:

We assume that the reasons for calls to the police station are independent. When $X=x$, $Y$ is a binomial random variable with parameters $(x, p)$. Thus, the joint PMF of $X$ and $Y$ is given by

$$\begin{aligned} p_{XY}(x,y) &= P[X=x, Y=y] = P[Y=y|X=x]P[X=x] \\ &= \binom{x}{y} p^y (1-p)^{x-y} \left\{ \frac{\lambda^x}{x!} e^{-\lambda} \right\} = \frac{e^{-\lambda}(\lambda p)^y [\lambda(1-p)]^{x-y}}{y!(x-y)!} \end{aligned}$$

Therefore,

$$p_Y(y) = \sum_x p_{XY}(x,y) = \frac{e^{-\lambda}(\lambda p)^y}{y!} \sum_{x \geq y} \frac{[\lambda(1-p)]^{x-y}}{(x-y)!}$$

Let $x-y=k$. Then we obtain

$$\begin{aligned} p_Y(y) &= \frac{e^{-\lambda}(\lambda p)^y}{y!} \sum_{k=0}^{\infty} \frac{[\lambda(1-p)]^k}{k!} = \frac{e^{-\lambda}(\lambda p)^y}{y!} e^{\lambda(1-p)} \\ &= \frac{(\lambda p)^y}{y!} e^{-\lambda p} \qquad y=0,1,\ldots \end{aligned}$$

This indicates that $Y$ has a Poisson distribution with mean $\lambda p$.

## EXAMPLE 5.4

A fair coin is tossed three times. Let $X$ be a random variable that takes the value 0 if the first toss is a tail and the value 1 if the first toss is a head. Also, let $Y$ be a random variable that defines the total number of heads in the three tosses.

a. Determine the joint PMF of $X$ and $Y$.
b. Are $X$ and $Y$ independent?

Solution:

Let H denote the event that a head appears on a toss and T the event that a tail appears on a toss. Table 5.1 shows the sample space and the values of the two random variables.

**Table 5.1** Sample Space and Values of Random Variables

| Sample Space | Value of $X$ | Value of $Y$ |
|---|---|---|
| HHH | 1 | 3 |
| HHT | 1 | 2 |
| HTH | 1 | 2 |
| HTT | 1 | 1 |
| THH | 0 | 2 |
| THT | 0 | 1 |
| TTH | 0 | 1 |
| TTT | 0 | 0 |

a. Each of these eight sample points is equally likely to be obtained. Thus, since $X$ takes values 0 and 1, and $Y$ takes values 0, 1, 2, and 3, the joint PMF of $X$ and $Y$ can then be constructed as follows:

$$p_{XY}(0,0) = P[X=0, Y=0] = P[TTT] = \frac{1}{8}$$
$$p_{XY}(0,1) = P[X=0, Y=1] = P[\{THT\} \cup \{TTH\}] = P[THT] + P[TTH] = \frac{1}{4}$$
$$p_{XY}(0,2) = P[X=0, Y=2] = P[TTH] = \frac{1}{8}$$
$$p_{XY}(0,3) = P[X=0, Y=3] = 0$$
$$p_{XY}(1,0) = P[X=1, Y=0] = 0$$
$$p_{XY}(1,1) = P[X=1, Y=1] = P[HTT] = \frac{1}{8}$$
$$p_{XY}(1,2) = P[X=1, Y=2] = P[\{HHT\} \cup \{HTH\}] = P[HHT] + P[HTH] = \frac{1}{4}$$
$$p_{XY}(1,3) = P[X=1, Y=3] = P[HHH] = \frac{1}{8}$$

b. If $X$ and $Y$ are independent, then for all $x$ and $y$ we have that $p_{XY}(x,y) = p_X(x)p_Y(y)$. Thus, to show that $X$ and $Y$ are not independent, all we have to do is to find a pair of $x$ and $y$ at which the joint PMF does not satisfy the above condition. Consider the point $(x,y) = (0,0)$.

$$p_X(0) = \sum_y p_{XY}(0,y) = p_{XY}(0,0) + p_{XY}(0,1) + p_{XY}(0,2) + p_{XY}(0,3) = \frac{1}{2}$$
$$p_Y(0) = \sum_x p_{XY}(x,0) = p_{XY}(0,0) + p_{XY}(1,0) = \frac{1}{8}$$

Since $p_X(0)p_Y(0) = \left(\frac{1}{2}\right) \times \left(\frac{1}{8}\right) = \frac{1}{16} \neq p_{XY}(0,0) = \frac{1}{8}$, we conclude that $X$ and $Y$ are not independent. A more extensive test will involve obtaining the marginal PMFs of $X$ and $Y$ and testing each pair of $x$-$y$ values. However, for this example, the above proof is sufficient.

## 5.4 CONTINUOUS BIVARIATE RANDOM VARIABLES

If both $X$ and $Y$ are continuous random variables, their joint PDF is given by

$$f_{XY}(x,y) = \frac{\partial^2}{\partial x \partial y} F_{XY}(x,y) \tag{5.5}$$

The joint PDF satisfies the following condition:

$$F_{XY}(x,y) = \int_{-\infty}^{x} \int_{-\infty}^{y} f_{XY}(u,v)\,dv\,du \tag{5.6}$$

The joint PDF also has the following properties:

a. For all $x$ and $y$, $f_{XY}(x,y) \geq 0$

b. $\int_{-\infty}^{\infty} \int_{-\infty}^{\infty} f_{XY}(x,y)\,dy\,dx = 1$

c. $f_{XY}(x,y)$ is continuous for all except possibly finitely many values of $x$ or of $y$

d. $P[x_1 < X \le x_2, y_1 < Y \le y_2] = \int_{x_1}^{x_2}\int_{y_1}^{y_2} f_{XY}(x,y)dydx$

The marginal PDFs are given by

$$f_X(x) = \int_{-\infty}^{\infty} f_{XY}(x,y)dy \tag{5.7a}$$

$$f_Y(y) = \int_{-\infty}^{\infty} f_{XY}(x,y)dx \tag{5.7b}$$

If $X$ and $Y$ are independent random variables, then

$$f_{XY}(x,y) = f_X(x)f_Y(y)$$

## EXAMPLE 5.5

$X$ and $Y$ are two continuous random variables whose joint PDF is given by

$$f_{XY}(x,y) = \begin{cases} e^{-(x+y)} & 0 \le x < \infty;\ 0 \le y < \infty \\ 0 & \text{otherwise} \end{cases}$$

Are $X$ and $Y$ independent?

Solution:

To answer the question, we first evaluate the marginal PDFs of $X$ and $Y$:

$$f_X(x) = \int_{-\infty}^{\infty} f_{XY}(x,y)dy = \int_0^{\infty} e^{-(x+y)}dy = e^{-x}\int_0^{\infty} e^{-y}dy = e^{-x} \quad x \ge 0$$
$$f_Y(y) = \int_{-\infty}^{\infty} f_{XY}(x,y)dx = \int_0^{\infty} e^{-(x+y)}dx = e^{-y}\int_0^{\infty} e^{-x}dx = e^{-y} \quad y \ge 0$$

Now, $f_X(x)f_Y(y) = e^{-x}e^{-y} = e^{-(x+y)} = f_{XY}(x,y)$, which means that $X$ and $Y$ are independent.

## EXAMPLE 5.6

Determine if random variables $X$ and $Y$ are independent when their joint PDF is given by

$$f_{XY}(x,y) = \begin{cases} 2e^{-(x+y)} & 0 \le x \le y;\ 0 \le y < \infty \\ 0 & \text{otherwise} \end{cases}$$

Solution:

We evaluate the marginal PDFs of $X$ and $Y$ by noting the relationship between $X$ and $Y$, thereby defining the region of interest as the shaded area in Figure 5.1. Note that the shaded area extends to $\infty$.

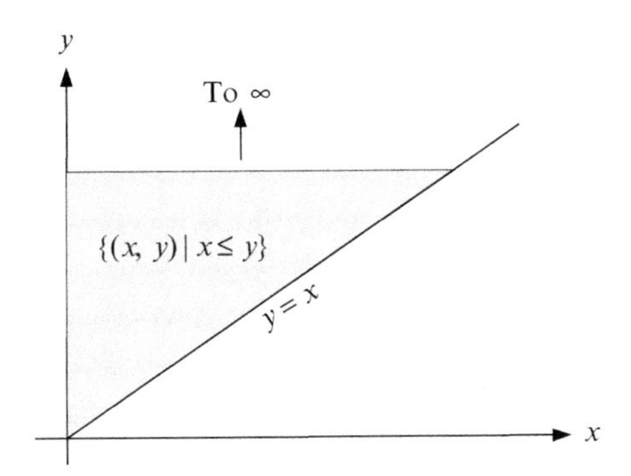


**FIGURE 5.1**
Relationship between $X$ and $Y$

Then we obtain

$$f_X(x) = \int_{-\infty}^{\infty} f_{XY}(x,y)dy = 2\int_x^{\infty} e^{-(x+y)}dy = 2e^{-x}\int_x^{\infty} e^{-y}dy = 2e^{-x}[-e^{-y}]_x^{\infty}$$
$$= 2e^{-2x}$$
$$f_Y(y) = \int_{-\infty}^{\infty} f_{XY}(x,y)dx = 2\int_0^{y} e^{-(x+y)}dx = 2e^{-y}\int_0^{y} e^{-x}dx = 2e^{-y}[-e^{-x}]_0^{y}$$
$$= 2e^{-y}\{1 - e^{-y}\}$$

Since $f_X(x)f_Y(y) = 4\{e^{-(2x+y)} - e^{-2(x+y)}\} \neq f_{XY}(x,y)$, $X$ and $Y$ are not independent.

## 5.5 DETERMINING PROBABILITIES FROM A JOINT CDF

Suppose that $X$ and $Y$ are given random variables and we are required to determine the probability of a certain event defined in terms of $X$ and $Y$ for which the joint CDF is known. We start by sketching the event in the $x$-$y$ plane. For example, assume we are required to find $P[a < X \leq b, c < Y \leq d]$. The region of interest is shown in Figure 5.2, which defines four partitions.

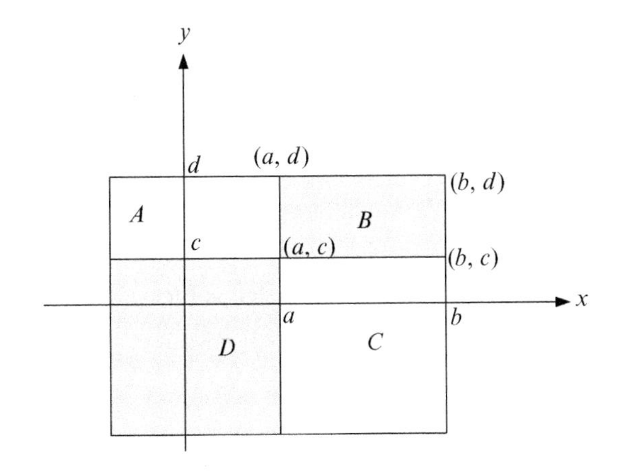


**FIGURE 5.2**
Domain Partitions

Consider the following events:

$$E_1 = \{(X \le b) \cap (Y \le d)\}$$
$$E_2 = \{(X \le b) \cap (Y \le c)\}$$
$$E_3 = \{(X \le a) \cap (Y \le d)\}$$
$$E_4 = \{(X \le a) \cap (Y \le c)\}$$
$$E_5 = \{(a < X \le b) \cap (c < Y \le d)\}$$

The region of interest is B, which corresponds to event $E_5$ that can be obtained as follows:

$$E_5 = E_1 - E_2 - E_3 + E_4$$

Thus,

$$P[a < X \le b, c < Y \le d] = F_{XY}(b, d) - F_{XY}(b, c) - F_{XY}(a, d) + F_{XY}(a, c)$$

Note that the probability is simply the joint CDF evaluated at the point where $X$ and $Y$ jointly have the larger of their two values plus the CDF evaluated at the point where they jointly have their smaller values minus the CDF evaluated at the two points where they have mixed smaller and larger values. When $X$ and $Y$ are independent random variables, the above result becomes

$$\begin{aligned} P[a < X \le b, c < Y \le d] &= F_{XY}(b, d) - F_{XY}(b, c) - F_{XY}(a, d) + F_{XY}(a, c) \\ &= F_X(b)F_Y(d) - F_X(b)F_Y(c) - F_X(a)F_Y(d) + F_X(a)F_Y(c) \\ &= F_X(b)\{F_Y(d) - F_Y(c)\} - F_X(a)\{F_Y(d) - F_Y(c)\} \\ &= \{F_X(b) - F_X(a)\}\{F_Y(d) - F_Y(c)\} \\ &= P[a < X \le b]P[c < Y \le d] \end{aligned}$$

Finally, for the case when $X$ and $Y$ are discrete random variables, the joint CDF can be obtained from the joint PMF as follows:

$$F_{XY}(x, y) = \sum_{m \le x} \sum_{n \le y} p_{XY}(m, n)$$

For example, if $X$ and $Y$ take only nonnegative values,

$$\begin{aligned} F_{XY}(1, 2) &= \sum_{m \le 1} \sum_{n \le 2} p_{XY}(m, n) \\ &= p_{XY}(0, 0) + p_{XY}(0, 1) + p_{XY}(0, 2) + p_{XY}(1, 0) + p_{XY}(1, 1) + p_{XY}(1, 2) \end{aligned}$$

## EXAMPLE 5.7

The joint CDF of two discrete random variables $X$ and $Y$ is given as follows:

$$F_{XY}(x,y)=\begin{cases}\frac{1}{8} & x=1,\ y=1\\ \frac{5}{8} & x=1,\ y=2\\ \frac{1}{4} & x=2,\ y=1\\ 1 & x=2,\ y=2\end{cases}$$

Determine the following:

a. Joint PMF of $X$ and $Y$
b. Marginal PMF of $X$
c. Marginal PMF of $Y$

**Solution:**

The joint PMF is obtained from the relationship

$$F_{XY}(x,y)=\sum_{m\le x}\sum_{n\le y} p_{XY}(m,n)$$

Thus,

$$\begin{aligned}
&F_{XY}(1,1)=p_{XY}(1,1)=1/8\\
&F_{XY}(1,2)=p_{XY}(1,1)+p_{XY}(1,2)=5/8\Rightarrow p_{XY}(1,2)=5/8-1/8=1/2\\
&F_{XY}(2,1)=p_{XY}(1,1)+p_{XY}(2,1)=1/4\Rightarrow p_{XY}(2,1)=1/4-1/8=1/8\\
&F_{XY}(2,2)=p_{XY}(1,1)+p_{XY}(1,2)+p_{XY}(2,1)+p_{XY}(2,2)=1\Rightarrow p_{XY}(2,2)=1/4
\end{aligned}$$

The joint PMF becomes

$$p_{XY}(x,y)=\begin{cases}\frac{1}{8} & x=1,\ y=1\\ \frac{1}{2} & x=1,\ y=2\\ \frac{1}{8} & x=2,\ y=1\\ \frac{1}{4} & x=2,\ y=2\end{cases}$$

The marginal PMF of $X$ is given by

$$p_X(x)=\begin{cases}p_{XY}(1,1)+p_{XY}(1,2)=\frac{5}{8} & x=1\\ p_{XY}(2,1)+p_{XY}(2,2)=\frac{3}{8} & x=2\end{cases}$$

The marginal PMF of $Y$ is given by

$$p_Y(y)=\begin{cases}p_{XY}(1,1)+p_{XY}(2,1)=\frac{1}{4} & y=1\\ p_{XY}(1,2)+p_{XY}(2,2)=\frac{3}{4} & y=2\end{cases}$$

## 5.6 CONDITIONAL DISTRIBUTIONS

Recall that for two events $A$ and $B$, the conditional probability of event $A$ given event $B$ is defined by

$$P[A|B] = \frac{P[A \cap B]}{P[B]}$$

which is defined when $P[B] > 0$. In this section we extend the same concept to two random variables $X$ and $Y$ governed by a joint CDF $F_{XY}(x, y)$.

### 5.6.1 Conditional PMF for Discrete Bivariate Random Variables

Consider two discrete random variables $X$ and $Y$ with the joint PMF $p_{XY}(x, y)$. The conditional PMF of $Y$, given $X=x$, is given by

$$p_{Y|X}(y|x) = \frac{P[X=x, Y=y]}{P[X=x]} = \frac{p_{XY}(x, y)}{p_X(x)} \tag{5.8}$$

provided $p_X(x) > 0$. Similarly, the conditional PMF of $X$, given $Y=y$, is given by

$$p_{X|Y}(x|y) = \frac{P[X=x, Y=y]}{P[Y=y]} = \frac{p_{XY}(x, y)}{p_Y(y)} \tag{5.9}$$

provided $p_Y(y) > 0$. If $X$ and $Y$ are independent random variables, $p_{X|Y}(x|y) = p_X(x)$ and $p_{Y|X}(y|x) = p_Y(y)$. Also, the conditional CDFs are defined by

$$F_{X|Y}(x|y) = P[X \leq x|Y=y] = \sum_{u \leq x} p_{X|Y}(u|y)$$

$$F_{Y|X}(y|x) = P[Y \leq y|X=x] = \sum_{v \leq y} p_{Y|X}(v|x)$$

---

### EXAMPLE 5.8

The joint PMF of two random variables $X$ and $Y$ is given by

$$p_{XY}(x, y) = \begin{cases} \frac{1}{30}(2x+y) & x=1,2; y=1,2,3 \\ 0 & \text{otherwise} \end{cases}$$

a. What is the conditional PMF of $Y$ given $X$?
b. What is the conditional PMF of $X$ given $Y$?

Solution:
From Example 5.2 we know that the marginal PMFs are given by

$$p_X(x) = \sum_y p_{XY}(x, y) = \frac{1}{30}\sum_{y=1}^{3}(2x+y) = \frac{1}{5}(x+1) \quad x=1, 2$$

$$p_Y(y) = \sum_x p_{XY}(x, y) = \frac{1}{30}\sum_{x=1}^{2}(2x+y) = \frac{1}{15}(y+3) \quad y=1, 2, 3$$

Thus, the conditional PMFs are given by

$$p_{X|Y}(x|y) = \frac{p_{XY}(x,y)}{p_Y(y)} = \frac{2x+y}{2(y+3)} \qquad x = 1,2$$
$$p_{Y|X}(y|x) = \frac{p_{XY}(x,y)}{p_X(x)} = \frac{2x+y}{6(x+1)} \qquad y = 1,2,3$$

### 5.6.2 Conditional PDF for Continuous Bivariate Random Variables

Consider two continuous random variables $X$ and $Y$ with the joint PDF $f_{XY}(x, y)$. The conditional PDF of $Y$, given $X=x$, is defined by

$$f_{Y|X}(y|x) = \frac{f_{XY}(x,y)}{f_X(x)} \tag{5.10}$$

provided $f_X(x) > 0$. Similarly, the conditional PDF of $X$, given $Y=y$, is given by

$$f_{X|Y}(x|y) = \frac{f_{XY}(x,y)}{f_Y(y)} \tag{511}$$

provided $f_Y(y) > 0$. If $X$ and $Y$ are independent, then $f_{X|Y}(x|y) = f_X(x)$ and $f_{Y|X}(y|x) = f_Y(y)$.

## EXAMPLE 5.9

Two random variables $X$ and $Y$ have the following joint PDF:

$$f_{XY}(x,y) = \begin{cases} xe^{-x(y+1)} & 0 \le x < \infty;\ 0 \le y < \infty \\ 0 & \text{otherwise} \end{cases}$$

Determine the conditional PDF of $X$ given $Y$ and the conditional PDF of $Y$ given $X$.

**Solution:**

To determine the conditional PDFs we first evaluate the marginal PDFs, which are given by

$$f_X(x) = \int_0^\infty f_{XY}(x,y)dy = \int_0^\infty xe^{-x(y+1)}dy = xe^{-x}\int_0^\infty e^{-xy}dy = xe^{-x}\left[-\frac{e^{-xy}}{x}\right]_{y=0}^{\infty}$$
$$= e^{-x} \qquad 0 \le x < \infty$$
$$f_Y(y) = \int_0^\infty f_{XY}(x,y)dx = \int_0^\infty xe^{-x(y+1)}dx$$

Let $u=x$, which means that $du=dx$; and let $dv = e^{-x(y+1)}dx$, which means that $v = -e^{-x(y+1)}/(y+1)$. Integrating by parts we obtain

$$f_Y(y) = \int_0^\infty xe^{-x(y+1)}dx = \left[-\frac{xe^{-x(y+1)}}{y+1}\right]_{x=0}^{\infty} + \frac{1}{y+1}\int_0^\infty e^{-x(y+1)}dx = 0 - \frac{1}{(y+1)^2}\left[e^{-x(y+1)}\right]_{x=0}^{\infty}$$
$$= \frac{1}{(y+1)^2} \qquad 0 \le y < \infty$$

Thus, the conditional PDFs are given by

$$f_{X|Y}(x|y)=\frac{f_{XY}(x,y)}{f_Y(y)}=\frac{xe^{-x(y+1)}}{1/(y+1)^2}=x(y+1)^2e^{-x(y+1)}\quad 0\leq x<\infty$$

$$f_{Y|X}(y|x)=\frac{f_{XY}(x,y)}{f_X(x)}=\frac{xe^{-x(y+1)}}{e^{-x}}=xe^{-xy}\quad 0\leq y<\infty$$

### 5.6.3 Conditional Means and Variances

If $X$ and $Y$ are discrete random variables with the joint PMF $p_{XY}(x,y)$, the conditional expected value of $Y$, given that $X=x$, is defined by

$$\mu_{Y|X}=E[Y|X]=\sum_y yp_{Y|X}(y|x) \tag{5.12}$$

The conditional variance of $Y$, given $X=x$, is given by

$$\begin{aligned}\sigma^2_{Y|X}&=E\left[\left(Y-\mu_{Y|X}\right)^2|X\right]=\sum_y\left(y-\mu_{Y|X}\right)^2p_{X|Y}(y|x)\\&=E\left[Y^2|X=x\right]-(E[Y|X=x])^2\end{aligned} \tag{5.13}$$

Similarly, the conditional expected value and variance of $X$, given $Y=y$, are given by

$$\mu_{X|Y}=E[X|Y]=\sum_x xp_{X|Y}(x|y) \tag{5.14a}$$

$$\begin{aligned}\sigma^2_{X|Y}&=E\left[\left(X-\mu_{X|Y}\right)^2|Y\right]=\sum_x\left(x-\mu_{X|Y}\right)^2p_{X|Y}(x|y)\\&=E\left[X^2|Y=y\right]-(E[X|Y=y])^2\end{aligned} \tag{5.14b}$$

If $X$ and $Y$ are continuous random variables with the joint PDF $f_{XY}(x,y)$, the conditional expected value and variance of $Y$, given $X=x$, are

$$\mu_{Y|X}=E[Y|X=x]=\int_{-\infty}^{\infty}yf_{Y|X}(y|x)dy \tag{5.15a}$$

$$\begin{aligned}\sigma^2_{Y|X}&=E\left[\left(Y-\mu_{Y|X}\right)^2|X\right]=\int_{-\infty}^{\infty}\left(y-\mu_{Y|X}\right)^2f_{Y|X}(y|x)dy\\&=E\left[Y^2|X=x\right]-(E[Y|X=x])^2\end{aligned} \tag{5.15b}$$

Finally, the conditional expected value and variance of $X$, given $Y=y$, are

$$\mu_{X|Y}=E[X|Y=y]=\int_{-\infty}^{\infty}xf_{X|Y}(x|y)dx \tag{5.16a}$$

$$\begin{aligned}\sigma^2_{X|Y}&=E\left[\left(X-\mu_{X|Y}\right)^2|Y\right]=\int_{-\infty}^{\infty}\left(x-\mu_{X|Y}\right)^2f_{X|Y}(x|y)dx\\&=E\left[X^2|Y=y\right]-(E[X|Y=y])^2\end{aligned} \tag{5.16b}$$

## EXAMPLE 5.10

Compute the conditional mean $E[X|Y=y]$ if the joint PDF of $X$ and $Y$ is given by

$$f_{XY}(x,y)=\begin{cases}\dfrac{e^{-x/y}e^{-y}}{y} & 0\le x<\infty;\ 0<y<\infty \\ 0 & \text{otherwise}\end{cases}$$

Solution:

We first compute the marginal PDF $f_Y(y)$ and the conditional PDF $f_{X|Y}(x|y)$, which are given by

$$f_Y(y)=\int_0^\infty f_{XY}(x,y)dx=\int_0^\infty \frac{e^{-x/y}e^{-y}}{y}dx=\frac{e^{-y}}{y}\int_0^\infty e^{-x/y}dx=e^{-y}$$

$$f_{X|Y}(x|y)=\frac{f_{XY}(x,y)}{f_Y(y)}=\frac{e^{-x/y}e^{-y}}{ye^{-y}}=\left(\frac{1}{y}\right)e^{-x/y}$$

Thus, the conditional mean is given by

$$E[X|Y=y]=\int_0^\infty xf_{X|Y}(x|y)dx=\left(\frac{1}{y}\right)\int_0^\infty xe^{-x/y}dx$$

Let $u=x$, which means that $du=dx$; and let $dv=e^{-x/y}dx$, which means that $v=-ye^{-x/y}$. Integrating by parts we obtain

$$\begin{aligned}E[X|Y=y]&=\left(\frac{1}{y}\right)\int_0^\infty xe^{-x/y}dx=\left(\frac{1}{y}\right)\left\{\left[-xye^{-x/y}\right]_{x=0}^{\infty}+y\int_0^\infty e^{-x/y}dx\right\}\\&=y\end{aligned}$$

### 5.6.4 Simple Rule for Independence

In many cases we are given the joint PDF of the random variables $X$ and $Y$ and are required to determine if they are independent random variables. It turns out that the determination can be made based on the nature of the PDF and the combined sample space of the PDF. The following is a general rule that applies when $X$ and $Y$ are independent:

If the joint PDF of $X$ and $Y$ is of the form:

$$f_{XY}(x,y)=\text{constant}\times\{\text{function of } x\}\times\{\text{function of } y\}$$

in a rectangular region (which can be finite or infinite) $a\le x\le b,\ \ c\le y\le d$, then $X$ and $Y$ are independent. Furthermore, if the joint PDF is not in the separable form shown above or the joint sample space is not rectangular, then $X$ and $Y$ are not independent.

The function of $x$ is $f_X(x)$ and the function of $y$ is $f_Y(y)$, provided the constant term is distributed in such a way that $f_X(x)$ and $f_Y(y)$ are true PDFs.

## EXAMPLE 5.11

Assume that the random variables $X$ and $Y$ have the joint PDF

$$f_{XY}(x,y)=\frac{1}{2}x^3y \qquad 0\le x\le 2,\ 0\le y\le 1$$

a. Determine if $X$ and $Y$ are independent.
b. What are the marginal PDFs?

Solution:

a. Applying the above rule we find that the joint PDF is separable and the joint sample space is rectangular. Therefore, $X$ and $Y$ are independent.
b. Thus, we have that

$$f_{XY}(x,y)=f_X(x)\times f_Y(y)=\frac{1}{2}\times x^3\times y \qquad 0\le x\le 2,\ 0\le y\le 1$$

where

$$\begin{aligned} f_X(x)&=Ax^3 \qquad 0\le x\le 2\\ f_Y(y)&=By \qquad 0\le y\le 1\\ \frac{1}{2}&=AB \end{aligned}$$

We can find the values of $A$ and $B$ as follows:

$$\int_0^2 f_X(x)dx=1=\int_0^2 Ax^3dx=A\left[\frac{x^4}{4}\right]_0^2=4A\Rightarrow A=\frac{1}{4}$$

Thus,

$$B=\frac{1/2}{A}=\frac{1/2}{1/4}=2$$

From these we obtain the marginal PDFs as

$$\begin{aligned} f_X(x)&=\frac{1}{4}x^3 \qquad 0\le x\le 2\\ f_Y(y)&=2y \qquad 0\le y\le 1 \end{aligned}$$

## 5.7 COVARIANCE AND CORRELATION COEFFICIENT

Consider two random variables $X$ and $Y$ with expected values $E[X]=\mu_X$ and $E[Y]=\mu_Y$, respectively, and variances $\sigma_X^2$ and $\sigma_Y^2$, respectively. The *covariance* of $X$ and $Y$, which is denoted by $\text{Cov}(X,Y)$ or $\sigma_{XY}$ is defined by

$$\begin{aligned} \text{Cov}(X,Y)=\sigma_{XY}&=E[(X-\mu_X)(Y-\mu_Y)]\\ &=E[XY-\mu_YX-\mu_XY+\mu_X\mu_Y]\\ &=E[XY]-\mu_YE[X]-\mu_XE[Y]+\mu_X\mu_Y\\ &=E[XY]-\mu_X\mu_Y-\mu_X\mu_Y+\mu_X\mu_Y\\ &=E[XY]-\mu_X\mu_Y \end{aligned} \tag{5.17}$$

If $X$ and $Y$ are independent, then $E[XY]=\mu_X\mu_Y$ and $\text{Cov}(X,Y)=0$. However, the converse is not true; that is, if the covariance of $X$ and $Y$ is zero, it does not mean that $X$ and $Y$ are independent random variables. If the covariance of two random variables is zero, we define the two random variables to be *uncorrelated*.

We define the *correlation coefficient* of X and Y, denoted by $\rho(X,Y)$ or $\rho_{XY}$, as follows:

$$\rho_{XY}=\frac{\text{Cov}(X,Y)}{\sqrt{\text{Var}(X)\text{Var}(Y)}}=\frac{\sigma_{XY}}{\sigma_X\sigma_Y} \tag{5.18}$$

The correlation coefficient has the property that

$$-1\leq\rho_{XY}\leq 1 \tag{5.19}$$

This can be proved as follows. Since the variance is always nonnegative, we have that if $X$ and $Y$ have variances given by $\sigma_X^2$ and $\sigma_Y^2$ respectively, then

$$0\leq\text{Var}\left(\frac{X}{\sigma_X}+\frac{Y}{\sigma_X}\right)=\frac{\text{Var}(X)}{\sigma_X^2}+\frac{\text{Var}(Y)}{\sigma_Y^2}+\frac{2\text{Cov}(X,Y)}{\sigma_X\sigma_Y}=2(1+\rho_{XY})$$

which implies that $-1\leq\rho_{XY}$. Also,

$$0\leq\text{Var}\left(\frac{X}{\sigma_X}-\frac{Y}{\sigma_X}\right)=\frac{\text{Var}(X)}{\sigma_X^2}+\frac{\text{Var}(Y)}{\sigma_Y^2}-\frac{2\text{Cov}(X,Y)}{\sigma_X\sigma_Y}=2(1-\rho_{XY})$$

which implies the $\rho_{XY}\leq 1$. Thus,

$$-1\leq\rho_{XY}\leq 1$$

The correlation coefficient $\rho_{XY}$ provides a measure of how good a linear prediction of the value of one of the two random variables can be formed based on an observed value of the other. Thus, if we represent the relationship between $X$ and $Y$ by the linear equation $Y=a+bX$, a value of $\rho_{XY}$ near $-1$ or $+1$ indicates a high degree of linearity between $X$ and $Y$. In particular, a positive $\rho_{XY}$ implies that $b>0$, and a negative $\rho_{XY}$ implies that $b<0$. That is, a positive $\rho_{XY}$ implies that as $X$ increases, $Y$ also tends to increase; and a negative $\rho_{XY}$ implies that as $X$ increases, $Y$ tends to decrease. A value of $\rho_{XY}=0$ means that there is no *linear correlation* between $X$ and $Y$. However, it does not mean that there is no correlation at all between them because there may still be a high *nonlinear correlation* between them. In general, $\rho_{XY}$ measures the goodness of fit of the equation that expresses $Y$ as a function of $X$ to actual (or measured) values of $Y$. That is, it indicates how closely the equation that expresses $Y$ as a function of $X$ matches measured (or observed) values of $Y$.

## EXAMPLE 5.12

The joint PDF of the random variables $X$ and $Y$ is defined as follows:

$$f_{XY}(x,y)=\begin{cases}25e^{-5y} & 0\le x<0.2;\ y\ge 0\\ 0 & \text{otherwise}\end{cases}$$

a. Find the marginal PDFs of $X$ and $Y$.
b. What is the covariance of $X$ and $Y$?

Solution:

a. The marginal PDFs are obtained as follows:

$$f_X(x)=\int_0^{\infty} f_{XY}(x,y)dy=\int_0^{\infty}25e^{-5y}dy=\begin{cases}5 & 0\le x<0.2\\ 0 & \text{otherwise}\end{cases}$$

$$f_Y(y)=\int_0^{0.2} f_{XY}(x,y)dx=\int_0^{0.2}25e^{-5y}dx=\begin{cases}5e^{-5y} & y\ge 0\\ 0 & \text{otherwise}\end{cases}$$

Thus, $X$ has a uniform distribution and $Y$ has an exponential distribution.

b. The expected values of $X$ and $Y$ are given by

$$E[X]=\mu_X=\frac{0+0.2}{2}=0.1$$

$$E[Y]=\mu_Y=\frac{1}{5}=0.2$$

Also,

$$\begin{aligned}E[XY]&=\int_{x=0}^{0.2}\int_{y=0}^{\infty}xyf_{XY}(x,y)dydx=\int_{x=0}^{0.2}\int_{y=0}^{\infty}25xye^{-5y}dydx\\&=\int_{x=0}^{0.2}x\left\{\int_{y=0}^{\infty}25ye^{-5y}dy\right\}dx=\int_{x=0}^{0.2}xdx=\left[\frac{x^2}{2}\right]_0^{0.2}\\&=0.02\end{aligned}$$

Thus, the covariance of $X$ and $Y$ is given by

$$\sigma_{XY}=E[XY]-\mu_X\mu_Y=0.02-(0.1)(0.2)=0$$

This means that $X$ and $Y$ are uncorrelated. Note that the reason why $\sigma_{XY}=0$ is because $X$ and $Y$ are independent. This follows from the fact that $f_{XY}(x,y)$ is separable into a function $x$ and a function of $y$, and the region of interest is rectangular. Thus, $f_{XY}(x,y)=f_X(x)f_Y(y)$.

## EXAMPLE 5.13

Hans and Ann planned to meet at their favorite restaurant on a date at about 6:30 pm. Both of them will arrive at the restaurant separately by train. They live in different parts of the city and so will be arriving on different trains that operate independently of each other's schedule. Hans' train will arrive at a stop by the restaurant at a time that is uniformly distributed between 6:00 pm and 7:00 pm. Ann's train will arrive at the same stop at a time that is uniformly distributed between 6:15 pm and 6:45 pm. They agreed that whoever arrives at the restaurant first will wait up to 5 minutes before leaving.

a. What is the probability that they meet?
b. What is the probability that Ann arrives before Hans?

Solution:
Let $X$ be the random variable that denotes Hans' arrival time, and let $Y$ be the random variable that denotes Ann's arriving time. As stated in the problem, $X$ and $Y$ are independent random variables. If we consider the time from 6:00 pm to 7:00 pm, we see that we can represent the PDFs of $X$ and $Y$ as follows:

$$f_X(x) = \begin{cases} \frac{1}{60} & 0 \leq x \leq 60 \\ 0 & \text{otherwise} \end{cases}$$

$$f_Y(y) = \begin{cases} \frac{1}{30} & 15 \leq x \leq 45 \\ 0 & \text{otherwise} \end{cases}$$

Thus, the joint PDF $f_{XY}(x, y)$, which is the product of the above marginal PDFs, has a uniform distribution over the rectangle shown in Figure 5.3.

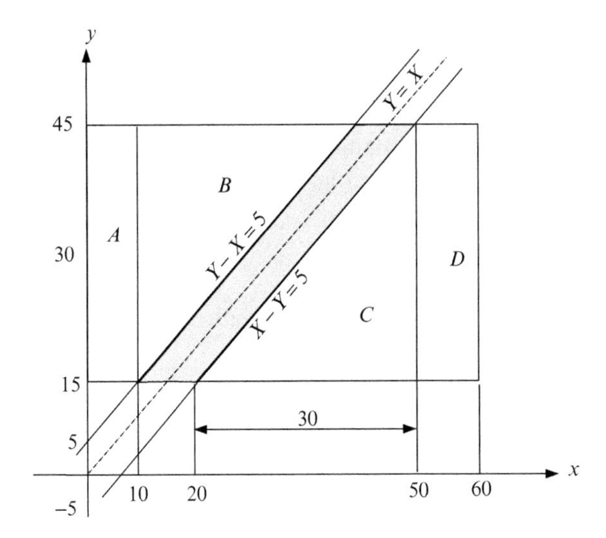


**FIGURE 5.3**
Domain of the Joint Distribution

a. The probability that they meet is given by $P[|X - Y| \leq 5]$, which is the probability of being in the shaded area of the rectangle. Now, the total area of the rectangle is $60 \times 30 = 1800$. The area of section A is $10 \times 30 = 300$, which is also the area of section D. The area of section B is $30 \times 30/2 = 450$, which is also the area of section C. Thus, the area of the shaded section is $1800 - 2(450 + 300) = 300$. This means that

$$p = \frac{300}{1800} = \frac{1}{6}$$

b. The probability that Ann arrives before Hans is $P[Y < X]$, which is the probability of being in the portion of the rectangle above the line $Y = X$. From the symmetry of the diagram, this can be seen to be equal to 1/2.

## 5.8 MULTIVARIATE RANDOM VARIABLES

In the previous sections we considered a system of two random variables. In this section we extend the concepts developed for two random variables to systems of more than two random variables.

Let $X_1, X_2, \ldots, X_n$ be a set of random variables that are defined on the same sample space. Their joint CDF is defined as

$$F_{X_1X_2\ldots X_n}(x_1, x_2, \ldots, x_n) = P[X_1 \leq x_1, X_2 \leq x_2, \ldots, X_n \leq x_n] \tag{5.20}$$

If all the random variables are discrete random variables, their joint PMF is defined by

$$p_{X_1X_2\ldots X_n}(x_1, x_2, \ldots, x_n) = P[X_1 = x_1, X_2 = x_2, \ldots, X_n = x_n] \tag{5.21}$$

The properties of the joint PMF include the following:

a. $0 \leq p_{X_1X_2\ldots X_n}(x_1, x_2, \ldots, x_n) \leq 1$
b. $\sum_{x_1}\sum_{x_2}\cdots\sum_{x_n} p_{X_1X_2\ldots X_n}(x_1, x_2, \ldots, x_n) = 1$
c. The marginal PMFs are obtained by summing the joint PMF over the appropriate ranges. For example, the marginal PMF of $X_n$ is given by

$$p_{X_n}(x_n) = \sum_{x_1}\sum_{x_2}\cdots\sum_{x_{n-1}} p_{X_1X_2\ldots X_n}(x_1, x_2, \ldots, x_n)$$

d. The conditional PMFs are similarly obtained. For example,

$$\begin{aligned} p_{X_n|X_1\ldots X_{n-1}}(x_n|x_1, \ldots, x_{n-1}) &= P[X_n = x_n | X_1 = x_1, \ldots, X_{n-1} = x_{n-1}] \\ &= \frac{p_{X_1X_2\ldots X_n}(x_1, x_2, \ldots, x_n)}{p_{X_1X_2\ldots X_{n-1}}(x_1, x_2, \ldots, x_{n-1})} \end{aligned}$$

provided $p_{X_1X_2\ldots X_{n-1}}(x_1, x_2, \ldots, x_{n-1}) > 0$. The random variables are defined to be mutually independent if

$$p_{X_1X_2\ldots X_n}(x_1, x_2, \ldots, x_n) = \prod_{k=1}^{n} p_{X_k}(x_k)$$

If all the random variables are continuous random variables, their joint PDF can be obtained from the joint CDF as follows:

$$f_{X_1X_2\ldots X_n}(x_1, x_2, \ldots, x_n) = \frac{\partial^n}{\partial x_1 \partial x_2 \ldots \partial x_n} F_{X_1X_2\ldots X_n}(x_1, x_2, \ldots, x_n) \tag{5.22}$$

The joint PDF has the following properties:

a. $f_{X_1X_2\ldots X_n}(x_1, x_2, \ldots, x_n) \geq 0$

b. $\int_{-\infty}^{\infty}\int_{-\infty}^{\infty}\cdots\int_{-\infty}^{\infty} f_{X_1X_2\ldots X_n}(x_1,x_2,\ldots,x_n)dx_1dx_2\ldots dx_n = 1$

c. The conditional PDFs are similarly obtained. For example,

$$f_{X_n|X_1\ldots X_{n-1}}(x_n|x_1,\ldots,x_{n-1}) = \frac{f_{X_1X_2\ldots X_n}(x_1,x_2,\ldots,x_n)}{f_{X_1X_2\ldots X_{n-1}}(x_1,x_2,\ldots,x_{n-1})}$$

provided $f_{X_1X_2\ldots X_{n-1}}(x_1,x_2,\ldots,x_{n-1}) > 0$. If the random variables are mutually independent, then

$$f_{X_1X_2\ldots X_n}(x_1,x_2,\ldots,x_n) = \prod_{k=1}^{n} f_{X_k}(x_k)$$

## EXAMPLE 5.14

A machine has $N$ identical components each of which has an exponentially distributed lifetime $T$ with PDF $\lambda e^{-\lambda t}, t \geq 0$. What is the probability that exactly $n$ of the components have failed by time $v \geq 0$?

Solution:

Let $p$ denote the probability that a component has failed by time $v$, which is the CDF of $T$ evaluated at $t = v$; that is,

$$p = P[T \leq v] = F_T(v) = 1 - e^{-\lambda v}$$

If we assume that the components fail independently, then if $q_n$ is the probability that exactly $n$ of them have failed by time $v$, we have that

$$\begin{aligned} q_n &= \binom{N}{n} p^n (1-p)^{N-n} = \binom{N}{n} \left[1 - e^{-\lambda v}\right]^n \left(e^{-\lambda v}\right)^{N-n} \\ &= \binom{N}{n} \left[1 - e^{-\lambda v}\right]^n e^{-\lambda (N-n) v} \end{aligned}$$

## 5.9 MULTINOMIAL DISTRIBUTIONS

The multinomial distribution is an extension of the binomial distribution, which was discussed in Chapter 4. It arises when a sequence of $n$ independent experiments is performed. Assume that each experiment can result in any one of $m$ possible outcomes with probabilities $p_1, p_2, \ldots, p_m$, where

$$\sum_{k=1}^{m} p_k = 1$$

Let $K_i$ denote the number of the $n$ experiments that result in outcome number $i$, where $i = 1, 2, \ldots, m$. Then

$$\begin{aligned} p_{K_1K_2\cdots K_m}(k_1,k_2,\cdots,k_m) &= P[K_1=k_1,K_2=k_2,\cdots,K_m=k_m] \\ &= \binom{n}{k_1\,k_2\cdots k_m} p_1^{k_1}p_2^{k_2}\cdots p_m^{k_m} \\ &= \frac{n!}{k_1!k_2!\cdots k_m!}p_1^{k_1}p_2^{k_2}\cdots p_m^{k_m} \end{aligned} \tag{5.23}$$

where $\sum_{i=1}^{m} k_i = n$, $k_i = 0,1,\ldots,n$ for $i=1,2,\ldots,m$. When $m=2$, we get the binomial distribution.

## EXAMPLE 5.15

A fair die is rolled 7 times. Find the probability that the number 1 appears twice and the number 2 appears once.

Solution:

The problem asks for the numbers 1 and 2. All the other four numbers (i.e., 3 through 6) are lumped into one event. Thus, instead of the normal 6 events, we have three events. Let $K_1$ be the number of times the number 1 appears in the 7 trials, $K_2$ the number times the number 2 appears in the 7 trials, and $K_3$ the number of times that all the other four numbers (3, 4, 5, 6) appear in the 7 trials. Similarly, let $p_1$ denote the probability that 1 appears in any roll of the die, $p_2$ the probability that 2 appears in any roll of the die and $p_3$ the probability that any number other than 1 and 2 appears in any roll of the die. Since the die is fair, we have that $p_1=p_2=1/6$ and $p_3=1-p_1-p_2=4/6$. Thus, the desired result is

$$\begin{aligned} p_{K_1K_2K_3}(2,1,4) &= \binom{7}{2\,1\,4}p_1^2p_2^1p_3^4 = \frac{7!}{2!1!4!}\left(\frac{1}{6}\right)^2\left(\frac{1}{6}\right)\left(\frac{4}{6}\right)^4 = \frac{(105)(256)}{6^7} \\ &= \frac{26{,}880}{279{,}936} = 0.0960 \end{aligned}$$

## EXAMPLE 5.16

The student population of a certain college in Massachusetts has been found to be made up as follows: 50% are from Massachusetts (or in-state students), 20% are from the other states of the United States (or out-of-state students), and 30% are from other countries (or foreign students). If 10 students from the college are randomly selected by a company conducting a survey, what is the probability that 6 of them will be in-state students, 2 are out-of-state students, and 2 are foreign students?

Solution:

Let $K_1$ be the number of in-state students among the 10 selected students, $K_2$ the number of out-of-state students among the 10 selected students, and $K_3$ the number of foreign students among the 10 selected students. Similarly, let $p_1$ denote the probability that an in-state student is randomly selected, $p_2$ the probability that an out-of-state student is randomly selected, and $p_3$ the probability that a foreign student is randomly selected. Since $p_1=0.5, p_2=0.2$ and $p_3=0.3$, the desired result is

$$\begin{aligned} p_{K_1K_2K_3}(6,2,2) &= \binom{10}{6\,2\,2}(0.5)^6(0.2)^2(0.3)^2 = \frac{10!}{6!2!2!}(0.5)^6(0.2)^2(0.3)^2 \\ &= 1260(0.5)^6(0.2)^2(0.3)^2 = 0.0709 \end{aligned}$$

## 5.10 CHAPTER SUMMARY

This chapter discussed problems that deal with two or more random variables simultaneously. It also discussed the concepts of covariance and correlation coefficient of two random variables. It discussed the multinomial distribution, which is an extension of the binomial distribution considered in Chapter 4. Sometimes multinomial distribution related problems can be reduced to binomial distribution problems, especially if we are only interested in obtaining results concerning only one member of the distribution.

## 5.11 PROBLEMS

### Section 5.3 Discrete Bivariate Random Variables

5.1 The joint PMF of two discrete random variables $X$ and $Y$ is given by

$$p_{XY}(x,y)=\begin{cases} kxy & x=1,2,3;\ y=1,2,3 \\ 0 & \text{otherwise} \end{cases}$$

a. Determine the value of the constant $k$.
b. Find the marginal PMFs of $X$ and $Y$.
c. Find $P[1 \le X \le 2, Y \le 2]$.

5.2 A fair coin is tossed three times. Let the random variable $X$ denote the number of heads in the first two tosses, and let the random variable $Y$ denote the number of heads in the third toss. Determine the joint PMF $p_{XY}(x,y)$ of $X$ and $Y$.

5.3 The joint PMF of two random variables $X$ and $Y$ is given by

$$p_{XY}(x,y)=\begin{cases} 0.10 & x=1, y=1 \\ 0.35 & x=2, y=2 \\ 0.05 & x=3, y=3 \\ 0.50 & x=1, y=1 \\ 0 & \text{otherwise} \end{cases}$$

a. Determine the joint CDF $F_{XY}(x,y)$
b. Find $P[1 \le X \le 2, Y \le 2]$.

5.4 Two discrete random variables $X$ and $Y$ have the joint CDF defined as follows:

$$F_{XY}(x,y)=\begin{cases} \frac{1}{12} & x=0, y=0 \\ \frac{1}{3} & x=0, y=1 \\ \frac{2}{3} & x=0, y=2 \\ \frac{1}{6} & x=1, y=0 \\ \frac{7}{12} & x=1, y=1 \\ 1 & x=1, y=2 \end{cases}$$

Determine the following:

a. $P[0 < X < 2, 0 < Y < 2]$
b. the marginal CDFs of $X$ and $Y$ (i.e., $F_X(x)$ and $F_Y(y)$)
c. $P[X = 1, Y = 1]$

5.5 Two discrete random variables $X$ and $Y$ have the joint PMF defined as follows:

$$p_{XY}(x,y) = \begin{cases} \frac{1}{12} & x=1, y=1 \\ \frac{1}{6} & x=1, y=2 \\ \frac{1}{12} & x=1, y=3 \\ \frac{1}{6} & x=2, y=1 \\ \frac{1}{4} & x=2, y=2 \\ \frac{1}{12} & x=2, y=3 \\ \frac{1}{12} & x=3, y=1 \\ \frac{1}{12} & x=3, y=2 \\ 0 & \text{otherwise} \end{cases}$$

Determine the following:

a. the marginal PMFs of $X$ and $Y$ (i.e., $p_X(x)$ and $p_Y(y)$)
b. $P[X < 2.5]$
c. the probability that $Y$ is odd

5.6 Two discrete random variables $X$ and $Y$ have the joint PMF given by

$$p_{XY}(x,y) = \begin{cases} 0.2 & x=1, y=1 \\ 0.1 & x=1, y=2 \\ 0.1 & x=2, y=1 \\ 0.2 & x=2, y=2 \\ 0.1 & x=3, y=1 \\ 0.3 & x=3, y=2 \\ 0 & \text{otherwise} \end{cases}$$

Determine the following:

a. the marginal PMFs of $X$ and $Y$ (i.e., $p_X(x)$ and $p_Y(y)$)
b. the conditional PMF of $X$ given $Y$, $p_{X|Y}(x|y)$
c. whether $X$ and $Y$ are independent.

## Section 5.4 Continuous Bivariate Random Variables

5.7 The joint PDF of two continuous random variables $X$ and $Y$ is given by

$$f_{XY}(x,y) = \begin{cases} kx & 0 < y \le x < 1 \\ 0 & \text{otherwise} \end{cases}$$

a. Determine the value of the constant $k$.
b. Find the marginal PDFs of $X$ and $Y$ (i.e., $f_X(x)$ and $f_Y(y)$)
c. Find $P\left[0<X<\frac{1}{2}, 0<Y<\frac{1}{4}\right]$

5.8 The joint CDF of two continuous random variables $X$ and $Y$ is given by

$$F_{XY}(x,y)=\begin{cases}1-e^{-ax}-e^{-by}+e^{-(ax+by)} & x\geq 0, y\geq 0\\ 0 & \text{otherwise}\end{cases}$$

a. Find the marginal PDFs of $X$ and $Y$
b. Carefully show why or why not $X$ and $Y$ are independent.

5.9 Two random variables $X$ and $Y$ have the joint PDF given by

$$f_{XY}(x,y)=\begin{cases}ke^{-(2x+3y)} & x\geq 0, y\geq 0\\ 0 & \text{otherwise}\end{cases}$$

Determine the following:
a. the value of the constant $k$ that makes $f_{XY}(x,y)$ a true joint PDF
b. the marginal PDFs of $X$ and $Y$
c. $P[X<1, Y<0.5]$

5.10 Two random variables $X$ and $Y$ have the joint PDF given by

$$f_{XY}(x,y)=\begin{cases}k(1-x^2y) & 0\leq x\leq 1; 0\leq y\leq 1\\ 0 & \text{otherwise}\end{cases}$$

Determine the following:
a. the value of the constant $k$ that makes $f_{XY}(x,y)$ a true joint PDF
b. the conditional PDFs of $X$ given $Y$, $f_{X|Y}(x|y)$, and $Y$ given $X$, $f_{Y|X}(y|x)$

5.11 The joint PDF of two random variables $X$ and $Y$ is given by

$$f_{XY}(x,y)=\frac{6}{7}\left(x^2+\frac{xy}{2}\right) \quad 0<x<1, 0<y<2$$

a. What is the CDF, $F_X(x)$, of $X$?
b. Find $P[X>Y]$
c. Find $P\left[Y>\frac{1}{2}|X<\frac{1}{2}\right]$

5.12 Two random variables $X$ and $Y$ have the joint PDF given by

$$f_{XY}(x,y)=\begin{cases}ke^{-(x+y)} & x\geq 0, y\geq x\\ 0 & \text{otherwise}\end{cases}$$

Determine the following:
a. the value of the constant $k$ that makes $f_{XY}(x,y)$ a true joint PDF
b. $P[Y<2X]$

5.13 The joint PDF of two random variables $X$ and $Y$ is given by

$$f_{XY}(x,y)=\frac{6x}{7} \quad 1\leq x+y\leq 2, x\geq 0, y\geq 0$$

a. Without actually performing the integration, obtain the integral that expresses the $P[Y>X^2]$. (That is, just give the exact limits of the integration.)
b. In a very convincing way, obtain the exact value of $P[X>Y]$.

5.14 Two random variables $X$ and $Y$ have the joint PDF given by

$$f_{XY}(x,y)=\begin{cases}\frac{1}{2}e^{-2y} & 0\leq x\leq 4, y\geq 0\\ 0 & \text{otherwise}\end{cases}$$

Determine the marginal PDFs of $X$ and $Y$.

## Section 5.6 Conditional Distributions

5.15 A box contains 3 red balls and 2 green balls. One ball is randomly selected from the box, its color is observed and it is put back into the box. A second ball is then selected and goes through the same process. Let the random variables $X$ and $Y$ be defined as follows: $X=0$ if the first ball is green, and $X=1$ if the first ball is red; $Y=0$ if the second ball is green and $Y=1$ if the second ball is red.
a. Find the joint PMF of $X$ and $Y$.
b. Find the conditional PMF of $X$ given $Y$.

5.16 Let the random variables $X$ and $Y$ have the joint PDF $f_{XY}(x,y)=2e^{-(x+2y)}$, $x\geq 0, y\geq 0$. Find the conditional expectation of
a. $X$ given $Y$
b. $Y$ given $X$

5.17 A fair coin is tossed four times. Let $X$ denote the number of heads obtained in the first two tosses, and let $Y$ denote the number of heads obtained in the last two tosses.
a. Find the joint PMF of $X$ and $Y$
b. Show that $X$ and $Y$ are independent random variables.

5.18 Two random variables $X$ and $Y$ have the joint PDF

$$f_{XY}(x,y)=xye^{-y^2/4} \qquad 0\leq x\leq 1,\ y\geq 0$$

a. Find the marginal PDFs of $X$ and $Y$.
b. Determine if $X$ and $Y$ are independent.

5.19 Two random variables $X$ and $Y$ have the joint PDF

$$f_{XY}(x,y)=\frac{6x}{7} \qquad 1\leq x+y\leq 2, x\geq 0, y\geq 0$$

a. Find the marginal PDFs of $X$ and $Y$.
b. Determine if $X$ and $Y$ are independent.

## Section 5.7 Covariance and Correlation Coefficient

5.20 Two discrete random variables $X$ and $Y$ have the joint PMF given by

$$p_{XY}(x,y) = \begin{cases} 0 & x=-1,\ y=0 \\ \frac{1}{3} & x=-1,\ y=1 \\ \frac{1}{3} & x=0,\ y=0 \\ 0 & x=0,\ y=1 \\ 0 & x=1,\ y=0 \\ \frac{1}{3} & x=1,\ y=1 \end{cases}$$

a. Are $X$ and $Y$ independent?
b. What is the covariance of $X$ and $Y$?

5.21 Two events A and B are such that $P[A]=\frac{1}{4}$, $P[B|A]=\frac{1}{2}$ and $P[A|B]=\frac{1}{4}$. Let the random variable $X$ be defined such that $X=1$ if event A occurs and $X=0$ if event A does not occur. Similarly, let the random variable $Y$ be defined such that $Y=1$ if event B occurs and $Y=0$ if event B does not occur.
a. Find $E[X]$ and the variance of $X$
b. Find $E[Y]$ and the variance of $Y$.
c. Find $\rho_{XY}$ and determine whether or not $X$ and $Y$ are uncorrelated.

5.22 A fair die is tossed three times. Let $X$ be the random variable that denotes the number of 1's and let $Y$ be the random variable that denotes the number of 3's. Find the correlation coefficient of $X$ and $Y$.

## Section 5.9 Multinomial Distributions

5.23 A box contains 10 chips from supplier A, 16 chips from supplier B, and 14 chips from supplier C. Assume that we perform the following experiment 20 times: We draw one chip from the box, note the supplier from where it came and put it back into the box. What is the probability that a chip from vendor B is drawn 9 times?

5.24 With reference to the previous problem, what is the probability that a chip from vendor A is drawn 5 times and a chip from vendor C is drawn 6 times?

5.25 The students in one college have the following rating system for their professors: excellent, good, fair, and bad. In a recent poll of the students, it was found that they believe that 20% of the professors are excellent, 50% are good, 20% are fair, and 10% are bad. Assume that 12 professors are randomly selected from the college.
a. What is the probability that 6 are excellent, 4 are good, 1 is fair, and 1 is bad?
b. What is the probability that 6 are excellent, 4 are good, and 2 are fair?
c. What is the probability that 6 are excellent and 6 are good?
d. What is the probability that 4 are excellent and 3 are good?
e. What is the probability that 4 are bad?
f. What is the probability that none is bad?

5.26 Studies on the toasters made by a company indicate that 50% are good, 35% are fair, 10% burn the toast, and 5% catch fire. If a store has 40 of these toasters in stock, determine the following probabilities:

a. 30 are good, 5 are fair, 3 burn the toast, and 2 catch fire
b. 30 are good and 4 are fair
c. None catches fire
d. None burns the toast and none catches fire

5.27 Ten pieces of candy are given out at random to a group that consists of 8 boys, 7 girls, and 5 adults. If anyone can get more than one piece of candy, find the following probabilities:

a. 4 pieces go to the girls and 2 go to the adults
b. 5 pieces go to the boys

CHAPTER 6

# Functions of Random Variables

## 6.1 INTRODUCTION

The previous chapters discussed basic properties of events defined in a given sample space and the random variables used to represent those events. The fundamental assumption that was made in those chapters is that events can always be defined by random variables. However, in many applications the events are functions of other events. For example, the time until a complex system fails is a function of the time to failure of the individual components that make up the system. This means that the random variable used to represent the time to failure of the complex system is a function of the random variables used to represent the times to failure of the component parts of the system. This chapter deals with functions of random variables. Because of the complexity involved in computing the CDFs and PDFs of functions of multiple random variables, we restrict our discussion to functions of at most two random variables.

## 6.2 FUNCTIONS OF ONE RANDOM VARIABLE

Let $X$ be a random variable, and let $Y$ be a new random variable that is a function of $X$. That is,

$$Y = g(X)$$

We are interested in computing the PDF or PMF of $Y$ when the PDF or PMF of $X$ is given. For example, let $g(X) = X + 5$. Then

$$F_Y(y) = P[Y \leq y] = P[X + 5 \leq y]$$

### 6.2.1 Linear Functions

Consider the function $g(X) = aX + b$, where $a$ and $b$ are constants. The CDF of $Y$ is given by



Fundamentals of Applied Probability and Random Processes. http://dx.doi.org/10.1016/B978-0-12-800852-2.00006-7

$$F_Y(y) = P[Y \leq y] = P[aX + b \leq y] = P\left[X \leq \frac{y-b}{a}\right] = F_X\left(\frac{y-b}{a}\right)$$

where $a$ is positive. The PDF of Y is given by

$$f_Y(y) = \frac{dF_Y(y)}{dy} = \frac{dF_X\left(\frac{y-b}{a}\right)}{dy} = \left(\frac{dF_X(u)}{du}\right)\left(\frac{du}{dy}\right)$$

where $u = \frac{y-b}{a}$ and $\frac{du}{dy} = \frac{1}{a}$. Thus,

$$f_Y(y) = \left(\frac{dF_X(u)}{du}\right)\left(\frac{du}{dy}\right) = f_X(u)\left(\frac{1}{a}\right) = \left(\frac{1}{a}\right)f_X\left(\frac{y-b}{a}\right)$$

If $a < 0$, we have that

$$\begin{aligned} F_Y(y) &= P[Y \leq y] = P[aX + b \leq y] = P[aX \leq y - b] = P\left[X \geq \frac{y-b}{a}\right] \\ &= 1 - \left\{P\left[X \leq \frac{y-b}{a}\right] - P\left[X = \frac{y-b}{a}\right]\right\} \end{aligned}$$

The change in sign on the second line arises from the fact that $a$ is negative. If $X$ is continuous, $P\left[X = \frac{y-b}{a}\right] = 0$. Thus, the CDF and PDF for the case of negative $a$ are given by

$$\begin{aligned} F_Y(y) &= 1 - P\left[X \leq \frac{y-b}{a}\right] = 1 - F_X\left(\frac{y-b}{a}\right) \\ f_Y(y) &= \frac{d}{dy}F_Y(y) = -\left(\frac{1}{a}\right)f_X\left(\frac{y-b}{a}\right) \end{aligned}$$

Therefore, the general PDF of $Y$ is given by

$$f_Y(y) = \frac{1}{|a|}f_X\left(\frac{y-b}{a}\right) \tag{6.1}$$

## EXAMPLE 6.1

Find the PDF of $Y$ in terms of the PDF of $X$ if $Y = 2X + 7$.

**Solution:**

From the results obtained above,

$$\begin{aligned} F_Y(y) &= F_X\left(\frac{y-7}{2}\right) \\ f_Y(y) &= \frac{d}{dy}F_Y(y) = \left(\frac{1}{2}\right)f_X\left(\frac{y-7}{2}\right) \end{aligned}$$

### 6.2.2 Power Functions

Consider the quadratic function $Y = aX^2$, where $a > 0$. The plot of $Y$ against $X$ is shown in Figure 6.1 where we see that for each value of $Y$ there are two corresponding values of $X$, namely $\sqrt{Y/a}$ and $-\sqrt{Y/a}$.

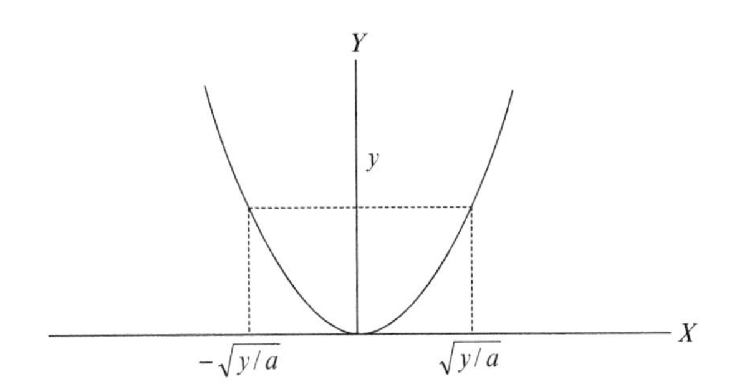


**FIGURE 6.1**
Plot of $Y = aX^2$

Thus, the CDF of $Y$ is given by

$$\begin{aligned} F_Y(y) &= P[Y \le y] = P[aX^2 \le y] = P\left[X^2 \le \frac{y}{a}\right] = P\left[|X| \le \sqrt{\frac{y}{a}}\right] \qquad y > 0 \\ &= P\left[-\sqrt{y/a} \le X \le \sqrt{y/a}\right] = F_X\left(\sqrt{y/a}\right) - F_X\left(-\sqrt{y/a}\right) \end{aligned}$$

The PDF of $Y$ is given by

$$f_Y(y) = \frac{d}{dy}F_Y(y) = \frac{d}{dy}\left\{F_X\left(\sqrt{y/a}\right) - F_X\left(-\sqrt{y/a}\right)\right\}$$

Let $u = \sqrt{y/a} = (y/a)^{1/2}$. Thus, $\dfrac{du}{dy} = \dfrac{1}{2(ay)^{1/2}} = \dfrac{1}{2\sqrt{ay}}$ and

$$\begin{aligned} f_Y(y) &= \frac{d}{dy}\left\{F_X\left(\sqrt{y/a}\right) - F_X\left(-\sqrt{y/a}\right)\right\} = \frac{dF_X(u)}{du}\frac{du}{dy} + \frac{dF_X(-u)}{du}\frac{du}{dy} \\ &= \frac{1}{2\sqrt{ay}}\left\{\frac{dF_X(u)}{du} + \frac{dF_X(-u)}{du}\right\} = \frac{1}{2\sqrt{ay}}\left\{f_X\left(\sqrt{y/a}\right) + f_X\left(-\sqrt{y/a}\right)\right\} \qquad (6.2) \\ &= \frac{f_X\left(\sqrt{y/a}\right) + f_X\left(-\sqrt{y/a}\right)}{2\sqrt{ay}} \qquad y/a > 0 \end{aligned}$$

If $f_X(x)$ is an even function, then $f_X(-x) = f_X(x)$ and $F_X(-x) = 1 - F_X(x)$. Thus, we have

$$f_Y(y) = \frac{f_X\left(\sqrt{y/a}\right) + f_X\left(-\sqrt{y/a}\right)}{2\sqrt{ay}} = \frac{2f_X\left(\sqrt{y/a}\right)}{2\sqrt{ay}} = \frac{f_X\left(\sqrt{y/a}\right)}{\sqrt{ay}} \qquad y/a > 0$$

## EXAMPLE 6.2

Find the PDF of the random variable $Y=aX^2$, where $X$ is the standard normal random variable and $a>0$.

**Solution:**
The PDF of $X$ is given by

$$f_X(x)=\frac{1}{\sqrt{2\pi}}e^{-x^2/2} \qquad -\infty<x<\infty$$

which is an even function. Thus, from the preceding result we have that

$$f_Y(y)=\frac{f_X\left(\sqrt{y/a}\right)}{\sqrt{ay}}=\frac{e^{-y/2a}}{\sqrt{2\pi ay}} \qquad y>0$$

## 6.3 EXPECTATION OF A FUNCTION OF ONE RANDOM VARIABLE

Let $X$ be a random variable and $g(X)$ a real-valued function of $X$. The expected value of $g(X)$ is defined by

$$E[g(X)]=\begin{cases}\sum_x g(x)p_X(x) & X \text{ discrete}\\ \int_{-\infty}^{\infty} g(x)f_X(x)dx & X \text{ continuous}\end{cases} \tag{6.3}$$

### 6.3.1 Moments of a Linear Function

Assume that $g(X)=aX+b$, where $X$ is a continuous random variable. Then

$$\begin{aligned}E[g(X)] = E[aX+b] &= \int_{-\infty}^{\infty} g(x)f_X(x)dx = \int_{-\infty}^{\infty}(ax+b)f_X(x)dx\\ &= a\int_{-\infty}^{\infty} xf_X(x)dx + b\int_{-\infty}^{\infty} f_X(x)dx\\ &= aE[X]+b\end{aligned} \tag{6.4}$$

This means that the expected value of a linear function of a single random variable is the linear function obtained by replacing the random variable by its expectation. When $X$ is a discrete random variable, we replace the summation by integration and obtain the same result.

The variance of $g(X)$ is given by

$$\begin{aligned}\text{Var}(g(X)) &= \text{Var}(aX+b) = E\left[(aX+b-E[aX+b])^2\right] \\ &= E[(aX+b-E[aX]-b])^2] = E\left[\{a(X-E[X])\}^2\right] \\ &= E\left[a^2\{X-E[X]\}^2\right] = a^2E\left[\{X-E[X]\}^2\right] \\ &\; a^2\sigma_X^2\end{aligned} \tag{6.5}$$

### 6.3.2 Expected Value of a Conditional Expectation

In Chapter 3 we saw that the conditional expectation of a random variable $X$, given that an event $A$ has occurred, is given by

$$E[X|A] = \begin{cases} \sum_k x_k p_{X|A}(x_k|A) & X \text{ discrete} \\ \int_{-\infty}^{\infty} x f_{X|A}(x|A)\,dx & X \text{ continuous} \end{cases}$$

where the conditional PMF $p_{X|A}(x|A)$ and the conditional PDF $f_{X|A}(x|A)$ are defined as follows:

$$p_{X|A}(x|A) = \frac{p_X(x)}{P[A]}$$

$$f_{X|A}(x|A) = \frac{f_X(x)}{P[A]}$$

where $x \in A$ and $P[A] > 0$ is the probability that event $A$ occurs. When $A$ is a random variable, say $A = Y$, the conditional expected value, $E[X|Y]$, is a random variable because it is a function of $Y$. Thus, it has an expected value, which for the case of continuous random variables is given by

$$\begin{aligned}E[E[X|Y]] &= \int_{-\infty}^{\infty} E[X|Y]f_Y(y)\,dy = \int_{-\infty}^{\infty}\left\{\int_{-\infty}^{\infty} x f_{X|Y}(x|y)dx\right\} f_Y(y)dy \\ &= \int_{-\infty}^{\infty} x\left\{\int_{-\infty}^{\infty} f_{X|Y}(x|y)f_Y(y)dy\right\}dx \\ &= \int_{-\infty}^{\infty} x\left\{\int_{-\infty}^{\infty} f_{XY}(x,y)dy\right\}dx = \int_{-\infty}^{\infty} x f_X(x)dx \\ &= E[X]\end{aligned} \tag{6.6}$$

## 6.4 SUMS OF INDEPENDENT RANDOM VARIABLES

Consider two independent continuous random variables $X$ and $Y$. We are interested in computing the CDF and PDF of their sum $g(X,Y) = U = X + Y$. The random variable $U$ can be used to model the reliability of systems with stand-by connections, as shown in Figure 6.2. In such systems, the component A whose time-to-failure is represented by the random variable $X$ is the primary component, and the component B whose time-to-failure is represented by the random variable $Y$ is the backup component that is brought into operation when the primary component fails. Thus, $U$ represents the time until the system fails, which is the sum of the lifetimes of both components.

The CDF of $U$ can be obtained as follows:

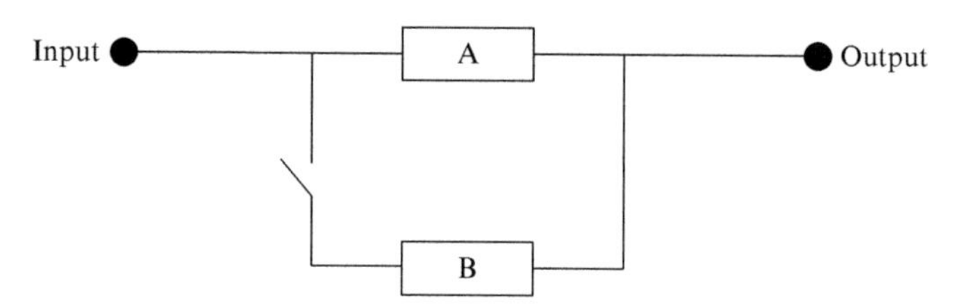


**FIGURE 6.2**
Stand-by Connection Modeled by the Random Variable $U$

$$F_U(u) = P[U \leq u] = P[X + Y \leq u] = \iint_D f_{XY}(x, y)dxdy$$

where $D$ is the set $D = \{(x, y) | x + y \leq u\}$, which is the area to the left of the line $u = x + y$ as shown in Figure 6.3.

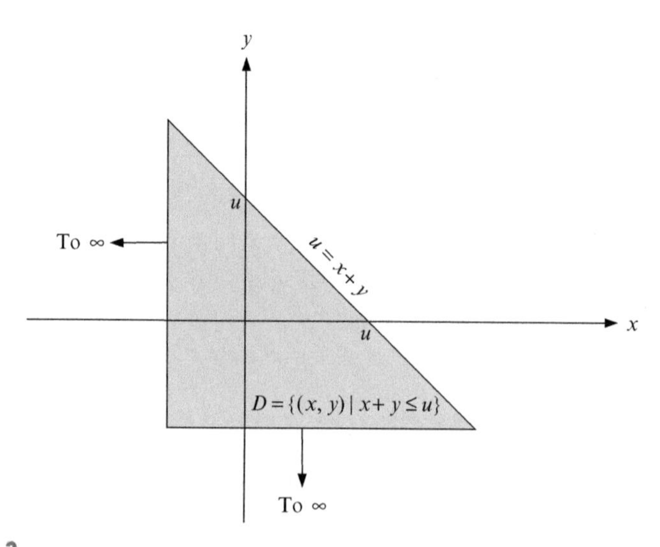


**FIGURE 6.3**
Domain of $D$

Thus,

$$F_U(u) = \int_{y=-\infty}^{\infty}\int_{x=-\infty}^{u-y} f_{XY}(x,y)dxdy = \int_{y=-\infty}^{\infty}\int_{x=-\infty}^{u-y} f_X(x)f_Y(y)dxdy$$
$$= \int_{y=-\infty}^{\infty}\left\{\int_{x=-\infty}^{u-y} f_X(x)dx\right\} f_Y(y)dy = \int_{y=-\infty}^{\infty} F_X(u-y)f_Y(y)dy$$

The PDF of $U$ is obtained by differentiating the CDF, as follows:

$$\begin{aligned} f_U(u) &= \frac{d}{du}F_U(u) = \frac{d}{du}\int_{y=-\infty}^{\infty} F_X(u-y)f_Y(y)dy = \int_{y=-\infty}^{\infty}\frac{d}{du}F_X(u-y)f_Y(y)dy \\ &= \int_{y=-\infty}^{\infty} f_X(u-y)f_Y(y)dy \end{aligned} \tag{6.7a}$$

where we have assumed that we can interchange differentiation and integration. The expression on the right-hand side of the last equality is a well-known result in signal analysis called the *convolution integral*. Thus, we find that the PDF of the sum $U$ of two independent random variables $X$ and $Y$ is the convolution of the PDFs of the two random variables; that is,

$$f_U(u) = f_X(u) * f_Y(u) \tag{6.7b}$$

## EXAMPLE 6.3

Find the PDF of the sum of $X$ and $Y$ if the two random variables are independent random variables with the common PDF

$$f_X(u) = f_Y(u) = \begin{cases} \frac{1}{4} & 0 < u < 4 \\ 0 & \text{otherwise} \end{cases}$$

**Solution:**

The limits of integration of the PDF of $U = X + Y$ can be computed with the aid of Figure 6.4. When $0 \le u \le 4$ (see Figure 6.4(a) where $f_Y(u-x)$ is shown in dashed lines), we have that

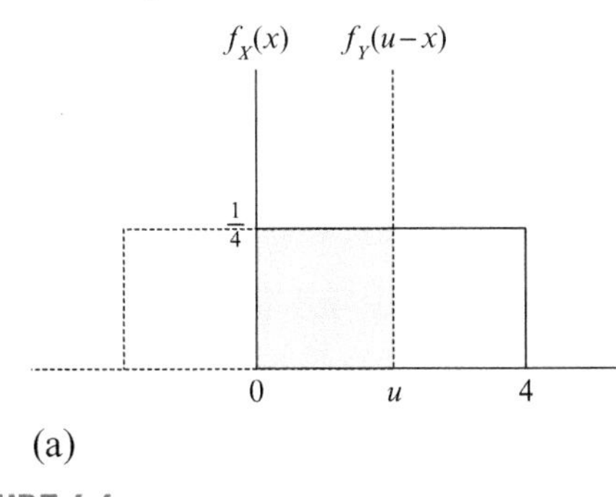


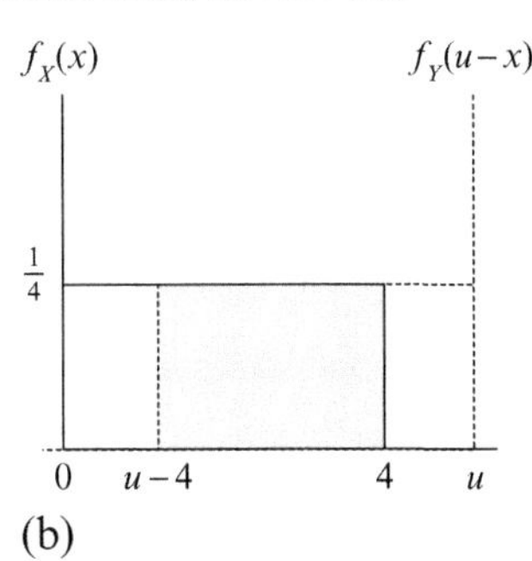


**FIGURE 6.4**
Convolution of PDFs (a) and (b)

$$f_U(u) = \int_{y=-\infty}^{\infty} f_X(u-y) f_Y(y) dy = \int_{y=0}^{u} \frac{1}{16} dy = \frac{u}{16}$$

For $4 < u < 8$ (see Figure 6.4(b)), we obtain

$$f_U(u) = \frac{1}{16} \int_{u-4}^{4} dy = \frac{8-u}{16}$$

Thus

$$f_U(u) = \begin{cases} \frac{u}{16} & 0 \le u < 4 \\ \frac{8-u}{16} & 4 \le u < 8 \\ 0 & \text{otherwise} \end{cases}$$

The PDF of $U$ is illustrated in Figure 6.5.

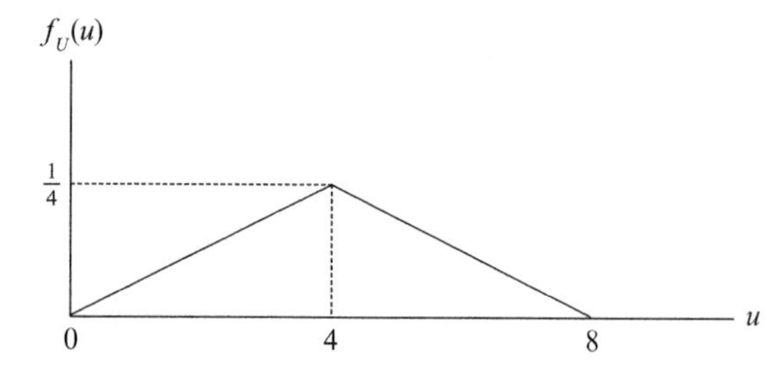


**FIGURE 6.5**
PDF of $U = X + Y$

## EXAMPLE 6.4

(A more general case of Example 6.3) Obtain the PDF of $Z = X + Y$, where $X$ and $Y$ are two independent random variables with the following PDFs:

$$f_X(x) = \begin{cases} \frac{1}{b-a} & a < x < b \\ 0 & \text{otherwise} \end{cases}$$

$$f_Y(y) = \begin{cases} \frac{1}{d-c} & c < y < d,\ d-c < b-a \\ 0 & \text{otherwise} \end{cases}$$

**Solution:**

The two PDFs are shown in Figure 6.6.

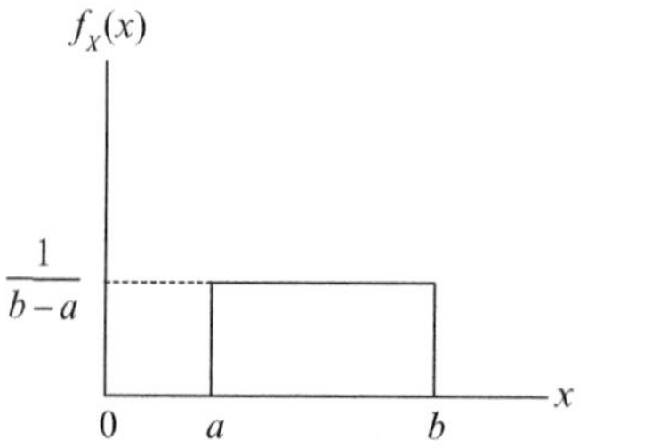


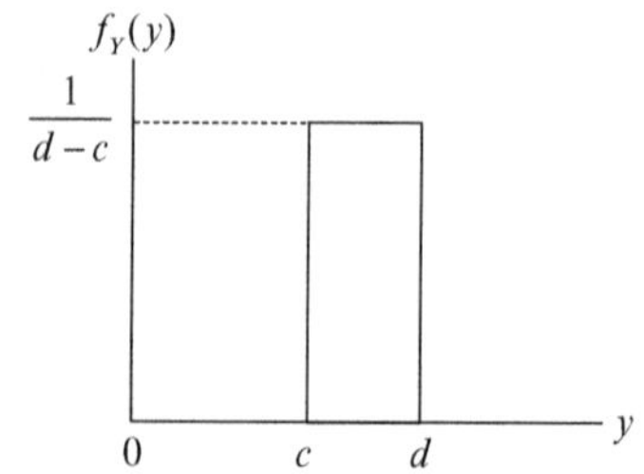


**FIGURE 6.6**
PDFs of $X$ and $Y$

To evaluate the limits of integration of the PDF of $Z$, we consider the following regions represented by the diagram shown in Figure 6.7.

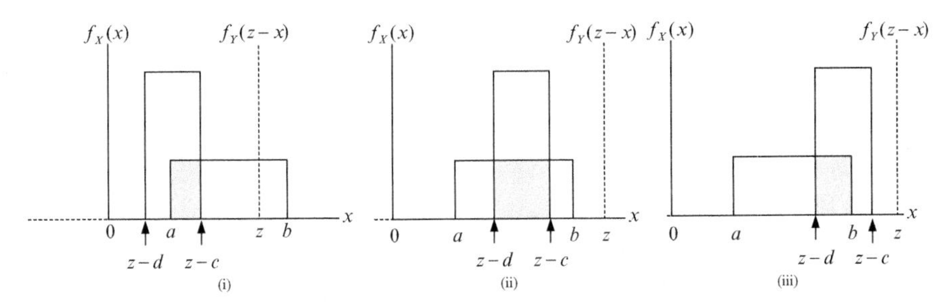


**FIGURE 6.7**
Convolution of the PDFs for Different Values of $z$

When $z<a+c, f_Z(z)=0$ because there is no overlap of the curves $f_X(x)$ and $f_Y(z-x)$. When $a+c \leq z \leq a+d$ (see Figure 6.7(*i*)), we obtain

$$f_Z(z)=\frac{1}{(b-a)(d-c)}\int_{a}^{z-c} dy=\frac{z-c-a}{(b-a)(d-c)}$$

When $a+d \leq z \leq b+c$ (see Figure 6.7(*ii*)), we obtain

$$f_Z(z)=\frac{1}{(b-a)(d-c)}\int_{z-d}^{z-c} dy=\frac{1}{b-a}$$

When $b+c \leq z \leq b+d$ (see Figure 6.7(*iii*)), we obtain

$$f_Z(z)=\frac{1}{(b-a)(d-c)}\int_{z-d}^{b} dy=\frac{b+d-z}{(b-a)(d-c)}$$

Finally, when $z>b+d, f_Z(z)=0$. Thus, the PDF of $Z$ is given by

$$f_Z(z)=\begin{cases} 0 & z<a+c \\ \dfrac{z-a-c}{(b-a)(d-c)} & a+c \leq z \leq a+d \\ \dfrac{1}{b-a} & a+d \leq z \leq b+c \\ \dfrac{b+d-z}{(b-a)(d-c)} & b+c \leq z \leq b+d \\ 0 & z>b+d \end{cases}$$

The PDF is graphically illustrated in Figure 6.8, which is a trapezoid. Note that when $b-a=d-c$, the PDF reduces to an isosceles triangle centered at $(a+c+b+d)/2$; similar to that in Figure 6.5. In the special case when $a=c$ and $b=d$, the isosceles triangle is centered at $z=a+b$.

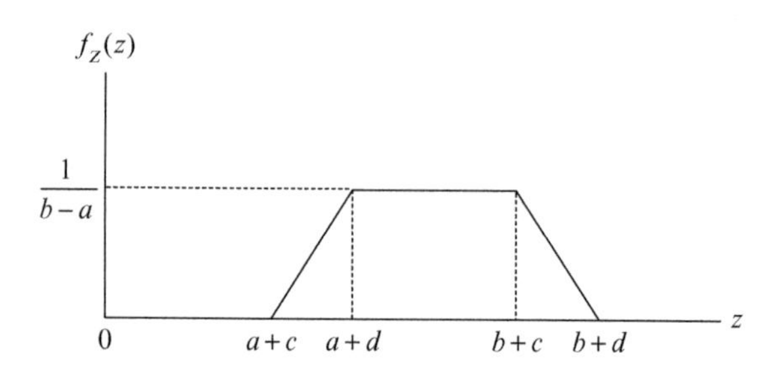


**FIGURE 6.8**
PDF of $Z=X+Y$

## EXAMPLE 6.5

The time $X$ between consecutive snowstorms in winter is a random variable with the PDF

$$f_X(x) = \begin{cases} \lambda e^{-\lambda x} & x \geq 0 \\ 0 & \text{otherwise} \end{cases}$$

Assume it has not snowed up until now. What is the PDF of the time $U$ until the second snowstorm?

**Solution:**
Let $X$ be the random variable that denotes the time until the first snowstorm from the reference time, and let $Y$ be the random variable that denotes the time between the first snowstorm and the second snowstorm. If we assume that the times between snowstorms are independent, then $X$ and $Y$ are independent and identically distributed random variables. That is, the PDF of $Y$ is given by

$$f_Y(y) = \begin{cases} \lambda e^{-\lambda y} & y \geq 0 \\ 0 & \text{otherwise} \end{cases}$$

Thus, $U=X+Y$, and the PDF of $U$ is given by

$$f_U(u) = \int_{x=0}^{\infty} f_X(x) f_Y(u-x)dx$$

The lower limit of the integration is zero because $f_X(x)=0$ when $x<0$. To obtain the upper limit of the integration we note that because $f_Y(y)=0$ when $y<0, f_Y(u-x)=0$ when $u-x<0$ (that is, when $x>u$). Thus, the range of interest in the integration is $0 \leq x \leq u$, and we obtain

$$\begin{aligned} f_U(u) &= \int_{x=0}^{u} f_X(x) f_Y(u-x)dx = \int_{x=0}^{u} \{\lambda e^{-\lambda x}\}\left\{\lambda e^{-\lambda(u-x)}\right\}dx = \lambda^2 e^{-\lambda u} \int_{x=0}^{u} dx \\ &= \lambda^2 u e^{-\lambda u} \qquad u \geq 0 \end{aligned}$$

This is the Erlang-2 distribution.

## EXAMPLE 6.6

Find the PDF of $U$, which is the sum of $X$ and $Y$ that are independent random variables with the following PDFs:

$$f_X(x) = \lambda e^{-\lambda x} \qquad x \geq 0$$

$$f_Y(y) = \lambda^2 y e^{-\lambda y} \qquad y \geq 0$$

**Solution:**
Since $X$ and $Y$ are independent and based on the argument developed in the previous example, the PDF of $U$ is given by

$$\begin{aligned} f_U(u) &= \int_{x=0}^{u} f_X(x) f_Y(u-x)dx = \int_{x=0}^{u} \{\lambda e^{-\lambda x}\}\left\{\lambda^2 (u-x) e^{-\lambda(u-x)}\right\}dx \\ &= \lambda^3 e^{-\lambda u} \int_{x=0}^{u} (u-x)dx = \lambda^3 e^{-\lambda u}\left[ux - \frac{x^2}{2}\right]_0^u = \lambda^3 e^{-\lambda u}\left[u^2 - \frac{u^2}{2}\right] \\ &= \frac{\lambda^3 u^2 e^{-\lambda u}}{2} = \frac{\lambda^3 u^2 e^{-\lambda u}}{2!} \qquad u \geq 0 \end{aligned}$$

This is the Erlang-3 distribution.

## EXAMPLE 6.7

Find the PDF of $W$, which is the sum of $X$ and $Y$ that are independent random variables with the following PDFs:

$$f_X(x) = \lambda e^{-\lambda x} \qquad x \geq 0$$

$$f_Y(y) = \mu e^{-\mu y} \qquad y \geq 0$$

where $\lambda \neq \mu$.

**Solution:**
Since $X$ and $Y$ are independent, the PDF of $W$ is given by

$$f_W(w) = \int_{-\infty}^{\infty} f_X(x) f_Y(w-x)dx$$

The limits of integration can be derived as follows based on the facts presented in Figure 6.9. The lower limit is zero because $f_X(x) = 0$ when $x < 0$. Also, because $f_Y(y) = 0$ when $y < 0$, we have that $f_Y(w-x) = 0$ when $w - x < 0$, which means that the integral is defined for $w - x \geq 0 \Rightarrow x \leq w$. Thus, the upper limit is $w$ and the range of interest in the integration is $0 \leq x \leq w$.

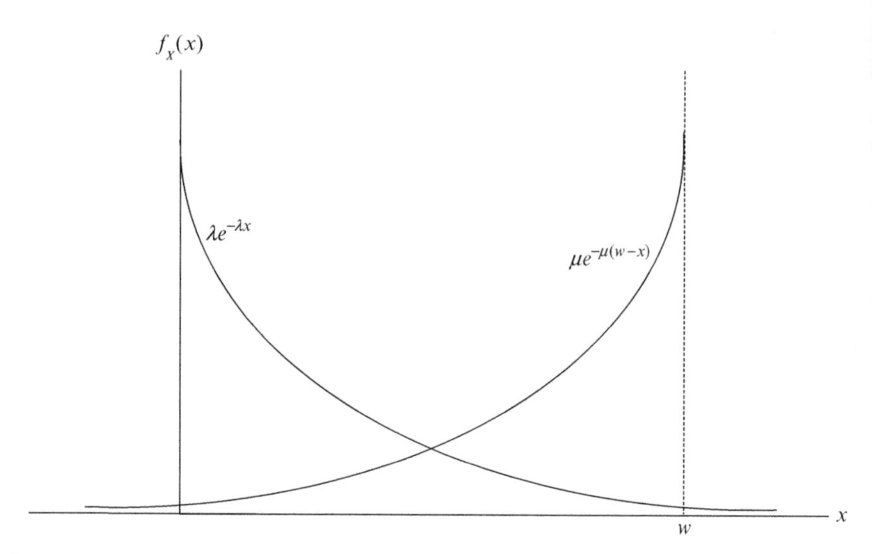


**FIGURE 6.9**
PDF of $W=X+Y$

Therefore, the PDF of $W$ is given by

$$f_W(w)=\int_{x=0}^{w} f_X(x)f_Y(w-x)dx=\int_{x=0}^{w}\{\lambda e^{-\lambda x}\}\{\mu e^{-\mu(w-x)}\}dx=\lambda\mu e^{-\mu w}\int_{x=0}^{w} e^{-(\lambda-\mu)x}dx$$
$$=\frac{\lambda\mu}{\lambda-\mu}e^{-\mu w}\left\{1-e^{-(\lambda-\mu)w}\right\}=\frac{\lambda\mu}{\lambda-\mu}\{e^{-\mu w}-e^{-\lambda w}\}$$

**Comment:**
The last three examples illustrate the fact that if $f_X(x)=0$ for $x<0$ and $f_Y(y)=0$ for $y<0$, then the PDF of the sum $U=X+Y$ is given by

$$f_U(u)=\int_{x=0}^{u} f_X(x)f_Y(u-x)dx=\int_{y=0}^{u} f_X(u-y)f_Y(y)dy$$

### 6.4.1 Moments of the Sum of Random Variables

Consider two continuous random variables $X$ and $Y$ with joint PDF $f_{XY}(x,y)$. Let the random variable $U$ be the sum of $X$ and $Y$; that is, $U=X+Y$. The mean of $U$ is given by

$$E[U]=E[X+Y]=\int_{y=-\infty}^{\infty}\int_{x=-\infty}^{\infty}(x+y)f_{XY}(x,y)dxdy$$
$$=\int_{x=-\infty}^{\infty}x\left\{\int_{y=-\infty}^{\infty}f_{XY}(x,y)dy\right\}dx+\int_{y=-\infty}^{\infty}y\left\{\int_{x=-\infty}^{\infty}f_{XY}(x,y)dx\right\}dy$$
$$=\int_{x=-\infty}^{\infty}xf_X(x)dx+\int_{y=-\infty}^{\infty}yf_Y(y)dy$$
$$=E[X]+E[Y]$$

Similarly, the variance of $U$ is given by

$$\begin{aligned}\sigma_U^2 &= E\left[\{U-E[U]\}^2\right] = E\left[\{X+Y-E[X]-E[Y]\}^2\right] \\ &= E\left[\{(X-E[X])+(Y-E[Y])\}^2\right] \\ &= E\left[(X-E[X])^2+(Y-E[Y])^2+2(X-E[X])(Y-E[Y])\right] \\ &= E\left[(X-E[X])^2\right]+E\left[(Y-E[Y])^2\right]+2E[(X-E[X])(Y-E[Y])] \\ &= \sigma_X^2+\sigma_Y^2+2\,\mathrm{Cov}(X,Y) = \sigma_X^2+\sigma_Y^2+2\sigma_{XY} \\ &= \sigma_X^2+\sigma_Y^2+2\rho_{XY}\sigma_X\sigma_Y\end{aligned} \tag{6.8}$$

where $\rho_{XY}$ is the correlation coefficient of $X$ and $Y$, $\sigma_X$ is the standard deviation of $X$ and $\sigma_Y$ is the standard deviation of $Y$. When $X$ and $Y$ are independent, $\mathrm{Cov}(X,Y)=0$, and we get $\sigma_U^2=\sigma_X^2+\sigma_Y^2$.

### 6.4.2 Sum of Discrete Random Variables

The preceding examples deal with continuous random variables. Let $U=X+Y$, where $X$ and $Y$ are discrete random variables. Then the PMF of $U$ is given by

$$p_U(u)=P[U=u]=P[X+Y=u]=\sum_{k\le u}P[X=k, Y=u-k]=\sum_{k\le u}p_{XY}(k,u-k)$$

If $X$ and $Y$ are independent random variables, then the PMF of $U$ is the convolution of the PMF of $X$ and the PMF of $Y$. That is,

$$p_U(u)=\sum_{k\le u}p_{XY}(k,u-k)=\sum_{k\le u}p_X(k)p_Y(u-k) \tag{6.9}$$

which is a discrete convolution of $p_X(x)$ and $p_Y(y)$.

## EXAMPLE 6.8

Assume that $X$ is a random variable with the PMF

$$p_X(x)=\begin{cases}\frac{1}{M} & x=0,1,2,\ldots,M-1 \\ 0 & \text{otherwise}\end{cases}$$

and $Y$ is the random variable with the PMF

$$p_Y(y)=\begin{cases}\frac{1}{N} & y=0,1,2,\ldots,N-1 \\ 0 & \text{otherwise}\end{cases}$$

where $N>M$. If $U=X+Y$, find the PMF of $U$.

**Solution:**

Since $X$ and $Y$ are independent,

$$p_U(u)=\sum_{k\le u}p_{XY}(k,u-k)=\sum_{k\le u}p_X(k)p_Y(u-k)$$

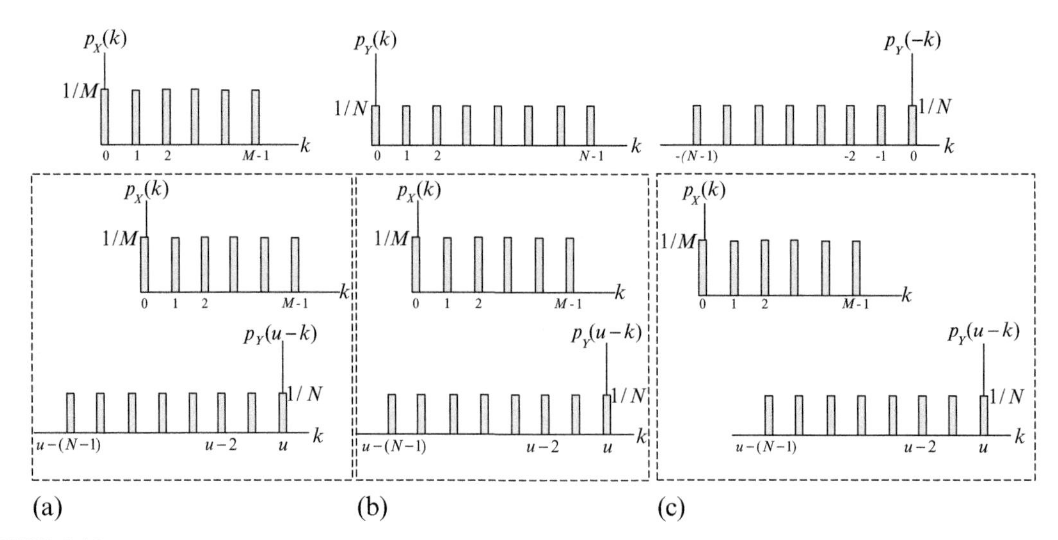


**FIGURE 6.10**
Derivation of the PMF of $U$

The plots of $p_X(k)$ and $p_Y(u-k)$ are shown in Figure 6.10.

When $u<0, p_U(u)=0$. When $0 \leq u \leq M-1$ (see Figure 6.10 (a)), $p_U(u)=(u+1)/NM$. When $M-1 \leq u \leq N-1$ (see Figure 6.10 (b)), $p_U(u)=1/N$. When $N-1 \leq u \leq M+N-2$ (see Figure 6.10 (c)), $p_U(u)=(N+M-1-u)/NM$. Finally, when $u>N+M-2, p_U(u)=0$. Thus, the PMF of U is given by the following trapezoid:

$$p_U(u)=\begin{cases} 0 & u<0 \\ \dfrac{u+1}{NM} & 0 \leq u \leq M-1 \\ \dfrac{1}{N} & M-1 \leq u \leq N-1 \\ \dfrac{N+M-1-u}{NM} & N-1 \leq u \leq N+M-2 \\ 0 & u>N+M-2 \end{cases}$$

Figure 6.11 shows the PMF of $U$.

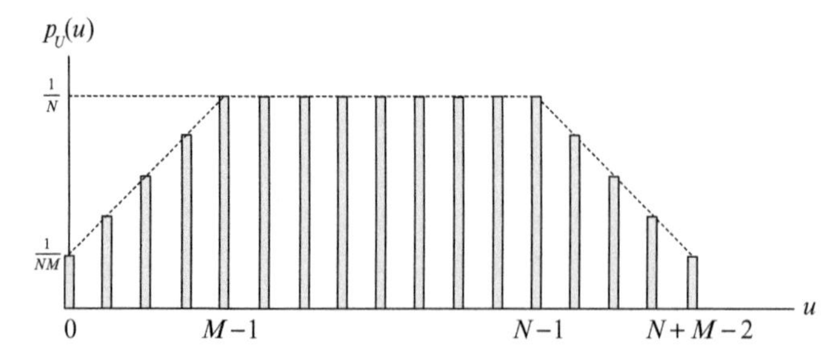


**FIGURE 6.11**
PMF of $U$

## EXAMPLE 6.9

What is the expected sum of the numbers that are obtained in 16 tosses of a fair die?

**Solution:**

Let $X$ be the random variable that denotes the sum of the numbers that appear in 16 tosses of the die. Let $X_k$ denote the number that appears on the $k$th toss, $1 \le k \le 16$. Then

$$X = X_1 + X_2 + \cdots + X_{16}$$

Thus, since the $X_k$ are identically distributed random variables,

$$E[X] = \sum_{k=1}^{16} E[X_k] = 16E[X_1]$$

Now, since each of the 6 numbers is equally likely to appear on a toss of the die,

$$p_{X_k}(x) = \begin{cases} \frac{1}{6} & x = 1, 2, \ldots, 6 \\ 0 & \text{otherwise} \end{cases}$$

Thus, $E[X_1] = (1+2+3+4+5+6)/6 = 7/2$. Therefore,

$$E[X] = 16E[X_1] = 16 \times \frac{7}{2} = 56$$

## EXAMPLE 6.10

A group of $N$ graduating seniors of a boys' high school threw their caps up into the air after their commencement exercise and walked away as the caps fell to the floor of the commencement hall. The janitor picked up the caps and put them in a room. If each student later came back and picked up a cap at random, what is the expected number of students who picked up their own caps?

**Solution:**

Let $X$ be the random variable that denotes the number of students who picked their very own caps. Let $X_k$ be the indicator random variable defined as follows:

$$X_k = \begin{cases} 1 & \text{if the } k\text{th student picked his own cap} \\ 0 & \text{otherwise} \end{cases}$$

where $1 \le k \le N$. Thus,

$$X = X_1 + X_2 + \cdots + X_N$$

If the caps are picked up randomly, then for each $k$, $E[X_k] = P[X_k = 1] = 1/N$. Thus,

$$E[X] = \sum_{k=1}^{N} E[X_k] = NE[X_1] = \frac{N}{N} = 1$$

## EXAMPLE 6.11

Students arrive at a dance hall in pairs of one boy and one girl. There are $N$ couples, and the building owner discovered that having $2N$ people in the hall violates the building code. To comply with the code, he must ask $m$ people to leave. If he selects these $m$ people in a random manner, what is the expected number of couples that would still remain in the hall?

**Solution:**

Let $X$ be the random variable that denotes the number of couples that remain in the hall after the $m$ people have been removed. Let $X_k$ be the indicator random variable defined as follows:

$$X_k = \begin{cases} 1 & \text{if the } k\text{th couple remains} \\ 0 & \text{otherwise} \end{cases}$$

where $1 \leq k \leq N$. Thus,

$$X = X_1 + X_2 + \cdots + X_N$$

For each $k$, $E[X_k] = P[X_k = 1]$. But the probability that a couple remains is the ratio of the number of ways of choosing $m$ people from $2N-2$ people to the number of ways of choosing $m$ people from $2N$ people, where the number $2N-2$ follows from the fact that the 2 people are not among those being considered for removal. That is,

$$E[X_k] = P[X_k = 1] = \frac{\binom{2N-2}{m}}{\binom{2N}{m}} = \frac{\dfrac{(2N-2)!}{(2N-m-2)!m!}}{\dfrac{(2N)!}{(2N-m)!m!}} = \frac{(2N-m)(2N-m-1)}{2N(2N-1)}$$

Thus,

$$E[X] = \sum_{k=1}^{N} E[X_k] = NE[X_1] = \frac{N(2N-m)(2N-m-1)}{2N(2N-1)} = \frac{(2N-m)(2N-m-1)}{2(2N-1)}$$

For example, if $N = 50$ and $m = 15$, then $E[X] = 36.06$, which means that 36 couples would still remain in the hall.

### 6.4.3 Sum of Independent Binomial Random Variables

Let $X$ and $Y$ be two independent binomial random variables with parameters $(n, p)$ and $(m, p)$, respectively. Let their sum be $V$; that is, $V = X + Y$. Then the PMF of $V$ is given by

$$\begin{aligned} p_V(v) &= P[V = v] = P[X + Y = v] = \sum_{k=0}^{n} P[X = k, Y = v - k] \\ &= \sum_{k=0}^{n} P[X = k] P[Y = v - k] \\ &= \sum_{k=0}^{n} \binom{n}{k} p^k (1-p)^{n-k} \binom{m}{v-k} p^{v-k} (1-p)^{m-v+k} \qquad (6.10) \\ &= p^v (1-p)^{m+n-v} \sum_{k=0}^{n} \binom{n}{k} \binom{m}{v-k} \\ &= \binom{n+m}{v} p^v (1-p)^{m+n-v} \end{aligned}$$

where we have used the combinatorial identity

$$\sum_{k=0}^{n} \binom{n}{k} \binom{m}{v-k} = \binom{n+m}{v}$$

This result shows that the sum of two independent binomial random variables with parameters $(n, p)$ and $(m, p)$ is a binomial random variable with parameter $(n+m,\ p)$.

### 6.4.4 Sum of Independent Poisson Random Variables

Let $X$ and $Y$ be two independent Poisson random variables with parameters $\lambda_X$ and $\lambda_Y$ and let $Z$ be their sum; that is, $Z=X+Y$. Consider the event $\{Z=z\}$, which can be written as a combination of two subevents: $\{Z=X+Y=z\}=\{X=k, Y=z-k\}$, $0 \le k \le z$. Thus, the PMF of $Z$ is given by

$$\begin{aligned}
p_Z(z) &= P[Z=z] = P[X+Y=z] = \sum_{k=0}^{z} P[X=k, Y=z-k] \\
&= \sum_{k=0}^{z} P[X=k]P[Y=z-k] = \sum_{k=0}^{z} \left\{ \frac{\lambda_X^k}{k!} e^{-\lambda_X} \right\} \left\{ \frac{\lambda_Y^{z-k}}{(z-k)!} e^{-\lambda_Y} \right\} \\
&= e^{-(\lambda_X+\lambda_Y)} \frac{1}{z!} \sum_{k=0}^{z} \frac{z!}{(z-k)!k!} \lambda_X^k \lambda_Y^{z-k} = \frac{e^{-(\lambda_X+\lambda_Y)}}{z!} \sum_{k=0}^{z} \binom{z}{k} \lambda_X^k \lambda_Y^{z-k} \\
&= \frac{(\lambda_X+\lambda_Y)^z}{z!} e^{-(\lambda_X+\lambda_Y)} \qquad z=0, 1, 2, \ldots
\end{aligned} \tag{6.11}$$

where we have used the binomial identity

$$\sum_{k=0}^{z} \binom{z}{k} \lambda_X^k \lambda_Y^{z-k} = (\lambda_X+\lambda_Y)^z$$

Thus, the sum of two independent Poisson random variables with parameters $\lambda_X$ and $\lambda_Y$ is a Poisson random variable with parameter $\lambda_X+\lambda_Y$.

### 6.4.5 The Spare Parts Problem

From Figure 6.2 we can see that finding the sum of independent random variables is equivalent to finding the lifetime of a system that achieves continuous operation by permitting instantaneous replacement of a component with a spare part at the time of the component's failure. One interesting issue is to find the probability that the life of the system exceeds a given value. For the case where only one spare part is available, we are basically dealing with the sum of two random variables. For the case where we have $n-1$ spare parts, we are dealing with the sum of $n$ random variables. For the case of $n=2$, we have that if the lifetime of the primary component is $X$ and the lifetime of the spare component is $Y$, where $X$ and $Y$ are independent, then the lifetime of the component $W$ and its PDF are given by

$$W = X + Y$$
$$f_W(w) = \int_0^w f_X(w-y) f_Y(y) dy$$

Thus, if we define the reliability function of the system by $R_W(w)$, the probability that the lifetime of the system exceeds the value $w_0$ is given by

$$P[W > w_0] = \int_{w_0}^{\infty} f_W(w) dw = 1 - F_W(w_0) = R_W(w_0)$$

If it is desired that $P[W > w_0] \geq \varphi$, where $0 \leq \varphi \leq 1$, then we could be required to find the parameters of $X$ and $Y$ that are necessary to achieve this goal. For example, if $X$ and $Y$ are independent and identically distributed exponential random variables with mean $1/\lambda$, we can find the smallest mean value of the random variables that can achieve this goal.

For the case of $n-1$ spare parts, the lifetime of the system $U$ is given by

$$U = X_1 + X_2 + \cdots + X_n$$

where $X_k$ is the life time of the $k$th component. If we assume that the $X_k$ are independent, the PDF of $U$ is given by the following $n$-fold convolution integral:

$$f_U(u) = \int_0^u \int_0^{x_1} \cdots \int_0^{x_{n-1}} f_{X_1}(u-x_1) f_{X_2}(x_1-x_2) \cdots f_{X_{n-1}}(x_{n-1}-x_n) \times f_{X_n}(x_n) dx_n dx_{n-1} \cdots dx_1$$

For the special case when the $X_k$ are identically distributed exponential random variables with mean $1/\lambda$, $U$ becomes an $n$th order Erlang random variable with the PDF, CDF, and reliability function given by

$$f_U(u) = \begin{cases} \dfrac{\lambda^n u^{n-1} e^{-\lambda u}}{(n-1)!} & n = 1, 2, \ldots; \ u \geq 0 \\ 0 & u < 0 \end{cases}$$
$$F_U(u) = 1 - \sum_{k=0}^{n-1} \frac{(\lambda u)^k e^{-\lambda u}}{k!} \tag{6.12}$$
$$R_U(u) = 1 - F_U(u) = \sum_{k=0}^{n-1} \frac{(\lambda u)^k e^{-\lambda u}}{k!}$$

## EXAMPLE 6.12

A system consists of one component whose lifetime is exponentially distributed with a mean of 50 hours. When the component fails, it is immediately replaced by a spare component whose lifetime is independent and identically distributed as that of the original component without the system suffering a downtime.

a. What is the probability that the system has not failed after 100 hours of operation?
b. If the mean lifetime of the component and its spare is increased by 10%, how does that affect the probability that the system exceeds a lifetime of 100 hours?

**Solution:**

Let $X$ be a random variable that denotes the lifetime of the component and let $U$ be the random variable that denotes the lifetime of the system. Then, $U$ is an Erlang-2 random variable whose PDF, CDF, and reliability function are given by

$$f_U(u) = \begin{cases} \lambda^2 u e^{-\lambda u} & u \geq 0 \\ 0 & u < 0 \end{cases}$$

$$F_U(u) = 1 - \sum_{k=0}^{1} \frac{(\lambda u)^k e^{-\lambda u}}{k!} = 1 - e^{-\lambda u}\{1 + \lambda u\}$$

$$R_U(u) = 1 - F_U(u) = \sum_{k=0}^{1} \frac{(\lambda u)^k e^{-\lambda u}}{k!} = e^{-\lambda u}\{1 + \lambda u\}$$

a. Since $1/\lambda = 50$, we have that

$$P[U > 100] = R_U(100) = e^{-100/50}\left\{1 + \frac{100}{50}\right\} = 3e^{-2} = 0.4060$$

b. When we increase the mean lifetime of the component by 10%, we obtain $1/\lambda = 50(1 + 0.1) = 55$. Thus, with the new $\lambda u = 100/55$, the corresponding value of $R_U(100)$ is

$$R_U(100) = e^{-100/55}\left\{1 + \frac{100}{55}\right\} = 2.8182e^{-1.82} = 0.4574$$

That is, the probability that the system lifetime exceeds 100 hours increases by approximately 13%.

## EXAMPLE 6.13

The time to failure of a component of a system is exponentially distributed with a mean of 100 hours. If the component fails, it is immediately replaced by an identical spare component whose time to failure is independent of that of the previous one and the system experiences no downtime in the process of component replacement. What is the smallest number of spare parts that must be used to guarantee continuous operation of the system for at least 300 hours with a probability of at least 0.95?

**Solution:**

Let $X$ be the random variable that denotes the lifetime of a component, and let the number of spare parts be $n-1$. Let $U$ be the random variable that denotes the lifetime of the system. Then $U = X_1 + X_2 + \cdots + X_n$, which is an Erlang-$n$ random variable whose reliability function is given by

$$R_U(u) = 1 - F_U(u) = \sum_{k=0}^{n-1} \frac{(\lambda u)^k e^{-\lambda u}}{k!} = e^{-\lambda u}\left\{1 + \lambda u + \frac{(\lambda u)^2}{2!} + \cdots + \frac{(\lambda u)^{n-1}}{(n-1)!}\right\} \geq 0.95$$

Since $1/\lambda = 100$, we have that

$$R_U(300) = e^{-3}\left\{1 + 3 + \frac{(3)^2}{2!} + \frac{(3)^3}{3!} + \frac{(3)^4}{4!} + \cdots + \frac{(3)^{n-1}}{(n-1)!}\right\} \geq 0.95$$

The following table shows the values of $R_U(300)$ for different values of $n$:

| $n-1$ | $R_U(300)$ |
|---|---|
| 0 | 0.0498 |
| 1 | 0.1991 |
| 2 | 0.4232 |
| 3 | 0.6472 |
| 4 | 0.8153 |
| 5 | 0.9161 |
| 6 | 0.9665 |

Thus, we see that with $n-1=5$ we cannot provide the required probability of operation, while with $n-1=6$ we can. This means that we need 6 spare components to achieve the goal.

## 6.5 MINIMUM OF TWO INDEPENDENT RANDOM VARIABLES

Consider two independent continuous random variables $X$ and $Y$. We are interested in a random variable $V$ that is the minimum of $X$ and $Y$; that is, $V = \min(X, Y)$. The random variable $V$ can be used to represent the reliability of systems with series connections, as shown in Figure 6.12. Such systems are operational as long as all components are operational. The first component to fail causes the system to fail; that is, the system has a single point of failure. Thus, if in the example shown in Figure 6.12 the times-to-failure are represented by the random variables $X$ and $Y$, then $V$ represents the time until the system fails, which is the minimum of the lifetimes of the two components.

**FIGURE 6.12**
Series Connection Modeled by the Random Variable $V$

The CDF of $V$ can be obtained as follows:

$$F_V(v) = P[V \leq v] = P[\min(X, Y) \leq v] = P[(X \leq v, X \leq Y) \cup (Y \leq v, X > Y)]$$

Since $P[A \cup B] = P[A] + P[B] - P[A \cap B]$, we have that

$$F_V(v) = P[X \leq v] + P[Y \leq v] - P[X \leq v, Y \leq v] = F_X(v) + F_Y(v) - F_{XY}(v, v)$$

Also, since $X$ and $Y$ are independent, we obtain the CDF and PDF of $U$ as follows:

$$\begin{aligned} F_V(v) &= F_X(v) + F_Y(v) - F_{XY}(v,v) = F_X(v) + F_Y(v) - F_X(v)F_Y(v) \\ f_V(v) &= \frac{d}{dv}F_V(v) = f_X(v) + f_Y(v) - f_X(v)F_Y(v) - F_X(v)f_Y(v) \\ &= f_X(v)\{1 - F_Y(v)\} + f_Y(v)\{1 - F_X(v)\} \end{aligned} \tag{6.13}$$

## EXAMPLE 6.14

Assume that $V = \min(X, Y)$, where $X$ and $Y$ are independent random variables with the respective PDFs

$$\begin{aligned} f_X(x) &= \lambda e^{-\lambda x} \quad x \geq 0 \\ f_Y(y) &= \mu e^{-\mu y} \quad y \geq 0 \end{aligned}$$

where $\lambda > 0$ and $\mu > 0$. What is the PDF of $V$?

**Solution:**
We first obtain the CDFs of $X$ and $Y$, which are as follows:

$$\begin{aligned} F_X(x) &= P[X \leq x] = \int_0^x \lambda e^{-\lambda u} du = 1 - e^{-\lambda x} \\ F_Y(y) &= P[Y \leq y] = \int_0^y \mu e^{-\mu w} dw = 1 - e^{-\mu y} \end{aligned}$$

Thus, the PDF of $V$ is given by

$$\begin{aligned} f_V(v) &= f_X(v)\{1 - F_Y(v)\} + f_Y(v)\{1 - F_X(v)\} = \lambda e^{-\lambda v} e^{-\mu v} + \mu e^{-\mu v} e^{-\lambda v} = \lambda e^{-(\lambda+\mu)v} + \mu e^{-(\lambda+\mu)v} \\ &= (\lambda + \mu) e^{-(\lambda+\mu)v} \qquad v \geq 0 \end{aligned}$$

Since $\lambda$ and $\mu$ are the failure rates of the components, the result indicates that the composite system behaves like a single unit whose failure rate is the sum of the two failure rates. More importantly, $V$ is an exponentially distributed random variable whose expected value is $E[V] = 1/(\lambda + \mu)$.

## 6.6 MAXIMUM OF TWO INDEPENDENT RANDOM VARIABLES

Consider two independent continuous random variables $X$ and $Y$. We are interested in the CDF and PDF of the random variable $W$ that is the maximum of the two random variables; that is, $W = \max(X, Y)$. The random variable $W$ can be used to represent the reliability of systems with parallel connections, as shown in Figure 6.13. In such systems, we are interested is passing a signal between the two endpoints through either the component labeled A or the component labeled B. Thus, as long as one or both components are operational, the system is operational. This implies that the system is declared to have failed when both paths become unavailable. That is, the reliability of the system depends on the reliability of the last component to fail.

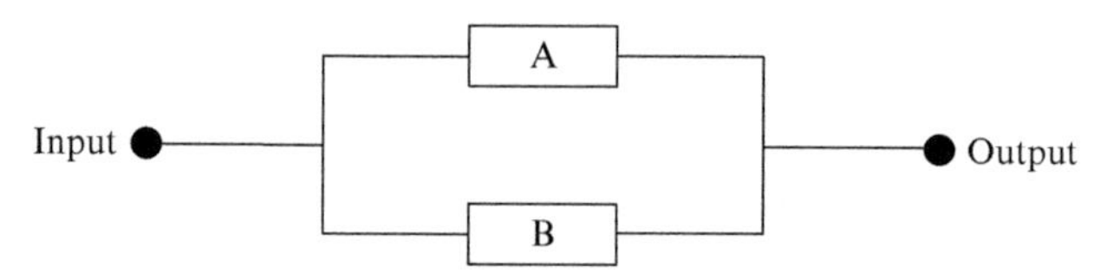


**FIGURE 6.13**
Parallel Connection Modeled by the Random Variable $W$

The CDF of $W$ can be obtained by noting that if the greater of the two random variables is less than or equal to $w$, then the smaller random variable must definitely also be less than or equal to $w$. Thus,

$$F_W(w) = P[W \le w] = P[\max(X, Y) \le w] = P[(X \le w) \cap (Y \le w)] = F_{XY}(w, w)$$

Since $X$ and $Y$ are independent, we obtain the CDF and PDF of $W$ as follows:

$$\begin{aligned} F_W(w) &= F_{XY}(w,w) = F_X(w)F_Y(w) \\ f_W(w) &= \frac{d}{dw}F_W(w) = f_X(w)F_Y(w) + F_X(w)f_Y(w) \end{aligned} \tag{6.14}$$

## EXAMPLE 6.15

Two components A and B have lifetimes $X$ and $Y$, respectively, that are independent random variables. The components are connected in parallel to create a system whose lifetime is $W$. Find the PDF of $W$ if the system needs at least one of the components to be operational and the PDFs of $X$ and $Y$ are given respectively by

$$\begin{aligned} f_X(x) &= \lambda e^{-\lambda x} \quad & x \ge 0 \\ f_Y(y) &= \mu e^{-\mu y} \quad & y \ge 0 \end{aligned}$$

where $\lambda > 0$ and $\mu > 0$.

**Solution:**
We have that $W = \max(X, Y)$. We first obtain the CDFs of $X$ and $Y$, which are as follows:

$$\begin{aligned} F_X(x) &= P[X \le x] = \int_0^x \lambda e^{-\lambda u} du = 1 - e^{-\lambda x} \\ F_Y(y) &= P[Y \le y] = \int_0^y \mu e^{-\mu w} dw = 1 - e^{-\mu y} \end{aligned}$$

Thus, the PDF of $W$ is given by

$$f_W(w) = f_X(w)F_Y(w) + F_X(w)f_Y(w) = \lambda e^{-\lambda w}(1 - e^{-\mu w}) + \mu e^{-\mu w}(1 - e^{-\lambda w})$$
$$= \lambda e^{-\lambda w} + \mu e^{-\mu w} - (\lambda + \mu)e^{-(\lambda+\mu)w} \qquad w \geq 0$$

Note that the expected value of $W$ is given by

$$E[W] = \frac{1}{\lambda} + \frac{1}{\mu} - \frac{1}{\lambda + \mu}$$

This result can be explained as follows. The mean time until the first failure is $\frac{1}{\lambda+\mu}$, according to the result in Example 6.14. After the first component has failed, the mean time until the second failure occurs is $\frac{1}{\lambda}$, if component B was the first to fail, and $\frac{1}{\mu}$, if component A was the first to fail. The probability that component A fails before component B is $\lambda/(\lambda+\mu)$, and the probability that component B fails before component A is $\mu/(\lambda+\mu)$. Thus, the mean life of the system is

$$E[W] = \frac{1}{\lambda + \mu} + \frac{1}{\lambda}\left(\frac{\mu}{\lambda + \mu}\right) + \frac{1}{\mu}\left(\frac{\lambda}{\lambda + \mu}\right) = \frac{\lambda\mu + \mu^2 + \lambda^2}{\lambda\mu(\lambda + \mu)} = \frac{2\lambda\mu + \mu^2 + \lambda^2 - \lambda\mu}{\lambda\mu(\lambda + \mu)}$$
$$= \frac{(\lambda + \mu)^2 - \lambda\mu}{\lambda\mu(\lambda + \mu)} = \frac{\lambda + \mu}{\lambda\mu} - \frac{1}{\lambda + \mu} = \frac{1}{\lambda} + \frac{1}{\mu} - \frac{1}{\lambda + \mu}$$

## 6.7 COMPARISON OF THE INTERCONNECTION MODELS

So far we have considered three models for interconnecting two components, namely the standby model, serial model, and parallel model. Figure 6.14 is a review of the three models. Let $X$ be the random variable that denotes the lifetime of component A, and $Y$ the random variable that denotes lifetime of component B.

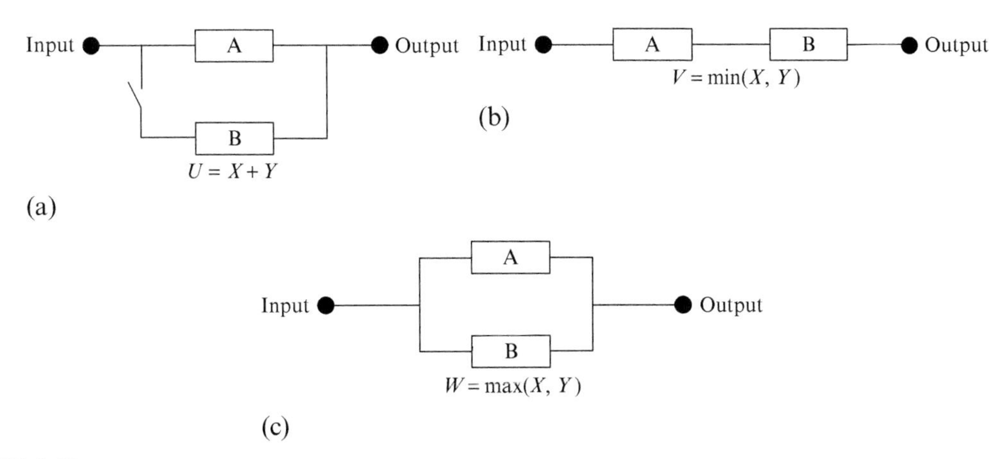


**FIGURE 6.14**
The Three Interconnection Models

We define the following random variables:

$$U = X + Y$$
$$V = \min(X, Y)$$
$$W = \max(X, Y)$$

Assume that the PDFs of $X$ and $Y$ are defined respectively as follows:

$$f_X(x) = \lambda e^{-\lambda x} \qquad x \geq 0, \lambda > 0$$
$$f_Y(y) = \mu e^{-\mu y} \qquad y \geq 0, \mu > 0$$

The following is a summary of the results obtained earlier of the PDFs of $U$, $V$ and $W$ in terms of those of $X$ and $Y$:

$$f_U(y) = \frac{\lambda\mu}{\lambda - \mu}\{e^{-\mu u} - e^{-\lambda u}\} \qquad u \geq 0$$
$$f_V(v) = (\lambda + \mu)e^{-(\lambda+\mu)v} \qquad v \geq 0$$
$$f_W(w) = \lambda e^{-\lambda w} + \mu e^{-\mu w} - (\lambda + \mu)e^{-(\lambda+\mu)w} \qquad w \geq 0$$

Similarly, the mean values of the random variables are given by

$$E[U] = \frac{1}{\lambda} + \frac{1}{\mu}$$
$$E[V] = \frac{1}{\lambda + \mu}$$
$$E[W] = \frac{1}{\lambda} + \frac{1}{\mu} - \frac{1}{\lambda + \mu}$$

From the above result, we find that $E[U] > E[W]$. We now compare $E[W]$ and $E[V]$. We have that

$$\begin{aligned} E[W] - E[V] &= \frac{1}{\lambda} + \frac{1}{\mu} - \frac{1}{\lambda+\mu} - \frac{1}{\lambda+\mu} = \frac{1}{\lambda} + \frac{1}{\mu} - \frac{2}{\lambda+\mu} \\ &= \frac{\mu(\lambda+\mu) + \lambda(\lambda+\mu) - 2\mu\lambda}{\lambda\mu(\lambda+\mu)} = \frac{\lambda^2 + \mu^2}{\lambda\mu(\lambda+\mu)} > 0 \end{aligned}$$

Thus, we have that $E[U] > E[W] > E[V]$. That is, the standby connection has the greatest mean lifetime, followed by the parallel connection, and the serial connection has the smallest mean lifetime. This result is not surprising because the failure rate of the serial connection is the sum of the failure rates of the two components, which means that the mean lifetime of the serial connection is smaller than the mean lifetime of either component. Similarly, the mean lifetime of the standby connection is the sum of the mean lifetimes of the individual components. Finally, the mean lifetime of the parallel connection is equal to the mean lifetime of the component that lasts the longer time, which means that it lies somewhere between those of the other two models.

## 6.8 TWO FUNCTIONS OF TWO RANDOM VARIABLES

Let $X$ and $Y$ be two random variables with a given joint PDF $f_{XY}(x, y)$. Assume that $U$ and $W$ are two functions of $X$ and $Y$; that is, $U = g(X, Y)$ and $W = h(X, Y)$. Sometimes it is necessary to obtain the joint PDF of $U$ and $W$, $f_{UW}(u, w)$, in terms of the PDFs of $X$ and $Y$.

It can be shown that if $(x_1, y_1), (x_2, y_2), \cdots, (x_n, y_n)$ are the real solutions to the equations $u = g(x, y)$ and $w = h(x, y)$ then $f_{UW}(u, w)$ is given by

$$f_{UW}(u, w) = \frac{f_{XY}(x_1, y_1)}{|J(x_1, y_1)|} + \frac{f_{XY}(x_2, y_2)}{|J(x_2, y_2)|} + \cdots + \frac{f_{XY}(x_n, y_n)}{|J(x_n, y_n)|} \tag{6.15}$$

where $J(x, y)$ is called the Jacobian of the transformation $\{u = g(x, y), w = h(x, y)\}$ and is defined by

$$J(x, y) = \begin{vmatrix} \frac{\partial g}{\partial x} & \frac{\partial g}{\partial y} \\ \frac{\partial h}{\partial x} & \frac{\partial h}{\partial y} \end{vmatrix} = \left(\frac{\partial g}{\partial x}\right)\left(\frac{\partial h}{\partial y}\right) - \left(\frac{\partial g}{\partial y}\right)\left(\frac{\partial h}{\partial x}\right) \tag{6.16}$$

### EXAMPLE 6.16

Let $U = g(X, Y) = X + Y$ and $W = h(X, Y) = X - Y$. Find $f_{UW}(u, w)$.

**Solution:**

The unique solution to the equations $u = x + y$ and $w = x - y$ is $x = (u + w)/2$ and $y = (u - w)/2$. Thus, there is only one set of solutions. Since

$$J(x, y) = \begin{vmatrix} \frac{\partial u}{\partial x} & \frac{\partial u}{\partial y} \\ \frac{\partial w}{\partial x} & \frac{\partial w}{\partial y} \end{vmatrix} = \begin{vmatrix} 1 & 1 \\ 1 & -1 \end{vmatrix} = -2$$

we obtain

$$f_{UW}(u, w) = \frac{f_{XY}(x, y)}{|J(x, y)|} = \frac{1}{|-2|} f_{XY}\left(\frac{u + w}{2}, \frac{u - w}{2}\right) = \frac{1}{2} f_{XY}\left(\frac{u + w}{2}, \frac{u - w}{2}\right)$$

### EXAMPLE 6.17

Find $f_{UW}(u, w)$ if $U = X^2 + Y^2$ and $W = X^2$.

**Solution:**

From the second equation we have that $x = \pm\sqrt{w}$. Substituting this value of $x$ in the first equation, we obtain $y = \pm\sqrt{(u - w)}$,which is real only when $u \geq w$. Also,

$$J(x,y)=\begin{vmatrix}\frac{\partial u}{\partial x} & \frac{\partial u}{\partial y}\\ \frac{\partial w}{\partial x} & \frac{\partial w}{\partial y}\end{vmatrix}=\begin{vmatrix}2x & 2y\\ 2x & 0\end{vmatrix}=-4xy$$

Thus,

$$\begin{aligned}f_{UW}(u,w)&=\frac{f_{XY}(\sqrt{w},\sqrt{u-w})}{4|\sqrt{w(u-w)}|}+\frac{f_{XY}(\sqrt{w},-\sqrt{u-w})}{4|-\sqrt{w(u-w)}|}+\frac{f_{XY}(-\sqrt{w},\sqrt{u-w})}{4|-\sqrt{w(u-w)}|}\\&\quad+\frac{f_{XY}(-\sqrt{w},-\sqrt{u-w})}{4|\sqrt{w(u-w)}|}\\&=\frac{f_{XY}(\sqrt{w},\sqrt{u-w})+f_{XY}(\sqrt{w},-\sqrt{u-w})+f_{XY}(-\sqrt{w},\sqrt{u-w})}{4|\sqrt{w(u-w)}|}\\&\quad+\frac{f_{XY}(-\sqrt{w},-\sqrt{u-w})}{4|\sqrt{w(u-w)}|}\end{aligned}$$

### 6.8.1 Application of the Transformation Method

Assume that $U=g(X,Y)$ and we are required to find the PDF of $U$. We can use the above transformation method by defining an auxiliary function $W=X$ or $W=Y$ so we can obtain the joint PDF $f_{UW}(u,w)$ of $U$ and $W$. Then we obtain the required marginal PDF $f_U(u)$ as follows:

$$f_U(u)=\int_{-\infty}^{\infty} f_{UW}(u,w)dw$$

## EXAMPLE 6.18

Find the PDF of the random variable $U=X+Y$, where the joint PDF of $X$ and $Y$, $f_{XY}(x,y)$, is given.

**Solution:**

We define the auxiliary random variable $W=X$. Then the unique solution to the two equations is $x=w$ and $y=u-w$, and the Jacobian of the transformation is

$$J(x,y)=\begin{vmatrix}\frac{\partial u}{\partial x} & \frac{\partial u}{\partial y}\\ \frac{\partial w}{\partial x} & \frac{\partial w}{\partial y}\end{vmatrix}=\begin{vmatrix}1 & 1\\ 1 & 0\end{vmatrix}=-1$$

Since there is only one solution to the equations, we have that

$$f_{UW}(u,w)=\frac{f_{XY}(x,y)}{|J(x,y)|}=\frac{1}{|-1|}f_{XY}(w,u-w)=f_{XY}(w,u-w)$$

$$f_U(u)=\int_{-\infty}^{\infty} f_{UW}(u,w)dw=\int_{-\infty}^{\infty} f_{XY}(w,u-w)dw$$

This reduces to the convolution integral we obtained earlier when $X$ and $Y$ are independent.

## EXAMPLE 6.19

Find the PDF of the random variable $U=X-Y$, where the joint PDF of $X$ and $Y$, $f_{XY}(x,y)$, is given.

**Solution:**

We define the auxiliary random variable $W=X$. Then the unique solution to the two equations is $x=w$ and $y=w-u$, and the Jacobian of the transformation is

$$J(x,y)=\begin{vmatrix}\frac{\partial u}{\partial x} & \frac{\partial u}{\partial y}\\ \frac{\partial w}{\partial x} & \frac{\partial w}{\partial y}\end{vmatrix}=\begin{vmatrix}1 & -1\\ 1 & 0\end{vmatrix}=1$$

Since there is only one solution to the equations, we have that

$$f_{UW}(u,w)=\frac{f_{XY}(x,y)}{|J(x,y)|}=f_{XY}(w,w-u)$$
$$f_U(u)=\int_{-\infty}^{\infty} f_{UW}(u,w)dw=\int_{-\infty}^{\infty} f_{XY}(w,w-u)dw$$

## EXAMPLE 6.20

The joint PDF of two random variables $X$ and $Y$ is given by $f_{XY}(x,y)$. If we define the random variable $U=XY$, determine the PDF of $U$.

**Solution:**

We define the auxiliary random variable $W=X$. Then the unique solution to the two equations is $x=w$ and $y=u/x=u/w$, and the Jacobian of the transformation is

$$J(x,y)=\begin{vmatrix}\frac{\partial u}{\partial x} & \frac{\partial u}{\partial y}\\ \frac{\partial w}{\partial x} & \frac{\partial w}{\partial y}\end{vmatrix}=\begin{vmatrix}y & x\\ 1 & 0\end{vmatrix}=-x=-w$$

Since there is only one solution to the equations, we have that

$$f_{UW}(u,w)=\frac{f_{XY}(x,y)}{|J(x,y)|}=\frac{1}{|w|}f_{XY}(w,u/w)$$
$$f_U(u)=\int_{-\infty}^{\infty} f_{UW}(u,w)dw=\int_{-\infty}^{\infty}\frac{1}{|w|}f_{XY}(w,u/w)dw$$

## EXAMPLE 6.21

The joint PDF of two random variables $X$ and $Y$ is given by $f_{XY}(x,y)$. If we define the random variable $V=X/Y$, determine the PDF of $V$.

**Solution:**

We define the auxiliary random variable $W=Y$. Then the unique solution to the two equations is $y=w$ and $x=vy=vw$, and the Jacobian of the transformation is

$$J(x,y)=\begin{vmatrix}\frac{\partial v}{\partial x} & \frac{\partial v}{\partial y}\\ \frac{\partial w}{\partial x} & \frac{\partial w}{\partial y}\end{vmatrix}=\begin{vmatrix}\frac{1}{y} & -\frac{x}{y^2}\\ 0 & 1\end{vmatrix}=\frac{1}{y}=\frac{1}{w}$$

Since there is only one solution to the equations, we have that

$$f_{VW}(v,w)=\frac{f_{XY}(x,y)}{|J(x,y)|}=|w|f_{XY}(vw,w)$$

$$f_V(v)=\int_{-\infty}^{\infty}f_{VW}(v,w)dw=\int_{-\infty}^{\infty}|w|f_{XY}(vw,w)dw$$

## 6.9 LAWS OF LARGE NUMBERS

There are two fundamental laws that deal with limiting behavior of probabilistic sequences. One law is called the "weak" law of large numbers, and the other is called the "strong" law of large numbers. The weak law describes how a sequence of probabilities converges, and the strong law describes how a sequence of random variables behaves in the limit. In this section we state and prove the weak law and only state the strong law.

**Proposition 6.1**

(Weak Law of Large Numbers) Let $X_1,X_2,\ldots,X_n$ be a sequence of mutually independent and identically distributed random variables each of which has a finite mean $E[X_k]=\mu_X<\infty, k=1,2,\ldots,n$. Let $S_n$ be the linear sum of the $n$ random variables; that is,

$$S_n=X_1+X_2+\cdots+X_n$$

Then for any $\varepsilon>0$,

$$\lim_{n\to\infty}P\left[\left|\frac{S_n}{n}-\mu_X\right|\geq\varepsilon\right]\to 0 \tag{6.17a}$$

Alternatively,

$$\lim_{n\to\infty}P\left[\left|\frac{S_n}{n}-\mu_X\right|<\varepsilon\right]\to 1 \tag{6.17b}$$

**Proof:**

By definition,

$$S_n = X_1 + X_2 + \cdots + X_n$$

$$\overline{S}_n = \frac{S_n}{n} = \frac{X_1 + X_2 + \cdots + X_n}{n} = \frac{n\mu_X}{n} = \mu_X$$

$$\begin{aligned}\text{Var}(\overline{S}_n) &= \text{Var}\left\{\frac{X_1 + X_2 + \cdots + X_n}{n}\right\} \\ &= \frac{1}{n^2}\{\text{Var}(X_1) + \text{Var}(X_2) + \cdots + \text{Var}(X_n)\} = \frac{n\sigma_X^2}{n^2} \\ &= \frac{\sigma_X^2}{n}\end{aligned}$$

From Chebyshev's inequality,

$$P[|\overline{S}_n - \mu_X| \geq \varepsilon] \leq \frac{\text{Var}(\overline{S}_n)}{\varepsilon^2} = \frac{\sigma_X^2}{n\varepsilon^2}$$

Thus,

$$\lim_{n\to\infty} P[|\overline{S}_n - \mu_X| \geq \varepsilon] = 0$$

which proves the theorem.

**Proposition 6.2:**
(Strong Law of Large Numbers) Let $X_1, X_2, \ldots, X_n$ be a sequence of mutually independent and identically distributed random variables each of which has a finite mean $E[X_k] = \mu_X < \infty, k = 1, 2, \ldots, n$. Let $S_n$ be the linear sum of the $n$ random variables; that is,

$$S_n = X_1 + X_2 + \cdots + X_n$$

Then for any $\varepsilon > 0$,

$$P\left[\lim_{n\to\infty} |\overline{S}_n - \mu_X| > \varepsilon\right] = 0 \tag{6.18a}$$

where $\overline{S}_n = S_n/n$. An alternative statement of the law is

$$P\left[\lim_{n\to\infty} |\overline{S}_n - \mu_X| \leq \varepsilon\right] = 1 \tag{6.18b}$$

**Comment:**
The weak law of large numbers essentially states that for any nonzero specified margin, no matter how small, there is a high probability that the average of a sufficiently large number of observations will be close to the expected value within the margin. That is,

$$\lim_{n\to\infty} \overline{S}_n \to \mu_X$$

Alternatively, the arithmetic average $\overline{S}_n$ of a sequence of independent observations of a random variable $X$ converges with probability 1 to the expected value $\mu_X$ of $X$. Thus, the weak law is a convergence statement about a sequence of probabilities; it states that the sequence of random variables $\{\overline{S}_n\}$ converges in probability to the population mean $\mu_X$ as $n$ becomes very large.

The strong law of large numbers states that with probability 1 the sequence of *sample* means $\overline{S}_n$ converges to a constant value $\mu_X$, which is the *population* mean of the random variables, as $n$ becomes very large. This validates the relative-frequency definition of probability.

## 6.10 THE CENTRAL LIMIT THEOREM

While the strong law of large numbers helps to validate the relative-frequency definition of probability, it says nothing about the limiting distribution of the sum $S_n$. The central limit theorem achieves this purpose. Let $X_1, X_2, \ldots, X_n$ be a sequence of mutually independent and identically distributed random variables each of which has a finite mean $\mu_X$ and a finite variance $\sigma_X^2$. Let

$$S_n = X_1 + X_2 + \cdots + X_n$$

The central limit theorem states that for large $n$ the distribution of $S_n$ is approximately normal, regardless of the form of the distribution of the $X_k$. That is, if we add a large number of independent and identically distributed variables together, the resulting sum will have a normal distribution, regardless of the distribution of the random variables that are added up. Now,

$$\overline{S}_n = E[S_n] = n\mu_X$$
$$\sigma_{S_n}^2 = n\sigma_X^2$$

Converting $S_n$ to standard normal random variable (i.e., zero mean and variance = 1) we obtain

$$Z_n = \frac{S_n - \overline{S}_n}{\sigma_{S_n}} = \frac{S_n - n\mu_X}{\sqrt{n\sigma_X^2}} = \frac{S_n - n\mu_X}{\sigma_X\sqrt{n}}$$

Then the central limit theorem states that if $F_{Z_n}(z)$ is the CDF of $Z_n$, then

$$\lim_{n\to\infty} F_{Z_n}(z) = \lim_{n\to\infty} P[Z_n \le z] = \frac{1}{\sqrt{2\pi}}\int_{-\infty}^{z} e^{-u^2/2}du = \Phi(z) \tag{6.19}$$

This means that $\lim_{n\to\infty} Z_n \sim N(0, 1)$. This is true for $n \ge 30$.

### EXAMPLE 6.22

Assume that the random variable $S_n$ is the sum of 48 independent experimental values of the random variable $X$ whose PDF is given by

$$f_X(x) = \begin{cases} \frac{1}{3} & 1 \le x \le 4 \\ 0 & \text{otherwise} \end{cases}$$

Find the probability that $S_n$ lies in the range $108 \le S_n \le 126$.

**Solution:**
The expected value and variance of $X$ are given by

$$E[X]=\frac{4+1}{2}=\frac{5}{2}$$
$$\sigma_X^2=\frac{(4-1)^2}{12}=\frac{3}{4}$$

Also, $S_n=X_1+X_2+\cdots+X_{48}$. Because the $X_i$ are independent and identically distributed, the mean and variance of $S_n$ are given by

$$E[S_n]=48E[X]=120$$
$$\sigma_{S_n}^2=48\sigma_X^2=36$$

Assuming that the sum approximates the normal random variable, which is usually true for $n\geq 30$, the CDF of the normalized random value of $S_n$ becomes

$$P[S_n\leq s]=F_{S_n}(s)=\Phi\left(\frac{s-E[S_n]}{\sigma_{S_n}}\right)=\Phi\left(\frac{s-120}{6}\right)$$

Therefore, the probability that $S_n$ lies in the range $108\leq S_n\leq 126$ is given by

$$\begin{aligned}P[108\leq S_n\leq 126]&=F_{S_n}(126)-F_{S_n}(108)=\Phi\left(\frac{126-120}{6}\right)-\Phi\left(\frac{108-120}{6}\right)\\&=\Phi(1)-\Phi(-2)=\Phi(1)-\{1-\Phi(2)\}=\Phi(1)+\Phi(2)-1\\&=0.8413+0.9772-1=0.8185\end{aligned}$$

where the values of $\Phi(1)$ and $\Phi(2)$ are obtained from Table 1 in Appendix 1.

## 6.11 ORDER STATISTICS

Consider an experiment in which we have $n$ identical light bulbs, labeled $1,2,\ldots,n$, that we turn on at the same time. We are interested in determining the time until each of the $n$ bulbs fails. Assume that the time $X$ until a bulb fails has a PDF $f_X(x)$ and a CDF $F_X(x)$. Let $X_1,X_2,\ldots,X_n$ denote the time until the bulbs failed. Assume we order the lifetimes of these bulbs after the experiment. Particularly, let the random variables $Y_k, k=1,2,\ldots,n$, be defined as follows.

$$Y_1=\max(X_1,X_2,\ldots,X_n)$$
$$Y_2=\text{second largest of } X_1,X_2,\ldots,X_n$$
$$\vdots$$
$$Y_n=\min(X_1,X_2,\ldots,X_n)$$

The random variables $Y_1,Y_2,\ldots,Y_n$ are called the *order statistics* corresponding to the random variables $X_1,X_2,\ldots,X_n$. In particular, $Y_k$ is called the $k$th order statistic. It is obvious that $Y_1 \geq Y_2 \geq \cdots \geq Y_n$, and in the case where the $X_k$ are continuous random variables, then $Y_1 > Y_2 > \cdots > Y_n$ with probability one.

The CDF of $Y_k, F_{Y_k}(y) = P[Y_k \leq y]$, can be computed as follows:

$$\begin{aligned}
F_{Y_k}(y) &= P[Y_k \leq y] = P[\text{at most } (k-1)\, X_i \geq y] \\
&= P[\{\text{all } X_i \leq y\} \cup \{[(n-1)\, X_i \leq y] \cap [1\, X_i \geq y]\} \cup \cdots \\
&\quad \cup \{[(n-k+1)\, X_i \leq y] \cap [(k-1)\, X_i \geq y]\}] \\
&= P[\text{all } X_i \leq y] + P[\{(n-1)\, X_i \leq y\} \cap \{1\, X_i \geq y\}] + \cdots \\
&\quad + P[\{(n-k+1)\, X_i \leq y\} \cap \{(k-1)\, X_i \geq y\}]
\end{aligned}$$

If we consider the events as results of $n$ Bernoulli trials, where in any trial we have that

$$\begin{aligned}
P[\text{success}] &= P[X_i \leq y] = F_X(y) \\
P[\text{failure}] &= P[X_i > y] = 1 - F_X(y)
\end{aligned}$$

then we obtain the result:

$$\begin{aligned}
F_{Y_k}(y) &= P[n \text{ successes}] + P[(n-1) \text{ successes}] + \cdots + P[(n-k+1) \text{ successes}] \\
&= [F_X(y)]^n + \binom{n}{n-1}[F_X(y)]^{n-1}[1 - F_X(y)] + \cdots \\
&\quad + \binom{n}{n-k+1}[F_X(y)]^{n-k+1}[1 - F_X(y)]^{k-1} \\
&= \sum_{m=0}^{k-1}\binom{n}{n-m}[F_X(y)]^{n-m}[1 - F_X(y)]^m
\end{aligned} \tag{6.20}$$

There are two ways we can obtain the PDF of $Y_k$. One way is to differentiate $F_{Y_k}(y)$ to obtain $f_{Y_k}(y)$. The other way, which we develop here, proceeds as follows:

$$\begin{aligned}
f_{Y_k}(y)dy &= P[Y_k \approx y] = P[1\, X_i \approx y, (k-1)\, X_i \geq y, (n-k)\, X_i \leq y] \\
&= \frac{n!}{1!(k-1)!(n-k)!}[f_X(y)dy]^1[1 - F_X(y)]^{k-1}[F_X(y)]^{n-k}
\end{aligned}$$

When we cancel out the $dy$'s we obtain

$$f_{Y_k}(y) = \frac{n!}{(k-1)!(n-k)!} f_X(y)[1 - F_X(y)]^{k-1}[F_X(y)]^{n-k} \tag{6.21}$$

## EXAMPLE 6.23

Assume that the random variables $X_1, X_2, \ldots, X_{10}$ are independent and identically distributed with the common PDF $f_X(x)$ and common CDF $F_X(x)$. Find the PDF and CDF of the following:

a. the 3rd largest random variable
b. the 5th largest random variable
c. the largest random variable
d. the smallest random variable

**Solution:**

a. The 3rd largest random variable is obtained by substituting $k=3$ and $n=10$ in the results in (6.20) and (6.21). Thus,

$$F_{Y_3}(y) = [F_X(y)]^{10} + \binom{10}{9}[F_X(y)]^9[1-F_X(y)] + \binom{10}{8}[F_X(y)]^8[1-F_X(y)]^2$$

$$= [F_X(y)]^{10} + 10[F_X(y)]^9[1-F_X(y)] + 45[F_X(y)]^8[1-F_X(y)]^2$$

$$f_{Y_3}(y) = \frac{10!}{2!7!} f_X(y)[1-F_X(y)]^2[F_X(y)]^7 = 360 f_X(y)[1-F_X(y)]^2[F_X(y)]^7$$

b. The 5th largest random variable is obtained by substituting $k=5$ and $n=10$, as follows:

$$F_{Y_5}(y) = [F_X(y)]^{10} + \binom{10}{9}[F_X(y)]^9[1-F_X(y)] + \binom{10}{8}[F_X(y)]^8[1-F_X(y)]^2$$

$$+ \binom{10}{7}[F_X(y)]^7[1-F_X(y)]^3 + \binom{10}{6}[F_X(y)]^6[1-F_X(y)]^4$$

$$= [F_X(y)]^{10} + 10[F_X(y)]^9[1-F_X(y)] + 45[F_X(y)]^8[1-F_X(y)]^2$$

$$+ 120[F_X(y)]^7[1-F_X(y)]^3 + 210[F_X(y)]^6[1-F_X(y)]^4$$

$$f_{Y_5}(y) = \frac{10!}{4!5!} f_X(y)[1-F_X(y)]^4[F_X(y)]^5 = 1260 f_X(y)[1-F_X(y)]^4[F_X(y)]^5$$

c. The largest random variable implies that $k=1$ in the formula with $n=10$. Thus, we obtain

$$F_{Y_1}(y) = [F_X(y)]^{10}$$

$$f_{Y_1}(y) = \frac{10!}{(1-1)!(10-1)!} f_X(y)[1-F_X(y)]^{1-1}[F_X(y)]^{10-1} = 10 f_X(y)[F_X(y)]^9$$

d. The smallest random variable implies that $k=10$. Thus, we obtain

$$F_{Y_{10}}(y) = \sum_{m=0}^{9}\binom{10}{10-m}[F_X(y)]^{10-m}[1-F_X(y)]^m$$

$$f_{Y_{10}}(y) = \frac{10!}{(10-1)!(10-10)!} f_X(y)[1-F_X(y)]^{10-1}[F_X(y)]^{10-10} = 10 f_X(y)[1-F_X(y)]^9$$

## EXAMPLE 6.24

Assume that the random variables $X_1, X_2, \ldots, X_{32}$ are independent and identically distributed with the common PDF $f_X(x)$ and common CDF $F_X(x)$. Find the PDF and CDF of the following:

a. the 4th largest random variable
b. the 27th largest random variable
c. the largest random variable
d. the smallest random variable
e. What is the probability that the 4th largest random variable has a value between 8 and 9?

**Solution:**

From (6.20) and (6.21) the CDF and PDF of the $k$th largest random variable are given respectively by

$$F_{Y_k}(y) = \sum_{m=0}^{k-1} \binom{n}{n-m} [F_X(y)]^{n-m}[1-F_X(y)]^m$$

$$f_{Y_k}(y) = \frac{n!}{(k-1)!(n-k)!} f_X(y)[1-F_X(y)]^{k-1}[F_X(y)]^{n-k}$$

a. The CDF and PDF of the 4th largest random variable are obtained by substituting $k=4$ and $n=32$ in the above equation; that is,

$$\begin{aligned} F_{Y_4}(y) &= [F_X(y)]^{32} + \binom{32}{31}[F_X(y)]^{31}[1-F_X(y)] + \binom{32}{30}[F_X(y)]^{30}[1-F_X(y)]^2 \\ &\quad + \binom{32}{29}[F_X(y)]^{29}[1-F_X(y)]^3 \\ &= [F_X(y)]^{32} + 32[F_X(y)]^{31}[1-F_X(y)] + 496[F_X(y)]^{30}[1-F_X(y)]^2 \\ &\quad + 4960[F_X(y)]^{29}[1-F_X(y)]^3 \end{aligned}$$

$$f_{Y_4}(y) = \frac{32!}{3!28!} f_X(y)[1-F_X(y)]^3[F_X(y)]^{28} = 14384060 f_X(y)[1-F_X(y)]^3[F_X(y)]^{28}$$

b. The CDF and PDF of the 27th largest random variable are obtained by substituting $k=27$ as follows:

$$F_{Y_{27}}(y) = \sum_{m=0}^{26} \binom{32}{32-m} [F_X(y)]^{32-m}[1-F_X(y)]^m$$

$$f_{Y_k}(y) = \frac{32!}{26!5!} f_X(y)[1-F_X(y)]^{26}[F_X(y)]^5 = 5437152 f_X(y)[1-F_X(y)]^{26}[F_X(y)]^5$$

c. The largest random variable implies that $k=1$. Thus, we obtain we obtain the CDF and PDF as follows:

$$F_{Y_1}(y) = [F_X(y)]^{32}$$

$$f_{Y_1}(y) = \frac{32!}{(1-1)!(32-1)!} f_X(y)[1-F_X(y)]^{1-1}[F_X(y)]^{32-1} = 32 f_X(y)[F_X(y)]^{31}$$

d. The smallest random variable implies that $k=32$. Thus, we obtain the CDF and PDF as follows:

$$F_{Y_{32}}(y)=\sum_{m=0}^{31}\binom{32}{32-m}[F_X(y)]^{32-m}[1-F_X(y)]^m$$

$$f_{Y_{32}}(y)=\frac{32!}{(32-1)!(32-32)!}f_X(y)[1-F_X(y)]^{32-1}[F_X(y)]^{32-32}=32f_X(y)[1-F_X(y)]^{31}$$

e. The probability that the 4th largest random variable lies between 8 and 9 is given by

$$P[8<Y_4<9]=F_{Y_4}(9)-F_{Y_4}(8)$$

where $F_{Y_4}(y)$ is as derived earlier in part (a).

## 6.12 CHAPTER SUMMARY

This chapter discussed how to model several functions of random variables. The concept of sums of random variable is related to the idea of systems with standby redundancy. The concept is extended to the so-called spare parts problem. Other functions discussed include the minimum of two random variables, which is used to model systems connected in series; the maximum of two random variables, which is used to model systems connected in parallel; the central limit theorem; the laws of large numbers; order statistics, which is concerned with arranging a set of observation data in an increasing order; and two functions of two random variables, which are analyzed using a transformation method.

## 6.13 PROBLEMS

### Section 6.2 Functions of One Random Variable

6.1 Suppose $X$ is a random variable and $Y=aX-b$, where $a$ and $b$ are constants. Find the PDF, expected value and variance of $Y$.

6.2 If $Y=aX^2+b$, where $a>0$ is a constant, $b$ is a constant and the PDF of $X$, $f_X(x)$, is known, find
   a. the PDF of $Y$
   b. the PDF of $Y$ when $f_X(x)$ is an even function

6.3 If $Y=aX^2$, where $a>0$ is a constant and the mean and other moments of $X$ are known, determine the following in terms of the moments of $X$:
   a. the mean of $Y$
   b. the variance of $Y$.

6.4 If $Y=|X|$ and the PDF of $X$, $f_X(x)$, is known, find the PDF of $Y$ in terms of $f_X(x)$.

6.5 The random variable $X$ has the following PDF:

$$f_X(x) = \begin{cases} \frac{1}{3} & -1 < x < 2 \\ 0 & \text{otherwise} \end{cases}$$

If we define $Y = 2X + 3$, what is the PDF of $Y$?

6.6 Assume that $Y = a^X$, where $a > 0$ is a constant and the PDF of $X$, $f_X(x)$, is known.

a. Determine the PDF of $Y$ in terms of the PDF of $X$

b. Find the PDF of $Y$ for the special case where $Y = e^X$ and the PDF of $X$ is given by

$$f_X(x) = \begin{cases} 1 & 0 < x < 1 \\ 0 & \text{otherwise} \end{cases}$$

6.7 Assume that $Y = \ln X$, where the PDF of $X$, $f_X(x)$, is known. Find the PDF of $Y$ in terms of the PDF of $X$.

## Section 6.4 Sums of Random Variables

6.8 A random variable $X$ has the PDF $f_X(x) = 2x, 0 \le x \le 1$, and 0 elsewhere. Independently of $X$, a random variable $Y$ is uniformly distributed between $-1$ and 1. Let the random variable $W = X + Y$. Determine the PDF of $W$.

6.9 $X$ and $Y$ are two independent random variables with PDFs

$$f_X(x) = 4e^{-4x} \quad x \ge 0$$
$$f_Y(y) = 2e^{-2y} \quad y \ge 0$$

If we define the random variable $U = X + Y$, find

a. the PDF of $U$

b. $P[U > 0.2]$

6.10 Suppose we roll two dice. Let the random variables $X$ and $Y$ denote the numbers that appear on the dice. What is the expected value of $X + Y$?

6.11 Assume that $X$ is a random variable that denotes the sum of the outcomes of two tosses of a fair coin. Denote the outcome "heads" by 1 and the outcome "tails" by 0. What is the expected value of $X$?

6.12 Suppose we select 4 students at random from a class of 10 boys and 12 girls. Let the random variable $X$ denote the number of boys selected, and let the random variable $Y$ denote the number of girls selected. What is $E[X - Y]$?

6.13 Suppose we put 8 balls randomly into 5 boxes. What is the expected number of empty boxes?

6.14 Two coins A and B are used in an experiment. Coin A is a biased coin that has a probability of heads equal to 1/4 and a probability of tails equal to 3/4. Coin B is a fair coin. Each coin is tossed four times. Let $X$ be the random variable that denotes the number of heads resulting from coin A, and let $Y$ be the random variable that denotes the number of heads resulting from coin B. Determine the following:

a. The probability that $X = Y$
b. The probability that $X > Y$
c. The probability that $X + Y \leq 4$

6.15 Two random variables $X$ and $Y$ have the joint PDF given by

$$f_{XY}(x, y) = 4xy \qquad 0 < x < 1, 0 < y < 1$$

If we define the random variable $U = X + Y$, find the PDF of $U$.

## Sections 6.4 and 6.5 Maximum and Minimum of Independent Random Variables

6.16 Suppose we roll two dice. Let the random variables $X$ and $Y$ denote the numbers that appear on the dice. What is the expected value of

a. $\max(X, Y)$
b. $\min(X, Y)$?

6.17 A system consists of two components A and B that are connected in series. If the lifetime of A is exponentially distributed with a mean 200 hours and the lifetime of B is exponentially distributed with a mean of 400 hours, what is the PDF of $X$, the time until the system fails?

6.18 A system consists of two components A and B that are connected in parallel. If the lifetime of A is exponentially distributed with a mean 200 hours and the lifetime of B is exponentially distributed with a mean of 400 hours, what is the PDF of $Y$, the time until the system fails?

6.19 The random variables $X_1, X_2, X_3, X_4, X_5$ are independent and identically distributed exponential random variables with parameter $\lambda$. Find the following probability: $P[\max(X_1, X_2, X_3, X_4, X_5) \leq a]$.

6.20 A system consists of three independent components $X$, $Y$, and $Z$ whose lifetimes are exponentially distributed with means $1/\lambda_X$, $1/\lambda_Y$ and $1/\lambda_Z$, respectively. Determine the PDF and expected value of $W$, the time until the system fails (or the lifetime of the system), under the following system configurations:

a. The components are connected in series
b. The components are connected in parallel
c. The components are connected in a backup mode with $X$ used first, then $Y$, and then $Z$.

## Section 6.8 Two Functions of Two Random Variables

6.21 Two independent random variables $X$ and $Y$ have variances $\sigma_X^2=9$ and $\sigma_Y^2=25$, respectively. If we define two new random variables $U$ and $V$ as follows:

$$U = 2X + 3Y$$
$$V = 4X - 2Y$$

a. Find the variances of $U$ and $V$
b. Find the correlation coefficient of $U$ and $V$
c. Find the joint PDF of $U$ and $V$ in terms of $f_{XY}(x,y)$.

6.22 Two random variables $X$ and $Y$ have variances $\sigma_X^2=16$ and $\sigma_Y^2=36$. If their correlation coefficient is 0.5, determine the following:
a. The variance of the sum of $X$ and $Y$
b. The variance of the difference of $X$ and $Y$.

6.23 The joint PDF of two continuous random variables $X$ and $Y$ is given by

$$f_{XY}(x,y) = \begin{cases} e^{-(x+y)} & x \geq 0, y \geq 0 \\ 0 & \text{otherwise} \end{cases}$$

If we define the random variable $W=X/Y$, find the PDF of $W$.

6.24 Let $X$ and $Y$ be two independent random variables that are uniformly distributed between 0 and 1. If we define $Z=XY$, find the PDF of $Z$.

6.25 Suppose $X$ and $Y$ are independent and identically distributed geometric random variables with success parameter $p$. Find the PMF of $S=X+Y$.

6.26 Three independent continuous random variables $X$, $Y$, and $Z$ are uniformly distributed between 0 and 1. If the random variable $S=X+Y+Z$, determine the PDF of $S$.

6.27 Suppose $X$ and $Y$ are two continuous random variables with the joint PDF $f_{XY}(x,y)$. Let the functions $U$ and $W$ be defined as follows: $U=2X+3Y$, and $W=X+2Y$. Find the joint PDF $f_{UW}(u,w)$.

6.28 Find $f_{UW}(u,w)$ in terms of $f_{XY}(x,y)$ if $U=X^2+Y^2$ and $W=X^2-Y^2$.

6.29 $X$ and $Y$ are independent normal random variables, where $X \sim N(\mu_X, \sigma_X^2)$ and $Y \sim N(\mu_Y, \sigma_Y^2)$. If we define $U=X+Y$ and $W=X-Y$, find the joint PDF of $U$ and $W$. (Note: Give the explicit expression for $f_{UW}(u,w)$.)

## Section 6.10 The Central Limit Theorem

6.30 If 30 fair dice are rolled, what is the probability that the sum obtained is between 95 and 125?

6.31 $X_1, X_2, \ldots, X_{35}$ are independent random variables each of which is uniformly distributed between 0 and 1. Let $S=X_1+X_2+\cdots+X_{35}$. What is the probability that $S>22$?

6.32 The random variable $X$ is uniformly distributed between 1 and 2. If $S$ is the sum of 40 independent experimental values of $X$, evaluate $P[55 < S \leq 65]$.

6.33 Consider the number of times $K$ that the number 4 appears in 600 tosses of a fair die. Determine the probability that the number appears 100 times using the following methods:

a. Stirling's formula, which states that $n! \approx \sqrt{2\pi n}\left(\frac{n}{e}\right)^n = n^n e^{-n}\sqrt{2\pi n}$

b. The Poisson approximation to the binomial distribution

c. The central limit theorem by replacing $K = 100$ with $99.5 < K < 100.5$

## Section 6.11 Order Statistics

6.34 A machine has 7 identical components that operate independently with respective lifetimes $X_1, X_2, \ldots, X_7$ hours. Their common PDF and CDF are $f_X(x)$ and $F_X(x)$, respectively. Find the probability that the machine lasts at most 5 hours if

a. It keeps going until all its components fail.

b. It fails as soon as one of its components fails.

c. It fails when it has only one component that has not failed.

6.35 A machine needs 4 out of its 6 identical independent components to operate. Let $X_1, X_2, \ldots, X_6$ denote the respective lifetimes of the components, and assume that each component's lifetime is exponentially distributed with a mean of $1/\lambda$ hours. Find

a. The CDF of the machine's lifetime.

b. The PDF of the machine's lifetime.

6.36 Assume that the random variables $X_1, X_2, \ldots, X_6$ are independent and identically distributed with the common PDF $f_X(x)$ and common CDF $F_X(x)$. Find the PDF and CDF of the following:

a. The 2nd largest random variable.

b. The maximum random variable.

c. The minimum random variable.

CHAPTER 7

# Transform Methods

## 7.1 INTRODUCTION

Different types of transforms are used in science and engineering. These include the z-transform, Laplace transform, and Fourier transform. One of the reasons for their popularity is that when they are introduced into the solution of many problems, the calculations become greatly simplified. For example, many solutions of equations that involve derivatives and integrals of functions are given as the convolution of two functions: $a(x) * b(x)$. As students of signal and systems know, the Fourier transform of a convolution is the product of the individual Fourier transforms. That is, if $F\{g(x)\}$ is the Fourier transform of the function $g(x)$, then

$$F\{a(x)*b(x)\} = A(w)B(w)$$

where $A(w)$ is the Fourier transform of $a(x)$ and $B(w)$ is the Fourier transform of $b(x)$. This means that the complicated convolution operation can be replaced by the much simpler multiplication operation. In fact, sometimes transform methods are the only tools available for solving some types of problems.

This chapter discusses how transform methods are used in probability theory. We consider three types of transforms: characteristic functions, the z-transform (or moment generating function) of PMFs, and the s-transform (or Laplace transform) of PDFs. The z-transform and the s-transform are particularly used when random variables take only nonnegative values. Thus, the s-transform is essentially the one-sided Laplace transform of a PDF. Examples of this class of random variables are frequently encountered in many engineering problems, such as the number of customers that arrive at the bank or the time until a component fails. We are interested in the s-transforms of PDFs and z-transforms of PMFs and not those of arbitrary functions. As a result these transforms satisfy certain conditions that relate to their probabilistic origin.

## 7.2 THE CHARACTERISTIC FUNCTION

Let $f_X(x)$ be the PDF of the continuous random variable $X$. The characteristic function of $X$ is defined by



Fundamentals of Applied Probability and Random Processes. http://dx.doi.org/10.1016/B978-0-12-800852-2.00007-9

$$\Phi_X(w) = E[e^{jwX}] = \int_{-\infty}^{\infty} e^{jwx} f_X(x)dx \tag{7.1}$$

where $j = \sqrt{-1}$. Note that because $e^{jwx} = \cos(wx) + j\sin(wx)$, $\Phi_X(w)$ is generally a complex function; that is, $\Phi_X(w) = U_X(w) + jV_X(w)$, where

$$U_X(w) = E[\cos(wx)] = \int_{-\infty}^{\infty} \cos(wx) f_X(x)dx$$
$$V_X(w) = E[\sin(wx)] = \int_{-\infty}^{\infty} \sin(wx) f_X(x)dx$$

We can obtain $f_X(x)$ from $\Phi_X(w)$ as follows:

$$f_X(x) = \frac{1}{2\pi}\int_{-\infty}^{\infty} e^{-jwx}\Phi_X(w)dw \tag{7.2}$$

If $X$ is a discrete random variable with PMF $p_X(x)$, the characteristic function is given by

$$\Phi_X(w) = \sum_{x=-\infty}^{\infty} e^{jwx} p_X(x) \tag{7.3}$$

Note that $\Phi_X(0) = 1$, which is a test of whether a given function of $w$ is a true characteristic function of a random variable.

### 7.2.1 Moment-Generating Property of the Characteristic Function

One of the primary reasons for studying the transform methods is to use them to derive the moments of the different probability distributions. By definition

$$\Phi_X(w) = \int_{-\infty}^{\infty} e^{jwx} f_X(x)dx$$

Taking the derivatives of $\Phi_X(w)$, we obtain

$$\frac{d}{dw}\Phi_X(w) = \frac{d}{dw}\int_{-\infty}^{\infty} e^{jwx} f_X(x)dx = \int_{-\infty}^{\infty} \frac{d}{dw} e^{jwx} f_X(x)dx = \int_{-\infty}^{\infty} jxe^{jwx} f_X(x)dx$$

$$\left.\frac{d}{dw}\Phi_X(w)\right|_{w=0} = \int_{-\infty}^{\infty} jx f_X(x)dx = jE[X]$$

$$\frac{d^2}{dw^2}\Phi_X(w) = \frac{d}{dw}\int_{-\infty}^{\infty} jxe^{jwx} f_X(x)dx = \int_{-\infty}^{\infty} j^2x^2e^{jwx} f_X(x)dx$$

$$\left.\frac{d^2}{dw^2}\Phi_X(w)\right|_{w=0} = \int_{-\infty}^{\infty} j^2x^2 f_X(x)dx = j^2E[X^2] = -E[X^2]$$

In general,

$$\frac{d^n}{dw^n}\Phi_X(w)\bigg|_{w=0} = j^n E[X^n] \tag{7.4}$$

## EXAMPLE 7.1

Find the mean and second moment of the random variable whose PDF has the characteristic function

$$\Phi_X(w) = \exp\left(jw\mu_X - \frac{w^2\sigma_X^2}{2}\right)$$

Solution:

The first and second derivatives of $\Phi_X(w)$ are given by

$$\frac{d}{dw}\Phi_X(w) = (j\mu_X - w\sigma_X^2)\exp\left(jw\mu_X - \frac{w^2\sigma_X^2}{2}\right)$$

$$\frac{d^2}{dw^2}\Phi_X(w) = \left\{(j\mu_X - w\sigma_X^2)^2 - \sigma_X^2\right\}\exp\left(jw\,\mu_X - \frac{w^2\sigma_X^2}{2}\right)$$

Thus, we obtain

$$E[X] = \frac{1}{j}\frac{d}{dw}\Phi_X(w)\bigg|_{w=0} = (j\mu_X - w\sigma_X^2)\exp\left(jw\,\mu_X - \frac{w^2\sigma_X^2}{2}\right)\bigg|_{w=0} = \mu_X$$

$$E\left[X^2\right] = -\frac{d^2}{dw^2}\Phi_X(w)\bigg|_{w=0} = -\left\{(j\mu_X - w\sigma_X^2)^2 - \sigma_X^2\right\}\exp\left(jw\,\mu_X - \frac{w^2\sigma_X^2}{2}\right)\bigg|_{w=0}$$

$$= \mu_X^2 + \sigma_X^2$$

### 7.2.2 Sums of Independent Random Variables

Let $Y = X_1 + X_2 + \cdots + X_n$, where the $X_i$ are independent and identically distributed random variables. Generally, the PDF of $Y$ is the $n$-fold convolution of the PDFs of the $X_i$. The characteristic function of $Y$ is given by

$$\begin{aligned}\Phi_Y(w) &= E\left[e^{jwY}\right] = E\left[e^{jw(X_1+X_2+\cdots+X_n)}\right] = E\left[e^{jwX_1}e^{jwX_2}\cdots e^{jwX_n}\right] \\ &= E\left[e^{jwX_1}\right]E\left[e^{jwX_2}\right]\cdots E\left[e^{jwX_n}\right] = \Phi_{X_1}(w)\Phi_{X_2}(w)\cdots\Phi_{X_n}(w) \\ &= [\Phi_X(w)]^n\end{aligned} \tag{7.5}$$

where the fourth equality is due to the fact that the $X_i$ are independent, and the last equality is due to the fact that they are identically distributed. This illustrates one of the advantages of using the characteristic function: It has reduced the complicated $n$-fold convolution operation to a simple multiplication, as we stated earlier.

### 7.2.3 The Characteristic Functions of Some Well-Known Distributions

We consider the characteristic functions of some of the distributions we considered in Chapter 4.

a. **Bernoulli Distribution**: Recall that the PMF of the Bernoulli random variable with parameter $p$ is given by

$$p_X(x) = \begin{cases} 1-p & x=0 \\ p & x=1 \end{cases}$$

Thus, the characteristic function is given by

$$\Phi_X(w) = \sum_{x=-\infty}^{\infty} e^{jwx} p_X(x) = 1 - p + pe^{jw} \tag{7.6}$$

b. **Binomial Distribution:** Recall that the PMF of the binomial random variable $X(n) \sim B(n,p)$ is

$$p_{X(n)}(x) = \binom{n}{x} p^x (1-p)^{n-x} \quad x = 0,1,\ldots,n$$

Since the binomial random variable $X(n)$ is the sum of $n$ independent and identically distributed Bernoulli random variables, we use the results in (7.5) and (7.6) to obtain the characteristic function of $X(n)$ as

$$\Phi_{X(n)}(w) = \left[1 - p + pe^{jw}\right]^n \tag{7.7}$$

c. **Geometric Distribution:** Recall that the PMF of the geometric random variable $X$ with parameter $p$ is given by

$$p_X(x) = p(1-p)^{x-1} \quad x = 1,2,\ldots$$

Thus, the characteristic function is given by

$$\begin{aligned} \Phi_X(w) &= \sum_{x=-\infty}^{\infty} e^{jwx} p_X(x) = \sum_{x=1}^{\infty} e^{jwx} p(1-p)^{x-1} \\ &= pe^{jw} \sum_{x=1}^{\infty} e^{jw(x-1)} (1-p)^{x-1} = pe^{jw} \sum_{x=1}^{\infty} \left\{e^{jw}(1-p)\right\}^{x-1} \\ &= \frac{pe^{jw}}{1 - e^{jw}(1-p)} \end{aligned} \tag{7.8}$$

d. **Pascal Distribution:** Recall that the PMF of the Pascal random variable of order $k$ (or Pascal-$k$ random variable), $X_k$, and parameter $p$ is given by

$$p_{X_k}(n) = \binom{n-1}{k-1} p^k (1-p)^{n-k} \quad k = 1,2,\ldots;\ n = k, k+1,\ldots$$

Because the Pascal-$k$ random variable is the sum of $k$ independent and identically distributed geometric random variables, we use the results of equations (7.5) and (7.8) to obtain the characteristic function of $X_k$ as

$$\Phi_{X_k}(w) = \left\{\frac{pe^{jw}}{1-e^{jw}(1-p)}\right\}^k \tag{7.9}$$

e. **Poisson Distribution:** Recall that the PMF of the Poisson distribution is given by

$$p_X(x) = \frac{\lambda^x e^{-\lambda}}{x!} \quad x = 0, 1, 2, \ldots; \ \lambda > 0$$

Thus, the characteristic function is given by

$$\begin{aligned}\Phi_X(w) &= \sum_{x=-\infty}^{\infty} e^{jwx} p_X(x) = \sum_{x=0}^{\infty} e^{jwx} \frac{\lambda^x e^{-\lambda}}{x!} = e^{-\lambda} \sum_{x=0}^{\infty} \frac{(\lambda e^{jw})^x}{x!} = e^{-\lambda} e^{\lambda e^{jw}} \\ &= e^{-\lambda(1-e^{jw})}\end{aligned} \tag{7.10}$$

f. **Exponential Distribution:** Recall that the PDF of the exponential random variable with parameter $\lambda > 0$ is

$$f_X(x) = \lambda e^{-\lambda x} \quad x \geq 0$$

Therefore, the characteristic function is given by

$$\begin{aligned}\Phi_X(w) &= E\left[e^{jwX}\right] = \int_0^{\infty} e^{jwx} \lambda e^{-\lambda x} dx = \lambda \int_0^{\infty} e^{-(\lambda - jw)x} dx \\ &= \frac{\lambda}{\lambda - jw}\end{aligned} \tag{7.11}$$

g. **Erlang Distribution:** Recall that the PDF of the Erlang random variable of order $k$ (or Erlang-$k$ random variable), $X_k$, is given by

$$f_{X_k}(x) = \frac{\lambda^k x^{k-1}}{(k-1)!} e^{-\lambda x} \quad x \geq 0$$

Because the Erlang-$k$ random variable is the sum of $k$ exponential random variables, we use the results of equations (7.5) and (7.11) to obtain

$$\Phi_{X_k}(w) = \left[\frac{\lambda}{\lambda - jw}\right]^k \tag{7.12}$$

h. **Uniform Distribution**: Recall that the PDF of the uniform random variable $X$ that takes on values between $a$ and $b$, where with parameter $a < b$, is given by

$$f_X(x)=\begin{cases}\dfrac{1}{b-a} & a\leq x\leq b\\ 0 & \text{otherwise}\end{cases}$$

Thus, the characteristic function of $X$ is given by

$$\begin{aligned}\Phi_X(w)&=\int_0^{\infty}e^{jwx}f_X(x)dx=\frac{1}{b-a}\int_a^b e^{jwx}dx=\left[\frac{e^{jwx}}{jw(b-a)}\right]_a^b\\&=\frac{e^{jbw}-e^{jaw}}{jw(b-a)}\end{aligned} \tag{7.13}$$

i. **Normal Distribution:** Recall that the PDF of the normal random variable $X$ is given by

$$f_X(x)=\frac{1}{\sqrt{2\pi\sigma_X^2}}e^{-(x-\mu_X)^2/2\sigma_X^2} \qquad -\infty<x<\infty$$

Thus, the characteristic function of $X$ is given by

$$\Phi_X(w)=E\left[e^{jwX}\right]=\int_{-\infty}^{\infty}e^{jwx}f_X(x)dx=\frac{1}{\sqrt{2\pi\sigma_X^2}}\int_{-\infty}^{\infty}e^{jwx}e^{-(x-\mu_X)^2/2\sigma_X^2}dx$$

Let $u=(x-\mu_X)/\sigma_X$, which implies that $x=u\sigma_X+\mu_X$ and $dx=\sigma_X du$. Thus, we can write

$$\Phi_X(w)=\frac{1}{\sqrt{2\pi}}\int_{-\infty}^{\infty}e^{jw(u\sigma_X+\mu_X)}e^{-u^2/2}du=\frac{e^{jw\mu_X}}{\sqrt{2\pi}}\int_{-\infty}^{\infty}e^{-(u^2-2jw\sigma_X u)/2}du$$

Now,

$$\frac{u^2-2jw\sigma_X u}{2}=\frac{u^2-2j\sigma_X u+(jw\sigma_X)^2}{2}-\frac{(jw\sigma_X)^2}{2}=\frac{(u-jw\sigma_X)^2}{2}+\frac{w^2\sigma_X^2}{2}$$

Thus,

$$\Phi_X(w)=\frac{e^{jw\mu_X}}{\sqrt{2\pi}}\int_{-\infty}^{\infty}e^{-(u-jw\sigma_X)^2/2}e^{-w^2\sigma_X^2/2}du=e^{\left(jw\mu_X-w^2\sigma_X^2/2\right)}\int_{-\infty}^{\infty}\frac{e^{-(u-jw\sigma_X)^2/2}}{\sqrt{2\pi}}du$$

Consider the function $g(u)=e^{-(u-jw\sigma_X)^2/2}/\sqrt{2\pi}$. If we substitute $v=u-jw\sigma_X$ we obtain $g(v)=e^{-v^2/2}/\sqrt{2\pi}$, which is the PDF of $N(0,1)$ random variable. This means that the integral in the preceding equation is 1. Thus, we obtain

$$\Phi_X(w)=e^{\left(jw\mu_X-w^2\sigma_X^2/2\right)}=\exp\left(jw\mu_X-\frac{w^2\sigma_X^2}{2}\right) \tag{7.14}$$

## 7.3 THE s-TRANSFORM

Let $f_X(x)$ be the PDF of the continuous random variable $X$ that takes only nonnegative values; that is, $f_X(x)=0$ for $x<0$. The s-transform of $f_X(x)$, denoted by $M_X(s)$, is defined by

$$M_X(s)=E[e^{-sX}]=\int_0^\infty e^{-sx}f_X(x)dx \tag{7.15}$$

One important property of an s-transform is that when it is evaluated at the point $s=0$, its value is equal to 1. That is,

$$M_X(s)|_{s=0}=\int_0^\infty f_X(x)dx=1$$

### EXAMPLE 7.2

For what value of $K$ is the function $A(s)=K/(s+5)$ a valid s-transform of a PDF?

Solution:

To be a valid s-transform of a PDF, $A(0)$ must be equal to 1. That is, we must have that

$$\frac{K}{0+5}=\frac{K}{5}=1 \Rightarrow K=5$$

### 7.3.1 Moment-Generating Property of the s-Transform

As stated earlier, one of the primary reasons for studying the transform methods is to use them to derive the moments of the different probability distributions. By definition

$$M_X(s)=\int_0^\infty e^{-sx}f_X(x)dx$$

Taking different derivatives of $M_X(s)$ and evaluating them at $s=0$, we obtain the following results:

$$\frac{d}{ds}M_X(s)=\frac{d}{ds}\int_0^\infty e^{-sx}f_X(x)dx=\int_0^\infty \frac{d}{ds}e^{-sx}f_X(x)dx=-\int_0^\infty xe^{-sx}f_X(x)dx$$

$$\left.\frac{d}{ds}M_X(s)\right|_{s=0}=-\int_0^\infty xf_X(x)dx=-E[X]$$

$$\frac{d^2}{ds^2}M_X(s)=\frac{d}{ds}(-1)\int_0^\infty xe^{-sx}f_X(x)dx=-\int_0^\infty \frac{d}{ds}xe^{-sx}f_X(x)dx$$

$$=\int_0^\infty x^2e^{-sx}f_X(x)dx$$

$$\left.\frac{d^2}{ds^2}M_X(s)\right|_{s=0}=\int_0^\infty x^2f_X(x)dx=E[X^2]$$

In general,

$$\left.\frac{d^n}{ds^n}M_X(s)\right|_{s=0} = (-1)^n E[X^n] \tag{7.16}$$

### 7.3.2 The s-Transform of the PDF of the Sum of Independent Random Variables

Let $X_1, X_2, \ldots, X_n$ be independent continuous random variables, and let their sum be

$$Y = X_1 + X_2 + \ldots + X_n$$

The s-transform of the PDF of $Y$ is given by

$$\begin{aligned} M_Y(s) &= E\left[e^{-sY}\right] = E\left[e^{-s(X_1+X_2+\ldots+X_n)}\right] = E\left[e^{-sX_1}e^{-sX_2}\cdots e^{-sX_n}\right] \\ &= E\left[e^{-sX_1}\right]E\left[e^{-sX_2}\right]\cdots E\left[e^{-sX_n}\right] \\ &= \prod_{k=1}^{n} M_{X_k}(s) \end{aligned} \tag{7.17}$$

where the fourth equality follows from the fact that the $X_k$ are independent. Thus, the s-transform of the PDF of $Y$ is the product of the s-transforms of the PDFs of the random variables in the sum. When the random variables are also identically distributed, the s-transform of the PDF of $Y$ becomes

$$M_Y(s) = [M_X(s)]^n \tag{7.18}$$

### 7.3.3 The s-Transforms of Some Well-Known PDFs

In this section we derive expressions for the s-transforms of the probability density functions of some of the random variables that are discussed in Chapter 4. These include the exponential distribution, the Erlang distribution, and the uniform distribution.

a. **Exponential Distribution**: Recall that the PDF of the exponential random variable with parameter $\lambda > 0$ is

$$f_X(x) = \lambda e^{-\lambda x} \quad x \geq 0$$

Therefore, the characteristic function is given by

$$\begin{aligned} M_X(s) &= E\left[e^{-sX}\right] = \int_0^\infty e^{-sx}\lambda e^{-\lambda x}dx = \lambda\int_0^\infty e^{-(s+\lambda)x}dx \\ &= \frac{\lambda}{s+\lambda} \end{aligned} \tag{7.19}$$

b. **Erlang Distribution:** Recall that the PDF of the Erlang random variable of order $k$ (or Erlang-$k$ random variable), $X_k$, with parameter $\lambda > 0$ is given by

$$f_{X_k}(x) = \frac{\lambda^k x^{k-1}}{(k-1)!} e^{-\lambda x} \quad x \geq 0$$

Because the Erlang-$k$ random variable is the sum of $k$ exponential random variables, we use the results of equations (7.18) and (7.19) to obtain

$$M_{X_k}(s) = \left[\frac{\lambda}{s+\lambda}\right]^k \tag{7.20}$$

c. **Uniform Distribution:** If $X$ is a uniformly distributed random variable that takes on values between $a$ and $b$, where with parameter $0 \leq a < b$, we know that its PDF is given by

$$f_X(x) = \begin{cases} \frac{1}{b-a} & 0 \leq a \leq x \leq b \\ 0 & \text{otherwise} \end{cases}$$

Thus, the s-transform of the PDF of $X$ is given by

$$\begin{aligned} M_X(s) &= \int_0^\infty e^{-sx} f_X(x)dx = \frac{1}{b-a}\int_a^b e^{-sx}dx = \left[-\frac{e^{-sx}}{s(b-a)}\right]_a^b \\ &= \frac{e^{-as} - e^{-bs}}{s(b-a)} \end{aligned} \tag{7.21}$$

## EXAMPLE 7.3

A communication channel is degraded beyond use in a random manner. A smart student figured out that the duration $Y$ of the intervals between consecutive periods of degradation has the PDF

$$f_Y(y) = \frac{0.2^4 y^3 e^{-0.2y}}{3!} \quad y \geq 0$$

What is the s-transform of the PDF of $Y$?

Solution:

First, we must realize that $Y$ is an Erlang random variable. To determine the order of the random variable, we put its PDF in the general form of the Erlang PDF as follows:

$$f_Y(y) = \frac{\lambda^k y^{k-1} e^{-\lambda y}}{(k-1)!} \quad y \geq 0$$

From the power of $y$ we see that $k-1=3$, which means that $k=4$ and so $Y$ is a fourth-order Erlang random variable. Let $X$ be the underlying exponential distribution. Since $\lambda = 0.2$, the PDF of $X$ and its s-transform are given as follows:

$$f_X(x) = 0.2e^{-0.2x} \quad x \geq 0$$
$$M_X(s) = \frac{0.2}{s+0.2}$$

Thus, $Y = X_1 + X_2 + X_3 + X_4$, where the $X_k$ are independent and identically distributed with the above PDF. Since the s-transform of the PDF of the sum of independent random variables is equal to the product of their s-transforms, we have that

$$M_Y(s) = [M_X(s)]^4 = \left[\frac{0.2}{s+0.2}\right]^4$$

## EXAMPLE 7.4

Determine in an efficient manner the fourth moment of the random variable $X$ whose PDF is given by

$$f_X(x) = 32x^2e^{-4x} \quad x \geq 0$$

**Solution:**

The easiest method of solving this problem is via the transform method. Thus, we first need to obtain the s-transform of the PDF. We shall use four methods to do this. The first method is a brute-force method, and the other three are "smart" methods.

a. **Brute-Force Method**

In this method we attempt to obtain $M_X(s)$ directly as follows:

$$M_X(s) = \int_0^\infty e^{-sx} f_X(x)dx = 32\int_0^\infty x^2 e^{-(s+4)x}dx$$

Let $u = x^2 \Rightarrow du = 2xdx$, and let $dv = e^{-(s+4)x}dx \Rightarrow v = -e^{-(s+4)x}/(s+4)$. Thus,

$$M_X(s) = 32\left[-\frac{x^2e^{-(s+4)x}}{s+4}\right]_0^\infty + \frac{32}{s+4}\int_0^\infty 2xe^{-(s+4)x}dx = \frac{64}{s+4}\int_0^\infty xe^{-(s+4)x}dx$$

Let $u = x \Rightarrow du = dx$, and let $dv = e^{-(s+4)x}dx \Rightarrow v = -e^{-(s+4)x}/(s+4)$. Thus,

$$M_X(s) = \frac{64}{s+4}\left\{\left[-\frac{xe^{-(s+4)x}}{s+4}\right]_0^\infty + \int_0^\infty \frac{e^{-(s+4)x}}{s+4}dx\right\} = \frac{64}{(s+4)^2}\int_0^\infty e^{-(s+4)x}dx$$
$$= \frac{64}{(s+4)^2}\left[-\frac{e^{-(s+4)x}}{s+4}\right]_0^\infty = \frac{64}{(s+4)^3} = \frac{4^3}{(s+4)^3} = \left(\frac{4}{s+4}\right)^3$$

b. **Smart Method 1**

In this method we exploit the properties of the moments of the exponential distribution in carrying out the integration. Thus, we proceed as follows:

$$M_X(s) = \int_0^\infty e^{-sx} f_X(x)dx = 32\int_0^\infty x^2 e^{-(s+4)x}dx$$

Let $\mu = s+4$. Then we have that

$$M_X(s)=32\int_0^\infty x^2e^{-\mu x}dx=\frac{32}{\mu}\int_0^\infty \mu x^2e^{-\mu x}dx$$

Now, the integration term is essentially the second moment of an exponential random variable $X$ with parameter $\mu$. From the results in Chapter 4 we know that for an exponential random variable,

$$E[X^n]=\frac{n!}{\mu^n} \quad n=1,2,\ldots$$

(See Equation 4.33.) Thus we have that

$$M_X(s)=\frac{32}{\mu}\left\{\frac{2!}{\mu^2}\right\}=\frac{64}{\mu^3}=\frac{64}{(s+4)^3}=\left(\frac{4}{s+4}\right)^3$$

c. **Smart Method 2**

In this method we realize that $X$ is an Erlang random variable. But to determine its order and parameter we need to put its PDF in the standard form of the Erlang PDF as follows:

$$f_X(x)=\frac{\lambda^k x^{k-1}e^{-\lambda x}}{(k-1)!}\equiv 32x^2e^{-4x} \quad x\geq 0$$

From the power of $x$ we see that $k-1=2$, which means that $k=3$. That is, $X$ is a third-order Erlang random variable. Similarly, from the exponential term we that $\lambda=4$. Let $Y$ be the underlying exponentially distributed random variable. Then the PDF of $Y$ and its s-transform are given by

$$f_Y(y)=4e^{-4y} \quad y\geq 0$$
$$M_Y(s)=\frac{4}{s+4}$$

Thus, $X=Y_1+Y_2+Y_3$, where the $Y_k$ are independent and identically distributed with the above PDF. Since the s-transform of the PDF of the sum of independent random variables is equal to the product of their s-transforms, we have that

$$M_X(s)=[M_Y(s)]^3=\left(\frac{4}{s+4}\right)^3$$

d. **Smart Method 3**

As in the previous methods, we start with

$$M_X(s)=32\int_0^\infty x^2e^{-(s+4)x}dx=32\int_0^\infty x^2e^{-\mu x}dx$$

Because the integrand (or integral kernel) looks like an Erlang-3 PDF, we rearrange it to obtain a true Erlang-3 PDF as follows:

$$x^2e^{-\mu x}=\left\{\frac{\mu^3x^2e^{-\mu x}}{2!}\right\}\left(\frac{2!}{\mu^3}\right)=\frac{2}{\mu^3}\left\{\frac{\mu^3x^2e^{-\mu x}}{2!}\right\}=\frac{2}{\mu^3}f_{X_3}(x)$$

This means that

$$\begin{aligned}M_X(s)&=32\int_0^\infty x^2e^{-\mu x}dx=\frac{(32)(2)}{\mu^3}\int_0^\infty \frac{\mu^3x^2}{2!}e^{-\mu x}dx=\frac{64}{\mu^3}\int_0^\infty \frac{\mu^3x^2}{2!}e^{-\mu x}dx\\&=\frac{64}{\mu^3}=\frac{4^3}{\mu^3}=\left(\frac{4}{\mu}\right)^3=\left(\frac{4}{s+4}\right)^3\end{aligned}$$

where the first equality of the second line follows from the fact that the integral is equal to 1 because a valid PDF is integrated over the entire range of the values of $x$.

**The Rest of the Solution**: Having obtained the s-transform of the PDF, the fourth moment of $X$ is given by

$$E[X^4] = (-1)^4 \frac{d^4}{ds^4} M_X(s)\Big|_{s=0} = \frac{d^4}{ds^4}\left(\frac{4}{s+4}\right)^3\Big|_{s=0}$$

$$= \frac{(-3)(-4)(-5)(-6)(4)^3}{(s+4)^7}\Big|_{s=0} = \frac{(360)(4)^3}{(s+4)^7}\Big|_{s=0} = 1.40625$$

## 7.4 THE z-TRANSFORM

Let $p_X(x)$ be the PMF of the nonnegative discrete random variable $X$; that is, $p_X(x) = 0$ for $x < 0$. The z-transform of $p_X(x)$, denoted by $G_X(z)$ is defined by

$$G_X(z) = E[z^X] = \sum_{x=0}^{\infty} z^x p_X(x) \tag{7.22}$$

The sum is guaranteed to converge and, therefore, the z-transform exists, when evaluated on or within the unit circle (where $|z| \le 1$). Note that

$$G_X(1) = \sum_{x=0}^{\infty} p_X(x) = 1$$

This means that a valid z-transform of a PMF reduces to unity when evaluated at $z = 1$. However, this is a necessary but not sufficient condition for a function to be the z-transform of a PMF. By definition,

$$G_X(z) = \sum_{x=0}^{\infty} z^x p_X(x) = p_X(0) + z p_X(1) + z^2 p_X(2) + z^3 p_X(3) + \cdots$$

This means that $P[X = k] = p_X(k)$ is the coefficient of $z^k$ in the series expansion. Thus, given the z-transform of a PMF, we can uniquely recover the PMF. The implication of this statement is that not every polynomial that has a value 1 when evaluated at $z = 1$ is a valid z-transform of a PMF. For example, consider the function $A(z) = 2z - 1$. Although $A(1) = 1$, the function contains invalid coefficients in the sense that these coefficients either have negative values or positive values that are greater than one. Specifically, the coefficient of $z$ is 2, which should be the value of $p_X(1)$, but the value is illegal because it is greater than 1. Similarly, the constant term, which is the coefficient of $z^0$ and thus the value of $p_X(0)$, is -1, another illegal probability value. Thus, for a function of $z$ to be a valid z-transform of a PMF, it must have a value of 1 when evaluated at

$z=1$, and all the coefficients of $z$ must be nonnegative numbers that cannot be greater than 1.

The individual terms of the PMF can also be determined as follows:

$$p_X(x)=\frac{1}{x!}\left[\frac{d^x}{dz^x}G_X(z)\right]_{z=0} \qquad x=0,1,2,\ldots \tag{7.23}$$

This feature of the z-transform is the reason it is sometimes called the *probability generating function*.

## EXAMPLE 7.5

Is the following function a valid z-transform of a PMF? If so, what is the PMF?

$$g(z)=\frac{1-a}{1-az} \qquad 0<a<1$$

Solution:
First, we note that $g(1)=1$, which means that the function $g(z)$ is potentially a valid z-transform of a PMF. Next, we test the coefficients of $z$ in the function. Now,

$$g(z)=(1-a)\sum_{k=0}^{\infty}(az)^k=(1-a)\{1+az+a^2z^2+a^3z^3+\cdots+a^kz^k+\cdots\}$$

Since $0<a<1$, we see that all the coefficients of $z$ are nonnegative quantities that are no greater than 1. Therefore, the function is a valid z-transform of a PMF. If $X$ is the random variable whose PMF has this z-transform, the PMF of $X$ is given by

$$p_X(x)=(1-a)a^x \qquad x=0,1,2,\ldots$$

## EXAMPLE 7.6

Explain why the function $F(z)=z^2+z-1$ is or is not a valid z-transform of the PMF of a random variable.

Solution:
One of the tests for a function of $z$ to be a valid z-transform of a PMF is that it must be equal to 1 when evaluated at $z=1$. As can be seen, $F(1)=1$, so the function has passed the first test. The second test is that the coefficients of $z$ must be nonnegative since, for example, the coefficient of $z^k$ is the probability that the random variable takes the value $k$. In the function above, the constant term, which represents the probability that the supposed random variable takes the value 0, is negative 1. This means that the function *cannot* be a valid z-transform of a PMF.

## EXAMPLE 7.7

The z-transform of the PMF of a discrete random variable $K$ is given by

$$G_K(z) = A\left[\frac{10+8z^2}{2-z}\right]$$

a. What is the expected value of $K$?
b. Find $p_K(1)$, the probability that $K$ has the value 1.

Solution:

Before answering both questions we need to obtain the numerical value of $A$. For $G_K(z)$ to be a valid z-transform, it must satisfy the condition $G_K(1)=1$. Thus, we have that

$$G_K(1) = A\left[\frac{10+8}{2-1}\right] = 18A = 1 \Rightarrow A = \frac{1}{18}$$

a. The expected value of $K$ is

$$E[K] = \frac{d}{dz}G_K(z)\bigg|_{z=1} = A\left[\frac{(2-z)16z-(10+8z^2)(-1)}{(2-z)^2}\right]_{z=1} = \frac{34}{18} = 1.9$$

b. To obtain the PMF of $K$, we can use two methods:

**Method 1:**

We observe that

$$G_K(z) = A\left[\frac{10+8z^2}{2-z}\right] = \frac{A}{2}\left[\frac{10+8z^2}{1-\frac{z}{2}}\right] = \frac{A(10+8z^2)}{2}\sum_{k=0}^{\infty}\left(\frac{z}{2}\right)^k$$
$$= \frac{A(10+8z^2)}{2}\left\{1+\frac{z}{2}+\frac{z^2}{4}+\frac{z^3}{8}+\frac{z^4}{16}+\cdots\right\}$$
$$= \frac{A}{2}\left\{10+z\left[\frac{10}{2}\right]+z^2\left[\frac{10}{4}+8\right]+z^3\left[\frac{10}{8}+4\right]+z^4\left[\frac{10}{16}+2\right]+\cdots\right\}$$

Thus, the probability that $K$ has a value 1 is the coefficient of $z$ in $G_K(z)$, which is

$$p_K(1) = \frac{A}{2}\times\frac{10}{2} = \frac{5}{36}$$

**Method 2:**

We can also obtain the value of $p_K(1)$ as follows:

$$p_K(1) = \frac{1}{1!}\frac{d}{dz}G_K(z)\bigg|_{z=0} = A\left\{\frac{(2-z)(16z)-(10+8z^2)(-1)}{(2-z)^2}\right\}\bigg|_{z=0}$$
$$= A\left\{\frac{10}{4}\right\} = \frac{1}{18}\times\frac{10}{4} = \frac{5}{36}$$

### 7.4.1 Moment-Generating Property of the z-Transform

As stated earlier, one of the major motivations for studying transform methods is their usefulness in computing the moments of the different random variables. Unfortunately the moment-generating capability of the z-transform is not as computationally efficient as that of the s-transform.

The moment-generating capability of the z-transform lies in the results obtained from evaluating the derivatives of the transform at $z=1$. For a discrete random variable $X$ with PMF $p_X(x)$, we have that

$$G_X(z)=\sum_{x=0}^{\infty} z^x p_X(x)$$

$$\frac{d}{dz}G_X(z)=\frac{d}{dz}\sum_{x=0}^{\infty} z^x p_X(x)=\sum_{x=0}^{\infty}\frac{d}{dz}z^x p_X(x)=\sum_{x=0}^{\infty} xz^{x-1}p_X(x)=\sum_{x=1}^{\infty} xz^{x-1}p_X(x)$$

$$\left.\frac{d}{dz}G_X(z)\right|_{z=1}=\sum_{x=1}^{\infty} xp_X(x)=\sum_{x=0}^{\infty} xp_X(x)=E[X]$$

That is,

$$E[X]=\sum_{x=0}^{\infty} xp_X(x)=\left.\frac{d}{dz}G_X(z)\right|_{z=1} \tag{7.24}$$

Similarly,

$$\frac{d^2}{dz^2}G_X(z)=\frac{d}{dz}\sum_{x=1}^{\infty} xz^{x-1}p_X(x)=\sum_{x=1}^{\infty} x\frac{d}{dz}z^{x-1}p_X(x)=\sum_{x=1}^{\infty} x(x-1)z^{x-2}p_X(x)$$

$$\begin{aligned}\left.\frac{d^2}{dz^2}G_X(z)\right|_{z=1}&=\sum_{x=1}^{\infty} x(x-1)p_X(x)=\sum_{x=0}^{\infty} x(x-1)p_X(x)=\sum_{x=0}^{\infty} x^2p_X(x)-\sum_{x=0}^{\infty} xp_X(x)\\&=E\left[X^2\right]-E[X]\end{aligned}$$

Thus,

$$E\left[X^2\right]=\left.\frac{d^2}{dz^2}G_X(z)\right|_{z=1}+\left.\frac{d}{dz}G_X(z)\right|_{z=1} \tag{7.25}$$

From this we obtain the variance as follows:

$$\sigma_X^2=E\left[X^2\right]-(E[X])^2=\left[\frac{d^2}{dz^2}G_X(z)+\frac{d}{dz}G_X(z)-\left\{\frac{d}{dz}G_X(z)\right\}^2\right]_{z=1} \tag{7.26}$$

### 7.4.2 The z-Transform of the PMF of the Sum of Independent Random Variables

Let $X_1, X_2, \ldots, X_n$ be independent discrete random variables, and let their sum be

$$Y = X_1 + X_2 + \ldots + X_n$$

The z-transform of the PMF of $Y$ is given by

$$\begin{aligned} G_Y(z) &= E\left[z^Y\right] = E\left[z^{X_1 + X_2 + \ldots + X_n}\right] = E\left[z^{X_1} z^{X_2} \cdots z^{X_n}\right] = E\left[z^{X_1}\right] E\left[z^{X_2}\right] \cdots E\left[z^{X_n}\right] \\ &= G_{X_1}(z) G_{X_2}(z) \cdots G_{X_n}(z) = \prod_{k=1}^{n} G_{X_k}(z) \end{aligned} \tag{7.27}$$

where the fourth equality is due to the independence of the random variables. In the case where the random variables are also identically distributed we obtain

$$G_Y(z) = [G_X(z)]^n \tag{7.28}$$

### 7.4.3 The z-Transform of Some Well-Known PMFs

In this section we consider the z-transforms of the Bernoulli, binomial, geometric, Pascal-$k$ and Poisson PMFs.

a. **Bernoulli Distribution**: Recall that the PMF of the Bernoulli random variable $X$ with parameter $p$ is given by

$$p_X(x) = \begin{cases} 1-p & x=0 \\ p & x=1 \end{cases}$$

Thus, the z-transform of the PMF of $X$ is given by

$$G_X(z) = \sum_{x=0}^{\infty} z^x p_X(x) = 1 - p + zp \tag{7.29}$$

b. **Binomial Distribution:** Recall that the PMF of the binomial random variable $X(n) \sim B(n, p)$ is

$$p_{X(n)}(x) = \binom{n}{x} p^x (1-p)^{n-x} \quad x = 0, 1, \ldots, n$$

Since the binomial random variable $X(n)$ is the sum of $n$ independent and identically distributed Bernoulli random variables, we use the results in equations (7.28) and (7.29) to obtain the z-transform of the PMF of $X(n)$ as

$$G_{X(n)}(z) = [1 - p + zp]^n \tag{7.30}$$

Note that we can also obtain the result by a direct method. Specifically,

$$G_{X(n)}(z) = \sum_{x=0}^{\infty} z^x p_{X(n)}(x) = \sum_{x=0}^{n} z^x \binom{n}{x} p^x (1-p)^{n-x} = \sum_{x=0}^{n} \binom{n}{x} (zp)^x (1-p)^{n-x}$$
$$= [zp + 1 - p]^n$$

which is the result we obtained earlier. The last equality follows from the binomial identity

$$(a+b)^n = \sum_{k=0}^{n} \binom{n}{k} a^k b^{n-k}$$

c. **Geometric Distribution:** Recall that the PMF of the geometric random variable $X$ with parameter $p$ is given by

$$p_X(x) = p(1-p)^{x-1} \quad x = 1, 2, \ldots$$

Thus, the z-transform of the PMF of $X$ is given by

$$\begin{aligned} G_X(z) &= \sum_{x=0}^{\infty} z^x p_X(x) = \sum_{x=1}^{\infty} z^x p(1-p)^{x-1} \\ &= zp \sum_{x=1}^{\infty} z^{x-1} (1-p)^{x-1} = zp \sum_{x=1}^{\infty} \{z(1-p)\}^{x-1} \\ &= \frac{zp}{1 - z(1-p)} \end{aligned} \tag{7.31}$$

d. **Pascal Distribution:** Recall that the PMF of the Pascal random variable of order $k$ (or Pascal-$k$ random variable), $X_k$, and parameter $p$ is given by

$$p_{X_k}(n) = \binom{n-1}{k-1} p^k (1-p)^{n-k} \quad k = 1, 2, \ldots; n = k, k+1, \ldots$$

Because the Pascal-$k$ random variable is the sum of $k$ independent and identically distributed geometric random variables, we use the results of equations (7.28) and (7.31) to obtain the z-transform of the PMF of $X_k$ as

$$G_{X_k}(z) = \left\{ \frac{zp}{1 - z(1-p)} \right\}^k \tag{7.32}$$

e. **Poisson Distribution:** Recall that the PMF of the Poisson distribution is given by

$$p_X(x) = \frac{\lambda^x e^{-\lambda}}{x!} \quad x = 0, 1, 2, \ldots; \ \lambda > 0$$

Thus, the z-transform of the PMF of $X$ is given by

$$\begin{aligned} G_X(z) &= \sum_{x=0}^{\infty} z^x p_X(x) = \sum_{x=0}^{\infty} z^x \frac{\lambda^x e^{-\lambda}}{x!} = e^{-\lambda} \sum_{x=0}^{\infty} \frac{(\lambda z)^x}{x!} = e^{-\lambda} e^{\lambda z} \\ &= e^{-\lambda(1-z)} = e^{\lambda(z-1)} \end{aligned} \tag{7.33}$$

## 7.5 RANDOM SUM OF RANDOM VARIABLES

Let $X$ be a continuous random variable with PDF $f_X(x)$ whose s-transform is $M_X(s)$. We know that if $Y$ is the sum of $n$ independent and identically distributed random variables with the PDF $f_X(x)$, then the s-transform of the PDF of $Y$ is given by

$$M_Y(s) = [M_X(s)]^n$$

The above result assumes that $n$ is a fixed number. However, there are certain situations when the number of random variables in a sum is itself a random variable. For this case, let $N$ denote a discrete random variable with PMF $p_N(n)$ whose z-transform is $G_N(z)$. Our goal is to find the s-transform of the PDF of $Y$ when the number of random variables is itself a random variable $N$.

Thus, we consider the sum

$$Y = X_1 + X_2 + \ldots + X_N$$

where $N$ has a known PMF, which in turn has a known z-transform. Now, let $N = n$. Then with $N$ fixed at $n$ we have that

$$\begin{aligned} Y|_{N=n} &= X_1 + X_2 + \ldots + X_n \\ M_{Y|N}(s|n) &= [M_X(s)]^n \end{aligned}$$

Thus,

$$\begin{aligned} M_Y(s) &= \sum_{n=0}^{\infty} M_{Y|N}(s|n) p_N(n) = \sum_{n=0}^{\infty} [M_X(s)]^n p_N(n) \\ &= G_N(M_X(s)) \end{aligned} \tag{7.34}$$

That is, the s-transform of the PDF of a random sum of independent and identically distributed random variables is the z-transform of the PMF of the number of variables evaluated at the s-transform of the PDF of the constituent random variables. Now, let $u = M_X(s)$. Then

$$\frac{d}{ds}M_Y(s) = \frac{d}{ds}G_N(M_X(s)) = \left\{\frac{dG_N(u)}{du}\right\}\left\{\frac{du}{ds}\right\}$$

$$\left.\frac{d}{ds}M_Y(s)\right|_{s=0} = \left[\left\{\frac{dG_N(u)}{du}\right\}\left\{\frac{du}{ds}\right\}\right]_{s=0}$$

Now, $u|_{s=0} = M_X(0) = 1$. Thus, we obtain

$$\left.\frac{d}{ds}M_Y(s)\right|_{s=0} = -E[Y] = \left[\left\{\frac{dG_N(u)}{du}\right\}\left\{\frac{du}{ds}\right\}\right]_{s=0} = \left.\frac{dG_N(u)}{du}\right|_{u=1}\left.\frac{dM_X(s)}{ds}\right|_{s=0}$$
$$= E[N]\{-E[X]\} = -E[N]E[X]$$

which gives

$$E[Y] = E[N]E[X] \tag{7.35}$$

Also,

$$\frac{d^2}{ds^2}M_Y(s) = \frac{d}{ds}\left[\left\{\frac{dG_N(u)}{du}\right\}\left\{\frac{du}{ds}\right\}\right]$$
$$= \left\{\frac{du}{ds}\right\}\frac{d}{ds}\left\{\frac{dG_N(u)}{du}\right\} + \left\{\frac{dG_N(u)}{du}\right\}\left\{\frac{d^2u}{ds^2}\right\}$$
$$= \left\{\frac{du}{ds}\right\}^2\left\{\frac{d^2G_N(u)}{du^2}\right\} + \left\{\frac{dG_N(u)}{du}\right\}\left\{\frac{d^2u}{ds^2}\right\}$$

$$\left.\frac{d^2}{ds^2}M_Y(s)\right|_{s=0} = E[Y^2] = \left[\left\{\frac{du}{ds}\right\}^2\left\{\frac{d^2G_N(u)}{du^2}\right\} + \left\{\frac{dG_N(u)}{du}\right\}\left\{\frac{d^2u}{ds^2}\right\}\right]_{s=0,\, u=1}$$
$$= \{-E[X]\}^2\{E[N^2] - E[N]\} + E[N]E[X^2]$$
$$= E[N^2]\{E[X]\}^2 + E[N]E[X^2] - E[N]\{E[X]\}^2$$

The variance of $Y$ is given by

$$\begin{aligned}\sigma_Y^2 &= E[Y^2] - \{E[Y]\}^2 \\ &= E[N^2]\{E[X]\}^2 + E[N]E[X^2] - E[N]\{E[X]\}^2 - \{E[N]E[X]\}^2 \\ &= E[N]\{E[X^2] - \{E[X]\}^2\} + \{E[X]\}^2\{E[N^2] - \{E[N]\}^2\} \\ &= E[N]\sigma_X^2 + \{E[X]\}^2\sigma_N^2\end{aligned} \tag{7.36}$$

If $X$ is also a discrete random variable, then we obtain

$$G_Y(z) = G_N(G_X(z)) \tag{7.37}$$

and the results for $E[Y]$ and $\sigma_Y^2$ still hold.

## EXAMPLE 7.8

Books are packed into cartons. The weight $W$ of a book is a continuous random variable with PDF

$$f_W(w) = \lambda e^{-\lambda w} \quad w \geq 0$$

The number $K$ of books in any carton is a random variable with the PMF

$$p_K(k) = \frac{\mu^k}{k!} e^{-\mu} \quad k = 0, 1, 2, \ldots$$

If we randomly select a carton and its weight is $X$, determine

a. the s-transform of the PDF of $X$.
b. $E[X]$
c. the variance of $X$.

Solution:

a. The s-transform of the PDF of $W$ is given by

$$M_W(s) = \frac{\lambda}{s + \lambda}$$

Similarly, the z-transform of the PMF of $K$ is given by

$$G_K(z) = e^{\mu(z-1)} = \exp(\mu(z-1))$$

Thus, the s-transform of the PDF if $X$ is given by

$$M_X(s) = G_K(M_W(s)) = \exp\left(\mu\left\{\frac{\lambda}{s+\lambda} - 1\right\}\right) = \exp\left(-\frac{\mu s}{s+\lambda}\right)$$

b. The expected weight of the randomly selected carton is

$$E[X] = E[K]E[W] = \mu\left(\frac{1}{\lambda}\right) = \frac{\mu}{\lambda}$$

c. The variance of $X$ is given by

$$\sigma_X^2 = E[K]\sigma_W^2 + \{E[W]\}^2\sigma_K^2 = \mu\left(\frac{1}{\lambda^2}\right) + \left(\frac{1}{\lambda^2}\right)\mu = \frac{2\mu}{\lambda^2}$$

## EXAMPLE 7.9

The number $K$ of parcels that the drivers of a parcel delivery service company can load in their trucks is a random variable with the PMF

$$p_K(k) = \frac{40^k e^{-40}}{k!} \quad k = 0, 1, 2, \ldots$$

The weight $W$ of a parcel in pounds is a continuous random variable with PDF

$$f_W(w) = \begin{cases} \frac{1}{6} & 3 \leq w \leq 9 \\ 0 & \text{otherwise} \end{cases}$$

Let $X$ denote the weight of a randomly selected loaded truck.

a. What is the s-transform of the PDF of $X$?
b. What is the expected value of $X$?
c. What is the variance of $X$?

Solution:

(a) Let the number of parcels in the truck be $K=k$, and let $W_i$ denote the weight of parcel $i$, $1 \leq i \leq k$. Then, since we have fixed $K$ at $k$, we have that

$$X|_{K=k} = W_1 + W_2 + \cdots + W_k$$
$$M_{X|K}(s|k) = [M_W(s)]^k$$

Thus, the s-transform of the PDF of $X$ is

$$M_X(s) = \sum_{k=0}^{\infty} M_{X|K}(s|k)p_K(k) = \sum_{k=0}^{\infty} [M_W(s)]^k p_K(k) = G_K(M_W(s))$$

where, $G_K(z) = \exp(-40\{1-z\})$ and from equation (7.21), $M_W(s)$ is given by:

$$M_W(s) = \frac{e^{-as} - e^{-bs}}{s(b-a)} = \frac{e^{-3s} - e^{-9s}}{6s}$$

(b) Since $K$ is a Poisson random variable, its expected value and variance are given by

$$E[K] = \sigma_K^2 = 40$$

Similarly, since $W$ has a uniform distribution, its mean and variance are given by

$$E[W] = \frac{3+9}{2} = 6$$
$$\sigma_W^2 = \frac{(9-3)^2}{12} = 3$$

Thus, $E[X] = E[K]E[W] = 240$.

(c) The variance of $X$ is given by

$$\sigma_X^2 = E[K]\sigma_W^2 + \{E[W]\}^2\sigma_K^2 = (40)(3) + \left(6^2\right)(40) = 1560$$

## EXAMPLE 7.10

The number $K$ of customers that arrive at Jay's supermarket in a given day has the PMF

$$p_K(k) = \frac{\lambda^k e^{-\lambda}}{k!} \qquad k = 0, 1, 2, \ldots$$

Independently of $K$, the number of items $N$ that any customer purchases from the supermarket has the PMF

$$p_N(n) = \frac{\mu^n e^{-\mu}}{n!} \qquad n = 0, 1, 2, \ldots$$

Determine the mean and the z-transform of the PMF of $Y$, the total number of items that the store sells on an arbitrary day.

Solution:
Let $K = k$, and let $N_i$ denote the number of items purchased by customer $i$, $1 \leq i \leq k$. Then

$$\begin{aligned}
Y|_{K=k} &= N_1 + N_2 + \cdots + N_k \\
G_{Y|K}(z|k) &= E\left[z^{Y|K}\right] = E\left[z^{N_1 + N_2 + \cdots + N_k}\right] = E\left[z^{N_1} z^{N_2} \cdots z^{N_k}\right] \\
&= E\left[z^{N_1}\right] E\left[z^{N_2}\right] \cdots E\left[z^{N_k}\right] = [G_N(z)]^k \\
G_Y(z) &= \sum_{k=0}^{\infty} G_{Y|K}(z|k) p_K(k) = \sum_{k=0}^{\infty} [G_N(z)]^k p_K(k) \\
&= G_K(G_N(z))
\end{aligned}$$

Since $G_K(z) = e^{\lambda(z-1)}$ and $G_N(z) = e^{\mu(z-1)}$, we have that

$$G_Y(z) = \exp\left(\lambda\left(e^{\mu(z-1)} - 1\right)\right)$$

## 7.6 CHAPTER SUMMARY

This chapter discussed three transform methods that are frequently used in the analysis of probabilistic problems. These are the characteristic function, the s-transform and the z-transform. Both the s-transform and the z-transform are used for random variables that take only nonnegative values, which include many random variables that are used to model practical systems. The moment-generating properties of the different transforms have also been demonstrated.

Table 7.1 is a summary of the different transforms of some of the well-known PMFs, and Table 7.2 is a summary of the different transforms of some of the well-known PDFs.

**Table 7.1** Summary of the Transforms of Well-known PMFs

| PMF, $p_X(x)$ | Characteristic Function, $\Phi_X(w)$ | z-Transform, $G_X(z)$ |
|---|---|---|
| Bernoulli, $X$ | $1 - p + pe^{jw}$ | $1 - p + zp$ |
| Binomial, $X(n)$ | $[1 - p + pe^{jw}]^n$ | $[1 - p + zp]^n$ |
| Geometric, $X$ | $\dfrac{pe^{jw}}{1 - e^{jw}(1 - p)}$ | $\dfrac{zp}{1 - z(1 - p)}$ |
| Pascal-$k$, $X_k$ | $\left\{\frac{pe^{jw}}{1 - e^{jw}(1 - p)}\right\}^k$ | $\left\{\frac{zp}{1 - z(1 - p)}\right\}^k$ |
| Poisson, $X$ | $e^{-\lambda(1 - e^{jw})}$ | $e^{-\lambda(1 - z)} = e^{\lambda(z - 1)}$ |

Table 7.2 Summary of the Transforms of Well-known PDFs

| PDF, $f_X(x)$ | Characteristic Function, $\Phi_X(w)$ | s-Transform, $M_X(s)$ |
|---|---|---|
| Exponential | $\frac{\lambda}{\lambda - jw}$ | $\frac{\lambda}{s+\lambda}$ |
| Erlang-$k$ | $\left[\frac{\lambda}{\lambda - jw}\right]^k$ | $\left[\frac{\lambda}{s+\lambda}\right]^k$ |
| Uniform | $\frac{e^{jbw} - e^{jaw}}{jw(b-a)}$ | $\frac{e^{-as} - e^{-bs}}{s(b-a)}$ |
| Normal | $\exp\left(jw\mu_X - \frac{w^2\sigma_X^2}{2}\right)$ | = |

# 7.7 PROBLEMS

## Section 7.2 Characteristic Functions

7.1 Find the characteristic function of the random variable $X$ with the following PDF:

$$f_X(x) = \begin{cases} \frac{1}{4} & 6 \le x \le 10 \\ 0 & \text{otherwise} \end{cases}$$

7.2 Find the characteristic function of the random variable $Y$ with the following PDF:

$$f_Y(y) = \begin{cases} 3e^{-3y} & y \ge 0 \\ 0 & \text{otherwise} \end{cases}$$

7.3 Find the characteristic function of the random variable $X$ with the following PDF:

$$f_X(x) = \begin{cases} 0 & x < -3 \\ \frac{x+3}{9} & -3 \le x < 0 \\ \frac{3-x}{9} & 0 \le x < 3 \\ 0 & x \ge 3 \end{cases}$$

7.4 The characteristic function of the random variable $X$ is given by $\Phi_X(w)$. If we define the random variable $Y = aX + b$, what is the characteristic function of $Y$?

## Section 7.3 s-Transforms

7.5 Explain why each of the following functions is or is not a valid s-transform of a PDF:

a. $A(s) = \frac{1 - e^{-5s}}{s}$

b. $B(s)=\dfrac{7}{4+3s}$

c. $C(s)=\dfrac{5}{5+3s}$

7.6 Assume that the s-transform of the PDF of the random variable $Y$ is given by

$$M_Y(s)=\frac{K}{s+2}$$

Determine the following:

a. the value of $K$ that makes the function a valid s-transform of a PDF

b. $E[Y^2]$

7.7 $X$ and $Y$ are independent random variables with the PDFs

$$f_X(x)=\begin{cases}\lambda e^{-\lambda x} & x\geq 0\\ 0 & \text{otherwise}\end{cases}$$

$$f_Y(y)=\begin{cases}\mu e^{-\mu y} & y\geq 0\\ 0 & \text{otherwise}\end{cases}$$

If the random variable $R$ is defined by $R=X+Y$, determine the following:

a. $M_R(s)$

b. $E[R]$

c. $\sigma_R^2$

7.8 The random variable $X$ has the following PDF:

$$f_X(x)=\begin{cases}2x & 0\leq x\leq 1\\ 0 & \text{otherwise}\end{cases}$$

Determine the numerical values of

a. $\left[\dfrac{d}{ds}[M_X(s)]^3\right]_{s=0}$

b. $\left[\dfrac{d^3}{ds^3}M_X(s)\right]_{s=0}$

7.9 The s-transform of the PDF of the random variable $X$ is given by

$$M_X(s)=\frac{\lambda^6}{(s+\lambda)^6}$$

Determine the following:

a. $E[X]$

b. $\sigma_X^2$

7.10 The s-transform of the PDF of the random variable $X$ is given as $M_X(s)$. If we define the random variable $Y=aX+b$, what is the s-transform of the PDF of $Y$?

7.11 The continuous random variables $X$ and $Y$ have the following PDFs:

$$f_X(x) = \begin{cases} 1 & 0 < x \leq 1 \\ 0 & \text{otherwise} \end{cases}$$

$$f_Y(y) = \begin{cases} 0.5 & 2 < y \leq 4 \\ 0 & \text{otherwise} \end{cases}$$

Assume that the function $L(s)$ is defined as follows:

$$L(s) = [M_X(s)]^3 [M_Y(s)]^2$$

Determine the value of the following quantity:

$$\left[\frac{d^2}{ds^2} L(s)\right]_{s=0} - \left\{\left[\frac{d}{ds} L(s)\right]_{s=0}\right\}^2$$

## Section 7.4 z-Transforms

7.12 The z-transform of the PMF of the random variable $X$ is given by

$$G_X(z) = \frac{1 + z^2 + z^4}{3}$$

Determine

a. $E[X]$

b. $p_X(E[X])$; that is, $P[X = E[X]]$

7.13 If the z-transform of the PMF of the random variable $X$ is given by

$$G_X(z) = A(1 + 3z)^3$$

determine the numerical values of the following:

a. $E[X^3]$

b. $p_X(2)$

7.14 If the z-transform of the PMF of the random variable $K$ is given by

$$G_K(z) = \frac{A(14 + 5z - 3z^2)}{2 - z}$$

determine the values of

a. $A$

b. $p_K(1)$

7.15 Explain why the function $C(z) = z^2 + 2z - 2$ is or is not a valid z-transform of the PMF of a random variable.

7.16 Consider the function:

$$D(z) = \frac{1}{2-z}$$

a. Is it a valid z-transform of the PMF of a random variable?
b. If it is, what PMF has the z-transform?

7.17 If the z-transform of the PMF of the random variable $N$ is given by

$$G_N(z) = 0.5z^5 + 0.3z^7 + 0.2z^{10}$$

determine
a. the PMF of $N$
b. $E[N]$
c. $\sigma_N^2$

7.18 The PMF of the random variable $X$ has the z-transform

$$G_X(z) = \left[\frac{zp}{1 - z(1-p)}\right]^6$$

Determine the following:
a. $E[X]$
b. $\sigma_X^2$

7.19 The z-transform of the PMF of the random variable $X$ is given as $G_X(z)$. If we define the random variable $Y = aX + b$, what is the z-transform of the PMF of $Y$?

## Section 7.5 Random Sum of Random Variables

7.20 People arrive at a restaurant by families. The number of families $X$ that arrive over the period of 1 hour is found to be a Poisson random variable with rate $\lambda$. If the number of people in each arriving family is a random variable $N$ whose PMF has the z-transform that is given by

$$G_N(z) = \frac{1}{2}z + \frac{1}{3}z^2 + \frac{1}{6}z^3$$

determine the following:
a. $G_M(z)$, the z-transform of the PMF of $M$, which is the total number of people arriving at the restaurant in an arbitrary hour
b. $E[Y]$, where $Y$ is the total number of people that arrive at the restaurant over a three-hour period

7.21 The number of customers, $K$, that shop at the neighborhood store in a day has the PMF

$$p_K(k) = \frac{\lambda^k e^{-\lambda}}{k!} \qquad k = 0, 1, 2, \ldots$$

Independently of $K$, the number of items $N$ that each customer purchases has the PMF

$$p_N(n) = \begin{cases} \frac{1}{4} & n=0 \\ \frac{1}{4} & n=1 \\ \frac{1}{3} & n=2 \\ \frac{1}{6} & n=3 \\ 0 & \text{otherwise} \end{cases}$$

What is the z-transform of the PMF of $Y$, the total number of items that the store sells on an arbitrary day?

7.22 Books are packed into cartons. The weight $W$ of a book in pounds is a continuous random variable with PDF

$$f_W(w) = \begin{cases} \frac{1}{4} & 1 \leq w \leq 5 \\ 0 & \text{otherwise} \end{cases}$$

The number $K$ of books in any carton is a random variable with the PMF

$$p_K(k) = \begin{cases} \frac{1}{4} & k=8 \\ \frac{1}{4} & k=9 \\ \frac{1}{3} & k=10 \\ \frac{1}{6} & k=12 \\ 0 & \text{otherwise} \end{cases}$$

If we randomly select a carton and its weight is $X$, determine

a. the s-transform of the PDF of $X$.

b. $E[X]$.

c. the variance of $X$.

CHAPTER 8

# Introduction to Descriptive Statistics

## 8.1 INTRODUCTION

The term "statistics" can be used in the singular sense or plural sense. In the singular sense, it refers to the procedures used to organize and interpret observed data. In this case we define statistics as a branch of mathematics that deals with collecting, organizing and summarizing data, and drawing conclusions about the environment from which the data was collected. In the plural sense, statistics are quantitative values that are used to describe a set of observed data. Thus, in this chapter we will sometimes say "statistics is" and "statistics are," depending on the context.

Statistics is different from and complementary to probability. Fully defined probability problems have unique and precise solutions. Also, probability laws apply across an entire population of interest. Statistics is concerned with the relationship between an observed segment of a population and the entire population. That is, in statistics we are interested in understanding an observation that is based on a segment of a population of interest and how the observation can apply to the entire population.

A statistician works by formatting the data in a way that makes sense and later postulating a probabilistic model for the system under investigation based on his or her knowledge of the physical mechanisms involved in the system and on personal experience. The statistician expects the model to exhibit a probabilistic behavior that is similar to that of the physical system.

There are two general branches of statistics: *descriptive statistics* and *inferential* (or *inductive*) *statistics*. Figure 8.1 shows the different aspects of both branches of statistics. Descriptive statistics is concerned with collecting, organizing, and summarizing the raw scores in more meaningful ways. These raw scores are measurement or observed values that are referred to as data. Similarly, inferential statistics deals with procedures or techniques that can be used to study a segment of the population called a sample, and make generalizations about the population from which the sample was obtained with the help of



Fundamentals of Applied Probability and Random Processes. http://dx.doi.org/10.1016/B978-0-12-800852-2.00008-0

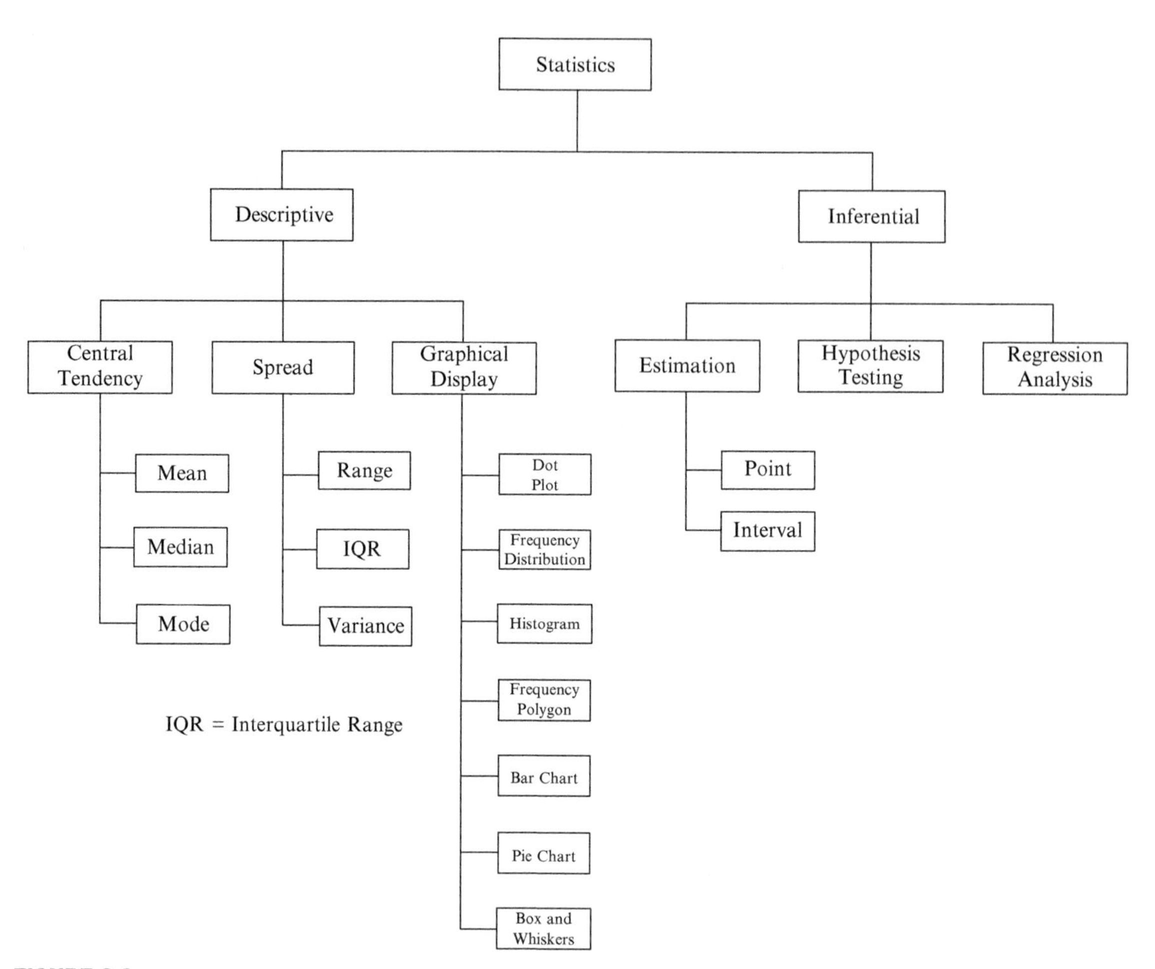


**FIGURE 8.1**
Different Branches of Statistics

probability theory. Thus, we can also say that inferential statistics enables us to infer from a given data set the generalizations that can be applied to a wider population.

A population is defined as the set of all individuals, items, or data of interest. This is the group about which inferential statistics attempts to generalize. A characteristic (usually numeric) that describes a population is referred to as a population *parameter*. Often, however, we do not have access to the whole population that we are interested in investigating. Under this condition the alternative is to select a portion or *sample* of members in the population. Thus, a sample is defined as a set of selected individuals, items, or data taken from a population of interest. Selecting a sample is more practical, and most scientific research is conducted on samples and not populations. A characteristic that describes a sample is called a sample *statistic* – this is similar to a parameter, except it describes characteristics in a sample and not a population. Inferential statistics uses the characteristics in a sample to infer what the unknown parameters are in a given population. In this way, a sample is selected from a population to learn more about the characteristics in the population of interest. It is, therefore, important that the sample accurately represents the population.

This chapter deals with descriptive statistics. We will discuss inferential statistics in Chapter 9.

## 8.2 DESCRIPTIVE STATISTICS

Descriptive statistics deals with collecting, grouping, and presenting data in a way that can be easily understood. Thus, it enables us to make sense of a set of observations. Descriptive statistics is very important because if we simply presented our raw data it would be hard to visualize what the data is showing, especially if there is a lot of it. Descriptive statistics therefore enables us to present the data in a more meaningful way, which allows simpler interpretation of the data. For example, if we had the grades of 100 students, we may be interested in the distribution or spread of the grades. Descriptive statistics allows us to do this. It works by organizing the data by a set of graphs, bar charts, tables, or frequency distributions. In this way we may see patterns emerging from the data. Descriptive statistics does not, however, allow us to make conclusions beyond the data we have analyzed; it is simply a way to describe a set of observed data by providing simple summaries about the sample and the measures.

There are three general methods that are used to describe a set of observation data:

a. *Measures of central tendency*, which are ways of describing the central position of a frequency distribution for a group of data.
b. *Measures of spread*, which are ways to summarize a group of data by describing how spread out the data values are.
c. *Graphical displays*, which are ways to visualize the data to see how it is distributed and if any patterns emerge from the data.

## 8.3 MEASURES OF CENTRAL TENDENCY

When we are given a set of raw data one of the most useful ways of summarizing the data is to find a way to describe the "center" of the data set. Central tendency describes the tendency of the observations to bunch around a particular value. Thus, measures of central tendency are numerical summaries that are used to summarize a data set with a single "typical" number. The three commonly used measures of central tendency are the mean, median, and mode. These are all measures of the 'average' of the distribution of the data set.

### 8.3.1 Mean

The mean, commonly called the average, is a mathematically computed value that represents a "typical value" of the set. The mean is computed by adding all the data values together and dividing by $N$, where $N$ represents the total number of data values. Thus, if the observation data set is $x_1, x_2, \ldots, x_N$, the mean is given by

$$\bar{x} = \frac{x_1 + x_2 + \cdots + x_N}{N} = \frac{1}{N}\sum_{n=1}^{N} x_n$$

For example, assume that we have the following observation data set:

66,54,88,56,34,12,48,50,80,50,90,65

Then, the mean is given by

$$\bar{x} = \frac{66 + 54 + 88 + 56 + 34 + 12 + 48 + 50 + 80 + 50 + 90 + 65}{12} = 57.75$$

Note that the mean may not be a member of the data set.

### 8.3.2 Median

If we divide the data set into two equal halves where each half contains 50% of the data, the numerical value where the data set is divided is called the median. To compute the median, three steps are required. First, the data are ordered by rank by arranging them in increasing order of magnitude. Second, the median data position is calculated. This requires examining the data set to determine if there is an even or odd number of data values. Let $R_M$ denote the rank-position of the median in the rank-ordered data set, and let $N$ denote the number of data values. The final step is the following: If $N$ is odd, we have that

$$R_M = \frac{N+1}{2} \tag{8.1}$$

and the median is $M$, which is the data value at the position $R_M$. If $N$ is even, then the value of the median is given

$$M = \frac{d_{\frac{N}{2}} + d_{\frac{N}{2}+1}}{2} \tag{8.2}$$

where $d_k$ is the data item in position $k$ of the rank-ordered data set and $M$ is the average of the value of $(\frac{N}{2})$th element and the value of the $(\frac{N}{2}+1)$th element in the rank-ordered data set. Thus, for an odd number of observations the median is a member of the data set, and for an even number of observations the median is not a member of the data set. For example, to find the median of the data set given earlier, we first reorder the set as in the following increasing order:

$$12,34,48,50,50,54,56,65,66,80,88,90 \tag{8.2a}$$

Second, there are 12 data values, which means that the median is not a member of the data set; it is the average of the sixth entry (54) and seventh entry (56) in the rank-ordered set. Thus, the median is

$$M=\frac{54+56}{2}=55$$

If we add 95 to the set, we obtain the new rank-ordered data set:

$$12,34,48,50,50,54,56,65,66,80,88,90,95 \tag{8.2b}$$

Since there are now 13 data values, the median is located in position $\frac{13+1}{2}=7$, which is 56 that is a member of the data set.

### 8.3.3 Mode

The mode is the data value that occurs the most frequently in a data set. For the data sets in (8.2a) and (8.2b), the value 50 occurs twice while every other value occurs once. Thus, the mode of the two data sets is 50 and the sets are said to be unimodal. Sometimes a data set can have more than one mode in which case it is said to be a multi-modal set.

## 8.4 MEASURES OF DISPERSION

Dispersion refers to the spread of the values around the central tendency. While measures of central tendency summarize a data set with a single "'typical" number, it is also useful to describe with a single number how "spread out" all the measurements are from that central number. Describing how a data set is distributed can be accomplished through one of the measures of dispersion: range, interquartile range, variance and standard deviation.

### 8.4.1 Range

The range of a data set is the measure that defines the difference between the largest data value and the smallest data value. This is the simplest measure of statistical dispersion or "spread." In the example in (8.2a), we have that

$$\text{Range}=\text{Maximum}-\text{Minimum}=90-12=78.$$

### 8.4.2 Quartiles and Percentiles

While the median divides a set of data into two halves, there are other division points that can also be used. The quartiles are used to divide an ordered data set into quarters. Similarly, the 100*p*th percentile of an ordered data set is a value such that at least 100*p*% of the observations are at or below this value and at least $100(1-p)\%$ are at or above this value, where $0<p<1$. Thus, we have that

i. First quartile $Q_1=25\text{th}$ percentile and $p=0.25$
ii. Second quartile $Q_2=50\text{th}$ percentile and $p=0.50$
iii. Third quartile $Q_3=75\text{th}$ percentile and $p=0.75$

The second quartile is the median. The *interquartile range* (IQR) is the difference between the third quartile and the first quartile; that is,

$$\text{IQR}=Q_3-Q_1 \tag{8.3}$$

Thus, the interquartile range describes the middle one-half (or 50%) of an ordered data set. This is illustrated in Figure 8.2. Like the median, the computation of quartiles is based on a data position in a rank-ordered data set and not on the data value itself.

The procedure for computing the 100*p*th percentile is as follows:

i. Order the $N$ observations from the smallest to the largest.
ii. Determine the product $Np$. If $Np$ is not an integer, round it to the next integer and find the corresponding ordered value. If $Np$ is an integer, say $k$, then calculate the mean of the $k$th and $(k+1)$th ordered data value.

For the example in (8.2a) where $N=12$, to obtain the first quartile where $p=0.25$, we compute $Np=(12)(0.25)=3$. Thus, the first quartile is the mean of the third and fourth entries in the rank-ordered set, which implies that $Q_1=(48+50)/2=49$. Similarly, for the third quartile we have that $Np=(12)(0.75)=9$. Thus, we find the mean of the ninth and tenth entries in the rank-ordered set, which is $Q_3=(66+80)/2=73$. Therefore, the interquartile range is $\text{IQR}=Q_3-Q_1=73-49=24$. Observe that the second quartile

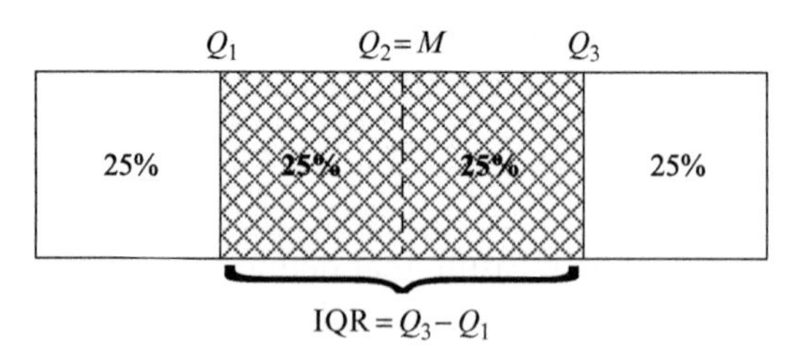


**FIGURE 8.2**
Relationship between IQR and the Quartiles

s obtained from $Np = (12)(0.5) = 6$. Thus, the median is the mean of the sixth and the seventh entries; that is, $Q_2 = (54+56)/2 = 55$, which is the value of the median that we obtained earlier.

A simple alternative method of computing IQR is as follows. First, find the median of the entire data set thereby dividing the data into two halves: the lower half, which is the set of values that are no greater than the median, and the upper half that is the set of values that are no less than the median. Then, find the median for the lower half, which is $Q_1$; and find the median for the upper half, which is $Q_3$. For example, in the example of (8.2a) we have the rank-ordered data set:

12,34,48,50,50,54,56,65,66,80,88,90

As discussed earlier, the median is $Q_2 = (54+56)/2 = 55$, which means that we have the lower half as 12, 34, 48, 50, 50, 54. Since there is an even number of entries, the median of this set is $Q_1$, which is average of the third and fourth entries; that is, $Q_1 = (48+50)/2 = 49$. Similarly, the upper half is 56, 65, 66, 80, 88, 90. Because the number of items is even, the median $Q_3$ is the average of the third and fourth entries; that is, $Q_3 = (66+80)/2 = 73$. These results agree with the earlier results.

### 8.4.3 Variance

If we subtract each data value from the mean, we obtain a value called a *deviation score* that tells us the numerical distance between the data value and the data set's "typical" value. The sum of all the deviation scores equals zero, as shown in Table 8.1. This follows from the fact that the data values above and below the mean have positive and negative deviation scores, respectively, that cancel each other out. To remove the negative values we can square the deviation scores and obtain the sum of the squared deviation scores. If we divide the sum of squares by the number of data values, which is 12 in our example, the resulting value is the variance. Thus, the variance is the average of the sum of squared deviation scores. In Table 8.1, the variance is obtained as

$$\sigma^2 = \frac{1}{12}\sum_{i=1}^{12}(x_i - \overline{x})^2 = 455.021$$

### 8.4.4 Standard Deviation

If we take the square root of the variance, the resulting number is called the standard deviation. The standard deviation is used to create bounds around the mean that describe data positions that are $\pm 1$, $\pm 2$, or $\pm 3$ standard deviations from the mean. For our example, the standard deviation is

**Table 8.1** Computation of the Variance

| Data Value $x$ | Deviation $x-\bar{x}$ | Squared Deviation $(x-\bar{x})^2$ |
|---|---|---|
| 66 | 8.25 | 68.0625 |
| 54 | -3.75 | 14.0625 |
| 88 | 30.25 | 915.0625 |
| 56 | -1.75 | 3.0625 |
| 34 | -23.75 | 564.0625 |
| 12 | -45.75 | 2093.0625 |
| 48 | -9.75 | 95.0625 |
| 50 | -7.75 | 60.0625 |
| 80 | 22.25 | 495.0625 |
| 50 | -7.75 | 60.0625 |
| 90 | 32.25 | 1040.0625 |
| 65 | 7.25 | 52.5625 |
| Sum | $\sum(x-\bar{x})=0$ | $\frac{1}{12}\sum(x-\bar{x})^2=455.021$ |

$\sigma=\sqrt{455.021}=21.33$. Figure 8.3 shows that for a normal distribution the probability that an observation value lies within one standard deviation from the mean is 0.6825, the probability that it lies within two standard deviations from the mean is 0.9544, and the probability that it lies within three standard deviations from the mean is 0.9974. That is,

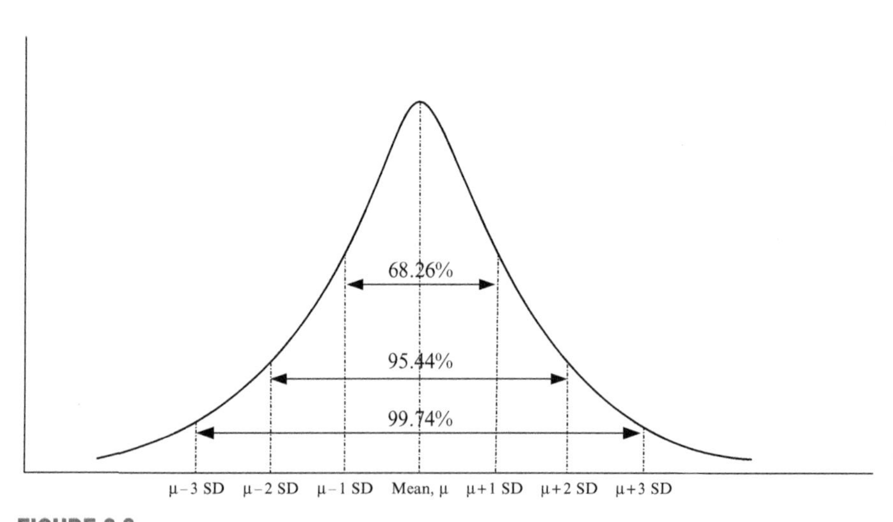


**FIGURE 8.3**
Areas Covered within Standard Deviations from the Mean

$$P[\mu_X - \sigma_X \le X \le \mu_X + \sigma_X] = 0.6826$$
$$P[\mu_X - 2\sigma_X \le X \le \mu_X + 2\sigma_X] = 0.9544$$
$$P[\mu_X - 3\sigma_X \le X \le \mu_X + 3\sigma_X] = 0.9974$$

## 8.5 GRAPHICAL AND TABULAR DISPLAYS

One of the methods of data analysis is to organize the data into a graphical or tabular form so that a trend, if any, emerging out of the data can be seen easily. Different graphical methods are used for this purpose and they include dot plots, frequency distribution, bar charts, histograms, frequency polygon, pie charts, and box and whiskers plot. These methods are discussed in this section.

### 8.5.1 Dot Plots

A dot plot, also called a *dot chart,* is used for relatively small data sets. The plot groups the data as little as possible and the identity of an individual observation is not lost. A dot plot uses dots to show where the data values (or scores) in a distribution are. The dots are plotted against their actual data values that are on the horizontal scale. If there are identical data values, the dots are "piled" on top of each other. Thus, to draw a dot plot, count the number of data points falling in each data value and draw a stack of dots that corresponds to the number of items in each data value. The plot makes it easy to see gaps and clusters in a data set as well as how the data spreads along the axis. For example, consider the following data set:

35,48,50,50,50,54,56,65,65,70,75,80

Since the value 50 occurs three times in the data set, there are three dots above 50. Similarly, since the value 65 occurs twice, there are two dots above 65. Thus, each dot in the plot represents a data item; there are as many dots on the plot as there are data items in the observation set. The dot plot for the data set is shown in Figure 8.4.

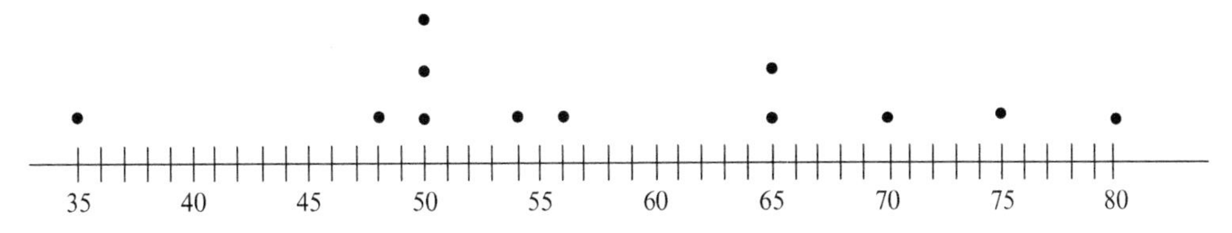


**FIGURE 8.4**
Example of a Dot Plot

### 8.5.2 Frequency Distribution

A frequency distribution is a table that lists the set of values in a data set and their frequencies (i.e., the number of times each occurs in the data set). For example, consider the following data set $X$:

35,48,50,50,50,54,56,65,65,70,75,80

The frequency distribution of the set is shown in Table 8.2.

Sometimes a set of data covers such a wide range of values that a list of all the $X$ values would be too long to be a "simple" presentation of the data. In this case we use a *grouped frequency distribution table* in which the $X$ column lists groups of data values, called *class intervals*, rather than individual data values. The width of each class can be determined by dividing the range of observations by the number of classes. It is advisable to have equal class widths, and the class intervals should be mutually exclusive and non-overlapping.

Class limits separate one class from another. The class width is the difference between the lower limits of two consecutive classes or the upper limits of two consecutive classes. A *class mark* (midpoint) is the number in the middle of the class. It is found by adding the upper and lower limits and dividing the sum by two. Table 8.3 shows how we can define the group frequency distribution of the following data set:

35,48,50,50,50,54,56,65,65,70,75,80

From the table we find that the class width is $41-34=47-40=7$. The table also shows the class marks (or the midpoints of the different classes). Observe that the class marks are also separated by the class width.

**Table 8.2** Example of Frequency Distribution

| X | Frequency |
|---|---|
| 35 | 1 |
| 48 | 1 |
| 50 | 3 |
| 54 | 1 |
| 56 | 1 |
| 65 | 2 |
| 70 | 1 |
| 75 | 1 |
| 80 | 1 |

Table 8.3 Example of Group Frequency Distribution

| Class | Frequency | Class Marks |
|---|---|---|
| 34 to 40 | 1 | 37 |
| 41 to 47 | 0 | 44 |
| 48 to 54 | 5 | 51 |
| 55 to 61 | 1 | 58 |
| 62 to 68 | 2 | 65 |
| 69 to 75 | 2 | 72 |
| 76 to 82 | 1 | 79 |

### 8.5.3 Histograms

A frequency histogram (or simply histogram) is used to graphically display the grouped frequency distribution. It consists of vertical bars drawn above the classes so that the height of a bar corresponds to the frequency of the class that it represents and the width of the bar extends to the real limits of the score class. Thus, the columns are of equal width, and there are no spaces between columns. For example, the histogram for the Table 8.3 is shown in Figure 8.5.

### 8.5.4 Frequency Polygons

A frequency polygon is a graph that is obtained by joining the class marks of a histogram with the two end points lying on the horizontal axis. It gives an idea of the shape of the distribution. It can be superimposed on the histogram by

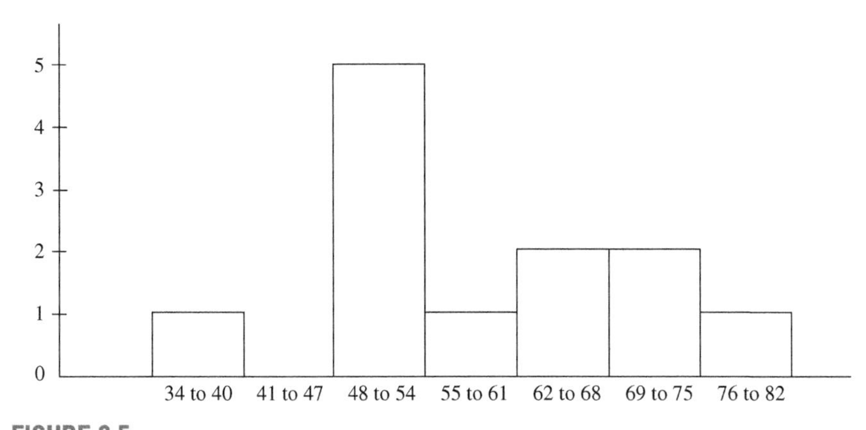


**FIGURE 8.5**
Histogram for Table 8.3

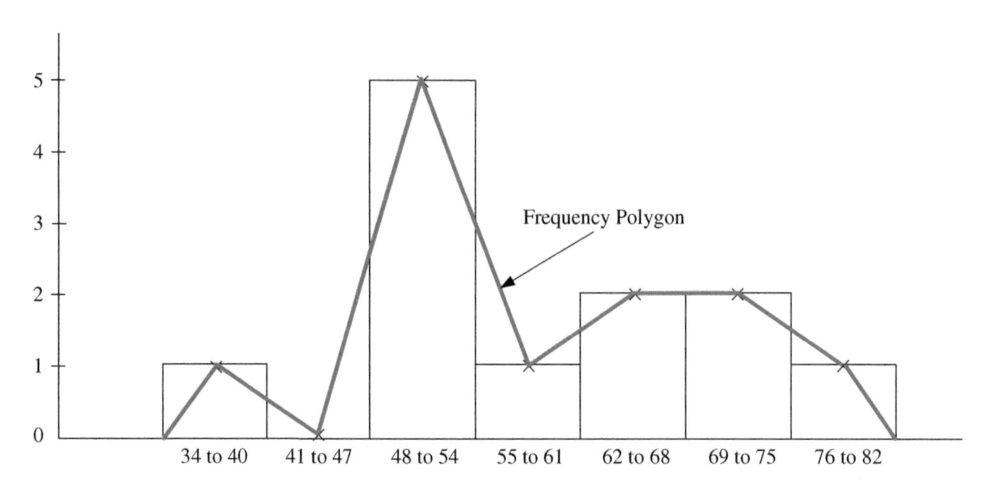


**FIGURE 8.6**
Frequency Polygon for Table 8.3

placing the dots on the class marks of the histogram, as shown in Figure 8.6, which is the frequency polygon of Figure 8.5.

### 8.5.5 Bar Graphs

A bar graph (or bar chart) is a type of graph in which each column (plotted either vertically or horizontally) represents a categorical variable. (A categorical variable is a variable that has two or more categories with no intrinsic ordering to the categories. For example, gender is a categorical variable with two categories: male and female.) A bar graph is used to compare the frequency of a category or characteristic with that of another category or characteristic. The bar height (if vertical) or length (if horizontal) shows the frequency for each category or characteristic.

For example, assume that data has been collected from a survey of 100 ECE students to determine how many of them indicated that Probability, Electronics, Electromechanics, Logic Design, Electromagnetics, or Signals and Systems is their best subject. Let the data show that 30 students indicated that Probability is their best subject, 20 students indicated that Electronics is their best subject, 15 students indicated that Electromechanics is their best subject, 15 students indicated that Logic Design is their best subject, 10 students indicated that Electromagnetics is their best subject, and 10 students indicated that Signals and Systems is their best subject. This result can be displayed in a bar graph as shown in Figure 8.7.

Because each column represents an individual category rather than intervals for a continuous measurement, gaps are included between the bars. Also, the bars can be arranged in any order without affecting the data.

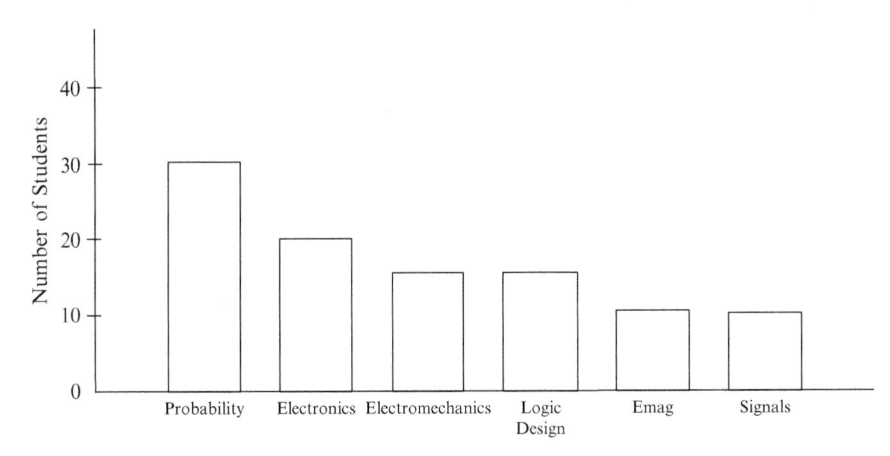


**FIGURE 8.7**
Example of a Bar Graph

Bar charts have a similar appearance as histograms. However, bar charts are used for categorical or qualitative data while histograms are used for quantitative data. Also, in histograms, classes (or bars) are of equal width and touch each other, while in bar charts the bars do not touch each other.

### 8.5.6 Pie Chart

A pie chart is a special chart that uses "pie slices" to show relative sizes of data. For example, consider the survey discussed earlier of ECE students to find out their favorite subjects. We noted that the survey result is as follows:

a. Probability 30%
b. Electronics 20%
c. Electromechanics 15%
d. Logic Design 15%
e. Electromagnetics 10%
f. Signals and Systems 10%

The result can be displayed in the pie chart shown in Figure 8.8. The size of each slice is proportional to the probability of the event that the slice represents.

Sometimes we are given a raw score of a survey and are required to display the result in a pie chart. For example, consider a survey of 25 customers who bought a particular brand of television. They were required to rate the TV as good, fair or bad. Assume that the following are their responses:

Good, good, fair, fair, fair, bad, fair, bad, bad, fair, good, bad, fair, good, fair, bad, fair, fair, good, bad, fair, good, fair, bad, and bad.

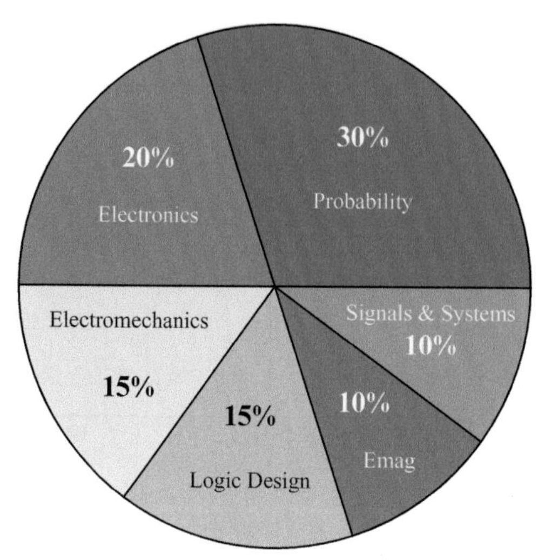


**FIGURE 8.8**
Example of a Pie Chart

To construct the pie chart we first create a list of the categories and tally each occurrence. Next, we add up the number of tallies to determine the frequency of each category. Finally, we obtain the relative frequency as the ratio of the frequency to the sum of the frequencies, where the sum is 25. This is illustrated in Table 8.4.

With this table we can then construct the pie chart as shown in Figure 8.9.

### 8.5.7 Box and Whiskers Plot

The box and whisker diagram (or box plot) is a way to visually organize data into fourths or quartiles. The diagram is made up of a "box," which lies between the first and third quartiles, and "whiskers" that are straight lines extending from the ends of the box to the maximum and minimum data values. Thus, the middle two-fourths are enclosed in a "box" and lower and upper fourths are drawn as whiskers. The procedure for drawing the diagram is as follows:

1. Arrange the data in increasing order
2. Find the median

**Table 8.4** Construction of Relative Frequencies

| Category | Tally | Frequency | Relative Frequency |
|---|---|---|---|
| Good | 𝍸 I | 6 | 0.24 |
| Fair | 𝍸 𝍸 I | 11 | 0.44 |
| Bad | 𝍸 III | 8 | 0.32 |

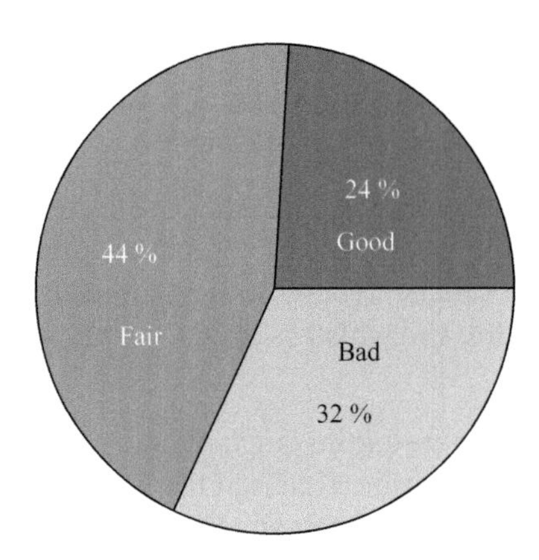


**FIGURE 8.9**
Pie Chart for TV Example

3. Find the first quartile $Q_1$, which is the median of the lower half of the data set; and the third quartile, $Q_3$, which is the median of the upper half of the data set.
4. On a line, mark points at the median, $Q_1$, $Q_3$, the minimum value of the data set and the maximum value of the data set.
5. Draw a box that lies between the first and third quartiles and thus represents the middle 50% of the data.
6. Draw a line from the first quartile to the minimum data value, and another line from the third quartile to the maximum data value. These lines are the whiskers of the plot.

Thus the box plot identifies the middle 50% of the data, the median, and the extreme points. The plot is illustrated in Figure 8.10 for the rank-order data set that we have used in previous sections:

12,34,48,50,50,54,56,65,66,80,88,90

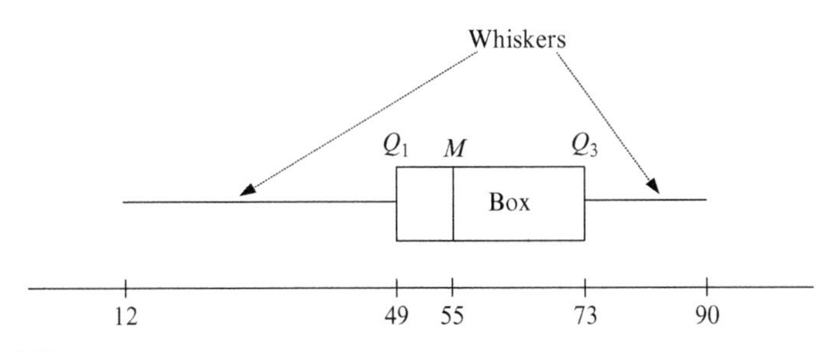


**FIGURE 8.10**
The Box and Whiskers Plot

As discussed earlier, the median is $M = Q_2 = 55$, the first quartile is $Q_1 = 49$, and the third quartile is $Q_3 = 73$. The minimum data value is 12, and the maximum data value is 90. These values are all indicated in the figure.

When collecting data, sometimes a result is collected that seems "wrong" because it is much higher or much lower than all of the other values. Such points are known as "outliers." These outliers are usually excluded from the whisker portion of the box and whiskers diagram. They are plotted individually and labeled as outliers.

As discussed earlier, the interquartile range, IQR, is the difference between the third quartile and the first quartile. That is, $IQR = Q_3 - Q_1$, which is the width of the box in the box and whiskers diagram. The IQR is one of the measures of dispersion, and statistics assumes that data values are clustered around some central value. The IQR can be used to tell when some of the other values are "too far" from the central value. In the box-and-whiskers diagram, an outlier is any data value that lies more than one and a half times the length of the box from either end of the box. That is, if a data point is below $Q_1 - 1.5 \times IQR$ or above $Q_3 + 1.5 \times IQR$, it is viewed as being too far from the central values to be reasonable. Thus, the values for $Q_1 - 1.5 \times IQR$ and $Q_3 + 1.5 \times IQR$ are the "fences" that mark off the "reasonable" values from the outlier values. That is, outliers are data values that lie outside the fences. Thus, we define

a. Lower fence $= Q_1 - 1.5 \times IQR$
b. Upper fence $= Q_3 + 1.5 \times IQR$

For the example in Figure 8.10, $IQR = Q_3 - Q_1 = 24 \Rightarrow IQR \times 1.5 = 36$. From this we have that the lower fence is at $Q_1 - 36 = 49 - 36 = 13$, and the upper fence is at $Q_3 + 36 = 73 + 36 = 109$. The only data value that is outside the fences is 12; all other data values are within the two fences. Thus, 12 is the only outliner in the data set.

The following example illustrates how to draw the box and whiskers plot with outliers. Supposed that we are given a new data set, which is 10, 12, 8, 1, 10, 13, 24, 15, 15, 24. First, we rank-order the data in an increasing order of magnitude to obtain:

1,8,10,10,12,13,15,15,24,24

Since there are 10 entries, the median is the average of the fifth and sixth numbers. That is, $M = (12 + 13)/2 = 12.5$. The lower half data set is 1, 8, 10, 10, 12 whose median is $Q_1 = 10$. Similarly, the upper half data set is 13, 15, 15, 24, 24 whose median is $Q_3 = 15$. Thus, $IQR = 15 - 10 = 5$, the lower fence is at $10 - (1.5)(5) = 2.5$ and the upper fence is at $15 + (1.5)(5) = 22.5$. Because the data values 1, 24 and 24 are outside the fences, they are outliers. The two values of 24 are stacked on top of each other. This is illustrated in Figure 8.11. The outliers are explicitly indicated in the diagram.

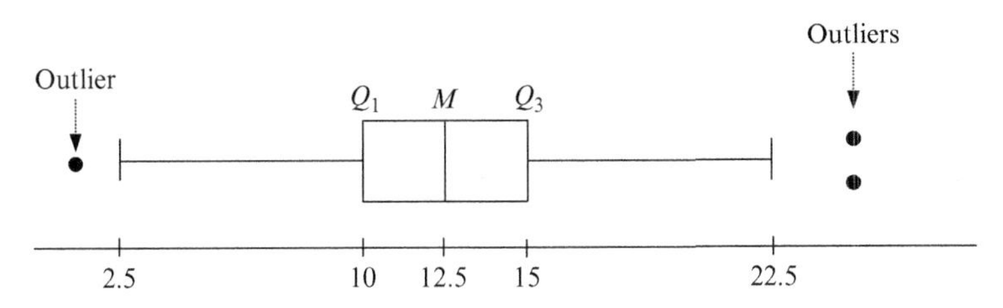


**FIGURE 8.11**
Example of Box and Whiskers Plot with Outliers

## 8.6 SHAPE OF FREQUENCY DISTRIBUTIONS: SKEWNESS

The shape of a curve of the frequency polygon of a data set is an important characteristic of the data set. One important distribution is the normal distribution, which as we have seen in earlier chapters has a bell-shaped curve that is symmetric about its highest point. Any distribution that is not symmetrical about a longitude is said to be *skewed*. If lower data values have higher frequencies, the distribution is said to be *positively skewed*. If the higher data values have higher frequencies, the distribution is *negatively skewed*. Thus a positively skewed distribution is a *right-tailed distribution*, and a negatively skewed distribution is a *left-tailed distribution*. This is illustrated in Figure 8.12.

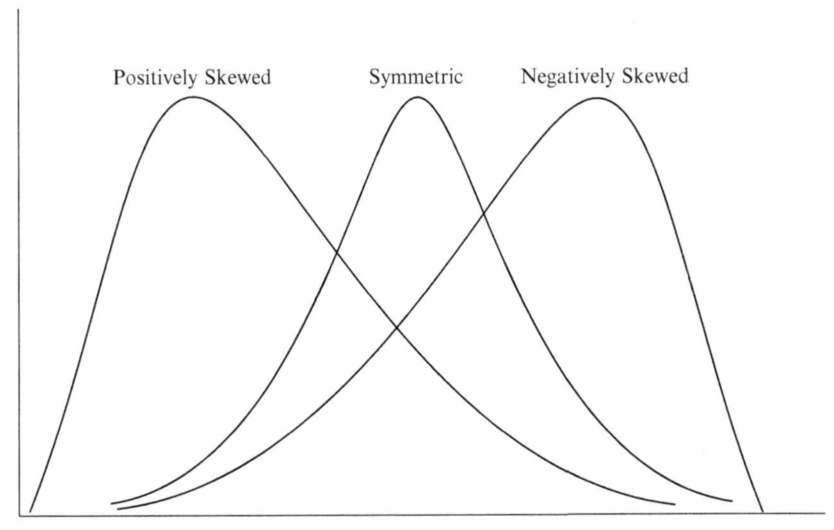


**FIGURE 8.12**
Skewness of Frequency Distribution

Quantitatively, the skewness of a random variable $X$ is denoted $\gamma_1$ and defined as

$$\gamma_1 = E\left[\left(\frac{X-\mu_X}{\sigma_X}\right)^3\right] = \frac{1}{(\sigma_X)^3}E\left[(X-\mu_X)^3\right] \tag{8.4}$$

where $\mu_X$ is the mean and $\sigma_X$ is the standard deviation. To compute the skewness of a data set of size $n$ we proceed as follows:

$$g_1 = \frac{\frac{1}{n}\sum_{i=1}^{n}(x_i-\overline{x})^3}{\left[\frac{1}{n-1}\sum_{i=1}^{n}(x_i-\overline{x})^2\right]^{3/2}} \tag{8.5}$$

However, this computation assumes that we have data for the whole population. When we have a sample of size $n$, the skewness is given by

$$G_1 = \frac{\sqrt{n(n-1)}}{n-2}g_1 \tag{8.6}$$

Sometimes skewness is measured by the *Pearson's coefficient of skewness*. There are two versions of this coefficient. Let $x_{Med}$ denote the median of a data set, and let $x_{Mode}$ denote the mode of the set. The *Pearson's mode* or *first skewness coefficient* is defined by:

$$\gamma_1 = \frac{\mu_X - x_{Mode}}{\sigma_X} \tag{8.7a}$$

The *Pearson's median* or *second skewness coefficient* is used when the mode of a distribution is not known and is given by

$$\gamma_1 = \frac{3(\mu_X - x_{Med})}{\sigma_X} \tag{8.7b}$$

The factor 3 in the preceding equation is due to the fact that empirical results indicate that

$$\text{Mean} - \text{Mode} \approx 3(\text{Mean} - \text{Median})$$

The Pearson's median coefficient is more popularly used. Thus, from this definition, we observe that when the mean is greater than the median, the distribution is positively skewed; and when the mean is less than the median, it is negatively skewed. When the mean and median are equal, then the skewness is zero and the distribution is symmetric. These facts are illustrated in Figure 8.13. We can remember the relationship between the mean, median and mode by the following simple rules:

a. For a positively skewed distribution, Mean > Median > Mode
b. For a negatively skewed distribution, Mean < Median < Mode
c. For a symmetric distribution, Mean = Median = Mode

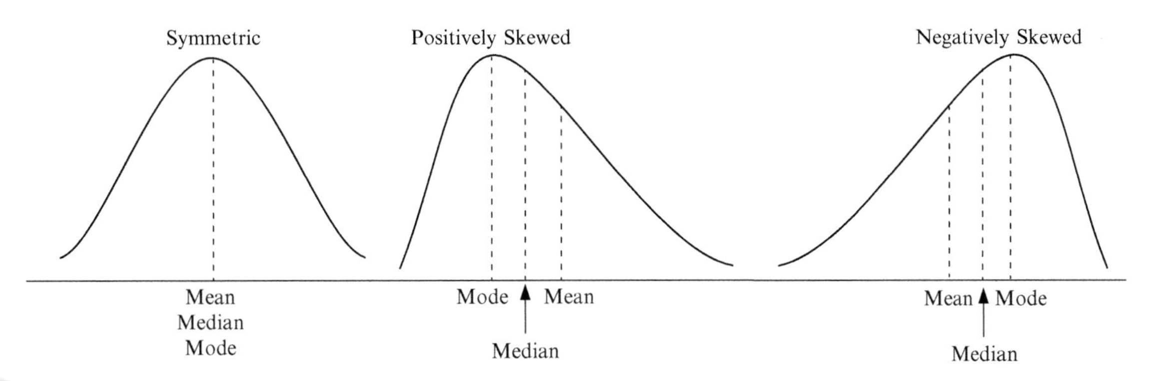


**FIGURE 8.13**
Relationship between Mean, Median and Mode

## 8.7 SHAPE OF FREQUENCY DISTRIBUTIONS: PEAKEDNESS

Another important property of frequency distributions is their *kurtosis*, which refers to their degree of peakedness. If a curve is more peaked than the normal distribution, it is said to be *leptokurtic*; and if it is less peaked than the normal distribution, it is said to be *platykurtic*. This is illustrated in Figure 8.14. The term "platy-" means "broad;" and in terms of shape, a platykurtic distribution has a

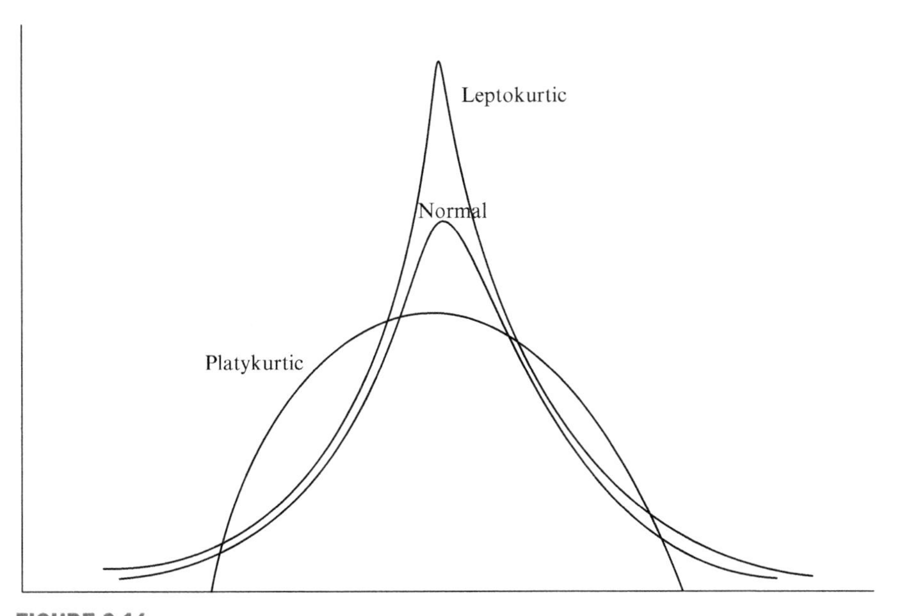


**FIGURE 8.14**
Peakedness of Frequency Distribution

lower, wider peak around the mean and thinner tails. Similarly, "lepto-" means "slender" or "skinny" and in terms of shape, a leptokurtic distribution has a more acute peak around the mean and fatter tails.

Quantitatively, the kurtosis of a random variable $X$ is denoted by $\beta_2$ and defined by

$$\beta_2 = E\left[\left(\frac{X-\mu_X}{\sigma_X}\right)^4\right] = \frac{1}{[\sigma_X]^4}E\left[(X-\mu_X)^4\right] = \frac{1}{[\sigma_X^2]^2}E\left[(X-\mu_X)^4\right] \tag{8.8}$$

Another term that is associated with the peakedness of a frequency distribution is *excess kurtosis*, $\gamma_2$, which is kurtosis minus 3. Thus, the excess kurtosis of a frequency distribution is given by

$$\gamma_2 = \beta_2 - 3 \tag{8.9}$$

The "minus 3" is used as a correction to make the kurtosis of the normal distribution equal to zero. Leptokurtic distributions have positive excess kurtosis while platykurtic distributions have negative excess kurtosis. The normal distribution has excess kurtosis of zero. Any distribution that is peaked the same way as the normal distribution is sometimes called a *mesokurtic* distribution. An example of a mesokurtic distribution is the binomial distribution with the value of $p$ close to 0.5.

When the data set has a size $n$, the computation of the kurtosis is given by

$$b_2 = \frac{\frac{1}{n}\sum_{i=1}^{n}(x_i-\overline{x})^4}{s^4} = \frac{\frac{1}{n}\sum_{i=1}^{n}(x_i-\overline{x})^4}{\left[\frac{1}{n-1}\sum_{i=1}^{n}(x_i-\overline{x})^2\right]^2} \tag{8.10}$$

As in the case of the skewness parameter, this computation applies to a population of $n$ members. For a sample of size $n$, the sample excess kurtosis is given by

$$g_2 = \frac{n-1}{(n-2)(n-3)}\{(n+1)b_2 + 6\} \tag{8.11}$$

## 8.8 CHAPTER SUMMARY

This chapter has discussed basic statistical methods. As stated earlier, statistics is a collection of methods for summarizing, presenting, analyzing and interpreting data and drawing conclusions from the data. The two general branches of statistics are descriptive statistics and inferential or inductive statistics.

Descriptive statistics, which is the subject of this chapter, deals with collecting, grouping, and presenting data in a way that can be easily understood. This task includes summarizing the available data by such variables as the mean, median, mode, and measures of the spread of the data (including range, interquartile range, variance, and standard deviation). It also includes describing the data by a set of graphs, bar charts, tables, frequency distributions, histograms, pie charts, and box and whiskers diagrams.

Inferential statistics uses probability theory to draw conclusions (or inferences) about, or estimate parameters of the environment from which the sample data came. Chapter 9 is devoted to a discussion on inferential statistics.

## 8.9 PROBLEMS

### Section 8.3 Measures of Central Tendency

8.1 Consider the following set of data: 15, 20, 21, 20, 36, 15, 25, 15.
  a. What is the sample mean?
  b. What is the median?
  c. What is the mode?

8.2 The following data represents the ages of 21 students in a class: 22, 24, 19, 17, 20, 27, 24, 23, 26, 17, 19, 22, 25, 21, 21, 22, 22, 21, 21, 20, 22.
  a. What is the mean of the data set?
  b. What is the median?
  c. What is the mode?

8.3 The scores in a class quiz are as follows: 58, 62, 62, 63, 65, 65, 65, 68, 69, 72, 72, 75, 76, 78, 79, 81, 84, 84, 85, 92, 94, 95, 98.
  a. What is the mean score in the quiz?
  b. What is the median score in the quiz?
  c. What is the mode of the quiz?

### Section 8.4 Measures of Dispersion

8.4 Consider the following set of data: 15, 20, 21, 20, 36, 15, 25, 15.
  d. What is the variance?
  e. What is the interquartile range?

8.5 The following data represents the ages of 21 students in a class: 22, 24, 19, 17, 20, 27, 24, 23, 26, 17, 19, 22, 25, 21, 21, 22, 22, 21, 21, 20, 22.
  a. What is the variance?
  b. What is the range?
  c. What is the interquartile range?

8.6 The scores in a class quiz are as follows: 58, 62, 62, 63, 65, 65, 65, 68, 69, 72, 72, 75, 76, 78, 79, 81, 84, 84, 85, 92, 94, 95, 98.
  a. What is the variance?
  b. What is the range?
  c. What is the interquartile range?

## Section 8.6 Graphical Displays

8.7 Bob found 15 pens in his new locker: 6 black, 4 blue, 3 red and 2 green.
  a. Draw a dot plot of the data set.
  b. Draw a bar graph of the data set.

8.8 The scores in a class quiz are as follows: 58, 62, 62, 63, 65, 65, 65, 68, 69, 72, 72, 75, 76, 78, 79, 81, 84, 84, 85, 92, 94, 95, 98. Draw a box and whiskers diagram of the data set.

8.9 Consider a survey concerning the ages of 800 students in a college. The results can be summarized as follows: 320 students were between 18 and 20 years old, 240 students were between 20 and 22 years old, 80 were between 22 and 24 years old, and 160 students were between 24 and 26 years old.
  a. Give the frequency distribution of the ages
  b. Plot the histogram of the ages
  c. Plot the frequency polygon of the ages.

8.10 Consider the following set of data: 2, 3, 5, 5, 6, 7, 7, 7, 8, 9, 9, 10, 10, 11, 12, 14, 14, 16, 18, 18, 22, 24, 24, 26, 28, 28, 32, 45, 50, and 55.
  a. Plot the box and whiskers diagram of the set.
  b. Determine if there are any outliers in the data set.

8.11 The following data were collected in a study in which 30 students were asked to rate their professors: very good, good, good, fair, excellent, good, good, very good, fair, good, good, excellent, very good, good, good, good, fair, very good, good, very good, excellent, very good, good, fair, fair, very good, very good, good, good, and excellent. Draw a pie chart to represent the data.

## Section 8.7 Shape of Frequency Distribution

8.12 The following data represents the ages of 21 students in a class: 22, 24, 19, 17, 20, 27, 24, 23, 26, 17, 19, 22, 25, 21, 21, 22, 22, 21, 21, 20, 22.
  a. Determine the skewness of the data
  b. Determine the kurtosis of the data
  c. Determine the excess kurtosis.

8.13 The scores in a class quiz are as follows: 58, 62, 62, 63, 65, 65, 65, 68, 69, 72, 72, 75, 76, 78, 79, 81, 84, 84, 85, 92, 94, 95, 98.
  a. What is the skewness of the data?
  b. What is the kurtosis of the data?
  c. What is the excess kurtosis?

8.14 Consider the following set of data: 15, 20, 21, 20, 36, 15, 25, 15.
  a. What is the skewness of the data?
  b. What is the kurtosis of the data?
  c. What is the excess kurtosis?

CHAPTER 9

# Introduction to Inferential Statistics

## 9.1 INTRODUCTION

As discussed in Chapter 8, a population is the set of all individuals, items, or data of interest. A characteristic (usually numeric) that describes a population is referred to as a population parameter. As also discussed in Chapter 8, we usually do not have access to the whole population that we are interested in investigating; we are limited to working with a sample of individuals in the population. Inferential statistics uses the characteristics in a sample to infer or draw conclusions on what the unknown parameters in a given population would be. Alternatively we say that inferential statistics uses the sample data to draw conclusions (or inferences) about, or estimate parameters of, the environment from which the data came. In this way, inferential statistics is concerned with making generalizations based on a set of data by going beyond information contained in the set. It uses probability theory to make decisions on the distributions of the data that are used in descriptive statistics.

Because a sample is selected from a population to learn more about the characteristics of the population of interest, it is important that the sample accurately represents the population. The process of ensuring that a sample fairly accurately represents the population is called *sampling*. Inferential statistics arises out of the fact that sampling naturally incurs sampling error and thus a sample is not expected to perfectly represent the population. The methods of inferential statistics that we consider in this chapter are:

a. *Sampling theory*, which deals with problems associated with selecting samples from some collection that is too large to be examined completely.
b. *Estimation theory*, which is concerned with making some prediction on or estimate of a population parameter based on the available data.
c. *Hypothesis testing*, which is an inferential procedure that uses sample data to evaluate the credibility of a hypothesis about a population. It works by choosing one model from several postulated (or hypothesized) models



Fundamentals of Applied Probability and Random Processes. http://dx.doi.org/10.1016/B978-0-12-800852-2.00009-2

of the physical system. Hypothesis testing is also called *detection theory* or *decision making*.

d. *Regression analysis*, which attempts to find mathematical expressions that best represent the collected data.

## 9.2 SAMPLING THEORY

As discussed earlier, the collection of data to be studied is called a population. A population can be finite or infinite. For example, a study on the number of students in the electrical engineering department of a college deals with a finite population. On the other hand, a study that involves the entire world population deals with what may be regarded as a countably infinite population.

For many statistical studies it is difficult and sometimes impossible to examine the entire population. In such cases the studies will be conducted with a small part of the population called a *sample*. Facts about the population can be inferred from the results obtained from the sample. The process of obtaining samples is called *sampling*. Thus, sampling is concerned with the selection of a subset of the members of a population to estimate characteristics of the whole population.

The reliability of conclusions drawn about the population depends on whether the sample is chosen to represent the population sufficiently well. One way to ensure that the sample sufficiently represents the population well is to ensure that each member of the population has the same chance of being in the sample. Samples constructed in this manner are called *random samples*.

We will be concerned with obtaining a sample of size $n$ that is described by the values $x_1, x_2, \ldots, x_n$ of a random variable $X$. We assume that each value is independent of the others. Thus, we can conceptualize these values as a sequence $X_1, X_2, \ldots, X_n$ of independent and identically distributed random variables, each of which has the same distribution as $X$. We will, therefore, define a random sample of size $n$ as a sequence of independent and identically distributed random variables $X_1, X_2, \ldots, X_n$. Once a sample has been taken, we denote the values obtained in the sample by $x_1, x_2, \ldots, x_n$.

Any quantity obtained from a sample for the purpose of estimating a population parameter is called a *sample statistic* (or simply *statistic*). An estimator $\hat{\theta}$ of a parameter $\theta$ of a random variable $X$ is a random variable that depends on a random sample $X_1, X_2, \ldots, X_n$. The two most common estimators are the sample mean and the sample variance.

### 9.2.1 The Sample Mean

Let $X_1, X_2, \ldots, X_n$ denote the random variables for a sample of size $n$. We define the sample mean $\overline{X}$ as the following random variable:

$$\overline{X} = \frac{X_1 + X_2 + \cdots + X_n}{n} = \frac{1}{n}\sum_{i=1}^{n} X_i \tag{9.1}$$

As stated earlier, the $X_i$ are random variables that are assumed to have the same PDF $f_X(x)$ (or PMF $p_X(x)$) as the population random variable $X$. When a particular sample with values $x_1, x_2, \ldots, x_n$ has been obtained, the sample mean is given by

$$\overline{x} = \frac{x_1 + x_2 + \cdots + x_n}{n}$$

Let $\mu_X$ denote the mean value of the population random variable $X$. The sample mean is a random variable because different sample sizes give different means. Thus, because it is a random variable, the sample mean has a mean value, which is given by

$$E\left[\overline{X}\right] = E\left[\frac{1}{n}\sum_{i=1}^{n} X_i\right] = \frac{1}{n}\sum_{i=1}^{n} E[X_i] = \frac{1}{n}\sum_{i=1}^{n} \mu_X = \mu_X \tag{9.2}$$

Since the mean value of the sample mean is equal to the true mean value of the population random variable, we define the sample mean to be an *unbiased estimate* of the population mean. The term "unbiased estimate" implies that the mean value of the estimate of a parameter is the same as the true mean of the parameter.

We can also compute the variance of the sample mean. If the population is infinite or the population is finite but sampling is done with replacement, the variance of the sample mean is given by

$$\begin{aligned}\sigma_{\overline{X}}^2 &= E\left[\left(\overline{X} - \mu_X\right)^2\right] = E\left[\left(\frac{X_1 + X_2 + \cdots + X_n}{n} - \mu_X\right)^2\right] \\ &= \frac{1}{n^2}\sum_{i=1}^{n} E\left[X_i^2\right] + \frac{1}{n^2}\sum_{\substack{i=1 \\ i \neq j}}^{n}\sum_{j=1}^{n} E\left[X_i X_j\right] - \frac{2\mu_X}{n}\sum_{i=1}^{n} E[X_i] + \mu_X^2\end{aligned}$$

Since the $X_i$ are independent random variables, $E[X_iX_j] = E[X_i]E[X_j] = \mu_X^2$. Therefore,

$$\begin{aligned}\sigma_{\overline{X}}^2 &= \frac{1}{n^2}\sum_{i=1}^{n} E\left[X_i^2\right] + \frac{1}{n^2}\sum_{\substack{i=1 \\ i \neq j}}^{n}\sum_{j=1}^{n} E\left[X_i X_j\right] - \frac{2\mu_X}{n}\sum_{i=1}^{n} E[X_i] + \mu_X^2 \\ &= \frac{E\left[X^2\right]}{n} + \left(\frac{n(n-1)}{n^2}\right)\mu_X^2 - \mu_X^2 = \frac{E\left[X^2\right]}{n} - \frac{\mu_X^2}{n} = \frac{E\left[X^2\right] - \mu_X^2}{n} \\ &= \frac{\sigma_X^2}{n}\end{aligned} \tag{9.3}$$

where $\sigma_X^2$ is the true variance of the population. When the population size is N and sampling is done without replacement, then if the sample size is $n \leq N$, the variance of the sample mean is given by

$$\sigma_{\overline{X}}^2 = \frac{\sigma_X^2}{n}\left(\frac{N-n}{N-1}\right) \tag{9.4}$$

Finally, since the sample mean is a random variable, we would like to find its PDF. Because the sample mean is derived from the sum of random variables, the central limit theorem says that it tends to be asymptotically normal regardless of the distribution of the random variables in the sample. In general this assumption of normal distribution is true when $n \geq 30$. If we define the standard normal score of the sample mean by

$$Z = \frac{\overline{X} - \mu_{\overline{X}}}{\sigma_{\overline{X}}} = \frac{\overline{X} - \mu_X}{\sigma_X/\sqrt{n}}$$

then when $n \geq 30$, we obtain

$$F_{\overline{X}}(x) = P\left[\overline{X} \leq x\right] = \Phi\left(\frac{x - \mu_X}{\sigma_X/\sqrt{n}}\right) \tag{9.5}$$

## EXAMPLE 9.1

A random variable $X$ is sampled 36 times obtaining a value $\overline{X}$ that is used as an estimate of $E[X]$. If the PDF of $X$ is $f_X(x) = 2e^{-2x}$, where $x \geq 0$,

a. Determine and $E\left[\overline{X}\right]$ and $E\left[\overline{X}^2\right]$.
b. What is the probability that the sample mean lies between 1/4 and 3/4?

**Solution:**

a. Since $X$ is an exponential random variable, the true mean and true variance are given by $E[X] = 1/2$ and $\sigma_X^2 = 1/4$. Thus, $E\left[\overline{X}\right] = E[X] = 1/2$. Since $n = 36$, the second moment of $\overline{X}$ is given by

$$E\left[\overline{X}^2\right] = \sigma_{\overline{X}}^2 + \left\{E\left[\overline{X}\right]\right\}^2 = \frac{\sigma_X^2}{n} + \{E[X]\}^2 = \frac{1/4}{36} + \left(\frac{1}{2}\right)^2 = \frac{37}{144}$$

b. Since $n = 36$, the variance of the sample mean becomes $\sigma_{\overline{X}}^2 = \sigma_X^2/36 = 1/144$. The probability that the sample mean lies between 1/4 and 3/4 is given by

$$P\left[\frac{1}{4} \leq \overline{X} \leq \frac{3}{4}\right] = F_{\overline{X}}(3/4) - F_{\overline{X}}(1/4) = \Phi\left(\frac{3/4 - 1/2}{1/12}\right) - \Phi\left(\frac{1/4 - 1/2}{1/12}\right)$$

$$= \Phi(3) - \Phi(-3) = \Phi(3) - \{1 - \Phi(3)\} = 2\Phi(3) - 1$$

$$= 2(0.9987) - 1 = 0.9974$$

where the values used in the last equality are taken from Table 1 of Appendix 1.

### 9.2.2 The Sample Variance

Since knowledge of the variance indicates the spread of values around the mean, it is desirable to obtain an estimate of the variance. The sample variance is denoted by $S^2$ and defined by

$$S^2 = \frac{1}{n}\sum_{i=1}^{n}(X_i - \overline{X})^2 \tag{9.6}$$

Since $X_i - \overline{X} = (X_i - E[X]) - (\overline{X} - E[X])$, we have that

$$(X_i - \overline{X})^2 = (X_i - E[X])^2 - 2(X_i - E[X])(\overline{X} - E[X]) + (\overline{X} - E[X])^2$$

$$\begin{aligned}\sum_{i=1}^{n}(X_i - \overline{X})^2 &= \sum_{i=1}^{n}(X_i - E[X])^2 - 2(\overline{X} - E[X])\sum_{i=1}^{n}(X_i - E[X]) \\ &\quad + \sum_{i=1}^{n}(\overline{X} - E[X])^2 \\ &= \sum_{i=1}^{n}(X_i - E[X])^2 - 2n(\overline{X} - E[X])^2 + n(\overline{X} - E[X])^2 \\ &= \sum_{i=1}^{n}(X_i - E[X])^2 - n(\overline{X} - E[X])^2\end{aligned}$$

From this we have that

$$S^2 = \frac{1}{n}\sum_{i=1}^{n}(X_i - \overline{X})^2 = \frac{1}{n}\sum_{i=1}^{n}(X_i - E[X])^2 - (\overline{X} - E[X])^2$$

Since the sample variance is also a random variable, its expected value is given by

$$\begin{aligned}E[S^2] &= E\left[\frac{1}{n}\sum_{i=1}^{n}(X_i - E[X])^2 - (\overline{X} - E[X])^2\right] \\ &= \frac{1}{n}\sum_{i=1}^{n}E[(X_i - E[X])^2] - E\left[(\overline{X} - E[X])^2\right] = \sigma_X^2 - \frac{\sigma_X^2}{n} \\ &= \frac{n-1}{n}\sigma_X^2\end{aligned} \tag{9.7}$$

where $\sigma_X^2$ is the variance of the population and the second to the last equality is due to equation (9.3). Since the mean of the sample variance is not equal to the population variance, the sample variance is a *biased estimate* of the variance. To obtain an unbiased estimate, we define a new random variable as follows:

$$\begin{aligned}\hat{S}^2 &= \frac{n}{n-1}S^2 = \frac{1}{n-1}\sum_{i=1}^{n}(X_i - \overline{X})^2 \\ E\left[\hat{S}^2\right] &= \sigma_X^2\end{aligned} \tag{9.8}$$

Thus, $\hat{S}^2$ is an unbiased estimate of the variance. The above results hold when sampling is from an infinite population or done with replacement in a finite population. If sampling is done without replacement from a finite population of size $N$, then the mean of the sample variance is given by

$$E[S^2] = \left(\frac{N}{N-1}\right)\left(\frac{n-1}{n}\right)\sigma_X^2 \tag{9.9}$$

### 9.2.3 Sampling Distributions

Assume that the population from which samples are taken has a mean $E[X]$ and variance $\sigma_X^2$. Then, when the sample size is $n$ and $n$ is large, the central limit theorem allows us to expect the sample mean $\overline{X}$ to be a normally distributed random variable with the standardized score given by

$$Z = \frac{\overline{X} - \mu_{\overline{X}}}{\sigma_{\overline{X}}} = \frac{\overline{X} - \mu_X}{\sigma_X/\sqrt{n}}$$

As stated earlier in the chapter, this claim of normal distribution is valid when $n \geq 30$. When $n < 30$, we use the Student's $t$ distribution by defining the normalized sample mean as

$$T = \frac{\overline{X} - E[X]}{\hat{S}/\sqrt{n}} = \frac{\overline{X} - E[X]}{S/\sqrt{(n-1)}} \tag{9.10}$$

When the sample size is $n$, the Student's $t$ distribution is said to have $n-1$ degrees of freedom. The PDF of the Student's $t$ distribution is given by

$$f_T(t) = \frac{\Gamma\left(\frac{v+1}{2}\right)}{\sqrt{v\pi}\,\Gamma\left(\frac{v}{2}\right)}\left(1 + \frac{t^2}{v}\right)^{-(v+1)/2} \tag{9.11}$$

where $\Gamma(x)$ is the gamma function of $x$ and $v = n-1$ is the number of degrees of freedom, which is the number of independent samples. $\Gamma(x)$ has the following properties:

$$\Gamma(k+1) = \begin{cases} k\Gamma(k) & \text{any } k \\ k! & k \text{ integer} \end{cases}$$

$$\Gamma(2) = \Gamma(1) = 1$$

$$\Gamma(1/2) = \sqrt{\pi}$$

Thus, for example,

$$\Gamma(2.5) = 1.5 \times \Gamma(1.5) = 1.5 \times 0.5 \times \Gamma(0.5) = 1.5 \times 0.5 \times \sqrt{\pi} = 1.3293$$

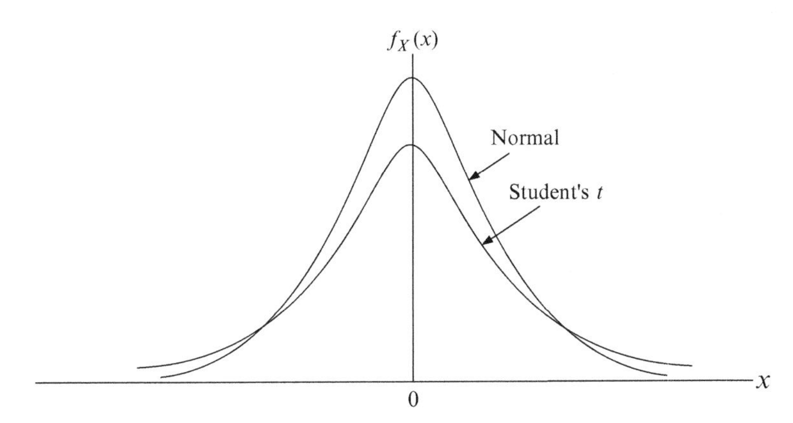


**FIGURE 9.1**
Student's $t$ and Normal Distributions

The Student's $t$ distribution is symmetrical about its mean value of 0 and is similar to the normal curve, as shown in Figure 9.1. The variance of the distribution depends on $v$; it is greater than 1 but approaches 1 as $n \rightarrow \infty$.

## 9.3 ESTIMATION THEORY

As discussed earlier, estimation theory is concerned with making some prediction or estimate on a population parameter based on the available data. In general the goal is to estimate a variable that is not directly observable but is observed only through some other measurable variables. For example, given a random variable $X$ whose distribution depends on some parameter $\theta$, a statistic $g(X)$ is called an estimator of $\theta$ if, for any observed value $x \in X$, $g(x)$ is considered to be an estimate of $\theta$. Thus, the estimation of the parameter can be defined as a rule or function that assigns a value to $\theta$ for each realization of $x$.

Estimation can also be defined as a method of fitting probability law/distribution to data, where the data consists of a collection of realized random variables generated by the probability distribution under consideration. The observation data is usually modeled as a realization of random variables $X_1, X_2, \ldots, X_n$ that are independent and identically distributed. The problem then becomes the following. Given observations of independent and identically distributed random variables $X_1, X_2, \ldots, X_n$ with a CDF $F(x|\theta)$, estimate $\theta$, where the notation $F(x|\theta)$ is used to indicate that the distribution depends on the parameter $\theta$ that is to be estimated. For example, the PMF of the Poisson distribution involves one parameter $\lambda$, and is given by

$$p(x|\lambda) = \frac{\lambda^x e^{-\lambda}}{x!} \qquad x = 0, 1, \ldots$$

Similarly, the PDF of the normal distribution involves two parameters, $\sigma$ and $\mu$, and is given by

$$f(x|\mu,\sigma) = \frac{1}{\sigma\sqrt{2\pi}} e^{-\frac{1}{2}(x-\mu)^2/\sigma^2} \qquad -\infty < x < \infty$$

In the same way, the exponential distribution involves one parameter, $\lambda$, and the PDF is given by

$$f(x|\lambda) = \lambda e^{-\lambda x} \qquad x \geq 0$$

Finally, the Erlang-$k$ distribution involves two parameters, $k$ and $\lambda$, and the PDF is given by

$$f(x|k,\lambda) = \frac{\lambda^k x^{k-1}}{(k-1)!} e^{-\lambda x} \qquad x \geq 0$$

Thus, the estimate of the parameter $\theta$ is the value of $\theta$ obtained for a given realization $x$. For example, the sample mean $\overline{x}$ is an estimator of the population mean $\mu$. Because $g(X)$ is a random variable, it is sometimes denoted by $\hat{\Theta}$ and the observed value $g(x)$ is denoted by $\hat{\theta}$.

The bias of an estimator $\hat{\theta}$ is defined by

$$B(\hat{\theta}) = E[\hat{\theta}] - \theta \tag{9.12}$$

An *unbiased estimator* of a population parameter is a statistic whose mean or expected value is equal to the parameter being estimated. Thus, an estimator $\hat{\theta}$ is defined to be unbiased if $E[\hat{\theta}] = \theta$; that is, if its expected value is equal to the parameter to be estimated. The corresponding value of the statistic is then called an *unbiased estimate*.

Another concept associated with an estimator is *efficiency*. If the sampling distributions of two statistics have the same mean, the statistic with the smaller variance is said to be a more *efficient estimator* of the mean. The corresponding value of the statistic is called an efficient estimator. Thus, if $\hat{X}_1$ and $\hat{X}_2$ are unbiased estimators of $X$, then $\hat{X}_1$ is a more efficient estimator of $X$ than $\hat{X}_2$ if $\sigma^2_{\hat{X}_1} < \sigma^2_{\hat{X}_2}$. Generally, we would like the estimates to be unbiased and efficient. But, in practice, biased or inefficient estimators are used due to the relative ease with which they are obtained.

Suppose we use an estimator of the sample mean in the form

$$\hat{X} = \frac{1}{n}\sum_{i=1}^{n} X_i$$

where $X_i, i=1,2,\ldots,n$, are the observed data used to estimate $\theta$. Such an estimator is defined to be a *consistent estimator* of $\theta$ if it converges in probability to $\theta$; that is,

$$\lim_{n\to\infty} P\left[|\hat{X}-\theta|\geq\varepsilon\right]=0$$

Thus, as the sample size $n$ increases, a consistent estimator gets closer to the true parameter; that is, it is *asymptotically unbiased*. If the observed data (or the samples) come from a population with finite mean and variance, we know that the variance of $\hat{X}$ is $\sigma_{\hat{X}}^2=\sigma_X^2/n$, which goes to zero as $n$ goes to infinity. Recall from Chapter 3 that the Chebyshev inequality states that for a random variable $Y$,

$$P[|Y-E[Y]|\geq a]\leq\frac{\sigma_Y^2}{a^2} \qquad a>0$$

Thus, we conclude that the sample mean is a consistent estimator of the population mean. In general, any unbiased estimator $\hat{X}$ of $\theta$ with the property that $\lim_{n\to\infty}\sigma_{\hat{X}}^2=0$ is a consistent estimator of $X$ due to the Chebyshev inequality.

### 9.3.1 Point Estimate, Interval Estimate, and Confidence Interval

The sample mean is called a *point estimate* because it assigns a single value to the estimate. Thus, a point estimate is a single value that best estimates a population parameter. Another type of estimate is the *interval estimate*, which is used to account for the fact that point estimates cannot be expected to coincide with the parameters they are estimating.

In interval estimation, the parameter being estimated lies within a certain interval, called the *confidence interval*, with a certain probability. Thus, a confidence interval is a range of values that can be expected to contain a given parameter $\theta$ of a population with a specified probability. It consists of two numbers between which the population parameter can be expected to lie with a certain degree of confidence. These end-points of a confidence interval are called the *confidence limits*.

A $q$-percent confidence interval is the interval within which the estimate will lie with a probability of $q/100$. This means that a $q$-percent confidence interval has a probability of $q/100$ of containing the population parameter. The parameter $q$ is also called the *confidence level*. Thus, a 95% confidence interval for the mean $\mu$ is a random interval that contains $\mu$ with probability 0.95. Confidence intervals are often used in conjunction with point estimates to highlight the uncertainty of estimates.

In the case of the sample mean, the $q$-percent confidence interval is defined as follows when sampling is done with replacement or from an infinite population:

$$\overline{X}-k\sigma_{\overline{X}} \leq \mu_X \leq \overline{X}+k\sigma_{\overline{X}} = \overline{X}-\frac{k\sigma_X}{\sqrt{n}} \leq \mu_X \leq \overline{X}+\frac{k\sigma_X}{\sqrt{n}} \qquad (9.13)$$

where $k$ is a constant that depends on $q$ and is called the *confidence coefficient* or *critical value*. The quantity $k$ defines the number of standard deviations on either side of the mean that the confidence interval is expected to cover. Thus, the confidence limits are $\overline{X} \pm k\sigma_{\overline{X}} = \overline{X} \pm k\sigma_X/\sqrt{n}$, and the *error of the estimate* is $k\sigma_{\overline{X}} = k\sigma_X/\sqrt{n}$.

When sampling is done without replacement from a finite population of size $N$, we obtain

$$\overline{X}-\frac{k\sigma_X}{\sqrt{n}}\sqrt{\frac{N-n}{N-1}} \leq \mu_X \leq \overline{X}+\frac{k\sigma_X}{\sqrt{n}}\sqrt{\frac{N-n}{N-1}} \qquad (9.14)$$

Values of $k$ that correspond to various confidence levels are shown in Table 9.1.

Figure 9.2 illustrates the confidence interval for the case of $q=95\%$. From Table 9.1 we can see that this corresponds to $k=1.96$.

**Table 9.1** Values of $k$ for Different Confidence Levels

| Confidence Level | 99.99% | 99.9% | 99% | 95% | 90% | 80% |
|---|---|---|---|---|---|---|
| $k$ | 3.89 | 3.29 | 2.58 | 1.96 | 1.64 | 1.28 |

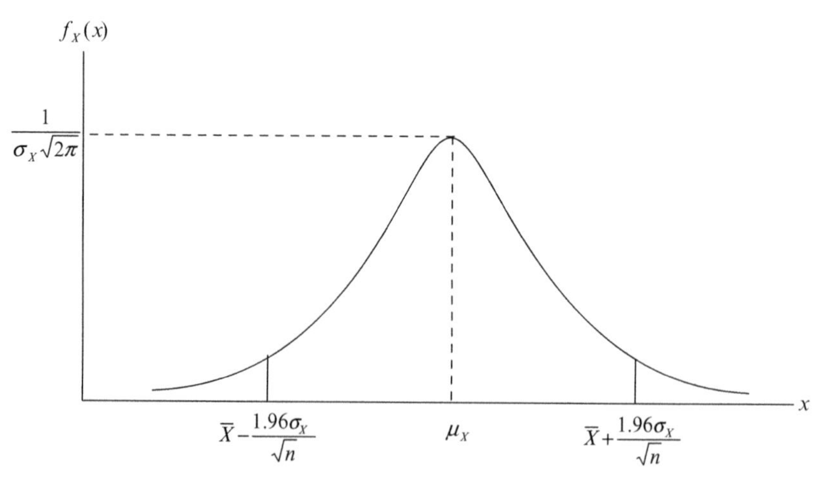


**FIGURE 9.2**
Confidence Limits for the 95% Confidence Level

In summary, the interval estimate is given by

$$\text{Interval Estimate} = \text{Point Estimate} \pm k_q \times \text{Sample Standard Deviation}$$

where $k_q$ is the value of the confidence coefficient $k$ for the specified confidence level $q$.

### EXAMPLE 9.2

One sample of a normally distributed random variable with standard deviation $\sigma_X$ is used to estimate the true mean $E[X]$. Determine the 95% confidence interval for the sample mean.

**Solution:**
Let $\overline{X}$ be the sample mean. Then 95% of the time the true mean, $E[X]$, falls within the interval given by $\overline{X} - k\sigma_X/\sqrt{n} \le E[X] \le \overline{X} + k\sigma_X/\sqrt{n}$. From Table 9.1 we find that when $q = 95\%$, $k = 1.96$. Since $n = 1$, the confidence interval is

$$\overline{X} - 1.96\sigma_X \le E[X] \le \overline{X} + 1.96\sigma_X$$

### EXAMPLE 9.3

If $\sigma_X = 1$, determine the number of observations required to ensure that at the 99% confidence level, $\overline{X} - 0.1 \le E[X] \le \overline{X} + 0.1$, where

$$\overline{X} = \frac{1}{n}\sum_{i=1}^{n} X_i$$

**Solution:**
The variance of the sample mean is $\sigma_{\overline{X}}^2 = \sigma_X^2/n$. From Table 9.1 the value of $k$ at the 99% confidence level is 2.58. Thus,

$$P\left[\overline{X} - 2.58\frac{\sigma_X}{\sqrt{n}} \le E[X] \le \overline{X} + 2.58\frac{\sigma_X}{\sqrt{n}}\right] = 0.99$$

and we are required to find the value of $n$ such that $2.58\sigma_X/\sqrt{n} = 0.1$. Since $\sigma_X = 1$, we have that

$$\frac{2.58}{\sqrt{n}} = 0.1 \Rightarrow \sqrt{n} = 25.8 \Rightarrow n = (25.8)^2 = 665.64$$

Since $n$ must be an integer, we have that $n = 666$.

### 9.3.2 Maximum Likelihood Estimation

Consider the following problem. A box contains a number of red and blue balls. Suppose that it is known that the ratio of the numbers is 3:1, but it is not known whether the red or blue balls are more numerous. Thus, the probability of drawing a red ball from the box is either 1/4 or 3/4. If we draw $m$ balls from the box with replacement, we know that the number $K$ of red balls drawn has the following binomial distribution:

$$p_K(k) = \binom{m}{k} p^k (1-p)^{m-k} \qquad k = 0, 1, 2, \ldots, m$$

where $p$ is the probability of drawing a red ball and has the value $p=1/4$ or $p=3/4$. The idea behind maximum likelihood estimation is to obtain the "best" estimate of $p$, which is the parameter of the binomial distribution. More formally, the maximum likelihood estimation of the parameters $\theta_1, \theta_2, \ldots, \theta_n$ that characterize a random variable $X$, chooses the value(s) of the parameters that make(s) the observed values $x_1, x_2, \ldots, x_n$ most probable.

Suppose $X$ is a random variable whose distribution depends on a single parameter $\theta$. Let $x_1, x_2, \ldots, x_n$ be an observed random sample. If $X$ is discrete with PMF $p_X(x)$, the probability that a random sample consists of exactly these values is given by

$$L(\theta) = L(\theta; x_1, x_2, \ldots, x_n) = p_X(x_1; \theta) p_X(x_2; \theta) \cdots p_X(x_n; \theta) \tag{9.15}$$

$L(\theta)$ is called the *likelihood function* and is a function of $\theta$; that is, its value depends on both the selected sample values and the choice of $\theta$. If $X$ is continuous with PDF $f_X(x)$, then the likelihood function is defined by

$$L(\theta) = L(\theta; x_1, x_2, \ldots, x_n) = f_X(x_1; \theta) f_X(x_2; \theta) \cdots f_X(x_n; \theta) \tag{9.16}$$

Generally, when a PDF or PMF is viewed as a function of an unknown parameter with a fixed data set, it is known as a likelihood function. The maximum likelihood estimate of $\theta$ is the value of $\theta$ that maximizes the value of $L(\theta)$. If $L(\theta)$ is a differentiable function, then a necessary condition for $L(\theta)$ to have a maximum value is that

$$\frac{\partial}{\partial \theta} L(\theta) = 0$$

The partial derivative is used because $L(\theta)$ depends on both $\theta$ and the sample values $x_1, x_2, \ldots, x_n$. If $\hat{\theta}$ is the value of $\theta$ that maximizes $L(\theta)$, then $\hat{\theta}$ is called the *maximum likelihood estimator*. Also, if the likelihood function contains $k$ parameters such that

$$L(\theta) = L(\theta_1, \theta_2, \ldots, \theta_k; x_1, x_2, \ldots, x_n) = \prod_{i=1}^{n} f_X(x_i; \theta_1, \theta_2, \cdots, \theta_k) \tag{9.17}$$

then the point where the likelihood function is a maximum is the solution to the following $k$ equations:

$$\frac{\partial}{\partial \theta_1} L(\theta_1, \theta_2, \ldots, \theta_k; x_1, x_2, \ldots, x_n) = 0$$

$$\frac{\partial}{\partial \theta_2} L(\theta_1, \theta_2, \ldots, \theta_k; x_1, x_2, \ldots, x_n) = 0$$

$$\vdots$$

$$\frac{\partial}{\partial \theta_k} L(\theta_1, \theta_2, \ldots, \theta_k; x_1, x_2, \ldots, x_n) = 0$$

In many cases it is more convenient to work with the logarithm of the likelihood function.

## EXAMPLE 9.4

Suppose a random sample of size $n$ is drawn from the Bernoulli distribution. What is the maximum likelihood estimate of $p$, the success probability?

**Solution:**

Let $X$ denote the Bernoulli random variable with a probability of success $p$. Then the PMF of $X$ is

$$p_X(x;p) = p^x(1-p)^{1-x} \quad x = 0,1;\ 0 \le p \le 1$$

The sample values $x_1, x_2, \ldots, x_n$ will be a sequence of 0s and 1s, and the likelihood function is

$$L(p) = L(p; x_1, x_2, \ldots, x_n) = \prod_{i=1}^{n} p^{x_i}(1-p)^{1-x_i} = p^{\sum x_i}(1-p)^{n-\sum x_i}$$

If we define

$$y = \sum_{i=1}^{n} x_i$$

we obtain the following, where the logarithm is taken to base $e$:

$$\log L(p) = y \log p + (n-y)\log(1-p)$$

Then, taking partial derivative we obtain

$$\frac{\partial}{\partial p} \log L(p) = \frac{y}{p} - \frac{n-y}{1-p} = 0$$

This gives the maximum likelihood estimate as

$$\hat{p} = \frac{y}{n} = \frac{1}{n}\sum_{i=1}^{n} x_i$$

Thus, $\hat{p}$ is the mean of the sample values.

## EXAMPLE 9.5

Consider a box that contains a mix of red and blue balls whose exact composition of red and blue balls is not known. If we draw $n$ balls from the box with replacement and obtain $k$ red balls, what is the maximum likelihood estimate of $p$, the probability of drawing a red ball?

**Solution:**

Let $K$ denote the number of red balls among the $n$ balls drawn from the box. We know that $K$ has the following binomial distribution:

$$p_K(k) = \binom{n}{k} p^k (1-p)^{n-k} \qquad k = 0, 1, 2, \ldots, n$$

where $p$ is the probability of drawing a red ball. Thus, the likelihood function is given by

$$L(p; k) = \binom{n}{k} p^k (1-p)^{n-k}$$

Taking the logarithm to base $e$ on both sides we obtain

$$\log L(p; k) = \log \binom{n}{k} + k \log p + (n-k) \log(1-p)$$

Then

$$\frac{\partial}{\partial p} \log L(p; k) = \frac{k}{p} - \frac{n-k}{1-p} = 0$$

This gives

$$\hat{p} = \frac{k}{n}$$

Thus, for example, if we perform the drawing $n = 15$ times and obtain red balls $k = 8$ times, then $\hat{p} = 8/15 = 0.533$.

## EXAMPLE 9.6

The campus police receives complaints about students' disturbances on campus according to a Poisson process. The police chief wants to estimate the average arrival rate $\lambda$ of these complaints from the random sample $x_1, x_2, \ldots, x_n$ of arrivals per one-hour interval. What result will the chief obtain?

**Solution:**

The PMF of a Poisson random variable $X$ with mean $\lambda$ is

$$p_X(x) = \frac{\lambda^x}{x!} e^{-\lambda} \qquad x = 0, 1, 2, \ldots$$

Thus, the maximum likelihood function for the random sample is given by

$$L(\lambda) = L(\lambda; x_1, x_2, \ldots, x_n) = \left(\frac{\lambda^{x_1}}{x_1!} e^{-\lambda}\right) \left(\frac{\lambda^{x_2}}{x_2!} e^{-\lambda}\right) \cdots \left(\frac{\lambda^{x_n}}{x_n!} e^{-\lambda}\right) = \frac{\lambda^y e^{-n\lambda}}{x_1! x_2! \cdots x_n!}$$

where $y = x_1 + x_2 + \ldots + x_n$. Thus, taking logarithms to base $e$ on both sides we obtain:

$$\log L(\lambda) = -n\lambda + y \log \lambda - \log(x_1! x_2! \cdots x_n!)$$

Then

$$\frac{\partial}{\partial \lambda} \log L(\lambda) = -n + \frac{y}{\lambda} = 0$$

This gives

$$\hat{\lambda} = \frac{y}{n} = \frac{x_1 + x_2 + \ldots + x_n}{n}$$

### 9.3.3 Minimum Mean Squared Error Estimation

Let $\hat{X}$ be an estimator of the random variable $X$. The *estimation error* $\varepsilon$ is defined as the difference between the true value $X$ and the estimated value $\hat{X}$; that is,

$$\varepsilon = X - \hat{X}$$

The error is a random variable that provides a measure of how well the estimator performs. There are several ways to define the goodness of an estimator, and they are all based on defining an appropriate cost function $C(\varepsilon)$ of $\varepsilon$. The goal is to choose the estimator in such a manner as to minimize the cost function. The choice of the cost function is subjective. For example, we may define the cost function as the squared error:

$$C_1(\varepsilon) = \varepsilon^2$$

Alternatively, we may define the cost function as the absolute error:

$$C_2(\varepsilon) = |\varepsilon|$$

These two cost functions are illustrated in Figure 9.3, where it can be seen that $C_1(\varepsilon)$ penalizes large errors severely and rewards errors that are less than 1, while $C_2(\varepsilon)$ treats errors on a linear basis.

*Mean squared error* (MSE) estimation, which is used to obtain estimates of random variables, modifies the cost function as follows:

$$C(\varepsilon) = E\left[\varepsilon^2\right] \tag{9.18}$$

MSE has the advantage that it is mathematically tractable. Also, when used in linear estimation, optimal estimates can be found in terms of the first-order and second-order moments, where by optimality we mean that errors are minimized.

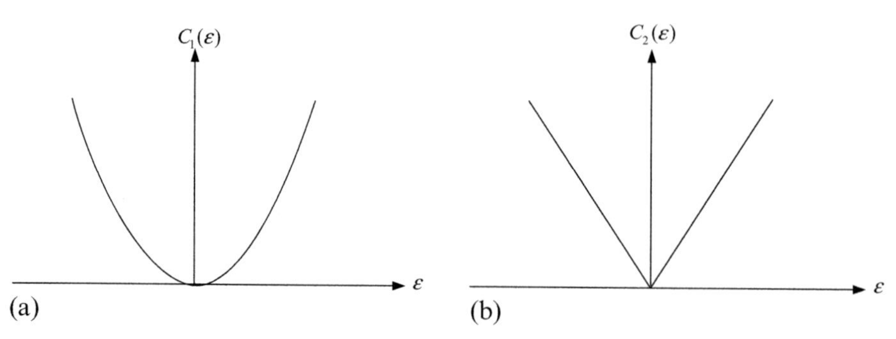


**FIGURE 9.3**
Squared Error and Absolute Error Cost Functions

## EXAMPLE 9.7

Assume that the random variable $Y$ is estimated from the random variable $X$ by the following linear function of $X$:

$$\hat{Y} = aX + b$$

Determine the values of $a$ and $b$ that minimize the mean squared error.

**Solution:**

The mean squared error is given by

$$e_{ms} = E\left[\left(Y - \hat{Y}\right)^2\right] = E\left[\{Y - (aX + b)\}^2\right]$$

The necessary conditions for $a$ and $b$ to minimize $e_{ms}$ are as follows:

$$\frac{\partial e_{ms}}{\partial a} = E[2(-X)\{Y - (aX + b)\}] = 0$$
$$\frac{\partial e_{ms}}{\partial b} = E[2(-1)\{Y - (aX + b)\}] = 0$$

Simplifying these two equations, we obtain

$$E[XY] = aE[X^2] + bE[X]$$
$$E[Y] = aE[X] + b$$

From these we obtain the optimal values $a^*$ and $b^*$ of $a$ and $b$, respectively, as follows:

$$a^* = \frac{E[XY] - E[X]E[Y]}{\sigma_X^2} = \frac{\text{Cov}(X, Y)}{\sigma_X^2} = \frac{\sigma_{XY}}{\sigma_X^2} = \frac{\rho_{XY}\sigma_X\sigma_Y}{\sigma_X^2} = \frac{\rho_{XY}\sigma_Y}{\sigma_X}$$

$$b^* = E[Y] - \frac{\sigma_{XY}E[X]}{\sigma_X^2} = E[Y] - \frac{\rho_{XY}\sigma_Y E[X]}{\sigma_X}$$

where $\rho_{XY}$ is the correlation coefficient of $X$ and $Y$. Finally, the minimum mean squared error becomes

$$e_{mms} = e_{ms}\big|_{\substack{a=a^* \\ b=b^*}} = \sigma_Y^2 - \frac{(\sigma_{XY})^2}{\sigma_X^2} = \sigma_Y^2 - \rho_{XY}^2\sigma_Y^2 = \sigma_Y^2(1 - \rho_{XY}^2)$$

When the joint PDF of $X$ and $Y$ is given, the different first-order and second-order moments are obtained from the specified PDF.

## 9.4 HYPOTHESIS TESTING

A hypothesis is a proposition that is assumed as a premise in an argument. It is a tentative proposition that is offered as an explanation for an observation, phenomenon or scientific problem that can be tested by further investigation. Hypothesis testing is a procedure for determining whether to accept or reject a certain statement (called *statistical hypothesis*) about the random variable determining the population, based on information obtained from a random sample of the population. Thus, a statistical hypothesis is an assumption made about a population and is generally stated as a proposition concerning the distribution of a random variable from that population. The hypothesis may be a statement about the values of one or more of the parameters of a given distribution, or it may be a statement about the form of the distribution. Examples of statistical hypotheses include the following:

a. The average waiting time in the checkout area of the cafeteria does not exceed 2 minutes.
b. The arrival pattern of customers at a bank is Poisson.
c. The time between bus arrivals at a students' bus station is exponentially distributed.

As stated earlier, hypothesis testing is also called detection theory. Detection is the process of deciding whether some phenomenon is present or not based on some observation. Also, because decision making involves choosing between a number of mutually exclusive alternatives called hypotheses, hypothesis testing is sometimes called decision making.

### 9.4.1 Hypothesis Test Procedure

A statistical hypothesis test is a formal step-by-step procedure. It starts by defining the so-called *null hypothesis*, which, as the name suggests, is a statement that there is no statistical difference between two procedures. For example, if we want to decide whether a given coin has been altered to give unequal probabilities of heads and tails, we may formulate the hypothesis that the coin is fair, which means the probability of heads on a toss of the coin is $p=0.5$. Similarly, if we want to decide whether a given drug is more effective than a placebo, we can make the statement that there is no difference between the drug and the placebo. The *null hypothesis* is denoted by $H_0$ and is sometimes aptly called the *no-difference hypothesis*. Any hypothesis that differs from a given null hypothesis is called an *alternative hypothesis*, which is denoted by $H_1$. There are different ways to formulate the alternative hypothesis. In the coin example mentioned above, the different forms of $H_1$ include $p \neq 0.5$, $p > 0.5$ or $p < 0.5$.

The test procedure consists of the following five steps:

1. State the hypotheses
   - $H_0$: The null hypothesis, which states that there is no difference between 2 means; any difference found is due to sampling error; that is, any significant difference found is not a true difference, but chance due to sampling error
   - $H_1$: The alternative hypothesis, which states that there is significant difference between the means; therefore $H_0$ is false. $H_1$ is essentially a statement of what the hypothesis test is interested in establishing.
2. Specify the level of significance, $\alpha$, which is the probability that sample means are different enough to reject $H_0$.
3. Calculate the test value, $z$, based on the observation data
4. Obtain the critical value $z_\alpha$ at the specified level of significance $\alpha$, which is used to compare against the calculated test value
5. Reject or accept $H_0$. The calculated test value is compared to the critical value to determine if the difference is significant enough to reject $H_0$ at the predetermined level of significance.

The critical region is determined by a parameter $\alpha$, which is called the *level of significance* of the test. The value of $\alpha$ is usually chosen to be 0.01 or 0.05 (or sometimes expressed as 1% or 5%), though any value between 0 and 1 can be selected. The level of significance is usually the confidence level subtracted from 100%. Thus, the rejection region is the region that lies outside the confidence interval; the confidence interval is the *acceptance region*. $H_0$ is accepted if $z$ lies in the acceptance region and it is rejected if $z$ lies in the rejection region. Note that choosing a high confidence level, and hence a small level of significance, makes it more likely that any given sample will result in accepting the null hypothesis, since a high confidence level will result in wider confidence interval (or greater acceptance region).

Figure 9.4 illustrates the rejection region and the confidence limits when the confidence level is $q=95\%$. The boundary between the confidence region and the rejection region is denoted by the symbol $z_\alpha$ and called the *critical value*. In Figure 9.4, $z_1=\overline{X}-1.96\sigma_X/\sqrt{n}$ is the lower confidence limit, and $z_2=\overline{X}+1.96\sigma_X/\sqrt{n}$ is the upper confidence limit. Thus, at the 95% confidence level we expect the mean to lie within the range $\overline{X}-1.96\sigma_X/\sqrt{n}\leq\mu_X\leq\overline{X}+1.96\sigma_X/\sqrt{n}$.

### 9.4.2 Type I and Type II Errors

Because hypothesis testing is based on sample means and not population means, there is a possibility of committing an error by making a wrong decision in rejecting or failing to reject $H_0$. There are two types of errors that can be committed: a *Type I error* and a *Type II error*.

A Type I error is committed if we reject $H_0$ when it is true; that is, it should have been accepted. The probability of committing this type of error is equal to $\alpha$. Thus, if $\alpha=0.05$, the probability of a Type I error is 0.05.

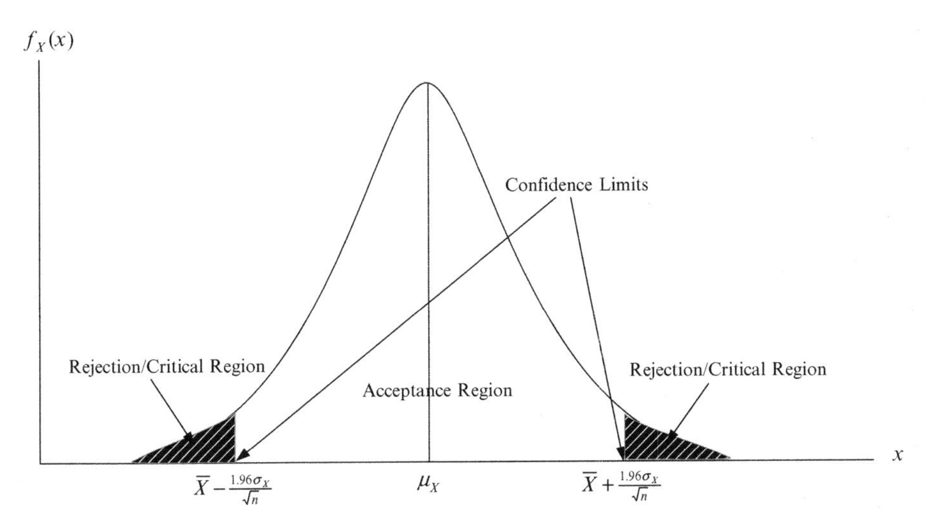


**FIGURE 9.4**
Relation of Confidence Limits and Level of Significance

A Type II error is committed if we accept $H_0$ when it should have been rejected. We can reduce the probability of this type of error by increasing $\alpha$.

Both types of error lead to a wrong decision. Which type of error is more serious than the other depends on the situation. The goal of any hypothesis test must be to minimize both types of error. One obvious way to reduce both types of error is to increase the sample size, which may not always be possible. As stated earlier, the level of significance, $\alpha$, represents the maximum probability with which we would be willing to commit a Type I error. These two types of error are summarized in Table 9.2.

### 9.4.3 One-Tailed and Two-Tailed Tests

Hypothesis tests are classified as either *one-tailed* (also called *one-sided*) tests or *two-tailed* (also called *two-sided*) tests. One tailed-tests are concerned with one side of a statistic, such as "the mean is greater than 10" or "the mean is less than 10." Thus, one-tailed tests deal with only one tail of the distribution, and the *z*-score is on only one side of the statistic.

**Table 9.2** Summary of the Types of Errors

| | | **Truth** | |
|---|---|---|---|
| | | **$H_0$ True** | **$H_0$ False** |
| Decision | Accept $H_0$ | OK, Correct Decision | False Acceptance, Type II Error |
| | Reject $H_0$ | False Rejection, Type I Error | OK, Correct Decision |

Two-tailed tests deal with both tails of the distribution, and the $z$-score is on both sides of the statistic. For example, Figure 9.4 illustrates a two-tailed test. A hypothesis like "the mean is not equal to 10" involves a two-tailed test because the claim is that the mean can be less than 10 or it can be greater than 10. Table 9.3 shows the critical values, $z_\alpha$, for both the one-tailed test and the two-tailed test in tests involving the normal distribution.

In a one-tailed test, the area under the rejection region is equal to the level of significance, $\alpha$. Also, the rejection region can be below (i.e., to the left of) the acceptance region or beyond (i.e., to the right of) the acceptance region depending on how $H_1$ is formulated. When the rejection region is below the acceptance region, we say that it is a *left-tail test*. Similarly, when the rejection region is above the acceptance region, we say that it is a *right-tail test*.

In the two-tailed test, there are two critical regions, and the area under each region is $\alpha/2$. As stated earlier, the two-tailed test is illustrated in Figure 9.4. Figure 9.5 illustrates the rejection region that is beyond the acceptance region for the one-tailed test, or more specifically the right-tail test.

**Table 9.3** Critical Points for Different Levels of Significance

| Level of Significance ($\alpha$) | 0.10 | 0.05 | 0.01 | 0.005 | 0.002 |
|---|---|---|---|---|---|
| $z_\alpha$ for 1-Tailed Tests | −1.28 or 1.28 | −1.645 or 1.645 | −2.33 or 2.33 | −2.58 or 2.58 | −2.88 or 2.88 |
| $z_\alpha$ for 2-Tailed Tests | −1.645 and 1.645 | −1.96 and 1.96 | −2.58 and 2.58 | −2.81 and 2.81 | −3.08 and 3.08 |

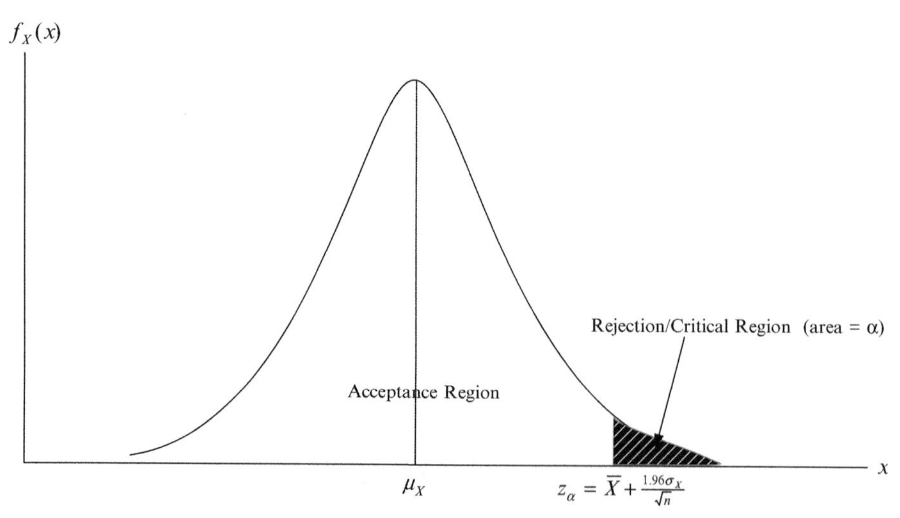


**FIGURE 9.5**
Critical Region for One-Tailed Tests

| One-Tailed Test (Left Tail) | Two-Tailed Test | One-Tailed Test (Right Tail) |
|---|---|---|
| $H_0 : \mu_X = \mu_0$<br>$H_1 : \mu_X < \mu_0$ | $H_0 : \mu_X = \mu_0$<br>$H_1 : \mu_X \neq \mu_0$ | $H_0 : \mu_X = \mu_0$<br>$H_1 : \mu_X > \mu_0$ |
| 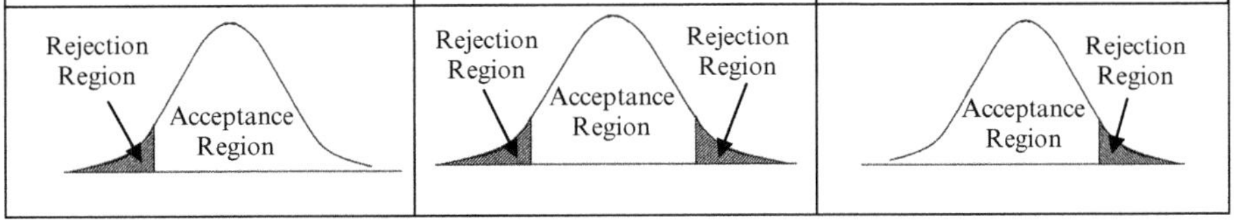 Rejection Region; Acceptance Region | Rejection Region; Acceptance Region; Rejection Region | Acceptance Region; Rejection Region |

**FIGURE 9.6**
Summary of the Different Tests

Note that in a one-tailed test, when $H_1$ involves values that are greater than $\mu_X$, we have a right-tail test. Similarly, when $H_1$ involves values that are less than $\mu_X$, we have a left-tail test. For example, an alternative hypothesis of the type $H_1 : \mu_X > 100$ is a right-tail test while an alternative hypothesis of the type $H_1 : \mu_X < 100$ is a left-tail test. Figure 9.6 is a summary of the different types of tests. In the figure, $\mu_0$ is the current value of the parameter.

## EXAMPLE 9.9

The mean lifetime $E[X]$ of the light bulbs produced by Lighting Systems Corporation is 1570 hours with a standard deviation of 120 hours. The president of the company claims that a new production process has led to an increase in the mean lifetimes of the light bulbs. If Joe tested 100 light bulbs made from the new production process and found that their mean lifetime is 1600 hours, test the hypothesis that $E[X]$ is not equal to 1570 hours using a level of significance of (a) 0.05 and (b) 0.01.

**Solution:**

The null hypothesis is

$H_0 : \mu_X = 1570$ hours

Similarly, the alternative hypothesis is

$H_1 : \mu_X \neq 1570$ hours

Since $\mu_X \neq 1570$ includes numbers that are both greater than and less than 1570, this is a two-tailed test. From the available data, the normalized value of the sample mean is

$$z = \frac{\overline{X} - \mu_X}{\sigma_{\overline{X}}} = \frac{\overline{X} - \mu_X}{\sigma_X / \sqrt{n}} = \frac{1600 - 1570}{120/\sqrt{100}} = \frac{30}{12} = 2.50$$

a. At a level of significance of 0.05, $z_\alpha = -1.96$ and $z_\alpha = 1.96$ for a two-tailed test. Thus, our acceptance region is $[-1.96, 1.96]$ of the standard normal distribution. The rejection and acceptance regions are illustrated in Figure 9.7.

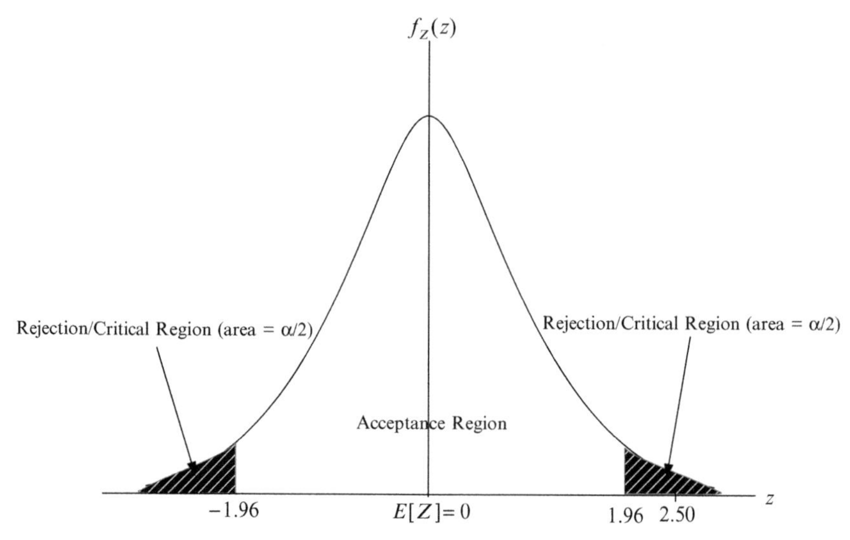


**FIGURE 9.7**
Critical Region for Problem 98.9(a)

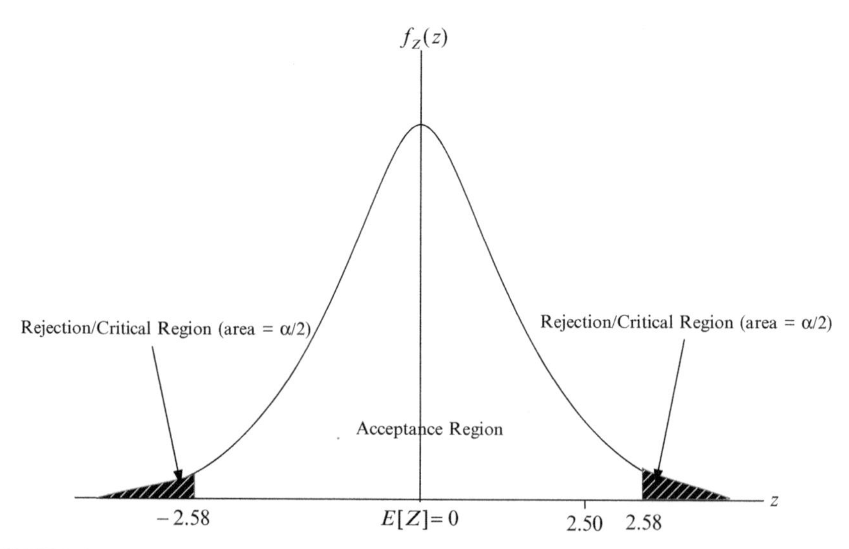


**FIGURE 9.8**
Critical Region for Problem 9.9(b)

Since $z=2.50$ lies outside the range $[-1.96, 1.96]$ (that is, it is in a rejection region), we reject $H_0$ at the 0.05 level of significance and accept $H_1$, which means that the difference in mean lifetimes is statistically significant.

b. At the 0.01 level of significance, $z_\alpha=-2.58$ and $z_\alpha=2.58$. The acceptance and rejection regions are shown in Figure 9.8. Since $z=2.50$ lies within the range $[-2.58, 2.58]$, which is the acceptance region, we accept $H_0$ at the 0.01 level of significance, which means that the difference in mean lifetimes is not statistically significant.

## EXAMPLE 9.10

For Example 9.9, test the hypothesis that the new mean lifetime is greater than 1570 hours using a level of significance of (a) 0.05 and (b) 0.01.

**Solution:**

Here we define the null hypothesis and alternative hypothesis as follows:

$H_0 : \mu_X = 1570$ hours
$H_1 : \mu_X > 1570$ hours

This is a one-tailed test. Since the $z$-score is the same as in Example 9.9, we only need to find the confidence limits for the two cases.

a. Because $H_1$ is concerned with values that are greater than 1570, we have a right-tail test, which means that we choose the rejection region that is above the acceptance region. Therefore, we choose $z_\alpha = 1.645$ for the 0.05 level of significance in Table 9.3. Since $z = 2.50$ lies in the rejection region (i.e., $2.50 > 1.645$), as illustrated in Figure 9.9, we reject $H_0$ at the 0.05 level of significance and thus accept $H_1$. This implies that the difference in mean lifetimes is statistically significant.
b. From Table 9.3, $z_\alpha = 2.33$ at the 0.01 level of significance, which is less than $z = 2.50$. Thus, we also reject $H_0$ at the 0.01 level of significance and accept $H_1$.

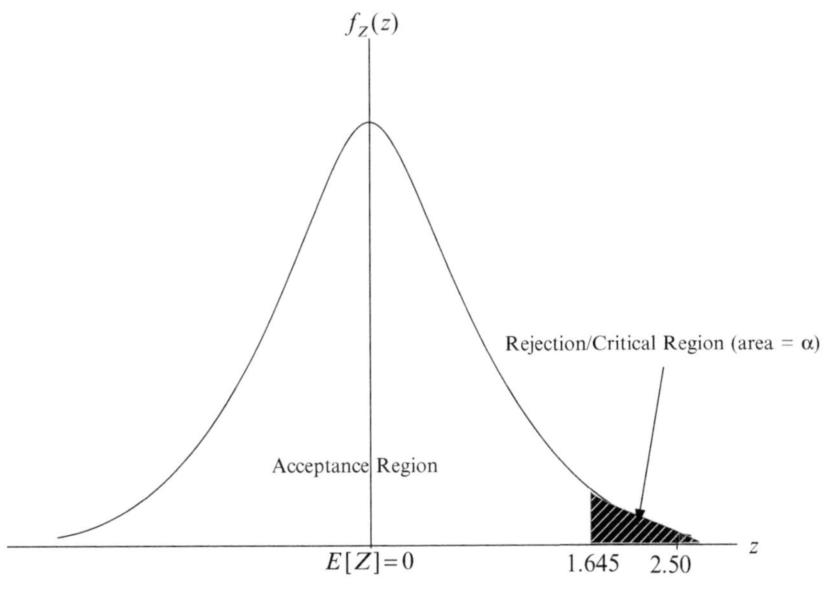


**FIGURE 9.9**
Critical Region for Problem 8.10(a)

Note that we had earlier accepted $H_0$ under the two-tailed test scheme at the 0.01 level of significance in Example 9.9. This means that decisions made under a one-tailed test do not necessarily agree with those made under a two-tailed test.

## EXAMPLE 9.11

A manufacturer of a migraine headache drug claimed that the drug is 90% effective in relieving migraines for a period of 24 hours. In a sample of 200 people who have migraine headache, the drug provided relief for 160 people for a period of 24 hours. Determine whether the manufacturer's claim is legitimate at the 0.05 level of significance.

**Solution:**
Since the success probability of the drug is $p=0.9$, the null hypothesis is

$$H_0: p=0.9$$

Also, since the drug is either effective or not, testing the drug on any individual is essentially a Bernoulli trial with claimed success probability of 0.9. Thus, the variance of the trial is

$$\sigma_p^2 = p(1-p) = 0.09$$

Because the drug provided relief for only 160 of the 200 people tested, the observed success probability is

$$\overline{p} = \frac{160}{200} = 0.8$$

We are interested in determining whether the proportion of people that the drug was effective in relieving their migraines is too low. Since $\overline{p} < 0.9$, we choose the alternative hypothesis as follows:

$$H_1: p < 0.9$$

Thus, we have a left-tail test. Now, the standard normal score of the observed proportion is given by

$$z = \frac{\overline{p}-p}{\sigma_{\overline{p}}} = \frac{\overline{p}-p}{\sigma_p/\sqrt{n}} = \frac{0.8-0.9}{\sqrt{0.09/200}} = -\frac{0.1}{0.0212} = -4.72$$

For a left-tail test at the 0.05 level of significance, the critical value is $z_\alpha = -2.33$. Since $z=-4.72$ falls within the rejection region, we reject $H_0$ and accept $H_1$; that is, the company's claim is false.

## 9.5 REGRESSION ANALYSIS

Sometimes we are required to use statistical data to reveal a mathematical relationship between two or more variables. Such information can be obtained from a *scatter diagram,* which is a graph of sample values that are plotted on the *xy*-plane. From the graph we can see whether the values fall into a linear or nonlinear pattern. An example of a scatter diagram is illustrated in Figure 9.10, which shows how a time function $X(t)$ varies with time. In this case, we see that the data points seem to fall into a straight line.

The general problem of finding a mathematical model to represent a given set of data is called regression analysis or *curve fitting*. The resulting model or curve is called a *regression curve* (or *regression line*), and the mathematical equation so developed is called a *regression equation*. The equation can be a linear or nonlinear one. However, in this book we consider only linear regression equations.

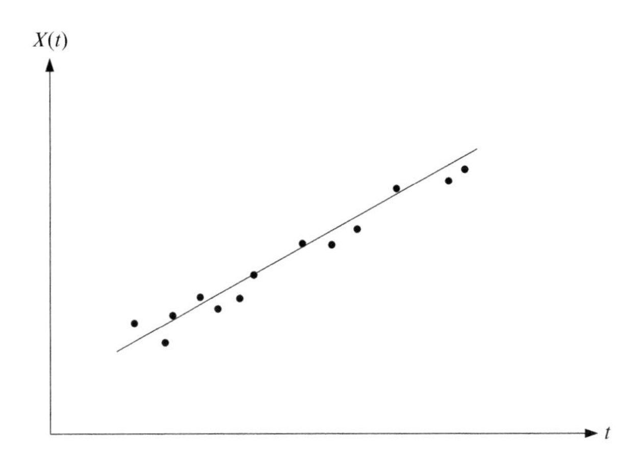


**FIGURE 9.10**
Example of Scatter Diagram

Since many regression lines can potentially be drawn through a scatter diagram, there is a need to find the "best" line. The problem is how to establish an acceptable criterion for defining "best." The general definition of the best regression equation is the so-called "least-squares" line. This is a regression line that has the property that the sum of the squares of the differences between the value predicted by the regression equation at any value of $x$ and the corresponding observed value of $y$ at that value of $x$ is a minimum. Thus, consider the scatter diagram in Figure 9.11. Assume that the regression line is a straight line given by the equation

$$y = a + bx$$

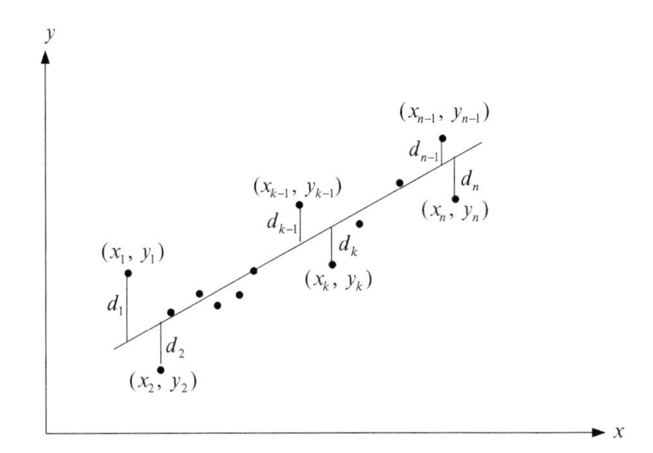


**FIGURE 9.11**
Difference Between Sample Values and Predicted Values

Let the points $(x_1, y_1), (x_2, y_2), \ldots, (x_n, y_n)$ represent the points on the scatter diagram. Then for any given $x_k$ the predicted value of $y$ on the regression line is $a + bx_k$. The difference between $y_k$ on the scatter diagram and the value $a + bx_k$ predicted by the line is defined by $d_k = y_k - (a + bx_k)$. The least-squares condition then states that the best line is the line for which $d_1^2 + d_2^2 + \cdots + d_n^2$ is a minimum.

The condition that $d_1^2 + d_2^2 + \cdots + d_n^2$ is a minimum implies that the sum

$$D = \sum_{i=1}^{n} [y_i - (a + bx_i)]^2$$

is a minimum. To find the parameters $a$ and $b$ that give $D$ its minimum value, we proceed as follows:

$$\frac{\partial D}{\partial a} = \frac{\partial}{\partial a} \sum_{i=1}^{n} [y_i - (a + bx_i)]^2 = \sum_{i=1}^{n} \frac{\partial}{\partial a} [y_i - (a + bx_i)]^2 = \sum_{i=1}^{n} -2[y_i - (a + bx_i)] = 0$$

This means that

$$\sum_{i=1}^{n} y_i = an + b \sum_{i=1}^{n} x_i \Rightarrow a = \frac{1}{n} \left\{ \sum_{i=1}^{n} y_i - b \sum_{i=1}^{n} x_i \right\} \tag{9.19}$$

Similarly,

$$\frac{\partial D}{\partial b} = \frac{\partial}{\partial b} \sum_{i=1}^{n} [y_i - (a + bx_i)]^2 = \sum_{i=1}^{n} \frac{\partial}{\partial b} [y_i - (a + bx_i)]^2 = \sum_{i=1}^{n} -2x_i[y_i - (a + bx_i)] = 0$$

This implies that

$$\sum_{i=1}^{n} x_i y_i = a \sum_{i=1}^{n} x_i + b \sum_{i=1}^{n} x_i^2 \tag{9.20}$$

From equations (9.19) and (9.20) we obtain the following conditions for $D$ to be a minimum:

$$b = \frac{n \sum_{i=1}^{n} x_i y_i - \sum_{i=1}^{n} x_i \sum_{i=1}^{n} y_i}{n \sum_{i=1}^{n} x_i^2 - \left( \sum_{i=1}^{n} x_i \right)^2} \tag{9.21}$$

$$a = \frac{1}{n} \left\{ \sum_{i=1}^{n} y_i - b \sum_{i=1}^{n} x_i \right\} \tag{9.22}$$

## EXAMPLE 9.12

(a) Fit a least-squares line to the data shown in Table 9.4. (b) Estimate the value of $y$ when $x=15$.

**Solution:**

a. We are looking for a least-squares line $y=a+bx$ such that $a$ and $b$ satisfy equations (9.21) and (9.22). Thus, we construct Table 9.5.
From the table we obtain the following results:

$$b=\frac{6\sum_{i=1}^{6}x_iy_i-\sum_{i=1}^{6}x_i\sum_{i=1}^{6}y_i}{6\sum_{i=1}^{6}x_i^2-\left(\sum_{i=1}^{6}x_i\right)^2}=\frac{6(226)-(42)(28)}{6(336)-(42)^2}=\frac{180}{252}=0.7143$$

$$a=\frac{1}{6}\left\{\sum_{i=1}^{6}y_i-0.7143\sum_{i=1}^{6}x_i\right\}=\frac{1}{6}\{28-(0.7143)(42)\}=-0.3334$$

Thus, the least-squares regression line is

$$y=a+bx=-0.3334+0.7143x$$

b. When $x=15$, we obtain

$$y=-0.3334+0.7143(15)=10.3811$$

**Table 9.4** Values for Example 9.12

| $x$ | 3 | 5 | 6 | 8 | 9 | 11 |
|---|---|---|---|---|---|---|
| $y$ | 2 | 3 | 4 | 6 | 5 | 8 |

**Table 9.5** Solution to 9.12

| $x$ | $y$ | $x^2$ | $xy$ |
|---|---|---|---|
| 3 | 2 | 9 | 6 |
| 5 | 3 | 25 | 15 |
| 6 | 4 | 36 | 24 |
| 8 | 6 | 64 | 48 |
| 9 | 5 | 81 | 45 |
| 11 | 8 | 121 | 88 |
| $\sum_i x_i=42$ | $\sum_i y_i=28$ | $\sum_i x_i^2=336$ | $\sum_i x_iy_i=226$ |

## 9.6 CHAPTER SUMMARY

This chapter has discussed inferential statistics, which uses probability theory to draw conclusions (or inferences) about, or estimate parameters of the environment from which the sample data came. Four different aspects of inferential statistics are discussed; these are:

1. *Sampling theory,* which deals with problems associated with selecting samples from some collection that is too large to be examined completely. These samples are selected in such a way that they are representative of the population.
2. *Estimation theory,* which is concerned with making some prediction or estimate based on the available data.
3. *Hypothesis testing,* which is also called detection theory, attempts to choose one model from several postulated (or hypothesized) models of the physical system.
4. *Regression analysis,* which attempts to find a mathematical expression that best represents the collected data. In this chapter we only considered linear relationships.

## 9.7 PROBLEMS

### Section 9.2 Sampling Theory

9.1 A sample size of 5 results in the sample values 9, 7, 1, 4, and 6.

a. What is the sample mean?
b. What is the sample variance?
c. What is the estimate of the sample variance?

9.2 The true mean of a quiz conducted in a class of 50 students is 70 points, and the true standard deviation is 12 points. It is desired to estimate the mean by sampling a subset of the scores, without replacement.

a. What is the standard deviation of the sample mean if only 10 scores are used?
b. How large should the sample size be for the standard deviation of the sample mean to be 1% of the true mean?

9.3 A random sample of size 81 is taken from a population that has a mean of 24 and variance 324. Use the central limit theorem to determine the probability that the sample mean lies between 23.9 and 24.2.

9.4 A random number generator produces three-digit random numbers that are uniformly distributed between 0.000 and 0.999.

a. If the generator produces the sequence of numbers 0.276, 0.123, 0.072, 0.324, 0.815, 0.312, 0.432, 0.283, 0.717, what is the sample mean?
b. What is the variance of the sample mean of numbers produced by the random number generator?
c. How large should the sample size be in order to obtain a sample mean whose standard deviation is no greater than 0.01?

9.5 Calculate the value of the Student's $t$ PDF for $t=2$ with (a) 6 degrees of freedom and (b) 12 degrees of freedom.

## Section 9.3 Estimation Theory

9.6 A large number of light bulbs were turned on continuously to determine the average number of days a bulb can last. The study revealed that the average lifetime of a bulb is 120 days with a standard deviation of 10 days. If the lifetimes are assumed to be independent normal random variables, find the confidence limits for a confidence level of 90% on the sample mean that is computed from a sample size of (a) 100 and (b) 25.

9.7 A random sample of 50 of the 200 electrical engineering students' grades in applied probability showed a mean of 75% and a standard deviation of 10%.

a. What are the 95% confidence limits for the estimate of the mean of the 200 grades?

b. What are the 99% confidence limits for the estimate of the mean of the 200 grades?

c. With what confidence can we say that the mean of all the 200 grades is $75 \pm 1$?

9.8 The mean of the grades of 36 freshmen is used to estimate the true average grade for the freshman class. If $\mu$ is the true mean, what is the probability that the estimate differs from the true mean by 3.6 marks if the standard deviation is known to be 24? (Note: The problem is asking for the probability that the true mean lies between $\mu - 3.6$ and $\mu + 3.6$.)

9.9 What is the increase in sample size that is required to increase the confidence level of a given confidence interval for a normal random variable from 90% to 99.9%?

9.10 A box contains red and white balls in an unknown proportion. A random sample of 60 balls selected with replacement from the box showed that 70% were red. Find the 95% confidence limits for the actual proportion of red balls in the box.

9.11 A box contains a mix of red and blue balls whose exact composition of red and blue balls is not known. If we draw 20 balls from the box with replacement and obtain 12 red balls, what is the maximum-likelihood estimate of $p$, the probability of drawing a red ball?

9.12 A box contains red and green balls whose exact composition is not known. An experimenter draws balls one by one with replacement until a green ball appears. Let $X$ denote the number of balls drawn until a green ball appears. This operation is repeated $n$ times to obtain the sample $X_1, X_2, \ldots, X_n$. Let $p$ be the fraction of green balls in the box. What is the maximum-likelihood estimate of $p$ on the basis of this sample?

## Section 9.4 Hypothesis Testing

9.13 A college provost claimed that 60% of the freshmen at his school receive their degrees within four years. A curious analyst followed the progress of a particular freshman class with 36 students and found that only 15 of the students received their degrees at the end of their fourth year. Determine whether this particular class performed worse than previous classes at a level of significance of (a) 0.05 and (b) 0.01.

9.14 An equipment manufacturing company claimed that at least 95% of the equipment it supplied to a factory conformed to specifications. An examination of a sample of 200 pieces of equipment revealed that 18 of them did not meet the specifications. Determine whether the company's claim is legitimate at a level of significance of (a) 0.01, and (b) 0.05.

9.15 A company claims that the boxes of detergent that it sells contain more than the current 500 grams of detergent each. From past experience the company knows that the amount of detergent in the boxes is normally distributed with a standard deviation of 75 grams. A worker takes a random sample of 100 boxes and finds that the average amount of detergent in a box is 510 grams. Test the company's claim at the 0.05 level of significance.

9.16 A government agency received many consumers' complaints that boxes of cereal sold by a company contain less than the advertised weight of 20 oz of cereal with a standard deviation of 5 oz. To check the consumers' complaints, the agency bought 36 boxes of the cereal and found that the average weight of cereal was 18 oz. If the amount of cereal in the boxes is normally distributed, test the consumers' complaint at the 0.05 level of significance.

## Section 9.5 Regression Analysis

9.17 Data were collected for a random variable $Y$ as a function of another random variable $X$. The recorded $(x, y)$ pairs are as follows:

$$(3, 2), (5, 3), (6, 4), (8, 6), (9, 5), (11, 8)$$

a. Plot the scatter diagram for these data.
b. Find the linear regression line of $y$ on $x$ that best fits these data.
c. Estimate the value of $y$ when $x = 15$.

9.18 Data were collected for a random variable $Y$ as a function of another random variable $X$. The recorded $(x, y)$ pairs are as follows:

$$(1, 11), (3, 12), (4, 14), (6, 15), (8, 17), (9, 18), (11, 19)$$

a. Plot the scatter diagram for these data.
b. Find the linear regression line of $y$ on $x$ that best fits these data.
c. Estimate the value of $y$ when $x = 20$.

9.19 The ages $x$ and systolic blood pressures $y$ of 12 people are shown in the following table:

| Age ($x$) | 56 | 42 | 72 | 36 | 63 | 47 | 55 | 49 | 38 | 42 | 68 | 60 |
|---|---|---|---|---|---|---|---|---|---|---|---|---|
| Blood Pressure ($y$) | 147 | 125 | 160 | 118 | 149 | 128 | 150 | 145 | 115 | 140 | 152 | 155 |

a. Find the least-squares regression line of $y$ on $x$.
b. Estimate the blood pressure of a person whose age is 45 years.

9.20 The following table shows a random sample of 12 couples who stated the number $x$ of children they planned to have at the time of their marriage and the number $y$ of actual children they have.

| Couple | 1 | 2 | 3 | 4 | 5 | 6 | 7 | 8 | 9 | 10 | 11 | 12 |
|---|---|---|---|---|---|---|---|---|---|---|---|---|
| Planned Number of Children ($x$) | 3 | 3 | 0 | 2 | 2 | 3 | 0 | 3 | 2 | 1 | 3 | 2 |
| Actual Number of Children ($y$) | 4 | 3 | 0 | 4 | 4 | 3 | 0 | 4 | 3 | 1 | 3 | 1 |

a. Find the least-squares regression line of $y$ on $x$.
b. Estimate the number of children that a couple who had planned to have 5 children actually had.

CHAPTER 10

# Introduction to Random Processes

## 10.1 INTRODUCTION

Chapters 1 to 7 are devoted to the study of probability theory. In those chapters we were concerned with outcomes of random experiments and the random variables used to represent them. This chapter deals with the concept of random processes that enlarges the random variable concept to include time. Thus, if $\Omega$ is the sample space, then instead of thinking of a random variable $X$ that maps an event $w \in \Omega$ to some number $X(w)$, we think of how the random variable maps the event to different numbers at different times. This implies that instead of the number $X(w)$ we deal with $X(t, w)$, where $t \in T$ and $T$ is called the *parameter set* of the process and is usually a set of times.

Random processes are widely encountered in such fields as communications, control, management science, and time series analysis. Examples of random processes include the population growth, an equipment failure, the price of a given stock over time, and the number of calls that arrive at a switchboard.

If we fix the sample point $w$, $X(t)$ is some real function of time. For each $w$, we have a function $X(t)$. Thus, $X(t, w)$ can be viewed as a collection of time functions, one for each sample point $w$, as shown in Figure 10.1.

On the other hand, if we fix $t$, we have a function $X(w)$ that depends only on $w$ and thus is a random variable. Thus, a random process becomes a random variable when time is fixed at some particular value. With many values of $t$ we obtain a collection of random variables. This means that we can define a random process as a family of random variables $\{X(t, w) | t \in T, w \in \Omega\}$ defined over a given probability space and indexed by the time parameter $t$.

A random process is also called a *stochastic process*. Consider a communication system example. Assume we have a set of possible messages that can be transmitted over a channel. The set of possible messages then constitutes our sample space. For each message $M$ generated by our source, we transmit an associated waveform $X(t, w)$ over the channel. The channel is not perfect; it selectively adds a noise waveform $N(t, w)$ to the original waveform so that what is seen at the



Fundamentals of Applied Probability and Random Processes. http://dx.doi.org/10.1016/B978-0-12-800852-2.00010-9

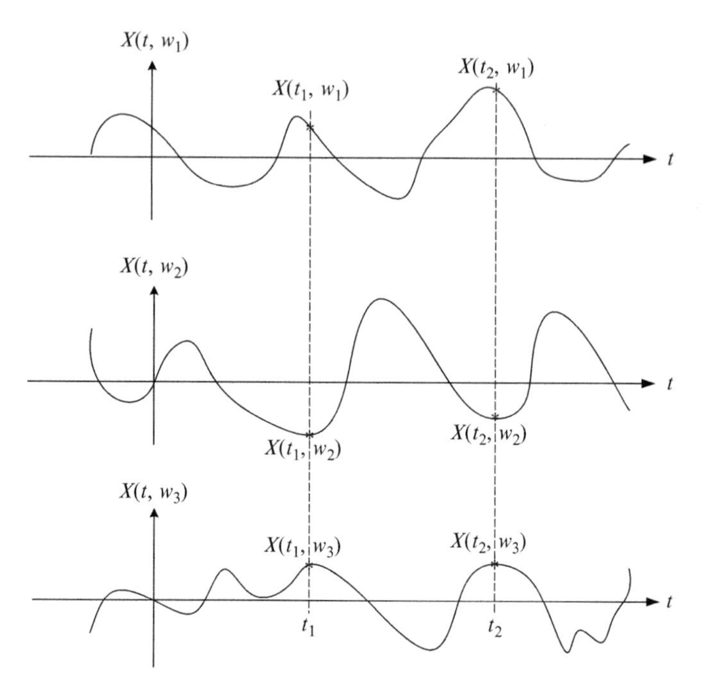


**FIGURE 10.1**
A Sample Random Process

receiver is a random signal $R(t, w)$ that is the sum of the transmitted waveform and the noise waveform. That is,

$$R(t, w) = X(t, w) + N(t, w)$$

Because the noise waveform is probabilistically selected by the channel, different noise waveforms can be associated with not only the same transmitted waveform but also different transmitted waveforms. Thus, the plot of $R(t, w)$ will be different for different values of $w$.

## 10.2 CLASSIFICATION OF RANDOM PROCESSES

A random process can be classified according to the nature of the time parameter and the values that $X(t, w)$ can take. As discussed earlier, $T$ is called the parameter set of the random process. If $T$ is an interval of real numbers and hence is continuous, the process is called a *continuous-time* random process. Similarly, if $T$ is a countable set and hence is discrete, the process is called a *discrete-time* random process. A discrete-time random process is also called a *random sequence*, which is denoted by $\{X[n] | n = 1, 2, \ldots\}$.

The values that $X(t, w)$ assumes are called the *states* of the random process. The set of all possible values of $X(t, w)$ forms the *state space, E,* of the random process. If $E$ is continuous, the process is called a *continuous-state* random process. Similarly, if $E$ is discrete, the process is called a *discrete-state* random process.

## 10.3 CHARACTERIZING A RANDOM PROCESS

In the remainder of the discussion we will represent the random process $X(t, w)$ by $X(t)$; that is, we will suppress $w$, the sample space parameter. A random process is completely described or characterized by the joint CDF. Since the value of a random process $X(t)$ at time $t_i$, $X(t_i)$, is a random variable, let

$$\begin{aligned} F_X(x_1, t_1) &= P[X(t_1) \leq x_1] \\ F_X(x_2, t_2) &= P[X(t_2) \leq x_2] \\ &\vdots \\ F_X(x_n, t_n) &= P[X(t_n) \leq x_n] \end{aligned}$$

where $0 < t_1 < t_2 < \cdots < t_n$. Then the joint CDF, which is defined by

$$F_X(x_1, x_2, \ldots, x_n; t_1, t_2, \ldots, t_n) = P[X(t_1) \leq x_1, X(t_2) \leq x_2, \ldots, X(t_n) \leq x_n]$$

for all $n$, completely characterizes the random process. If $X(t)$ is a continuous-time random process, then it is specified by a collection of PDFs:

$$f_X(x_1, x_2, \ldots, x_n; t_1, t_2, \ldots, t_n) = \frac{\partial^n}{\partial x_1 \partial x_2 \ldots \partial x_n} F_X(x_1, x_2, \ldots, x_n; t_1, t_2, \ldots, t_n)$$

Similarly, if $X(t)$ is a discrete-time random process, then it is specified by a collection of PMFs:

$$p_X(x_1, x_2, \ldots, x_n; t_1, t_2, \ldots, t_n) = P[X(t_1) = x_1, X(t_2) = x_2, \ldots, X(t_n) = x_n]$$

### 10.3.1 Mean and Autocorrelation Function

The mean of $X(t)$ is a function of time called the *ensemble average* and is denoted by

$$\mu_X(t) = E[X(t)] \tag{10.1}$$

The autocorrelation function provides a measure of similarity between two observations of the random process $X(t)$ at different points in time $t$ and $s$. The autocorrelation function of $X(t)$ and $X(s)$ is denoted by $R_{XX}(t, s)$ and defined as follows:

$$R_{XX}(t, s) = E[X(t)X(s)] = E[X(s)X(t)] = R_{XX}(s, t) \tag{10.2a}$$

$$R_{XX}(t, t) = E[X^2(t)] \tag{10.2b}$$

It is common to define $s = t + \tau$, which gives the autocorrelation function as

$$R_{XX}(t, t+\tau) = E[X(t)X(t+\tau)] \tag{10.3}$$

The parameter $\tau$ is sometimes called the *delay time* (or *lag time*). The autocorrelation function of a deterministic periodic function of period $T$ is given by

$$R_{XX}(t, t+\tau) = \frac{1}{2T}\int_{-T}^{T} f_X(t) f_X(t+\tau)\,dt \tag{10.4}$$

Similarly, for an aperiodic function the autocorrelation function is given by

$$R_{XX}(t, t+\tau) = \int_{-\infty}^{\infty} f_X(t) f_X(t+\tau)\,dt \tag{10.5}$$

Basically the autocorrelation function defines how much a signal is similar to a time-shifted version of itself. A random process $X(t)$ is called a *second order process* if $E[X^2(t)] < \infty$ for each $t \in T$.

### 10.3.2 The Autocovariance Function

The autocovariance function of the random process $X(t)$ is another quantitative measure of the statistical coupling between $X(t)$ and $X(s)$. It is denoted by $C_{XX}(t, s)$ and defined as follows:

$$\begin{aligned} C_{XX}(t, s) &= \text{Cov}\{X(t), X(s)\} = E[\{X(t) - \mu_X(t)\}\{X(s) - \mu_X(s)\}] \\ &= E[X(t)X(s)] - \mu_X(s)E[X(t)] - \mu_X(t)E[X(s)] + \mu_X(t)\mu_X(s) \\ &= R_{XX}(t, s) - \mu_X(t)\mu_X(s) \end{aligned} \tag{10.6}$$

If $X(t)$ and $X(s)$ are independent, then $R_{XX}(t, s) = \mu_X(t)\mu_X(s)$, and we have $C_{XX}(t, s) = 0$, which means that there is no coupling between $X(t)$ and $X(s)$. This is equivalent to saying that $X(t)$ and $X(s)$ are uncorrelated, but the reverse is not true. That is, if $C_{XX}(t, s) = 0$, it does not mean that $X(t)$ and $X(s)$ are independent.

---

## EXAMPLE 10.1

A random process is defined by

$$X(t) = K\cos(wt) \qquad t \geq 0$$

where $w$ is a constant and $K$ is uniformly distributed between 0 and 2. Determine the following:

a. $E[X(t)]$
b. The autocorrelation function of $X(t)$
c. The autocovariance function of $X(t)$

**Solution:**

The expected value and variance of $K$ are given by $E[K]=(2+0)/2=1$ and $\sigma_K^2=(2-0)^2/12=1/3$. Thus, $E[K^2]=\sigma_K^2+(E[K])^2=4/3$.

a. The mean of $X(t)$ is given by $E[X(t)]=E[K\cos(wt)]=E[K]\cos(wt)=\cos(wt)$

b. The autocorrelation function of $X(t)$ is given by

$$
\begin{aligned}
R_{XX}(t,s) &= E[X(t)X(s)] = E\left[K^2\cos(wt)\cos(ws)\right] = E\left[K^2\right]\cos(wt)\cos(ws) \\
&= \frac{4}{3}\cos(wt)\cos(ws)
\end{aligned}
$$

c. The autocovariance function of $X(t)$ is given by

$$
\begin{aligned}
C_{XX}(t,s) &= R_{XX}(t,s) - E[X(t)]E[X(s)] = \frac{4}{3}\cos(wt)\cos(ws) - \cos(wt)\cos(ws) \\
&= \frac{1}{3}\cos(wt)\cos(ws)
\end{aligned}
$$

## 10.4 CROSSCORRELATION AND CROSSCOVARIANCE FUNCTIONS

Let $X(t)$ and $Y(t)$ be two random processes defined on the same probability space and with means $\mu_X(t)$ and $\mu_Y(t)$, respectively. The crosscorrelation function $R_{XY}(t,s)$ of the two random processes is defined by

$$
R_{XY}(t,s) = E[X(t)Y(s)] = R_{YX}(s,t) \tag{10.7}
$$

for all $t$ and $s$. The crosscorrelation function essentially measures how similar two different processes (or *signals*) are when one of them is shifted in time relative to the other. If $R_{XY}(t,s)=0$ for all $t$ and $s$, we say that $X(t)$ and $Y(t)$ are *orthogonal processes*. If the two processes are statistically independent, the crosscorrelation function becomes

$$
R_{XY}(t,s) = E[X(t)]E[Y(s)] = \mu_X(t)\mu_Y(s)
$$

The crosscovariance functions of $X(t)$ and $Y(t)$, denoted by $C_{XY}(t,s)$, is defined by

$$
\begin{aligned}
C_{XY}(t,s) &= \text{Cov}\{X(t),Y(s)\} = E[\{X(t)-\mu_X(t)\}\{Y(s)-\mu_Y(s)\}] \\
&= E[X(t)Y(s)] - \mu_Y(s)E[X(t)] - \mu_X(t)E[Y(s)] + \mu_X(t)\mu_Y(s) \\
&= R_{XY}(t,s) - \mu_X(t)\mu_Y(s)
\end{aligned} \tag{10.8}
$$

The random processes $X(t)$ and $Y(t)$ are said to be uncorrelated if $C_{XY}(t,s)=0$ for all $t$ and $s$. That is, $X(t)$ and $Y(t)$ are said to be uncorrelated if for all $t$ and $s$, we have that

$$
R_{XY}(t,s) = \mu_X(t)\mu_Y(s)
$$

In many situations the random process $Y(t)$ is the sum of the random process $X(t)$ and a statistically independent noise process $N(t)$.

## EXAMPLE 10.2

A random process $Y(t)$ consists of the sum of the random process $X(t)$ and a statistically independent noise process $N(t)$. Find the crosscorrelation function of $X(t)$ and $Y(t)$.

**Solution:**

By definition the crosscorrelation function is given by

$$\begin{aligned} R_{XY}(t,s) &= E[X(t)Y(s)] = E[X(t)\{X(s)+N(s)\}] \\ &= E[X(t)X(s)] + E[X(t)N(s)] = R_{XX}(t,s) + E[X(t)]E[N(s)] \\ &= R_{XX}(t,s) + \mu_X(t)\mu_N(s) \end{aligned}$$

where the fourth equality follows from the fact that $X(t)$ and $N(s)$ are independent. Using the results obtained earlier, the crosscovariance function of $X(t)$ and $Y(t)$ is given by

$$\begin{aligned} C_{XY}(t,s) &= \mathrm{Cov}\{X(t),Y(s)\} = E[\{X(t)-\mu_X(t)\}\{Y(s)-\mu_Y(s)\}] \\ &= R_{XY}(t,s) - \mu_X(t)\mu_Y(s) = R_{XX}(t,s) + \mu_X(t)\mu_N(s) - \mu_X(t)\mu_Y(s) \\ &= R_{XX}(t,s) + \mu_X(t)\mu_N(s) - \mu_X(t)\{\mu_X(s)+\mu_N(s)\} \\ &= R_{XX}(t,s) - \mu_X(t)\mu_X(s) = C_{XX}(t,s) \end{aligned}$$

Thus, the crosscovariance function is identical to the autocovariance function of $X(t)$.

### 10.4.1 Review of Some Trigonometric Identities

Some of the problems that we will encounter in this chapter will deal with trigonometric functions. As a result we summarize some of the relevant trigonometric identities. These identities are derived from expansions of $\sin(A \pm B)$ and $\cos(A \pm B)$. Specifically, we know that

$$\sin(A+B) = \sin A \cos B + \cos A \sin B \quad \text{(a)}$$

$$\sin(A-B) = \sin A \cos B - \cos A \sin B \quad \text{(b)}$$

Adding the two equations we obtain the following identity:

$$\sin A \cos B = \frac{1}{2}\{\sin(A+B) + \sin(A-B)\}$$

By subtracting equation (b) from equation (a) we obtain:

$$\cos A \sin B = \frac{1}{2}\{\sin(A+B) - \sin(A-B)\}$$

Similarly,

$$\cos(A-B) = \cos A \cos B + \sin A \sin B \quad \text{(c)}$$

$$\cos(A+B) = \cos A \cos B - \sin A \sin B \quad \text{(d)}$$

Adding the two equations we obtain the following identity:

$$\cos A \cos B = \frac{1}{2}\{\cos(A-B) + \cos(A+B)\}$$

Finally, subtracting equation (d) from equation (c) we obtain the following identity:

$$\sin A \sin B = \frac{1}{2}\{\cos(A-B) - \cos(A+B)\}$$

Also, when $A = B$ we obtain

$$\sin(2A) = 2\sin A \cos A \quad \text{(e)}$$

$$\cos(2A) = \cos^2 A - \sin^2 A \quad \text{(f)}$$

$$1 = \cos^2 A + \sin^2 A \quad \text{(g)}$$

From equations (f) and (g) we obtain

$$\cos^2 A = \frac{1+\cos(2A)}{2}$$
$$\sin^2 A = \frac{1-\cos(2A)}{2}$$

Also, the derivatives are given by

$$\frac{d}{dx}\sin(x) = \cos(x)$$

$$\frac{d}{dx}\cos(x) = -\sin(x)$$

## EXAMPLE 10.3

Two random processes $X(t)$ and $Y(t)$ are defined as follows:

$$X(t) = A\cos(wt+\Theta)$$
$$Y(t) = B\sin(wt+\Theta)$$

where $A$, $B$, and $w$ are constants and $\Theta$ is a random variable that is uniformly distributed between 0 and $2\pi$. Find the crosscorrelation function of $X(t)$ and $Y(t)$.

**Solution:**
The crosscorrelation function of $X(t)$ and $Y(t)$ is given by

$$\begin{aligned} R_{XY}(t,s) &= E[X(t)Y(s)] = E[A\cos(wt+\Theta)B\sin(ws+\Theta)] \\ &= E[AB\cos(wt+\Theta)\sin(ws+\Theta)] = ABE[\cos(wt+\Theta)\sin(ws+\Theta)] \\ &= ABE\left[\frac{1}{2}\{\sin(wt+ws+2\Theta) - \sin(wt-ws)\}\right] \\ &= \frac{AB}{2}E[\sin(wt+ws+2\Theta) - \sin(wt-ws)] \\ &= \frac{AB}{2}\{E[\sin(wt+ws+2\Theta)] - \sin[w(t-s)]\} \end{aligned}$$

Now, since $f_\Theta(\theta) = 1/2\pi$ for $0 \le \theta \le 2\pi$, we have that

$$\begin{aligned} E[\sin(wt+ws+2\Theta)] &= \int_{-\infty}^{\infty} \sin(wt+ws+2\theta) f_\Theta(\theta)d\theta = \frac{1}{2\pi}\int_0^{2\pi} \sin(wt+ws+2\theta)d\theta \\ &= \frac{1}{2\pi}\left[-\frac{\cos(wt+ws+2\theta)}{2}\right]_0^{2\pi} \\ &= \frac{1}{4\pi}\{\cos(wt+ws) - \cos(wt+ws+4\pi)\} \\ &= \frac{1}{4\pi}\{\cos(wt+ws) - \cos(wt+ws)\} = 0 \end{aligned}$$

Thus,

$$R_{XY}(t,s) = \frac{AB}{2}[0 - \sin\{w(t-s)\}] = -\frac{AB}{2}\sin\{w(t-s)\}$$

If we define $s = t + \tau$, then

$$R_{XY}(t,s) = R_{XY}(t,t+\tau) = -\frac{AB}{2}\sin\{w(-\tau)\} = \frac{AB}{2}\sin(w\tau)$$

## 10.5 STATIONARY RANDOM PROCESSES

There are several ways to define a stationary random process. At a high level, it is a process whose statistical properties do not vary with time. In this book we consider only two types of stationary processes. These are the *strict-sense stationary* processes and the *wide-sense stationary* processes.

### 10.5.1 Strict-Sense Stationary Processes

A random process is defined to be a strict-sense stationary process if its CDF is invariant to a shift in the time origin. This means that the process $X(t)$ with the CDF $F_X(x_1, x_2, \ldots, x_n; t_1, t_2, \ldots, t_n)$ is a strict-sense stationary process if its CDF is identical to that of for any arbitrary $X(t+\varepsilon)$. Thus, we have that being a strict-sense stationary process implies that for any arbitrary $\varepsilon$,

$$F_X(x_1, x_2, \ldots, x_n; t_1, t_2, \ldots, t_n) = F_X(x_1, x_2, \ldots, x_n; t_1+\varepsilon, t_2+\varepsilon, \ldots, t_n+\varepsilon)$$

for all $n$. When the CDF is differentiable, the equivalent condition for strict-sense stationarity is that the PDF is invariant to a shift in the time origin; that is,

$$_X(x_1, x_2, \ldots, x_n; t_1, t_2, \ldots, t_n) = f_X(x_1, x_2, \ldots, x_n; t_1 + \varepsilon, t_2 + \varepsilon, \ldots, t_n + \varepsilon)$$

or all $n$. If $X(t)$ is a strict-sense stationary process, then the CDF $F_{X_1X_2}(x_1, x_2; t_1, t_1 + \tau)$ does not depend on $t$ but it may depend on $\tau$. Thus, if $t_2 = t_1 + \tau$, then $F_{X_1X_2}(x_1, x_2; t_1, t_2)$ may depend on $t_2 - t_1$, but not on $t_1$ and $t_2$ individually. This means that if $X(t)$ is a strict-sense stationary process, then the autocorrelation and autocovariance functions do not depend on $t$. Thus, we have that for all $\tau \in T$:

$$\begin{aligned} \mu_X(t) &= \mu_X(0) \\ R_{XX}(t, t+\tau) &= R_{XX}(0, \tau) \\ C_{XX}(t, t+\tau) &= C_{XX}(0, \tau) \end{aligned}$$

If the condition $\mu_X(t) = \mu_X(0)$ holds for all $t$, the mean is constant and denoted by $\mu_X$. Similarly, if $R_{XX}(t, t+\tau)$ does not depend on $t$ but is a function of $\tau$, we write $R_{XX}(t, t+\tau) = R_{XX}(\tau)$. Finally, whenever the condition $C_{XX}(t, t+\tau) = C_{XX}(0, \tau)$ holds for all $t$, we write $C_{XX}(t, t+\tau) = C_{XX}(\tau)$.

### 10.5.2 Wide-Sense Stationary Processes

Many practical problems that we encounter require that we deal with only the mean and autocorrelation function of a random process. Solutions to these problems are simplified if these quantities do not depend on absolute time. Random processes in which the mean and autocorrelation function do not depend on absolute time are called wide-sense stationary (WSS) processes. Thus, for a wide-sense stationary process $X(t)$,

$$\begin{aligned} E[X(t)] &= \mu_X \qquad \text{(constant)} \\ R_{XX}(t, t+\tau) &= R_{XX}(\tau) \end{aligned}$$

Note that a strict-sense stationary process is also a wide-sense stationary process. However, in general the converse is not true; that is, a WSS process is not necessarily stationary in the strict sense.

## EXAMPLE 10.4

A random process $X(t)$ is defined by

$$X(t) = A\cos(t) + B\sin(t) \qquad -\infty < t < \infty$$

where $A$ and $B$ are independent random variables each of which has a value $-2$ with probability 1/3 and a value 1 with probability 2/3. Show that $X(t)$ is a wide-sense stationary process.

**Solution:**

The PMF of $A$ and $B$ is shown in Figure 10.2.

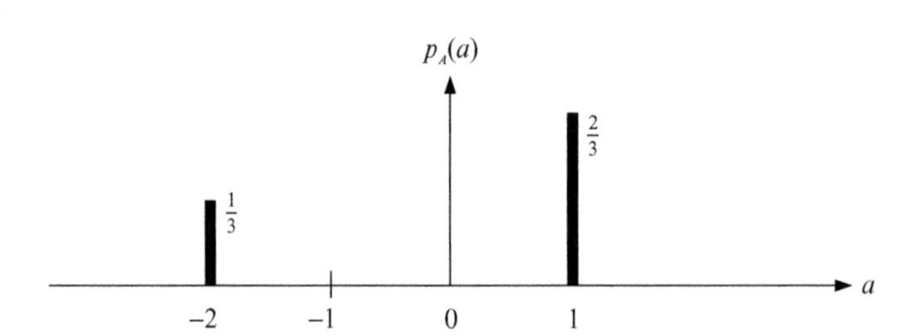


**FIGURE 10.2**
PMF of $A$ and $B$

$$E[A] = E[B] = \frac{1}{3}(-2) + \frac{2}{3}(1) = 0$$
$$E\left[A^2\right] = E\left[B^2\right] = \frac{1}{3}(-2)^2 + \frac{2}{3}(1)^2 = 2$$

Thus, $E[X(t)] = E[A]\cos(t) + E[B]\sin(t) = 0$, which is a constant. Since $A$ and $B$ are independent, $E[AB] = E[A]E[B] = 0$. Thus,

$$\begin{aligned}
R_{XX}(t,s) &= E[X(t)X(s)] = E[\{A\cos(t) + B\sin(t)\}\{A\cos(s) + B\sin(s)\}] \\
&= E\left[A^2\cos(t)\cos(s) + AB\cos(t)\sin(s) + AB\sin(t)\cos(s) + B^2\sin(t)\sin(s)\right] \\
&= E\left[A^2\right]\cos(t)\cos(s) + E[AB]\{\cos(t)\sin(s) + \sin(t)\cos(s)\} + E\left[B^2\right]\sin(t)\sin(s) \\
&= 2\{\cos(t)\cos(s) + \sin(t)\sin(s)\} \\
&= 2\cos(s-t)
\end{aligned}$$

Since the mean is constant and the autocorrelation function is a function of the difference between the two times, we conclude that the random process $X(t)$ is wide-sense stationary.

## EXAMPLE 10.5

Assume that $X(t)$ is a random process defined as follows:

$$X(t) = A\cos(2\pi t + \Phi)$$

where $A$ is a zero-mean normal random variable with variance $\sigma_A^2 = 2$ and $\Phi$ is a uniformly distributed random variable over the interval $[-\pi, \pi]$. $A$ and $\Phi$ are statistically independent. Let the random variable $Y$ be defined as follows:

$$Y = \int_0^1 X(t)dt$$

Determine

1. the mean $E[Y]$ of $Y$
2. the variance of $Y$.

**Solution:**
The mean of $X(t)$ is given by

$$E[X(t)] = E[A\cos(2\pi t + \Phi)] = E[A]E[\cos(2\pi t + \Phi)] = 0$$

Similarly the variance of $X(t)$ is given by

$$\begin{aligned}
\sigma^2_{X(t)} &= E\left[\{X(t) - E[X(t)]\}^2\right] = E\left[X^2(t)\right] = E\left[\{A\cos(2\pi t + \Phi)\}^2\right] \\
&= E\left[A^2\right]E\left[\cos^2(2\pi t + \Phi)\right] = 2E\left[\frac{1 + \cos(4\pi t + 2\Phi)}{2}\right] = E[1 + \cos(4\pi t + 2\Phi)] \\
&= 1 + \int_{-\pi}^{\pi} \cos(4\pi t + 2\phi) f_\Phi(\phi) d\phi = 1 + \frac{1}{2\pi}\int_{-\pi}^{\pi} \cos(4\pi t + 2\phi) d\phi \\
&= 1
\end{aligned}$$

1. The mean of $Y$ is given by

$$E[Y] = E\left[\int_0^1 X(t)dt\right] = \int_0^1 E[X(t)]dt = 0$$

2. The variance of $Y$ is given by

$$\begin{aligned}
\sigma^2_Y &= E\left[\{Y = E[Y]\}^2\right] = E\left[Y^2\right] \\
&= E\left[\left(\int_0^1 X(t)dt\right)^2\right] = E\left[\left(\int_0^1 A\cos(2\pi t + \Phi)dt\right)^2\right] = E\left[\left\{\frac{A\sin(2\pi t + \Phi)}{2\pi}\bigg|_0^1\right\}^2\right] \\
&= \frac{1}{4\pi^2}E\left[\{A\sin(2\pi + \Phi) - A\sin(\Phi)\}^2\right] = \frac{1}{4\pi^2}E\left[\{A\sin(\Phi) - A\sin(\Phi)\}^2\right] \\
&= 0
\end{aligned}$$

---

Note that

$$Y = \int_0^1 X(t)dt = \int_0^1 A\cos(2\pi t + \Phi)dt = \frac{A[\sin(2\pi + \Phi) - \sin(\Phi)]}{2\pi} = 0$$

which is why we got the results for the mean and variance of $Y$.

#### 10.5.2.1 *Properties of Autocorrelation Functions for WSS Processes*

As defined earlier, the autocorrelation function of a wide-sense stationary random process $X(t)$ is defined as

$$R_{XX}(t, t + \tau) = R_{XX}(\tau)$$

The properties of autocorrelation functions of wide-sense stationary processes include the following:

1. $|R_{XX}(\tau)| \le R_{XX}(0)$, which means that $R_{XX}(\tau)$ is bounded by its value at the origin (or the largest value of $R_{XX}(\tau)$ occurs at $\tau = 0$)
2. $R_{XX}(\tau) = R_{XX}(-\tau)$, which means that $R_{XX}(\tau)$ is an even function
3. $R_{XX}(0) = E[X^2(t)]$, which means that the largest value of the autocorrelation function, $R_{XX}(0)$ (according property 1 above), is equal to the second moment of the random process. $E[X^2(t)]$ is usually referred to as the *mean-square value*.
4. If $X(t)$ has no periodic components and is ergodic (where the concept of "ergodic processes" is discussed later), and $E[X(t)] = \mu_X \neq 0$, then

   $$\lim_{|\tau| \to \infty} R_{XX}(\tau) = \mu_X^2$$

5. If $X(t)$ has a direct-current (dc) component or mean value, then $R_{XX}(\tau)$ will have a constant component. Thus, if $X(t) = K + N(t)$, where $K$ is a constant, then $R_{XX}(\tau) = K^2 + R_{NN}(\tau)$.
6. If $X(t)$ has a periodic component, then $R_{XX}(\tau)$ will have a periodic component with the same period.
7. $R_{XX}(\tau)$ cannot have an arbitrary shape; this means that any arbitrary function cannot be an autocorrelation function.

## EXAMPLE 10.6

Compute the variance of the random process $X(t)$ whose autocorrelation function is given by

$$R_{XX}(\tau) = 25 + \frac{4}{1 + 6\tau^2}$$

**Solution:**

By property 4, the square of the mean is given by:

$$\mu_X^2 = \lim_{|\tau| \to \infty} R_{XX}(\tau) = 25$$

Thus, $\mu_X = \sqrt{25} = \pm 5$. Note that the property yields only the magnitude of the mean but not its sign. Also, from property 3,

$$E\left[X^2(t)\right] = R_{XX}(0) = 25 + 4 = 29$$

Thus, the variance is given by

$$\sigma_{X(t)}^2 = E\left[X^2(t)\right] - \mu_X^2 = 29 - 25 = 4$$

## EXAMPLE 10.7

A random process has the autocorrelation function

$$R_{XX}(\tau) = \frac{4\tau^2 + 6}{\tau^2 + 1}$$

Find the mean-square value, the mean value and the variance of the process.

**Solution:**

We first decompose the function to obtain its dc component as follows:

$$R_{XX}(\tau) = \frac{4\tau^2 + 6}{\tau^2 + 1} = \frac{4(\tau^2 + 1) + 6 - 4}{\tau^2 + 1} = \frac{4(\tau^2 + 1) + 2}{\tau^2 + 1} = 4 + \frac{2}{\tau^2 + 1}$$

Thus,

$$E\left[X^2(t)\right] = R_{XX}(0) = 6$$

$$E[X(t)] = \pm\sqrt{\lim_{|\tau| \to \infty} R_{XX}(\tau)} = \pm\sqrt{4} = \pm 2$$

$$\sigma_X^2 = E\left[X^2(t)\right] - \{E[X(t)]\}^2 = 6 - 4 = 2$$

#### 10.5.2.2 *Autocorrelation Matrices for WSS Processes*

Consider a WSS random signal $X(t)$ that is sampled at periodic time instants and assume that we accumulate $N$ such samples. If the sampling times are $t_1, t_2, \ldots, t_N$, the vector $\mathbf{X}$ representing the different samples of $X(t)$ is given by

$$\mathbf{X} = \begin{bmatrix} X(t_1) \\ X(t_2) \\ \vdots \\ X(t_N) \end{bmatrix}$$

Since each sample is a random variable, we can define an $N \times N$ autocorrelation matrix that gives the autocorrelation function for every pair of random variables in the vector $\mathbf{X}$. Also, if the interval between two consecutive samples is $\Delta t$, we have that

$$t_2 = t_1 + \Delta t$$

$$t_3 = t_2 + \Delta t = t_1 + 2\Delta t$$

$$\vdots$$

$$t_N = t_{N-1} + \Delta t = t_1 + (N-1)\Delta t$$

Thus, the autocorrelation matrix becomes

$$R_{XX} = E\left[XX^T\right] = E\left[\begin{bmatrix} X(t_1)X(t_1) & X(t_1)X(t_2) & \cdots & X(t_1)X(t_N) \\ X(t_2)X(t_1) & X(t_2)X(t_2) & \cdots & X(t_2)X(t_N) \\ \cdots & \cdots & \cdots & \cdots \\ X(t_N)X(t_1) & X(t_N)X(t_2) & \cdots & X(t_N)X(t_N) \end{bmatrix}\right]$$

$$= \begin{bmatrix} R_{XX}(t_1,t_1) & R_{XX}(t_1,t_2) & \cdots & R_{XX}(t_1,t_N) \\ R_{XX}(t_2,t_1) & R_{XX}(t_2,t_2) & \cdots & R_{XX}(t_2,t_N) \\ \cdots & \cdots & \cdots & \cdots \\ R_{XX}(t_N,t_1) & R_{XX}(t_N,t_2) & \cdots & R_{XX}(t_N,t_N) \end{bmatrix}$$

$$= \begin{bmatrix} R_{XX}(0) & R_{XX}(\Delta t) & \cdots & R_{XX}([N-1]\Delta t) \\ R_{XX}(\Delta t) & R_{XX}(0) & \cdots & R_{XX}([N-2]\Delta t) \\ \cdots & \cdots & \cdots & \cdots \\ R_{XX}([N-1]\Delta t) & R_{XX}([N-2]\Delta t) & \cdots & R_{XX}(0) \end{bmatrix}$$

where $X^T$ is the transpose of **X**, and we have taken advantage of the fact that $R_{XX}(-\tau) = R_{XX}(\tau)$. Thus, for a wide-sense stationary process, $R_{XX}$ is a symmetric matrix. In a similar manner we can obtain the autocovariance matrix $C_{XX}$, which is defined as follows:

$$C_{XX} = E\left[\left(X - \overline{X}\right)\left(X^T - \overline{X}^T\right)\right] = R_{XX} - \overline{X}\overline{X}^T$$

## EXAMPLE 10.8

Determine the missing elements denoted by *xx* in the following autocorrelation matrix of a WSS random process $Y(t)$:

$$R_{YY} = \begin{bmatrix} 2.0 & 1.3 & 0.4 & xx \\ xx & 2.0 & 1.2 & 0.8 \\ 0.4 & 1.2 & xx & 1.1 \\ 0.9 & xx & xx & 2.0 \end{bmatrix}$$

**Solution:**

Since the autocorrelation matrix of a WSS process is a symmetric matrix, the real matrix is as follows:

$$R_{YY} = \begin{bmatrix} 2.0 & 1.3 & 0.4 & 0.9 \\ 1.3 & 2.0 & 1.2 & 0.8 \\ 0.4 & 1.2 & 2.0 & 1.1 \\ 0.9 & 0.8 & 1.1 & 2.0 \end{bmatrix}$$

#### 10.5.2.3 Properties of Crosscorrelation Functions for WSS Processes

As defined earlier, the crosscorrelation function $R_{XY}(t,s)$ of the two random processes $X(t)$ and $Y(t)$ is defined by

$$R_{XY}(t,s) = E[X(t)Y(s)]$$

If we set $s = t + \tau$, we may write

$$R_{XY}(t,t+\tau) = E[X(t)Y(t+\tau)]$$

We say that $X(t)$ and $Y(t)$ are jointly wide-sense stationary if $R_{XY}(t,t+\tau)$ is independent of the absolute time. That is, $X(t)$ and $Y(t)$ are jointly wide-sense stationary random processes if

$$R_{XY}(t,t+\tau) = E[X(t)Y(t+\tau)] = R_{XY}(\tau)$$

Generally the crosscorrelation function is not an even function, as is true for the autocorrelation function. Also, it does not necessarily have a maximum value at the origin as is true for the autocorrelation function. Some of the properties of $R_{XY}(\tau)$ include the following:

1. $R_{XY}(\tau) = R_{YX}(-\tau)$
2. $|R_{XY}(\tau)| \le \sqrt{R_{XX}(0)R_{YY}(0)}$
3. $|R_{XY}(\tau)| \le [R_{XX}(0) + R_{YY}(0)]/2$

## 10.6 ERGODIC RANDOM PROCESSES

One desirable property of a random process is the ability to estimate its parameters from measurement data. Consider a random process $X(t)$ whose observed samples are $x(t)$. The time average of a function of $x(t)$ is defined by

$$\bar{x} = \lim_{T\to\infty} \frac{1}{2T}\int_{-T}^{T} x(t)dt \tag{10.9}$$

The statistical average of the random process $X(t)$ is the expected value $E[X(t)]$ of the process. The expected value is also called the *ensemble average*. An ergodic random process is a stationary process in which every member of the ensemble exhibits the same statistical behavior as the ensemble. This implies that it is possible to determine the statistical behavior of the ensemble by examining only one typical sample function. Thus, for an ergodic random process, the mean values and moments can be determined by time averages as well as by ensemble averages (or expected values), which are equal. That is,

$$E[X^n] = \int_{-\infty}^{\infty} x^n f_X(x)dx = \lim_{T\to\infty} \frac{1}{2T}\int_{-T}^{T} x^n(t)dt = \overline{x^n}$$

A random process $X(t)$ is defined to be *mean-ergodic* (or *ergodic in the mean*) if $E[X(t)] = \overline{x}$.

## EXAMPLE 10.9

A random process has sample functions of the form

$$X(t) = A\cos(wt + \Theta)$$

where $w$ is constant, $A$ is a random variable that has a magnitude of +1 and −1 with equal probability, and $\Theta$ is a random variable that is uniformly distributed between 0 and $2\pi$. Assume that the random variables $A$ and $\Theta$ are independent.

a. Is $X(t)$ a wide-sense stationary process?
b. Is $X(t)$ a mean-ergodic process?

**Solution:**

The PMF of $A$ and the PDF of $\Theta$ are shown in Figure 10.3.

$$E[A] = \frac{1}{2}(-1) + \frac{1}{2}(1) = 0$$

$$\sigma_A^2 = \frac{1}{2}(-1)^2 + \frac{1}{2}(1)^2 = 1 = E\left[A^2\right]$$

$$E[\Theta] = \frac{0 + 2\pi}{2} = \pi$$

$$\sigma_\Theta^2 = \frac{(2\pi - 0)^2}{12} = \frac{4\pi^2}{12} = \frac{\pi^2}{3}$$

a. Since $A$ and $\Theta$ are independent,

$$E[X(t)] = E[A]E[\cos(wt + \Theta)] = 0$$

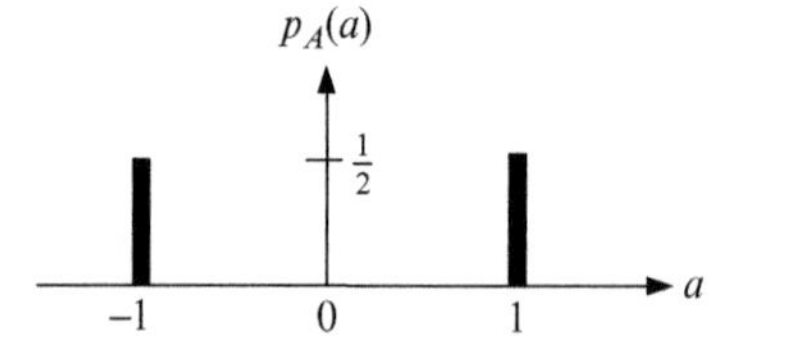


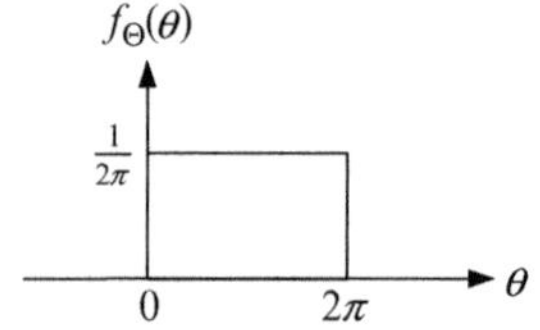


**FIGURE 10.3**
PMF of $A$ and PDF of $\Theta$ for Example 10.9

which is a constant. Also, the autocorrelation function of $X(t)$ is given by

$$
\begin{aligned}
R_{XX}(t,t+\tau) &= E[X(t)X(t+\tau)] = E[A\cos(wt+\Theta)A\cos(wt+w\tau+\Theta)] \\
&= E\left[A^2\right]E[\cos(wt+\Theta)\cos(wt+w\tau+\Theta] \\
&= \frac{1}{2}E[\cos(-w\tau)+\cos(2wt+w\tau+2\Theta)] \\
&= \frac{1}{2}E[\cos(-w\tau)]+\frac{1}{2}E[\cos(2wt+w\tau+2\Theta)] \\
&= \frac{1}{2}\cos(w\tau)+\frac{1}{2}E[\cos(2wt+w\tau+2\Theta)] \\
&= \frac{1}{2}\cos(w\tau)+\frac{1}{2}\int_0^{2\pi}\frac{\cos(2wt+w\tau+2\theta)}{2\pi}d\theta \\
&= \frac{1}{2}\cos(w\tau)+\frac{1}{8\pi}[\sin(2wt+w\tau+2\theta)]_0^{2\pi} \\
&= \frac{1}{2}\cos(w\tau)+\frac{1}{8\pi}\{\sin(2wt+w\tau+4\pi)-\sin(2wt+w\tau)\} \\
&= \frac{1}{2}\cos(w\tau)
\end{aligned}
$$

Since the mean is constant and the autocorrelation function depends only on the difference between the two times and not on $t$, we conclude that the process is wide-sense stationary.

b.

$$
\begin{aligned}
\lim_{T\to\infty}\frac{1}{2T}\int_{-T}^{T}X(t)dt &= \lim_{T\to\infty}\frac{1}{2T}\int_0^{2\pi}A\cos(wt+\Theta)dt = \lim_{T\to\infty}\frac{A}{2wT}[\sin(wt+\Theta)]_0^{2\pi} \\
&= \lim_{T\to\infty}\frac{A}{2wT}[\sin(2\pi w+\Theta)-\sin(\Theta)] = 0
\end{aligned}
$$

Thus, $X(t)$ is mean-ergodic.

## 10.7 POWER SPECTRAL DENSITY

So far we have been able to characterize a random process by its mean, autocorrelation function, and covariance function. All these functions deal with the time domain; we have not said anything about the spectral (or frequency domain) properties of the process. For a deterministic signal $y(t)$, it is well known that its spectral properties are contained in its Fourier transform $Y(w)$, which is given by

$$Y(w) = \int_{-\infty}^{\infty} y(t)e^{-jwt}dt$$

Conversely, given $Y(w)$ we can recover $y(t)$ by means of the inverse Fourier transform:

$$y(t) = \frac{1}{2\pi}\int_{-\infty}^{\infty} Y(w)e^{jwt}dw$$

Thus, $Y(w)$ provides a complete description of $y(t)$ and vice versa. Unfortunately the same argument cannot be applied to a random process $X(t)$ because the Fourier transform may not exist for most sample functions of the process One of the conditions for the function $y(t)$ to be Fourier transformable is that it must be absolutely integrable, which means that

$$\int_{-\infty}^{\infty} |y(t)|dt < \infty$$

Recall that for wide-sense stationary processes the autocorrelation function, $R_{XX}(\tau)$, is bounded: $|R_{XX}(\tau)| \le R_{XX}(0) = E[X^2(t)]$. Thus, instead of working directly with the random process $X(t)$, we work with its autocorrelation function, which is bounded and hence absolutely integrable.

For a wide-sense stationary process, the Fourier transform of the autocorrelation function is called the *power spectral density*, $S_{XX}(w)$, of the random process. Thus,

$$S_{XX}(w) = \int_{-\infty}^{\infty} R_{XX}(\tau)e^{-jw\tau}d\tau \tag{10.10}$$

We can recover $R_{XX}(\tau)$ via the inverse Fourier transform operation as follows:

$$R_{XX}(\tau) = \frac{1}{2\pi}\int_{-\infty}^{\infty} S_{XX}(w)e^{jw\tau}dw \tag{10.11}$$

The statement that the autocorrelation function of a random process and the power spectral density of the process constitute a Fourier transform pair is called the *Wiener-Khintchin theorem*. Note that the mean-square value of the random process, $E[X^2(t)]$, which is also called the *average power*, is given by

$$E\left[X^2(t)\right] = R_{XX}(0) = \frac{1}{2\pi}\int_{-\infty}^{\infty} S_{XX}(w)dw \tag{10.12}$$

Thus, the properties of the power spectral density include the following:

a. $S_{XX}(w) \ge 0$, which means that is a nonnegative function
b. $S_{XX}(-w) = S_{XX}(w)$, which means that is an even function
c. The power spectral density is a real function if $X(t)$ is real because we have that

$$\begin{aligned} S_{XX}(w) &= \int_{-\infty}^{\infty} R_{XX}(\tau)e^{-jw\tau}d\tau = \int_{-\infty}^{\infty} R_{XX}(\tau)\{\cos(w\tau) - j\sin(w\tau)\}d\tau \\ &= \int_{-\infty}^{\infty} R_{XX}(\tau)\cos(w\tau)d\tau - j\int_{-\infty}^{\infty} R_{XX}(\tau)\sin(w\tau)d\tau \end{aligned}$$

We know from the second property above that $S_{XX}(w)$ is an even function. Since $R_{XX}(\tau)\cos(w\tau)$ is an even function of $\tau$ and $R_{XX}(\tau)\sin(w\tau)$ is an odd function of $\tau$, the imaginary part in the above equation vanishes. Thus, we have that

$$S_{XX}(w) = \int_{-\infty}^{\infty} R_{XX}(\tau)\cos(w\tau)d\tau = 2\int_{0}^{\infty} R_{XX}(\tau)\cos(w\tau)d\tau$$

where the second equality follows from the fact that $R_{XX}(\tau)\cos(w\tau)$ is an even function and $S_{XX}(w)$ is also an even function.

d. As stated earlier, the average power of $X(t)$ is given by

$$E[X^2(t)] = R_{XX}(0) = \frac{1}{2\pi}\int_{-\infty}^{\infty} S_{XX}(w)dw$$

e. $S_{XX}^*(w) = S_{XX}(w)$, where $S_{XX}^*(w)$ is the complex conjugate of $S_{XX}(w)$. This means that $S_{XX}(w)$ cannot be a complex function; it must be a real function.

f. If $\int_{-\infty}^{\infty}|R_{XX}(\tau)|d\tau < \infty$, then $S_{XX}(w)$ is a continuous function of $w$.

Table 10.1 shows some of the common Fourier transform pairs used in random processes analysis.

**Table 10.1** Some Common Fourier Transform Pairs

| $x(\tau)$ | $X(w)$ |
|---|---|
| $e^{-a\lvert\tau\rvert}$, $a>0$ | $\frac{2a}{a^2+w^2}$ |
| $e^{-a\tau}$, $a>0$, $\tau\geq 0$ | $\frac{1}{a+jw}$ |
| $e^{b\tau}$, $b>0$, $\tau<0$ | $\frac{1}{b-jw}$ |
| $\tau e^{-a\tau}$, $a>0$, $\tau\geq 0$ | $\frac{1}{(a+jw)^2}$ |
| 1 | $2\pi\delta(w)$ |
| $\delta(\tau)$ | 1 |
| $e^{jw_0\tau}$ | $2\pi\delta(w-w_0)$ |
| $\begin{cases} 1 & -T/2<\tau<T/2 \\ 0 & \text{otherwise} \end{cases}$ | $T\frac{\sin(wT/2)}{(wT/2)}$ |
| $\begin{cases} 1-\lvert\tau\rvert/T & \lvert\tau\rvert<T \\ 0 & \text{otherwise} \end{cases}$ | $T\left[\frac{\sin(wT/2)}{(wT/2)}\right]^2$ |
| $\cos(w_0\tau)$ | $\pi\{\delta(w-w_0)+\delta(w+w_0)\}$ |
| $\sin(w_0\tau)$ | $-j\pi\{\delta(w-w_0)-\delta(w+w_0)\}$ |
| $e^{-a\lvert\tau\rvert}\cos(w_0\tau)$ | $\frac{a}{a^2+(w-w_0)^2}+\frac{a}{a^2+(w+w_0)^2}$ |

Note that because of the fact that the power spectral density must be an even, non-negative, real function, some of the entries for $x(\tau)$ in Table 10.1 cannot be autocorrelation functions of wide-sense stationary processes. In particular, the functions $e^{-a\tau}$,$\tau e^{-a\tau}$ and $\sin(w_0\tau)$ cannot be autocorrelation functions of wide-sense stationary processes because their Fourier transforms are complex functions.

For two random processes $X(t)$ and $Y(t)$ that are jointly wide-sense stationary, the Fourier transform of their crosscorrelation function $R_{XY}(\tau)$ is called the *cross-power spectral density*, $S_{XY}(w)$, of the two random processes. Thus,

$$S_{XY}(w) = \int_{-\infty}^{\infty} R_{XY}(\tau)e^{-jw\tau}d\tau \tag{10.13}$$

The cross-power spectral density is generally a complex function even when both $X(t)$ and $Y(t)$ are real. Thus, since $R_{YX}(\tau) = R_{XY}(-\tau)$, we have that

$$S_{YX}(w) = S_{XY}(-w) = S^*_{XY}(w) \tag{10.14}$$

where $S^*_{XY}(w)$ is the complex conjugate of $S_{XY}(w)$.

## EXAMPLE 10.10

Determine the autocorrelation function of the random process with the power spectral density given by

$$S_{XX}(w) = \begin{cases} S_0 & |w| < w_0 \\ 0 & \text{otherwise} \end{cases}$$

**Solution:**

$S_{XX}(w)$ is plotted in Figure 10.4.

$$R_{XX}(\tau) = \frac{1}{2\pi}\int_{-\infty}^{\infty} S_{XX}(w)e^{jw\tau}dw = \frac{1}{2\pi}\int_{-w_0}^{w_0} S_0e^{jw\tau}dw$$

$$= \frac{S_0}{2j\pi\tau}\left[e^{jw\tau}\right]_{w=-w_0}^{w_0} = \frac{S_0}{2j\pi\tau}\left\{e^{jw_0\tau} - e^{-jw_0\tau}\right\} = \frac{S_0}{\pi\tau}\left[\frac{e^{jw_0\tau} - e^{-jw_0\tau}}{2j}\right]$$

$$= \frac{S_0}{\pi\tau}\sin(w_0\tau)$$

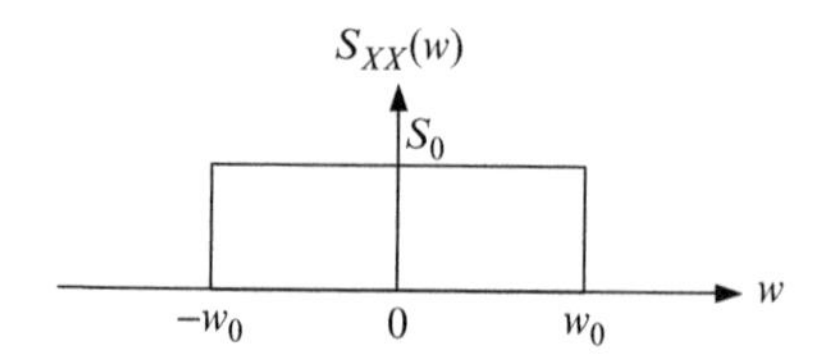


**FIGURE 10.4**
Plot of $S_{XX}(w)$ for Example 10.10

## EXAMPLE 10.11

A stationary random process $X(t)$ has the power spectral density

$$S_{XX}(w) = \frac{24}{w^2+16}$$

Find the mean-square value of the process.

**Solution:**

We will use two methods to solve the problem, as follows:

**Method 1 (Brute-force Method)**: The mean-square value is given by

$$E\left[X^2(t)\right] = R_{XX}(0) = \frac{1}{2\pi}\int_{-\infty}^{\infty} S_{XX}(w)dw = \frac{1}{2\pi}\int_{-\infty}^{\infty}\left\{\frac{24}{w^2+16}\right\}dw$$
$$= \frac{1}{2\pi}\int_{-\infty}^{\infty}\frac{24}{16\left[1+(w/4)^2\right]}dw$$

Let $w/4 = \tan(\theta) \Rightarrow dw = 4\sec^2(\theta)d\theta$, and

$$1+(w/4)^2 = 1+\tan^2(\theta) = \sec^2(\theta)$$

Also, when $w=-\infty$, $\theta=-\pi/2$; and when $w=\infty$, $\theta=\pi/2$. Thus, we obtain

$$E\left[X^2(t)\right] = \frac{24}{32\pi}\int_{-\pi/2}^{\pi/2}\frac{4\sec^2(\theta)}{\sec^2(\theta)}d\theta = \frac{3}{\pi}\int_{-\pi/2}^{\pi/2}d\theta = \frac{3}{\pi}[\theta]_{-\pi/2}^{\pi/2}$$
$$= \frac{3}{\pi}\left[\frac{\pi}{2}-\left(-\frac{\pi}{2}\right)\right] = \frac{3}{\pi}\left[\frac{\pi}{2}+\frac{\pi}{2}\right]$$
$$= 3$$

**Method 2 (Smart Method)**: From Table 10.1 we observe that

$$e^{-a|\tau|} \leftrightarrow \frac{2a}{a^2+w^2}$$

That is, $e^{-a|\tau|}$ and $2a/(a^2+w^2)$ are Fourier transform pairs. Thus, if we can identify the parameter $a$ in the given problem, we can readily obtain the autocorrelation function. Rearranging the power spectral density, we obtain

$$S_{XX}(w) = \frac{24}{w^2+16} = \frac{24}{w^2+4^2} = \frac{6(4)}{w^2+4^2} = 3\left\{\frac{2(4)}{w^2+4^2}\right\} \equiv 3\left\{\frac{2a}{w^2+a^2}\right\}$$

This means that $a=4$, and the autocorrelation function is

$$R_{XX}(\tau) = 3e^{-4|\tau|}$$

Therefore, the mean-square value of the process is

$$E\left[X^2(t)\right] = R_{XX}(0) = 3$$

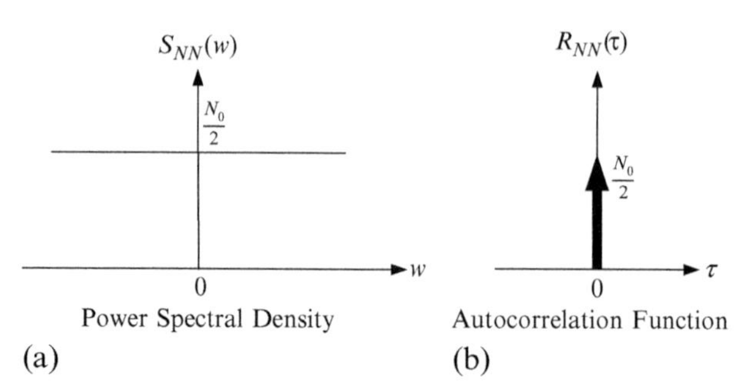


**FIGURE 10.5**
Power Spectral Density and Autocorrelation Function of White Noise

## 10.7.1 White Noise

White noise is the term used to define a random function whose power spectral density is constant for all frequencies. Thus, if $N(t)$ denotes white noise, then

$$S_{NN}(w) = \frac{1}{2}N_0 \tag{10.15}$$

where $N_0$ is a real positive constant. The inverse Fourier transform of $S_{NN}(w)$ gives the autocorrelation function of $N(t)$, $R_{NN}(\tau)$, as follows:

$$R_{NN}(\tau) = \frac{1}{2}N_0\delta(\tau) \tag{10.16}$$

where $\delta(\tau)$ is the impulse function. The two functions are shown in Figure 10.5.

## EXAMPLE 10.12

Let $Y(t) = X(t) + N(t)$ be a wide-sense stationary process where $X(t)$ is the actual signal and $N(t)$ is a zero-mean noise process with variance $\sigma_N^2$ and is independent of $X(t)$. Find the power spectral density of $Y(t)$.

**Solution:**

Since $X(t)$ and $N(t)$ are independent random processes, the autocorrelation function of $Y(t)$ is given by

$$\begin{aligned}
R_{YY}(\tau) &= E[Y(t)Y(t+\tau)] = E[\{X(t)+N(t)\}\{X(t+\tau)+N(t+\tau)\}] \\
&= E[X(t)X(t+\tau) + X(t)N(t+\tau) + N(t)X(t+\tau) + N(t)N(t+\tau)] \\
&= E[X(t)X(t+\tau)] + E[X(t)]E[N(t+\tau)] + E[N(t)]E[X(t+\tau)] \\
&\quad + E[N(t)N(t+\tau)] \\
&= R_{XX}(\tau) + R_{NN}(\tau) = R_{XX}(\tau) + \sigma_N^2\delta(\tau)
\end{aligned}$$

Thus, the power spectral density of $Y(t)$ is given by

$$S_{YY}(w) = S_{XX}(w) + \sigma_N^2$$

## 10.8 DISCRETE-TIME RANDOM PROCESSES

All the discussion thus far assumes that we are dealing with continuous-time random processes. In this section we extend the discussion to discrete-time random processes $\{X[n],\ n=0,1,2,\ldots\}$, which are also called random sequences. A discrete-time random process can be obtained by sampling a continuous-time random process. Thus, if the sampling interval is $T_S$, then for such a case we have that

$$X[n]=X(nT_S) \qquad n=0,\pm 1,\ \pm 2,\ \ldots$$

We provide a summary of the key results as follows.

### 10.8.1 Mean, Autocorrelation Function and Autocovariance Function

The mean of $X[n]$ is given by

$$\mu_X[n]=E[X[n]]$$

The autocorrelation function is given by

$$R_{XX}[n,\,n+m]=E[X[n]X^*[n+m]]$$

where $X^*[n]$ is the complex conjugate of $X[n]$. The random process is wide-sense stationary if $\mu_X[n]=\mu$ a constant, and $R_{XX}[n,\ n+m]=R_{XX}[m]$.

Finally, the autocovariance function of $X[n]$, $C_{XX}[n_1,n_2]$, which is one measure of the coupling between $X[n_1]$ and $X[n_2]$, is defined by

$$\begin{aligned}C_{XX}[n_1,n_2]&=E[\{X[n_1]-\mu_X[n_1]\}\{X[n_2]-\mu_X[n_2]\}]\\&=E[X[n_1]X[n_2]]-\mu_X[n_1]\mu_X[n_2]\\&=R_{XX}[n_1,n_2]-\mu_X[n_1]\mu_X[n_2]\end{aligned}$$

If $X[n_1]$ and $X[n_2]$ are independent, then $R_{XX}[n_1,n_2]=\mu_X[n_1]\mu_X[n_2]$, and we have $C_{XX}[n_1,n_2]=0$, which means that $X[n_1]$ and $X[n_2]$ are uncorrelated.

A discrete-time random process is called a white noise if the random variables $X[n_k]$ are uncorrelated. If the white noise is a Gaussian wide-sense stationary process, then $X[n]$ consists of a sequence of independent and identically distributed random variables with variance $\sigma^2$ and the autocorrelation function is given by

$$R_{XX}[m]=\sigma^2\delta[m]$$

$$\delta[m]=\begin{cases}1 & m=0\\0 & m\neq 0\end{cases}$$

### 10.8.2 Power Spectral Density of a Random Sequence

The power spectral density of $X[n]$ is given by the following discrete-time Fourier transform of its autocorrelation function:

$$S_{XX}(\Omega) = \sum_{m=-\infty}^{\infty} R_{XX}[m]e^{-j\Omega m} \tag{10.17}$$

where $\Omega$ is the discrete frequency. Note that $e^{-j\Omega m}$ is periodic with period $2\pi$. That is,

$$e^{-j(\Omega+2\pi)m} = e^{-j\Omega m}e^{-j2\pi m} = e^{-j\Omega m}$$

because $e^{-j2\pi m} = 1$. Thus, $S_{XX}(\Omega)$ is periodic with period $2\pi$, and it is sufficient to define $S_{XX}(\Omega)$ only in the range $(-\pi, \pi)$. This means that the autocorrelation function is given by

$$R_{XX}[m] = \frac{1}{2\pi}\int_{-\pi}^{\pi} S_{XX}(\Omega)e^{j\Omega m}d\Omega \tag{10.18}$$

The properties of $S_{XX}(\Omega)$ include the following:

a. $S_{XX}(\Omega+2\pi) = S_{XX}(\Omega)$, which means that $S_{XX}(\Omega)$ is periodic with period $2\pi$, as stated earlier.
b. $S_{XX}(-\Omega) = S_{XX}(\Omega)$, which means that $S_{XX}(\Omega)$ is an even function.
c. $S_{XX}(\Omega)$ is real, which means that $S_{XX}(\Omega) \geq 0$.
d. The average power of the process is given by

$$E[X^2[n]] = R_{XX}[0] = \frac{1}{2}\int_{-\pi}^{\pi} S_{XX}(\Omega)d\Omega$$

## EXAMPLE 10.13

Assume that $X[n]$ is a real process, which means that $R_{XX}[-m] = R_{XX}[m]$. Find the power spectral density $S_{XX}(\Omega)$.

**Solution:**

The power spectral density is the discrete-time Fourier transform of the autocorrelation function and is given by

$$\begin{aligned}
S_{XX}(\Omega) &= \sum_{m=-\infty}^{\infty} R_{XX}[m]e^{-j\Omega m} = \sum_{m=-\infty}^{-1} R_{XX}[m]e^{-j\Omega m} + \sum_{m=0}^{\infty} R_{XX}[m]e^{-j\Omega m} \\
&= \sum_{k=1}^{\infty} R_{XX}[-k]e^{j\Omega k} + \sum_{m=1}^{\infty} R_{XX}[m]e^{-j\Omega m} + R_{XX}[0] \\
&= R_{XX}[0] + 2\sum_{m=1}^{\infty} R_{XX}[m]\left\{\frac{e^{j\Omega m} + e^{-j\Omega m}}{2}\right\} \\
&= R_{XX}[0] + 2\sum_{m=1}^{\infty} R_{XX}[m]\cos(m\Omega)
\end{aligned}$$

where the fourth equality follows from the fact that $R_{XX}[-m] = R_{XX}[m]$.

## EXAMPLE 10.14

Find the power spectral density of a random sequence $X[n]$ whose autocorrelation function is given by $R_{XX}[m]=a^{|m|}$.

**Solution:**

The power spectral density is given by

$$\begin{aligned}S_{XX}(\Omega)&=\sum_{m=-\infty}^{\infty}R_{XX}[m]e^{-j\Omega m}=\sum_{m=-\infty}^{\infty}a^{|m|}e^{-j\Omega m}=\sum_{m=-\infty}^{-1}a^{-m}e^{-j\Omega m}+\sum_{m=0}^{\infty}a^{m}e^{-j\Omega m}\\&=\sum_{k=1}^{\infty}a^{k}e^{j\Omega k}+\sum_{m=0}^{\infty}a^{m}e^{-j\Omega m}=\sum_{k=1}^{\infty}\left\{ae^{j\Omega}\right\}^{k}+\sum_{m=0}^{\infty}\left\{ae^{-j\Omega}\right\}^{m}\\&=\frac{1}{1-ae^{j\Omega}}-1+\frac{1}{1-ae^{-j\Omega}}=\frac{1-a^2}{1+a^2-2\cos(\Omega)}\end{aligned}$$

### 10.8.3 Sampling of Continuous-Time Processes

As discussed earlier, one of the methods that can be used to generate discrete-time processes is by sampling a continuous-time process. Thus, if $X(t)$ is a continuous-time process that is sampled at constant intervals of $T_S$ time units (that is, $T_S$ is the sampling period), then the samples constitute the discrete-time process defined by

$$X[n]=X(nT_S) \qquad n=0,\pm 1,\pm 2,\ldots$$

If $\mu_X(t)$ and $R_{XX}(t_1,t_2)$ are the mean and autocorrelation function, respectively, of $X(t)$, the mean and autocorrelation function of $X[n]$ are given by

$$\begin{aligned}\mu_X[n]&=\mu_X(nT_S)\\R_{XX}[n_1,n_2]&=R_{XX}(n_1T_S,n_2T_S)\end{aligned}$$

It can be shown that if $X(t)$ is a wide-sense stationary process, then $X[n]$ is also a wide-sense stationary process with mean $\mu_X[n]=\mu_X$ and autocorrelation function $R_{XX}[m]=R_{XX}(mT_S)$. If $X(t)$ is a wide-sense stationary process, then the power spectral density of $X[n]$ is given by

$$\begin{aligned}S_{XX}(\Omega)&=\sum_{m=-\infty}^{\infty}R_{XX}[m]e^{-j\Omega m}=\sum_{m=-\infty}^{\infty}R_{X_CX_C}(mT_S)e^{-j\Omega m}\\&=\frac{1}{T_S}\sum_{m=-\infty}^{\infty}S_{X_CX_C}\left(\frac{\Omega-2\pi m}{T_S}\right)\end{aligned} \tag{10.19}$$

where $S_{X_CX_C}(w)$ and $R_{X_CX_C}(\tau)$ are the power spectral density and autocorrelation function, respectively, of $X(t)$.

**An Aside**: The last result in (10.19) is obtained from the concept of signal sampling. Consider a continuous-time signal $x_C(t)$ that is sampled at intervals of $T_S$

time units. $T_S$ is called the sampling period. We define $f_S = 1/T_S$ as the sampling frequency; the unit of $f_S$ is Hz when $T_S$ is measured in seconds. Similarly, we define the angular sampling frequency $w_S = 2\pi f_S = 2\pi/T_S$ radians/second. Thus, the sampled signal $x_S(t)$ is obtained by multiplying the continuous-time signal with an impulse train; that is,

$$x_S(t) = x_C(t) \sum_{n=-\infty}^{\infty} \delta(t - nT_S) \tag{10.20}$$

Because the impulse train $\sum_{n=-\infty}^{\infty} \delta(t - nT_S)$ is periodic with a period $T_S$, its Fourier series is $\frac{1}{T_S} \sum_{k=-\infty}^{\infty} e^{jkw_S t}$. Thus, we can rewrite (10.20) as

$$x_S(t) = x_C(t) \sum_{n=-\infty}^{\infty} \delta(t - nT_S) = x_C(t) \frac{1}{T_S} \sum_{k=-\infty}^{\infty} e^{jkw_S t} = \frac{1}{T_S} \sum_{k=-\infty}^{\infty} x_C(t) e^{jkw_S t}$$

Thus, taking the Fourier transform we obtain

$$\begin{aligned} X_S(w) &= \int_{-\infty}^{\infty} x_S(t) e^{-jwt} dt = \int_{-\infty}^{\infty} \left\{ \frac{1}{T_S} \sum_{k=-\infty}^{\infty} x_C(t) e^{jkw_S t} \right\} e^{-jwt} dt \\ &= \frac{1}{T_S} \sum_{k=-\infty}^{\infty} \int_{-\infty}^{\infty} x_C(t) e^{-j(w - kw_S)t} dt = \frac{1}{T_S} \sum_{k=-\infty}^{\infty} X_C(w - kw_S) \\ &= \frac{1}{T_S} \sum_{k=-\infty}^{\infty} X_C(w - 2\pi k f_S) = \frac{1}{T_S} \sum_{k=-\infty}^{\infty} X_C\left(w - \frac{2\pi k}{T_S}\right) \end{aligned} \tag{10.21}$$

Note that from (10.20) we have that

$$x_S(t) = x_C(t) \sum_{n=-\infty}^{\infty} \delta(t - nT_S) = \sum_{n=-\infty}^{\infty} x_C(t) \delta(t - nT_S) = \sum_{n=-\infty}^{\infty} x_C(nT) \delta(t - nT_S)$$

Taking the Fourier transform we obtain

$$\begin{aligned} X_S(w) &= \int_{-\infty}^{\infty} \left\{ \sum_{k=-\infty}^{\infty} x_C(kT_S) \delta(t - kT_S) \right\} e^{-jwt} dt \\ &= \sum_{k=-\infty}^{\infty} \int_{-\infty}^{\infty} x_C(kT_S) \delta(t - kT_S) e^{-jwt} dt = \sum_{k=-\infty}^{\infty} x_C(kT_S) e^{-jwT_S k} \\ &\equiv \sum_{k=-\infty}^{\infty} x_C[k] e^{-j\Omega k} = X_C(\Omega) \end{aligned} \tag{10.22}$$

where $\Omega = wT_S$. For equations (10.21) and (10.22) to be identical, we must have that $w = \Omega/T_S$. Thus, $X_C(\Omega) = X_S(\Omega/T_S)$ and the discrete Fourier transform of the sampled signal is

$$X_S(\Omega) = \frac{1}{T_S} \sum_{k=-\infty}^{\infty} X_C\left(\frac{\Omega}{T_S} - \frac{2\pi k}{T_S}\right) = \frac{1}{T_S} \sum_{k=-\infty}^{\infty} X_C\left(\frac{\Omega - 2\pi k}{T_S}\right)$$

## EXAMPLE 10.15

A wide-sense stationary continuous-time process $X_C(t)$ has the autocorrelation function given by

$$R_{X_CX_C}(\tau)=e^{-4|\tau|}$$

If $X_C(t)$ is sampled with a sampling period 20 seconds to produce the discrete-time process $X[n]$, find the power spectral density of $X[n]$.

**Solution:**
The discrete-time process $X[n]=X_C(20n)$. From Table 10.1 we see that the power spectral density of the continuous-time process is given by

$$S_{X_CX_C}(w)=\frac{2(4)}{4^2+w^2}=\frac{8}{16+w^2}$$

Thus, the power spectral density of the discrete-time process $X[n]$ is given by

$$\begin{aligned}S_{XX}(\Omega)&=\frac{1}{T_S}\sum_{m=-\infty}^{\infty}S_{X_CX_C}\left(\frac{\Omega-2\pi m}{T_S}\right)=\frac{1}{20}\sum_{m=-\infty}^{\infty}\frac{8}{16+\left[\frac{\Omega-2\pi m}{20}\right]^2}\\&=\sum_{m=-\infty}^{\infty}\frac{160}{6400+[\Omega-2\pi m]^2}\end{aligned}$$

## 10.9 CHAPTER SUMMARY

This chapter has presented an introduction to random (or stochastic) processes. It provided different classifications of random processes including discrete-state random processes, continuous-state random processes, discrete-time random processes, and continuous-time random processes. It also discussed two types of stationarity for random processes. A random process whose CDF is invariant to a shift in the time origin is defined to be a strict-sense stationary random process. In many practical situations, this stringent condition is not required. For these situations two conditions are required: the mean value of the process must be a constant, and the autocorrelation function must be dependent only on the difference between the two observation times and not on the absolute time. Such processes are said to be stationary in the wide sense.

The power spectral density of a continuous-time wide-sense stationary random process is defined as the Fourier transform of its autocorrelation function. Thus, the autocorrelation function and the power spectral density are Fourier transform pairs, which means that given one of them the other can be obtained by an inverse transformation. Similarly, the power spectral density of a discrete-time wide-sense stationary random process is defined as the discrete-time Fourier transform of its autocorrelation function.

## 10.10 PROBLEMS

### Section 10.3 Mean, Autocorrelation Function and Autocovariance Function

10.1 Calculate the autocorrelation function of the rectangular pulse shown in Figure 10.6; that is,

$$X(t) = A \qquad 0 \le t \le T$$

where $A$ and $T$ are constants.

10.2 Calculate the autocorrelation function of the periodic function $X(t) = A\sin(wt+\varphi)$, where the period $T = 2\pi/w$, and $A$, $\varphi$ and $w$ are constants.

10.3 The random process $X(t)$ is given by

$$X(t) = Y\cos(2\pi t) \qquad t \ge 0$$

where $Y$ is a random variable that is uniformly distributed between 0 and 2. Find the expected value and autocorrelation function of $X(t)$.

10.4 The sample function $X(t)$ of a stationary random process $Y(t)$ is given by

$$X(t) = Y(t)\sin(wt + \Theta)$$

where $w$ is a constant, $Y(t)$ and $\Theta$ are statistically independent, and $\Theta$ is uniformly distributed between 0 and $2\pi$. Find the autocorrelation function of $X(t)$ in terms of $R_{YY}(\tau)$.

10.5 The sample function $X(t)$ of a stationary random process $Y(t)$ is given by

$$X(t) = Y(t)\sin(wt + \Theta)$$

where $w$ is a constant, $Y(t)$ and $\Theta$ are statistically independent, and $\Theta$ is uniformly distributed between 0 and $2\pi$. Find the autocovariance function of $X(t)$.

10.6 The random process $X(t)$ is given by

$$X(t) = A\cos(wt) + B\sin(wt)$$

where $w$ is a constant, and $A$ and $B$ are independent standard normal random variables (i.e., zero mean and variance of 1). Find the autocovariance function of $X(t)$.

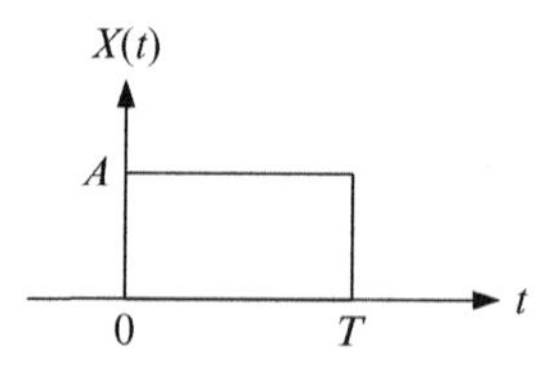


**FIGURE 10.6**
Figure for Problem 10.1

10.7 Assume that $Y$ is a random variable that is uniformly distributed between 0 and 2. If we define the random process $X(t)=Y\cos(2\pi t)$, $t\geq 0$, find the autocovariance function of $X(t)$.

10.8 A random process $X(t)$ is given by

$$X(t)=A\cos(t)+(B+1)\sin(t) \qquad -\infty<t<\infty$$

where $A$ and $B$ are independent random variables with $E[A]=E[B]=0$ and $E[A^2]=E[B^2]=1$. Find the autocovariance function of $X(t)$.

10.9 Determine the missing elements of the following autocovariance matrix of a zero-mean wide-sense stationary random process $X(t)$, where the missing elements are denoted by $xx$.

$$C_{XX}=\begin{bmatrix} 1.0 & xx & 0.4 & xx \\ 0.8 & xx & 0.6 & 0.4 \\ xx & 0.6 & 1.0 & 0.6 \\ 0.2 & xx & xx & 1.0 \end{bmatrix}$$

10.10 The random process $X(t)$ is defined as follows:

$$X(t)=A+e^{-B|t|}$$

where $A$ and $B$ are independent random variables. $A$ is uniformly distributed over the range $-1\leq A\leq 1$, and $B$ is uniformly distributed over the range $0\leq B\leq 2$. Find the following:

a. the mean of $X(t)$

b. the autocorrelation function of $X(t)$

10.11 The random process $X(t)$ has the autocorrelation function $R_{XX}(\tau)=e^{-2|\tau|}$. The random process $Y(t)$ is defined as follows:

$$Y(t)=\int_0^t X^2(u)\,du$$

Find $E[Y(t)]$.

## Section 10.4 Crosscorrelation and Crosscovariance Functions

10.12 Two random processes $X(t)$ and $Y(t)$ are both zero-mean and wide-sense stationary processes. If we define the random process $Z(t)=X(t)+Y(t)$, determine the autocorrelation function of $Z(t)$ under the following conditions:

a. $X(t)$ and $Y(t)$ are jointly wide-sense stationary

b. $X(t)$ and $Y(t)$ are orthogonal.

10.13 Two random processes $X(t)$ and $Y(t)$ are defined as follows:

$$X(t) = A\cos(wt) + B\sin(wt)$$
$$Y(t) = B\cos(wt) - A\sin(wt)$$

where $w$ is a constant, and $A$ and $B$ zero-mean and uncorrelated random variables with variances $\sigma_A^2 = \sigma_B^2 = \sigma^2$. Find the crosscorrelation function $R_{XY}(t, t+\tau)$.

10.14 Two random processes $X(t)$ and $Y(t)$ are defined as follows:

$$X(t) = A\cos(wt + \Theta)$$
$$Y(t) = B\sin(wt + \Theta)$$

where $w$, $A$ and $B$ are constants, and $\Theta$ is a random variable that is uniformly distributed between 0 and $2\pi$.

a. Find the autocorrelation function $R_{XX}(t, t+\tau)$, and show that $X(t)$ is a wide-sense stationary process
b. Find the autocorrelation function $R_{YY}(t, t+\tau)$, and show that $Y(t)$ is a wide-sense stationary process
c. Find the crosscorrelation function $R_{XY}(t, t+\tau)$, and show that $X(t)$ and $Y(t)$ are jointly wide-sense stationary.

## Section 10.5 Wide-Sense Stationary Processes

10.15 Two random processes $X(t)$ and $Y(t)$ are defined as follows:

$$X(t) = A\cos(w_1 t + \Theta)$$
$$Y(t) = B\sin(w_2 t + \Phi)$$

where $w_1, w_2$, $A$ and $B$ are constants, and $\Theta$ and $\Phi$ are statistically independent random variables, each of which is uniformly distributed between 0 and $2\pi$.

a. Find the crosscorrelation function $R_{XY}(t, t+\tau)$, and show that $X(t)$ and $Y(t)$ are jointly wide-sense stationary.
b. If $\Theta = \Phi$, show that $X(t)$ and $Y(t)$ are not jointly wide-sense stationary.
c. If $\Theta = \Phi$, under what condition are $X(t)$ and $Y(t)$ jointly wide-sense stationary?

10.16 Explain why the following matrices can or cannot be valid autocorrelation matrices of a zero-mean wide-sense stationary random process $X(t)$.

a.

$$G = \begin{bmatrix} 1.0 & 1.2 & 0.4 & 1.0 \\ 1.2 & 1.0 & 0.6 & 0.9 \\ 0.4 & 0.6 & 1.0 & 1.3 \\ 1.0 & 0.9 & 1.3 & 1.0 \end{bmatrix}$$

b.

$$H=\begin{bmatrix} 2.0 & 1.2 & 0.4 & 1.0 \\ 1.2 & 2.0 & 0.6 & 0.9 \\ 0.4 & 0.6 & 2.0 & 1.3 \\ 1.0 & 0.9 & 1.3 & 2.0 \end{bmatrix}$$

c.

$$K=\begin{bmatrix} 1.0 & 0.7 & 0.4 & 0.8 \\ 0.5 & 1.0 & 0.6 & 0.9 \\ 0.4 & 0.6 & 1.0 & 0.3 \\ 1.0 & 0.9 & 0.3 & 1.0 \end{bmatrix}$$

10.17 Two jointly stationary random processes $X(t)$ and $Y(t)$ are defined as follows:

$$X(t)=2\cos(5t+\Phi)$$
$$Y(t)=10\sin(5t+\Phi)$$

where $\Phi$ is a random variable that is uniformly distributed between 0 and $2\pi$. Find the crosscorrelation functions $R_{XY}(\tau)$ and $R_{YX}(\tau)$.

10.18 State why each of the functions, $F(\tau)$, $G(\tau)$ and $H(\tau)$, shown in Figure 10.7 can or cannot be a valid autocorrelation function of a wide-sense stationary process.

10.19 A random process $Y(t)$ is given by

$$Y(t)=A\sin(Wt+\Phi)$$

where $A$, $W$ and $\Phi$ are independent random variables. Assume that $A$ has a mean of 3 and a variance of 9, $\Phi$ is uniformly distributed between $-\pi$ and $\pi$, and $W$ is uniformly distributed between $-6$ and 6. Determine if the process is stationary in the wide sense.

10.20 A random process $X(t)$ is given by

$$X(t)=A\cos(t)+(B+1)\sin(t) \qquad -\infty<t<\infty$$

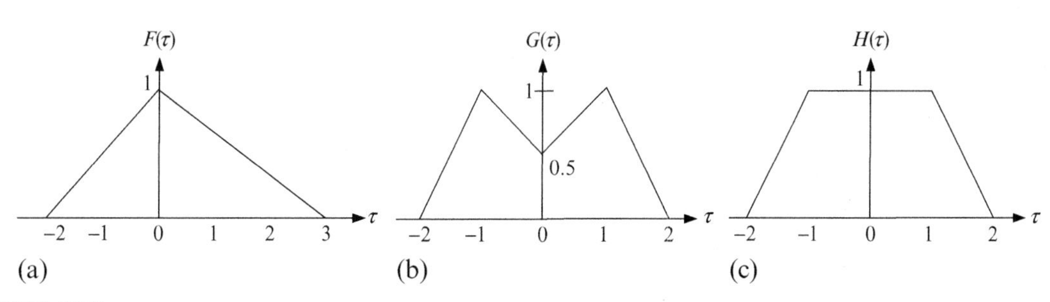


**FIGURE 10.7**
Figure for Problem 10.18

where $A$ and $B$ are independent random variables with $E[A]=E[B]=0$ and $E[A^2]=E[B^2]=1$. Is $X(t)$ wide-sense stationary?

10.21 A random process has the autocorrelation function

$$R_{XX}(\tau)=\frac{16\tau^2+28}{\tau^2+1}$$

Find the mean-square value, the mean value and the variance of the process.

10.22 A wide-sense stationary random process $X(t)$ has a mean-square value (or average power) $E[X^2(t)]=11$. Give reasons why the functions given below can or cannot be its autocorrelation function.

a. $R_{XX}(\tau)=\frac{11\sin(2\tau)}{1+\tau^2}$

b. $R_{XX}(\tau)=\frac{11\tau}{1+3\tau^2+4\tau^4}$

c. $R_{XX}(\tau)=\frac{\tau^2+44}{\tau^2+4}$

d. $R_{XX}(\tau)=\frac{11\cos(\tau)}{1+3\tau^2+4\tau^4}$

e. $R_{XX}(\tau)=\frac{11\tau^2}{1+3\tau^2+4\tau^4}$

10.23 An ergodic random process $X(t)$ has the autocorrelation function

$$R_{XX}(\tau)=\frac{36\tau^2+40}{\tau^2+1}$$

Determine the mean value, mean-square value, and variance of $X(t)$.

10.24 Assume that $X(t)$ is the sum of a deterministic quantity $Q$ and a wide-sense stationary noise process $N(t)$. Determine the following:

a. the mean of $X(t)$
b. the autocorrelation function of $X(t)$
c. the autocovariance function of $X(t)$

10.25 Two statistically independent and zero-mean random processes $X(t)$ and $Y(t)$ have the following autocorrelation functions, respectively:

$$R_{XX}(\tau)=e^{-|\tau|}$$

$$R_{YY}(\tau)=\cos(2\pi\tau)$$

Determine the following:

a. the autocorrelation function of the process $U(t)=X(t)+Y(t)$
b. the autocorrelation function of the process $V(t)=X(t)-Y(t)$
c. the crosscorrelation function of $U(t)$ and $V(t)$

## Section 10.6 Ergodic Random Processes

10.26 A random process $Y(t)$ is given by

$$Y(t) = A\cos(wt + \Phi)$$

where $w$ is a constant, and $A$ and $\Phi$ are independent random variables. The random variable $A$ has a mean of 3 and a variance of 9, and $\Phi$ is uniformly distributed between $-\pi$ and $\pi$. Determine if the process is a mean-ergodic process.

10.27 A random process $X(t)$ is given by

$$X(t) = A$$

where $A$ is a random variable with a finite mean of $\mu_A$ and finite variance $\sigma_A^2$. Determine if $X(t)$ is a mean-ergodic process.

## Section 10.7 Power Spectral Density

10.28 Assume that $V(t)$ and $W(t)$ are both zero-mean wide-sense stationary random processes and let the random process $M(t)$ be defined as follows:

$$M(t) = V(t) + W(t)$$

a. If $V(t)$ and $W(t)$ are jointly wide-sense stationary, determine the following in terms of those of $V(t)$ and $W(t)$:
   i. the autocorrelation function of $M(t)$
   ii. the power spectral density of $M(t)$
b. If $V(t)$ and $W(t)$ are orthogonal, determine the following in terms of those of $V(t)$ and $W(t)$:
   i. the autocorrelation function of $M(t)$
   ii. the power spectral density of $M(t)$

10.29 A stationary random process $X(t)$ has an autocorrelation function given by

$$R_{XX}(\tau) = 2e^{-|\tau|} + 4e^{-4|\tau|}$$

Find the power spectral density of the process.

10.30 A random process $X(t)$ has a power spectral density given by

$$S_{XX}(w) = \begin{cases} 4 - \dfrac{w^2}{9} & |w| \le 6 \\ 0 & \text{otherwise} \end{cases}$$

Determine (a) the average power and (b) the autocorrelation function of the process.

10.31 A random process $Y(t)$ has the power spectral density

$$S_{XX}(w)=\frac{9}{w^2+64}$$

Find (a) the average power in the process and (b) the autocorrelation function.

10.32 A random process $Z(t)$ has the autocorrelation function given by

$$R_{ZZ}(\tau)=\begin{cases}1+\frac{\tau}{\tau_0} & -\tau_0\le\tau\le 0\\ 1-\frac{\tau}{\tau_0} & 0\le\tau\le\tau_0\\ 0 & \text{otherwise}\end{cases}$$

where $\tau_0$ is a constant. Calculate the power spectral density of the process.

10.33 Give reasons why the functions given below can or cannot be the power spectral density of a wide-sense stationary random process.

a. $S_{XX}(w)=\frac{\sin(w)}{w}$

b. $S_{XX}(w)=\frac{\cos(w)}{w}$

c. $S_{XX}(w)=\frac{8}{w^2+16}$

d. $S_{XX}(w)=\frac{5w^2}{1+3w^2+4w^4}$

e. $S_{XX}(w)=\frac{5w}{1+3w^2+4w^4}$

10.34 A bandlimited white noise has the power spectral density defined by

$$S_{XX}(w)=\begin{cases}0.01 & 400\pi\le|w|\le 500\pi\\ 0 & \text{otherwise}\end{cases}$$

Find the mean-square value of the process.

10.35 A wide-sense stationary process $X(t)$ has an autocorrelation function

$$R_{XX}(\tau)=ae^{-4|\tau|}$$

where $a$ is a constant. Determine the power spectral density.

10.36 Two random processes $X(t)$ and $Y(t)$ are defined as follows:

$$X(t)=A\cos(w_0t)+B\sin(w_0t)$$
$$Y(t)=B\cos(w_0t)-A\sin(w_0t)$$

where $w_0$ is a constant, and $A$ and $B$ zero-mean and uncorrelated random variables with variances $\sigma_A^2=\sigma_B^2=\sigma^2$. Find the cross-power spectral density of $X(t)$ and $Y(t)$, $S_{XY}(w)$. (Note that $S_{XY}(w)$ is the Fourier transform of the crosscorrelation function $R_{XY}(\tau)$.)

10.37 Two random processes $X(t)$ and $Y(t)$ are both zero-mean and wide-sense stationary processes. If we define the random process $Z(t)=X(t)+Y(t)$, determine the power spectral density of $Z(t)$ under the following conditions:

a. $X(t)$ and $Y(t)$ are jointly wide-sense stationary
b. $X(t)$ and $Y(t)$ are orthogonal.

10.38 Two jointly stationary random processes $X(t)$ and $Y(t)$ have the crosscorrelation function given by:

$$R_{XY}(\tau)=2e^{-2\tau} \quad \tau \geq 0$$

Determine the following:

a. the cross-power spectral density $S_{XY}(w)$
b. the cross-power spectral density $S_{YX}(w)$

10.39 Two jointly stationary random processes $X(t)$ and $Y(t)$ have the cross-power spectral density given by

$$S_{XY}(w)=\frac{1}{-w^2+j4w+4}$$

Find the corresponding crosscorrelation function.

10.40 Two zero-mean independent wide-sense stationary random processes $X(t)$ and $Y(t)$ have the following power spectral densities

$$S_{XX}(w)=\frac{4}{w^2+4}$$

$$S_{YY}(w)=\frac{4}{w^2+4}$$

respectively. A new random process $W(t)$ is defined as follows: $W(t)=X(t)+Y(t)$. Determine the following:

a. the power spectral density of $W(t)$
b. the cross-power spectral density $S_{XW}(w)$
c. the cross-power spectral density $S_{YW}(w)$

10.41 Two zero-mean independent wide-sense stationary random processes $X(t)$ and $Y(t)$ have the following power spectral densities:

$$S_{XX}(w)=\frac{4}{w^2+4}$$

$$S_{YY}(w)=\frac{w^2}{w^2+4}$$

respectively. Two new random processes $V(t)$ and $W(t)$ are defined as follows:

$$V(t) = X(t) + Y(t)$$
$$W(t) = X(t) - Y(t)$$

respectively. Determine the cross-power spectral density $S_{VW}(w)$.

10.42 A zero-mean wide-sense stationary random process $X(t)$, $-\infty < t < \infty$, has the following power spectral density:

$$S_{XX}(w) = \frac{2}{1 + w^2} \qquad -\infty < w < \infty$$

The random process $Y(t)$ is defined by

$$Y(t) = \sum_{k=0}^{2} X(t+k)$$

a. Find the mean of $Y(t)$.
b. Find the variance of $Y(t)$.

10.43 Consider two individual wide-sense stationary processes $X(t)$ and $Y(t)$. Consider the random process $Z(t) = X(t) + Y(t)$.

a. Show that the autocorrelation function of $Z(t)$ is given by

$$R_{ZZ}(t, t+\tau) = R_{XX}(\tau) + R_{YY}(\tau) + R_{XY}(t, t+\tau) + R_{YX}(t, t+\tau)$$

b. If $X(t)$ and $Y(t)$ are jointly wide-sense stationary, show that the autocorrelation function of $Z(t)$ is given by

$$R_{ZZ}(t, t+\tau) = R_{XX}(\tau) + R_{YY}(\tau) + R_{XY}(\tau) + R_{YX}(\tau)$$

c. If $X(t)$ and $Y(t)$ are jointly wide-sense stationary, find the power spectral density of $Z(t)$.
d. If $X(t)$ and $Y(t)$ are uncorrelated, find the power spectral density of $Z(t)$.
e. If $X(t)$ and $Y(t)$ are orthogonal, find the power spectral density of $Z(t)$.

## Section 10.8 Discrete-time Random Processes

10.44 Find the power spectral density of a random sequence $X[n]$ whose autocorrelation function is given by $R_{XX}[m] = a^m, m = 0, 1, 2, \ldots,$ where $|a| < 1$.

10.45 A wide-sense stationary continuous-time process $X(t)$ has the autocorrelation function given by

$$R_{X_C X_C}(\tau) = e^{-2|\tau|} \cos(w_0 \tau)$$

where $w_0$ is a constant. If $X(t)$ is sampled with a sampling period 10 seconds to produce the discrete-time process $X[n]$, find the power spectral density of $X[n]$. [Hint: Use Table 10.1 to find $S_{X_cX_c}(w)$]

0.46 Periodic samples of the autocorrelation function of white noise $N(t)$ with period $T$ are defined by

$$R_{NN}(kT) = \begin{cases} \sigma_N^2 & k=0 \\ 0 & k \neq 0 \end{cases}$$

Find the power spectral density of the discrete-time random process.

10.47 The autocorrelation function of a discrete-time process is given by

$$R_{XX}[k] = \begin{cases} \sigma_X^2 & k=0 \\ \dfrac{4\sigma_X^2}{k^2\pi^2} & k \neq 0,\ k \text{ odd} \\ 0 & k \neq 0,\ k \text{ even} \end{cases}$$

Find the power spectral density $S_{XX}(\Omega)$ of the process.

CHAPTER 11

# Linear Systems with Random Inputs

## 11.1 INTRODUCTION

In Chapter 10 the concept of random processes was introduced. Different parameters that are associated with random processes were also discussed. These parameters include autocorrelation function, autocovariance function, crosscorrelation function, crosscovariance function, and power spectral density. The goal of this chapter is to determine the response or output of a linear system when the input is a random signal instead of a deterministic signal. To set the stage for the discussion, we first begin with a brief review of linear systems with deterministic inputs. Then we examine the response of linear systems to random inputs.

## 11.2 OVERVIEW OF LINEAR SYSTEMS WITH DETERMINISTIC INPUTS

Consider a system with a deterministic input signal $x(t)$ and a deterministic response $y(t)$. The system is usually represented either in terms of its *impulse function* (or *impulse response*) $h(t)$ or its *system response* $H(w)$, which is the Fourier transform of the impulse function. This is illustrated in Figure 11.1. The system response is also called the *transfer function* of the system.

The system is defined to be linear if its response to a sum of inputs $x_k(t)$, $k=1,2,\ldots,K$, is equal to the sum of the responses taken separately. Also, the system is said to be a time-invariant system if the form of the impulse response does not depend on the time the impulse is applied. For linear time-invariant systems, the response of the system to an input $x(t)$ is the convolution of $x(t)$ and $h(t)$. That is,

$$y(t)=x(t)*h(t)=\int_{-\infty}^{\infty} x(\tau)h(t-\tau)d\tau \tag{11.1}$$

where the last equation is called the *convolution integral* of $x(t)$ and $h(t)$. If we define $u=t-\tau$, we see that



Fundamentals of Applied Probability and Random Processes. http://dx.doi.org/10.1016/B978-0-12-800852-2.00011-0

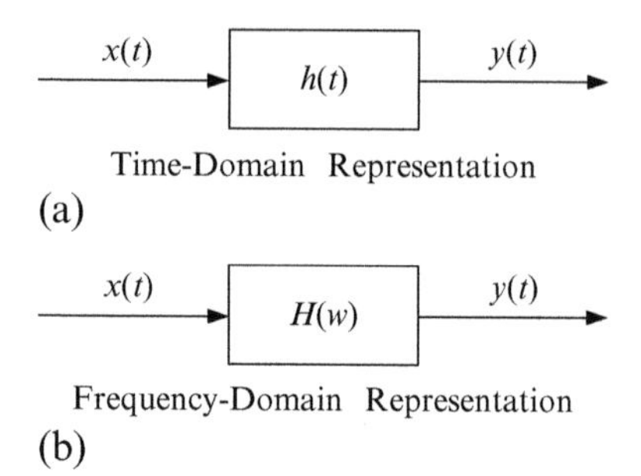


**FIGURE 11.1**
Time-Domain and Frequency-Domain Representations of Linear System

$$y(t) = \int_{-\infty}^{\infty} x(t-u)h(u)du = \int_{-\infty}^{\infty} h(u)x(t-u)du = h(t) * x(t)$$

Thus, the convolution operation between two functions is commutative, which implies that the convolution equation can be written in one of two forms:

$$y(t) = x(t) * h(t) = h(t) * x(t) = \int_{-\infty}^{\infty} x(\tau)h(t-\tau)d\tau = \int_{-\infty}^{\infty} h(\tau)x(t-\tau)d\tau$$

In the frequency domain, we can compute the Fourier transform of $y(t)$ as follows:

$$Y(w) = F[y(t)] = \int_{-\infty}^{\infty} y(t)e^{-jwt}dt = \int_{-\infty}^{\infty}\left\{\int_{-\infty}^{\infty} x(\tau)h(t-\tau)d\tau\right\}e^{-jwt}dt$$

where $F[y(t)]$ denotes the Fourier transform of $y(t)$. Interchanging the order of integration and since $x(\tau)$ does not depend on $t$, we obtain

$$Y(w) = \int_{-\infty}^{\infty} x(\tau)\left\{\int_{-\infty}^{\infty} h(t-\tau)e^{-jwt}dt\right\}d\tau$$

But the inner integral is the Fourier transform of $h(t-\tau)$, which is $e^{-jw\tau}H(w)$. Therefore,

$$Y(w) = \int_{-\infty}^{\infty} x(\tau)e^{-jw\tau}H(w)d\tau = H(w)\int_{-\infty}^{\infty} x(\tau)e^{-jw\tau}d\tau = H(w)X(w) \tag{11.2}$$

Thus, the Fourier transform of the output signal is the product of the system response and the Fourier transform of the input signal.

A system is said to be *causal* if its output at time $t_0$, $y(t_0)$, resulting from input $x(t)$ does not depend on $x(t)$ for any time $t > t_0$, for all $t_0$ and all inputs $x(t)$. Thus, the output of a causal system cannot anticipate or precede the input signal to produce the output; it only depends on values of the input at the present time and n the past. This means that for a causal system, $h(t)=0$ for $t<0$, which is why a ausal system is sometimes referred to as a *non-anticipative* system because the ystem output does not anticipate future values of the input. For a causal linear ime-invariant system, the output process is given by

$$(t)=\int_{-\infty}^{t} x(\tau)h(t-\tau)d\tau=\int_{0}^{\infty} h(\tau)x(t-\tau)d\tau$$

## 11.3 LINEAR SYSTEMS WITH CONTINUOUS-TIME RANDOM INPUTS

We can now extend this principle to the case of continuous-time random nputs. Recall that the random process $X(t)$ is not guaranteed to be Fourier ransformable, which is the reason we deal with its autocorrelation function n the frequency domain. The problem we want to address then is this: Given that $X(t)$ is the input to a linear time-invariant system with impulse response $h(t)$ and $Y(t)$ is the corresponding output of the system, can we determine the mean and autocorrelation function of $Y(t)$ if those of $X(t)$ are known? From our earlier discussion, the output process is given by

$$Y(t)=\int_{-\infty}^{\infty} X(\tau)h(t-\tau)d\tau$$

Thus, the mean of $Y(t)$ is given by

$$\begin{aligned}\mu_Y(t)=E[Y(t)]=E\left[\int_{-\infty}^{\infty} X(\tau)h(t-\tau)d\tau\right]&=\int_{-\infty}^{\infty} E[X(\tau)]h(t-\tau)d\tau\\&=\int_{-\infty}^{\infty} \mu_X(\tau)h(t-\tau)d\tau=\mu_X(t)*h(t)\end{aligned} \tag{11.3}$$

Since we can also write

$$Y(t)=\int_{-\infty}^{\infty} h(\tau)X(t-\tau)d\tau$$

we see that the mean is

$$\mu_Y(t)=\int_{-\infty}^{\infty} h(\tau)\mu_X(t-\tau)d\tau=h(t)*\mu_X(t)$$

This means that

$$\mu_Y(t)=\mu_X(t)*h(t)=h(t)*\mu_X(t) \tag{11.4}$$

Thus, the mean of the output process is the convolution of the mean of the input process and the impulse response.

The crosscorrelation function between the input process $X(t)$ and the output process $Y(t)$ is given by

$$R_{XY}(t,t+\tau)=E[X(t)Y(t+\tau)]=E\left[X(t)\int_{-\infty}^{\infty}h(u)X(t+\tau-u)du\right]$$
$$=\int_{-\infty}^{\infty}E[X(t)X(t+\tau-u)]h(u)du=\int_{-\infty}^{\infty}R_{XX}(t,t+\tau-u)h(u)du$$

If $X(t)$ is a wide-sense stationary process, this reduces to

$$R_{XY}(\tau)=\int_{-\infty}^{\infty}R_{XX}(\tau-u)h(u)du=R_{XX}(\tau)*h(\tau) \tag{11.5}$$

If we take the Fourier transforms of both sides, we obtain the *cross-power spectral density* between $X(t)$ and $Y(t)$ as

$$S_{XY}(w)=H(w)S_{XX}(w) \tag{11.6}$$

Thus, the transfer function of the system is given by

$$H(w)=\frac{S_{XY}(w)}{S_{XX}(w)} \tag{11.7}$$

Similarly, the crosscorrelation function between the output process $Y(t)$ and a wide-sense stationary input process $X(t)$ is given by

$$R_{YX}(\tau)=E[Y(t)X(t+\tau)]=\int_{-\infty}^{\infty}R_{XX}(\tau-u)h(-u)du=R_{XX}(\tau)*h(-\tau) \tag{11.8}$$

And if we take the Fourier transforms of both sides, we obtain the cross-power spectral density between $Y(t)$ and $X(t)$ as

$$S_{YX}(w)=H^{*}(w)S_{XX}(w) \tag{11.9}$$

where $H^{*}(w)$ is the complex conjugate of $H(w)$ and is given by

$$H^{*}(w)=\frac{S_{YX}(w)}{S_{XX}(w)} \tag{11.10}$$

Finally, the autocorrelation function of the output of a linear time-invariant system with a wide-sense stationary process input is given by

$$R_{YY}(t,t+\tau)=E[Y(t)Y(t+\tau)]=E\left[Y(t)\int_{-\infty}^{\infty}h(u)X(t+\tau-u)du\right]$$
$$=\int_{-\infty}^{\infty}h(u)E[Y(t)X(t+\tau-u)]du=\int_{-\infty}^{\infty}h(u)R_{YX}(t,t+\tau-u)du$$
$$=\int_{-\infty}^{\infty}h(u)R_{YX}(\tau-u)du=R_{YX}(\tau)*h(\tau)$$
$$=R_{XX}(\tau)*h(-\tau)*h(\tau)=h(-\tau)*h(\tau)*R_{XX}(\tau)$$

where the last equality follows from the fact that the order in which convolutions are performed does not change the answer. Thus, the output process is also wide-sense stationary, and we have that

$$R_{YY}(t,t+\tau)=R_{YY}(\tau)=R_{XX}(\tau)*h(-\tau)*h(\tau)=h(-\tau)*h(\tau)*R_{XX}(\tau) \qquad (11.11)$$

Finally, we can obtain the power spectral density of the output process of a linear time-invariant system with a wide-sense stationary input by noting that the Fourier transform of a convolution is the product of the Fourier transforms. Thus, we obtain

$$S_{YY}(w)=H^*(w)H(w)S_{XX}(w)=|H(w)|^2S_{XX}(w) \qquad (11.12)$$

The quantity $|H(w)|^2$ is called the *power transfer function* of the system and is given by

$$|H(w)|^2=\frac{S_{YY}(w)}{S_{XX}(w)} \qquad (11.13)$$

## EXAMPLE 11.1

Assume that the input $X(t)$ to a linear time-invariant system is white noise. Determine the power spectral density of the output process $Y(t)$ if the system response $H(w)$ is given by the following:

$$H(w)=\begin{cases}1 & w_1<|w|<w_2\\ 0 & \text{otherwise}\end{cases}$$

**Solution:**

The plot of $H(w)$ is illustrated in Figure 11.2, which is an ideal bandpass filter.

Since $S_{XX}(w)=S_{NN}(w)=N_0/2$ for $-\infty<w<\infty$, we obtain

$$S_{YY}(w)=S_{XX}(w)|H(w)|^2=\frac{N_0}{2}|H(w)|^2=\begin{cases}\dfrac{N_0}{2} & w_1<|w|<w_2\\ 0 & \text{otherwise}\end{cases}$$

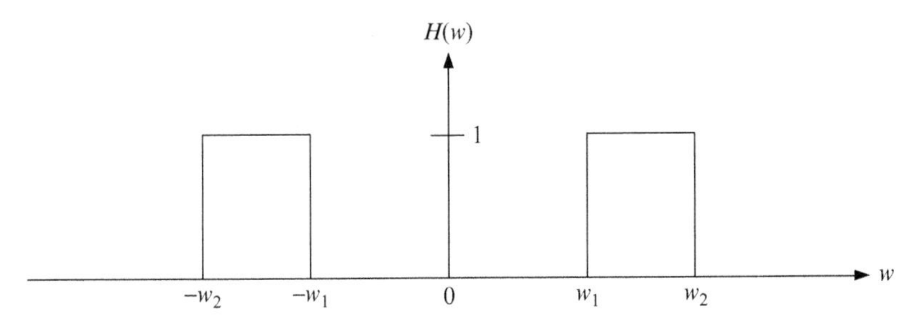


**FIGURE 11.2**
The Bandpass Filter

## EXAMPLE 11.2

A random process $X(t)$ is the input to a linear system whose impulse response is $h(t)=2e^{-t}$, $t\geq 0$. If the autocorrelation function of the process is $R_{XX}(\tau)=e^{-2|\tau|}$, find the power spectral density of the output process $Y(t)$.

**Solution:**

The spectral density of the input process is

$$\begin{aligned} S_{XX}(w) &= \int_{-\infty}^{\infty} R_{XX}(\tau)e^{-jw\tau}d\tau = \int_{-\infty}^{\infty} e^{-2|\tau|}e^{-jw\tau}d\tau \\ &= \int_{-\infty}^{0} e^{2\tau}e^{-jw\tau}d\tau + \int_{0}^{\infty} e^{-2\tau}e^{-jw\tau}d\tau = \int_{-\infty}^{0} e^{(2-jw)\tau}d\tau + \int_{0}^{\infty} e^{-(2+jw)\tau}d\tau \\ &= \frac{1}{2-jw}+\frac{1}{2+jw}=\frac{4}{w^2+4} \end{aligned}$$

The transfer function of the linear system is given by

$$\begin{aligned} H(w) &= \int_{-\infty}^{\infty} h(t)e^{-jwt}dt = \int_{0}^{\infty} 2e^{-t}e^{-jwt}dt = \int_{0}^{\infty} 2e^{-(1+jw)t}dt \\ &= \frac{2}{1+jw} \end{aligned}$$

Thus, the power spectral density of the output process is given by

$$S_{YY}(w)=S_{XX}(w)|H(w)|^2=\left\{\frac{4}{w^2+4}\right\}\left\{\left|\frac{2}{1+jw}\right|^2\right\}=\frac{16}{(w^2+1)(w^2+4)}$$

Note that we can also use the results in Table 10.1 to solve the problem as follows:

$$e^{-a|\tau|}\leftrightarrow\frac{2a}{a^2+w^2},\ e^{-a\tau}\leftrightarrow\frac{1}{a+jw}$$

## EXAMPLE 11.3

A random process $X(t)$ is the input to a linear system whose impulse response is $h(t)=2e^{-t}$, $t\geq 0$. If the autocorrelation function of the process is $R_{XX}(\tau)=e^{-2|\tau|}$, determine the following:

a. The crosscorrelation function $R_{XY}(\tau)$ between the input process $X(t)$ and the output process $Y(t)$
b. The crosscorrelation function $R_{YX}(\tau)$ between the output process $Y(t)$ and the output process $X(t)$.

**Solution:**

a. The crosscorrelation function between the input process and the output process is given by

$$R_{XY}(\tau)=\int_{-\infty}^{\infty} R_{XX}(\tau-u)h(u)du=R_{XX}(\tau)*h(\tau)$$

While we can use a brute-force method to obtain $R_{XY}(\tau)$, a better approach is via the cross-power spectral density $S_{XY}(w)$ associated with $R_{XY}(\tau)$. From Table 10.1 we have that

$$e^{-a|\tau|} \leftrightarrow \frac{2a}{a^2+w^2}, \quad e^{-a\tau} \leftrightarrow \frac{1}{a+jw}$$

Therefore,

$$S_{XY}(w) = S_{XX}(w)H(w) = \left\{\frac{4}{w^2+4}\right\}\left\{\frac{2}{1+jw}\right\} = \frac{8}{(2+jw)(2-jw)(1+jw)}$$
$$\equiv \frac{a}{2+jw} + \frac{b}{2-jw} + \frac{c}{1+jw}$$

where

$$a = (2+jw)S(w)|_{jw=-2} = -2$$
$$b = (2-jw)S(w)|_{jw=2} = 2/3$$
$$c = (1+jw)S(w)|_{jw=-1} = 8/3$$

Therefore,

$$S_{XY}(w) = -\frac{2}{2+jw} + \frac{2/3}{2-jw} + \frac{8/3}{1+jw}$$

Now, from Table 10.1 we see that the Fourier transform of the function $f(t) = e^{-at}$ for $t \geq 0$ and $a > 0$ is $F(w) = 1/(a+jw)$. Similarly, the Fourier transform of the function $g(t) = e^{bt}$ for $t < 0$ and $b > 0$ is $G(w) = 1/(b-jw)$. Thus, if we define the unit step function

$$u(\tau) = \begin{cases} 1 & \tau \geq 0 \\ 0 & \text{otherwise} \end{cases}$$

we obtain the crosscorrelation function between the input process and the output process as

$$R_{XY}(\tau) = \left\{\frac{8}{3}e^{-\tau} - 2e^{-2\tau}\right\}u(\tau) + \frac{2}{3}e^{2\tau}u(-\tau)$$

b. To find $R_{YX}(\tau)$, we use the relationship $R_{YX}(\tau) = R_{XY}(-\tau)$ to obtain

$$R_{YX}(\tau) = \left\{\frac{8}{3}e^{\tau} - 2e^{2\tau}\right\}u(-\tau) + \frac{2}{3}e^{-2\tau}u(\tau)$$

## EXAMPLE 11.4

Consider the linear system shown in Figure 11.3 with an input process $X(t)$ and a zero-mean noise process $N(t)$, where the two input processes are wide-sense stationary processes that are mutually uncorrelated and have power spectral densities $S_{XX}(w)$ and $S_{NN}(w)$, respectively. Find the power spectral density of the output process $Y(t)$.

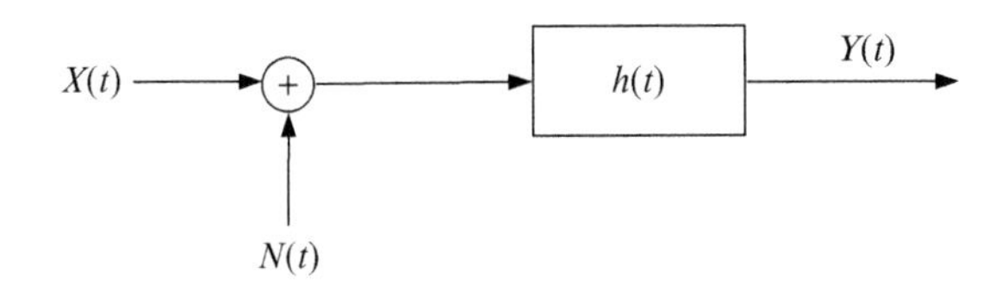


**FIGURE 11.3**
Figure for Example 11.4

**Solution:**
Let the input process be $V(t)=X(t)+N(t)$. Then the autocorrelation function of $V(t)$ is

$$\begin{aligned}R_{VV}(\tau)&=E[V(t)V(t+\tau)]=E[\{X(t)+N(t)\}\{X(t+\tau)+N(t+\tau)\}]\\&=E[X(t)X(t+\tau)]+E[X(t)N(t+\tau)]+E[N(t)X(t+\tau)]+E[N(t)N(t+\tau)]\\&=R_{XX}(\tau)+R_{NN}(\tau)\end{aligned}$$

where the last equality follows from the fact that $X(t)$ and $N(t)$ are mutually uncorrelated and the mean of $N(t)$ is zero. Thus, $S_{VV}(w)=S_{XX}(w)+S_{NN}(w)$. Let the system response be $H(w)$; then we have the following results:

$$S_{YY}(w)=|H(w)|^2S_{VV}(w)=|H(w)|^2\{S_{XX}(w)+S_{NN}(w)\}$$

## 11.4 LINEAR SYSTEMS WITH DISCRETE-TIME RANDOM INPUTS

The analysis of the linear time-invariant system with continuous-time random processes can easily be extended to discrete-time random processes (or random sequences). Let the input process of a linear time-invariant system with impulse response $h[n]$ be the random sequence $\{X[n], n=1,2,\ldots\}$. The response of the system to an input random sequence $X[n]$ is given by

$$Y[n]=\sum_{k=-\infty}^{\infty}X[k]h_k[n]$$

where $h_k[n]$ is the system response for $X[k]$. Since the system is time-invariant, we may write $h_k[n]=h[n-k]$. Thus, for a linear time-invariant system we obtain

$$Y[n]=\sum_{k=-\infty}^{\infty}X[k]h[n-k] \tag{11.14}$$

which is usually called the *convolutional sum* and written as $Y[n]=X[n]*h[n]$. As in the continuous-time system, it can also be shown that

$$Y[n]=\sum_{k=-\infty}^{\infty}h[k]X[n-k]=h[n]*X[n] \tag{11.15}$$

which means that the convolution operation is also commutative, as in the continuous-time case. Given that the mean of $X[n]$ is $\mu_X$, the mean of $Y[n]$ is given by

$$\begin{aligned}\mu_Y[n]=E[Y[n]]&=E\left[\sum_{k=-\infty}^{\infty}X[k]h[n-k]\right]=\sum_{k=-\infty}^{\infty}E[X[k]]h[n-k]\\&=\mu_X\sum_{k=-\infty}^{\infty}h[n-k]\end{aligned} \tag{11.16}$$

Unlike the continuous-time system that uses the Fourier transform for spectral analysis, the discrete-time system uses the discrete-time Fourier transform that is based on the *discrete frequency* $\Omega$. Thus, if the autocorrelation sequence of a discrete sequence $X[n]$ is $R_{XX}[n]$ the power spectral density of $X[n]$ is obtained through the following discrete-time Fourier transform:

$$S_{XX}(\Omega) = \sum_{n=-\infty}^{\infty} R_{XX}[n]e^{-j\Omega n}$$

As stated in Chapter 10, $S_{XX}(\Omega)$ is periodic with period $2\pi$, which enables us to recover the autocorrelation sequence as follows:

$$R_{XX}[n] = \frac{1}{2\pi}\int_{-\pi}^{\pi} S_{XX}(\Omega)e^{j\Omega n}d\Omega$$

The crosscorrelation function of $X[n]$ and the output discrete sequence $Y[n]$ is given by

$$\begin{aligned} R_{XY}[n, n+k] &= E[X[n]Y[n+k]] = E\left[X[n]\sum_{l=-\infty}^{\infty} h[l]X[n+k-l]\right] \\ &= \sum_{l=-\infty}^{\infty} h[l]E[X[n]X[n+k-l]] = \sum_{k=-\infty}^{\infty} h[l]R_{XX}[n, n+k-l] \end{aligned}$$

If $X[n]$ is wide-sense stationary, then we have that

$$R_{XY}[n, n+k] = \sum_{k=-\infty}^{\infty} h[l]R_{XX}[k-l] = h[k] * R_{XX}[k] = R_{XY}[k] \tag{11.17}$$

For the case where $X[n]$ is a wide-sense stationary random sequence, the cross-power spectral density is given by

$$S_{XY}(\Omega) = \sum_{k=-\infty}^{\infty} R_{XY}[k]e^{-j\Omega k} = \sum_{k=-\infty}^{\infty} \{h[k] * R_{XX}[k]\}e^{-j\Omega k} = H(\Omega)S_{XX}(\Omega) \tag{11.18}$$

The autocorrelation function of the output discrete sequence $Y[n]$ is given by

$$\begin{aligned} R_{YY}[n, n+k] &= E[Y[n]Y[n+k]] = E\left[\sum_{m=-\infty}^{\infty} h[m]X[n-m]\sum_{l=-\infty}^{\infty} h[l]X[n+k-l]\right] \\ &= \sum_{m=-\infty}^{\infty}\sum_{l=-\infty}^{\infty} h[m]h[l]E[X[n-m]X[n+k-l]] \\ &= \sum_{m=-\infty}^{\infty}\sum_{l=-\infty}^{\infty} h[m]h[l]R_{XX}[n-m, n+k-l] \end{aligned}$$

If $X[n]$ is a wide-sense stationary discrete sequence, then we obtain

$$R_{YY}[n, n+k] = \sum_{m=-\infty}^{\infty} \sum_{l=-\infty}^{\infty} h[m]h[l]R_{XX}[k+m-l] = R_{YY}[k] \tag{11.19}$$

In a manner similar to the continuous-time case, it can be shown that the power spectral density of $Y[n]$ is given by

$$S_{YY}(\Omega) = |H(\Omega)|^2 S_{XX}(\Omega) \tag{11.20}$$

## EXAMPLE 11.5

The impulse response of a discrete linear time-invariant system is given by

$$h[n] = a^n u[n]$$

where $|a| < 1$, and $u[n]$ is the unit step sequence defined by

$$u[n] = \begin{cases} 1 & n \geq 0 \\ 0 & n < 0 \end{cases}$$

If the input sequence $X[n]$ is a discrete-time white noise with power spectral density $N_0/2$, find the power spectral density of the output $Y[n]$.

**Solution:**
The system response is given by

$$H(\Omega) = \sum_{n=-\infty}^{\infty} h[n]e^{-j\Omega n} = \sum_{n=0}^{\infty} a^n e^{-j\Omega n} = \sum_{n=0}^{\infty} \left(ae^{-j\Omega}\right)^n = \frac{1}{1-ae^{-j\Omega}}$$

Since $S_{XX}(\Omega) = N_0/2$, we have that

$$\begin{aligned} S_{YY}(\Omega) &= |H(\Omega)|^2 S_{XX}(\Omega) = H^*(\Omega)H(\Omega)S_{XX}(\Omega) = \frac{N_0}{2}\left(\frac{1}{1-ae^{-j\Omega}}\right)\left(\frac{1}{1-ae^{j\Omega}}\right) \\ &= \frac{N_0}{2[1-a(e^{j\Omega}+e^{-j\Omega})+a^2]} = \frac{N_0}{2[1-2a\cos(\Omega)+a^2]} \end{aligned}$$

## 11.5 AUTOREGRESSIVE MOVING AVERAGE PROCESS

The autoregressive moving average (ARMA) process is frequently used in time series analysis. It consists of two parts: the moving average process and the autoregressive process. Let $\{W[n], n \geq 0\}$ be a wide-sense stationary random input process with zero mean and variance $\sigma_W^2$. In general the $W[n]$ are assumed to be uncorrelated. An example of such a process is noise. The different processes are defined as follows.

### 11.5.1 Moving Average Process

A moving average process is a process whose current value $Y[n]$ depends linearly on the $q$ past values of the random input process. Thus, given a set of constants $\beta_0, \beta_1, \ldots, \beta_q$, the output process defined by

$$\begin{aligned} Y[n] &= \beta_0 W[n] + \beta_1 W[n-1] + \beta_2 W[n-2] + \cdots + \beta_q W[n-q] \\ &= \sum_{k=0}^{q} \beta_k W[n-k] \qquad n \geq 0 \end{aligned} \tag{11.21}$$

is called a moving average process of order $q$, MA($q$). The moving average process is a special case of the purely feedforward system called the *finite impulse response* (FIR) system with input $X[n]$ that has nonzero mean. The general structure of FIR systems is shown in Figure 11.4, where D indicates a unit delay.

Since the $W[n]$ are uncorrelated, the mean, variance, and autocorrelation function of MA($q$) are given by

$$E[Y[n]] = E\left[\sum_{k=0}^{q} \beta_k W[n-k]\right] = \sum_{k=0}^{q} \beta_k E[W[n-k]] = 0$$

$$\begin{aligned} \sigma_Y^2[n] &= E[\{Y[n] - E[Y[n]]\}^2] = E[Y^2[n]] = E\left[\left(\sum_{k=0}^{q} \beta_k W[n-k]\right)^2\right] \\ &= E\left[\sum_{k=0}^{q} \beta_k W[n-k] \sum_{m=0}^{q} \beta_m W[n-m]\right] = \sum_{k=0}^{q}\sum_{m=0}^{q} \beta_k \beta_m E[W[n-k]W[n-m]] \end{aligned}$$

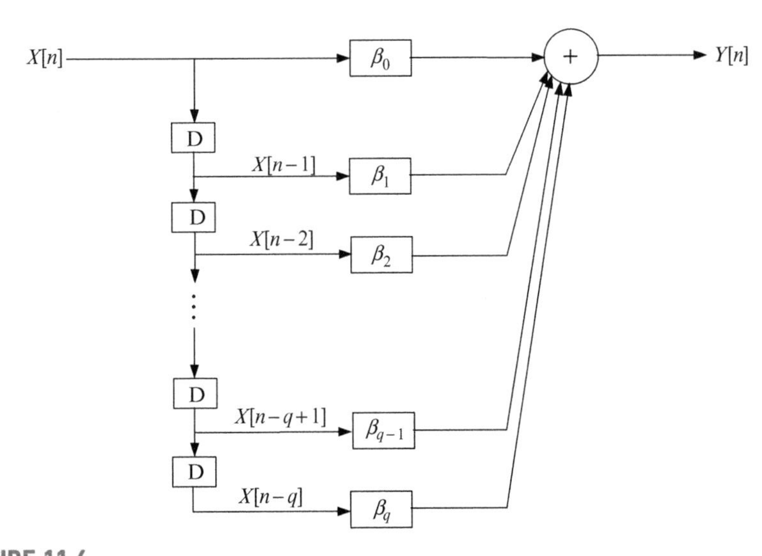


**FIGURE 11.4**

Structure of a Finite Impulse Response System

Since we know that

$$E[W[m]W[k]] = \begin{cases} \sigma_W^2 & m = k \\ 0 & m \neq k \end{cases}$$

the quantity $E[W[n-k]W[n-m]]$ will have a nonzero value $\sigma_W^2$ when $k=m$ and a zero value elsewhere. Thus, we have that

$$\sigma_Y^2[n] = \sigma_W^2 \sum_{k=0}^{q} \beta_k^2 \tag{11.22}$$

In the same way, the autocorrelation function is given by

$$\begin{aligned} R_{YY}[n, n+m] &= E[Y[n]Y[n+m]] = E\left[\sum_{k=0}^{q} \beta_k W[n-k] \sum_{l=0}^{q} \beta_l W[n+m-l]\right] \\ &= \sum_{k=0}^{q} \sum_{l=0}^{q} \beta_k \beta_l E[W[n-k]\, W[n+m-l]] \end{aligned}$$

Since the quantity $E[W[n-k]\, W[n+m-l]]$ has a nonzero value $\sigma_W^2$ when $k=l-m$ (that is, when $l=k+m$) and a zero value elsewhere, we have that

$$R_{YY}[n, n+m] = \sum_{k=0}^{q-m} \beta_k \beta_{k+m} E\left[W^2[n-k]\right] = \sigma_W^2 \sum_{k=0}^{q-m} \beta_k \beta_{k+m} \qquad 0 \leq m \leq q$$

For $m > q$, the autocorrelation function is zero because there is no overlap in the products. Note that the autocorrelation function depends only on $m$ and not on $n$. Therefore, $Y[n]$ is a wide-sense stationary process.

Note also that if $S_{XX}(\Omega)$ is the discrete-time Fourier transform of $R_{XX}[m]$, then $S_{XX}(\Omega)e^{-j\Omega m_0}$ is the discrete-time Fourier transform of $R_{XX}[m-m_0]$, where $m_0$ is a constant. Thus, the crosscorrelation function of $W[n]$ and $Y[n]$ is given by

$$\begin{aligned} R_{WY}[n, n+m] &= E[W[n]Y[n+m]] = E\left[W[n] \sum_{k=0}^{q} \beta_k W[n+m-k]\right] \\ &= \sum_{k=0}^{q} \beta_k E[W[n]W[n+m-k]] = \sum_{k=0}^{q} \beta_k R_{WW}[m-k] = R_{WY}[m] \end{aligned}$$

This means that the cross-power spectral density is given by

$$\begin{aligned} S_{WY}(\Omega) &= \sum_{m=-\infty}^{\infty} R_{WY}[m] e^{-j\Omega m} = \sum_{m=-\infty}^{\infty} \sum_{k=0}^{q} \beta_k R_{WW}[m-k] e^{-j\Omega m} \\ &= \sum_{k=0}^{q} \beta_k e^{-j\Omega k} \sum_{m=-\infty}^{\infty} R_{WW}[m-k] e^{-j\Omega(m-k)} \\ &= S_{WW}(\Omega) \sum_{k=0}^{q} \beta_k e^{-j\Omega k} \end{aligned} \tag{11.23}$$

From this we see that the transfer function of the linear system defined by the MA($q$) is given by

$$H(\Omega) = \frac{S_{WY}(\Omega)}{S_{WW}(\Omega)} = \sum_{k=0}^{q} \beta_k e^{-j\Omega k} \tag{11.24}$$

Thus, the MA(q) process has a transfer function that is similar to that of the finite impulse response (or non-recursive) filter. This is not surprising, since the moving average process equation is a non-recursive equation; that is, we do not recursively use previously known values of the output process to compute the present value of the output process.

## EXAMPLE 11.6

Find the variance and autocorrelation function of the first-order moving average process MA(1).

Solution:

The first-order moving average process is given by

$$Y[n] = \beta_0 W[n] + \beta_1 W[n-1] \qquad n \geq 0$$

Thus, the variance and autocorrelation function are given by

$$\sigma_Y^2[n] = \sigma_W^2 \sum_{k=0}^{1} \beta_k^2 = \sigma_W^2 \{\beta_0^2 + \beta_1^2\}$$

$$R_{YY}[n, n+m] = \sigma_W^2 \sum_{k=0}^{1-m} \beta_k \beta_{k+m}$$

where $0 \leq m \leq 1$. This gives

$$R_{YY}[n, n] = R_{YY}[0] = \sigma_W^2 \{\beta_0^2 + \beta_1^2\}$$

$$R_{YY}[n, n+1] = R_{YY}[1] = \sigma_W^2 \beta_0 \beta_1$$

$$R_{YY}[n, n+k] = 0 \qquad k > 1$$

Thus, the power spectral density of the MA(1) is given by

$$S_{YY}(\Omega) = \sum_{m=-\infty}^{\infty} R_{YY}[m] e^{-j\Omega m} = \sigma_W^2 \{\beta_0^2 + \beta_1^2\} + \sigma_W^2 \beta_0 \beta_1 e^{-j\Omega}$$

### 11.5.2 Autoregressive Process

Given a set of constants $\beta_0, \alpha_1, \ldots, \alpha_p$, the output process defined by

$$\begin{aligned} Y[n] &= \alpha_1 Y[n-1] + \alpha_2 Y[n-2] + \cdots + \alpha_p Y[n-p] + \beta_0 W[n] \\ &= \sum_{k=1}^{p} \alpha_k Y[n-k] + \beta_0 W[n] \qquad n \geq 0 \end{aligned} \tag{11.25}$$

is called an autoregressive process of order $p$, AR($p$), where, as stated earlier, $\{W[n], \ n \geq 0\}$ is a wide-sense stationary random input process with zero mean and variance $\sigma_W^2$. It is called "autoregressive" because $Y[n]$ regresses on itself, which means that the value of $Y[n]$ can be written as a linear function of its own past $p$ values rather than the past values of the input random process. The autoregressive process is a special case of the feedback system called the *infinite impulse response* (IIR) system with input $X[n]$ that has nonzero mean. The general structure of IIR systems is shown in Figure 11.5, where D indicates a unit delay.

One example of an AR($p$) process is the behavior of a company's stock in the stock market. Suppose we can approximate the stock price by the following model:

Stock price on day $k$ = Stock price on day $(k-1)$ + Random events on day $k$

Then we can model it as an AR(1) process. If the stock price depends on the previous 3 days' prices, for example, then it can be modeled as an AR(3) process.

The mean of $Y[n]$ is given by

$$E[Y[n]] = E\left[\sum_{k=1}^{p} \alpha_k Y[n-k] + \beta_0 W[n]\right] = \sum_{k=1}^{p} \alpha_k E[Y[n-k]] + \beta_0 E[W[n]] \ n \geq 0$$

Since $E[W[n]] = 0$ and $Y[n] = 0$ for $n < 0$, we can solve the above equation recursively to obtain

$$E[Y[n]] = 0$$

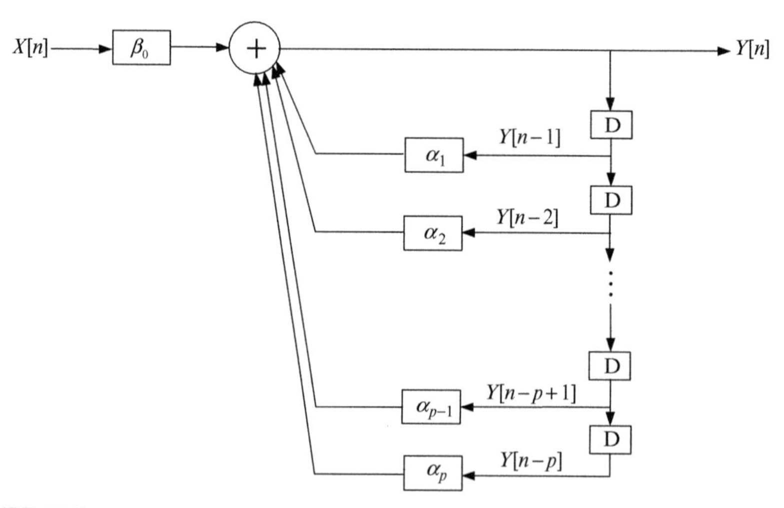


**FIGURE 11.5**

Structure of an Infinite Impulse Response System

Similarly, it can be shown that the variance is given by

$$\sigma_Y^2[n] = \sum_{m=0}^{p}\sum_{k=0}^{p} \alpha_k \alpha_m R_{YY}[n-k, n-m] + \beta_0^2 \sigma_W^2 \tag{11.26}$$

Thus, the variance is a function of various preceding autocorrelation functions at various time indices. The autocorrelation function at $n$ and $n$ can be shown to be equal to

$$R_{YY}[n, n] = \sum_{k=0}^{p} \alpha_k R_{YY}[n, n-k] + \beta_0^2 \sigma_W^2 \tag{11.27}$$

Finally, it can be shown that the transfer function of the linear system defined by the AR($p$) is given by

$$H(\Omega) = \frac{\beta_0}{1 - \sum_{k=0}^{p} \alpha_k e^{-j\Omega k}} \tag{11.28}$$

Thus, the transfer function is similar to that of a recursive (or infinite response) filter. This is so because the equation for the AR($p$) process is a recursive equation. Consequently it has an impulse response of infinite duration as long as at least one of the $\alpha_k$, $k = 1, 2, \ldots, p$, is nonzero.

## EXAMPLE 11.7

Find the autocorrelation function and variance for the first-order autoregressive process, AR(1).

Solution:

AR(1) is defined by

$$\begin{aligned} Y[n] &= \alpha_1 Y[n-1] + \beta_0 W[n] = \alpha_1\{\alpha_1 Y[n-2] + \beta_0 W[n-1]\} + \beta_0 W[n] \\ &= \alpha_1^2\{\alpha_1 Y[n-3] + \beta_0 W[n-2]\} + \beta_0 \alpha_1 W[n-1] + \beta_0 W[n] \\ &= \beta_0 W[n] + \beta_0 \alpha_1 W[n-1] + \beta_0 \alpha_1^2 W[n-2] + \cdots \\ &= \beta_0 \sum_{k=0}^{\infty} \alpha_1^k W[n-k] \end{aligned}$$

From this we obtain

$$E[Y[n]] = E\left[\beta_0 \sum_{k=0}^{\infty} \alpha_1^k W[n-k]\right] = \beta_0 \sum_{k=0}^{\infty} \alpha_1^k E[W[n-k]] = 0$$

$$\sigma_Y^2[n] = \sigma_W^2 \beta_0^2 \{1 + \alpha_1^2 + \alpha_1^4 + \cdots\}$$

Thus, the variance is finite provided $|\alpha_1| < 1$. Under this condition we have that

$$\sigma_Y^2[n] = \frac{\sigma_W^2 \beta_0^2}{1 - \alpha_1^2}$$

The autocorrelation function is given by

$$R_{YY}[n,n+m]=E[Y[n]Y[n+m]]=E\left[\beta_0\sum_{k=0}^{\infty}\alpha_1^k W[n-k]\beta_0\sum_{l=0}^{\infty}\alpha_1^l W[n+m-l]\right]$$
$$=\beta_0^2\sum_{k=0}^{\infty}\sum_{l=0}^{\infty}\alpha_1^k\alpha_1^l E[W[n-k]W[n+m-l]]$$

Since $E[W[m]W[n]]=0$ if $m\neq n$ and $E[W[m]W[n]]=\sigma_W^2$ if $m=n$, we have that

$$E[W[n-k]W[n+m-l]]=\begin{cases}\sigma_W^2 & \text{if } l=m+k\\ 0 & \text{otherwise}\end{cases}$$

Thus, we have that

$$R_{YY}[n,n+m]=\sigma_W^2\beta_0^2\sum_{k=0}^{\infty}\alpha_1^k\alpha_1^{m+k}=\sigma_W^2\beta_0^2\alpha_1^m\sum_{k=0}^{\infty}\alpha_1^{2k}=\frac{\sigma_W^2\beta_0^2}{1-\alpha_1^2}\alpha_1^m$$
$$=\sigma_Y^2[n]\alpha_1^m \qquad m\geq 0$$

Therefore, the power spectral density of AR(1) is given by

$$S_{YY}(\Omega)=\sum_{m=-\infty}^{\infty}R_{YY}[m]e^{-j\Omega m}=\frac{\sigma_W^2\beta_0^2}{1-\alpha_1^2}\sum_{k=0}^{\infty}\alpha_1^m e^{-j\Omega m}=\frac{\sigma_W^2\beta_0^2}{(1-\alpha_1^2)(1-\alpha_1 e^{-j\Omega})}$$

### 11.5.3 ARMA Process

The ARMA process of order $(p, q)$ is obtained by combining an MA($q$) process and an AR($p$) processes. That is, it contains $p$ AR terms and $q$ MA terms and is given by

$$Y[n]=\sum_{k=1}^{p}\alpha_k Y[n-k]+\sum_{k=0}^{q}\beta_k W[n-k] \quad n\geq 0 \tag{11.29}$$

A structural representation of the ARMA process is a combination of the structures shown in Figure 11.4 and Figure 11.5, as shown in Figure 11.6.

One of the advantages of ARMA is that a stationary random sequence (or time series) may be more adequately modeled by an ARMA model involving fewer parameters than a pure MA or AR process alone. Since $E[W[n-k]]=0$ for $k=0,1,2,\ldots,q$, it is easy to show that $E[Y[n]]=0$. Similarly, it can be shown that the variance of $Y[n]$ is given by

$$\sigma_Y^2[n]=\sum_{k=1}^{p}\alpha_k R_{YY}[n,n-k]+\sum_{k=0}^{q}\beta_k R_{YW}[n,n-k] \tag{11.30}$$

Thus, the variance is obtained as the weighted sum of the autocorrelation function evaluated at different times and the weighted sum of various

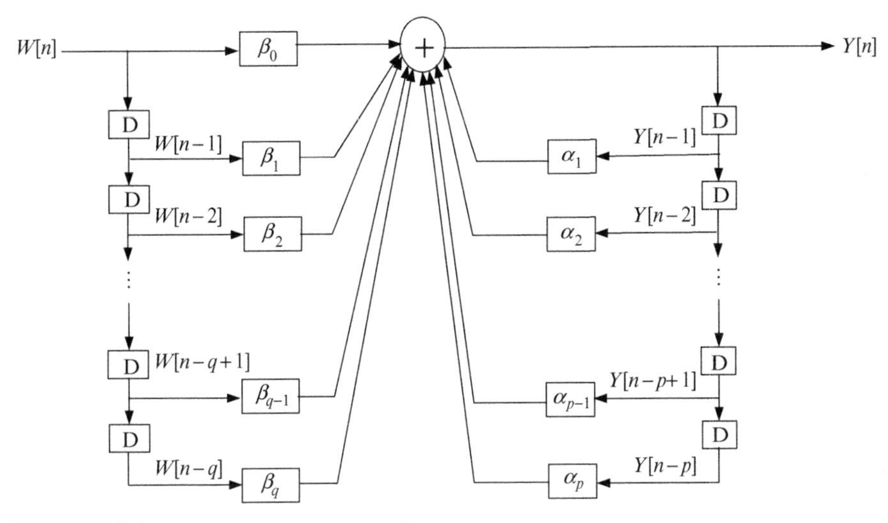


**FIGURE 11.6**
Structure of an ARMA Process

crosscorrelation functions at different times. Finally, it can be shown that the transfer function of the linear system defined by the ARMA($p$, $q$) is given by

$$H(\Omega) = \frac{\sum_{k=0}^{q} \beta_k e^{-j\Omega k}}{1 - \sum_{k=0}^{p} \alpha_k e^{-j\Omega k}} \tag{11.31}$$

## 11.6 CHAPTER SUMMARY

This chapter has dealt with the response of linear time-invariant systems to random inputs. Both continuous-time and discrete-time systems were considered. Both the output power spectral density and the cross-power spectral density between the input and output processes were obtained. The power spectral density $S_{XX}(\Omega)$ of a discrete-time input process was shown to be periodic with a period of $2\pi$. This means that it is sufficient to define $S_{YY}(\Omega)$ only in the range $(-\pi, \pi)$. The chapter also introduced the autoregressive process, the moving average process, and the autoregressive moving average process.

## 11.7 PROBLEMS

### Section 11.2 Linear Systems with Deterministic Input

11.1 Find the Fourier transform of the following "sawtooth" function $x(t)$ defined in the interval $[-T, T]$:

$$x(t)=\begin{cases}1+\dfrac{t}{T} & -T\leq t\leq 0\\ 1-\dfrac{t}{T} & 0\leq t\leq T\end{cases}$$

11.2 Consider a system that performs a *differentiation* of the input function. This means that when the input function is $x(t)$, the output function is

$$y(t)=\frac{d}{dt}x(t)$$

Find the Fourier transform of $y(t)$ in terms of the Fourier transform of $x(t)$.

11.3 Consider a system that performs a *complex modulation* on an input function. This means that when the input function is $x(t)$, the output function is

$$y(t)=e^{jw_0t}x(t)$$

where $w_0$ is a constant. Find the Fourier transform of $y(t)$ in terms of the Fourier transform of $x(t)$.

11.4 Consider a system that introduces a *delay* of $t_0$ to an input function. This means that when the input function is $x(t)$, the output function is

$$y(t)=x(t-t_0)$$

where $t_0>0$ is a constant. Find the Fourier transform of $y(t)$ in terms of the Fourier transform of $x(t)$.

11.5 Consider a system that performs a *scaling* operation on an input function. This means that when the input function is $x(t)$, the output function is

$$y(t)=x(at)$$

where $a>0$ is a constant. Find the Fourier transform of $y(t)$ in terms of the Fourier transform of $x(t)$.

## Section 11.3 Linear Systems with Continuous Random Input

11.6 A stationary zero-mean random signal $X(t)$ is the input to two filters, as shown in Figure 11.7.

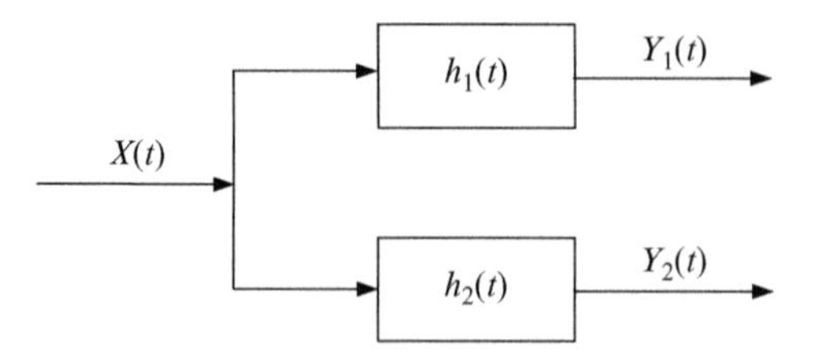


**FIGURE 11.7**
Figure for Problem 11.6

The power spectral density of $X(t)$ is $S_{XX}(w) = N_0/2$, and the filter impulse responses are given by

$$h_1(t) = \begin{cases} 1 & 0 \le t < 1 \\ 0 & \text{otherwise} \end{cases}$$

$$h_2(t) = \begin{cases} 2e^{-t} & t \ge 0 \\ 0 & \text{otherwise} \end{cases}$$

Determine the following:

a. The mean $E[Y_i(t)]$ and second moment $E[Y_i^2(t)]$ of the output signal $Y_i(t)$ for $i = 1, 2$

b. The crosscorrelation function $R_{Y_1Y_2}(t, t+\tau)$

11.7 A wide-sense stationary process $X(t)$ is the input to a linear system whose impulse response is $h(t) = 2e^{-7t}$, $t \ge 0$. If the autocorrelation function of the process is $R_{XX}(\tau) = e^{-4|\tau|}$ and the output process is $Y(t)$, find the following:

a. The power spectral density of $Y(t)$

b. The cross-spectral power density $S_{XY}(w)$

c. The crosscorrelation function $R_{XY}(\tau)$

11.8 A linear system has a transfer function given by

$$H(w) = \frac{w}{w^2 + 15w + 50}$$

Determine the power spectral density of the output when the input function is

a. a stationary random process $X(t)$ with an autocorrelation function. $R_{XX}(\tau) = 10e^{-|\tau|}$

b. white noise that has a mean-square value of 1.2 $V^2/Hz$

11.9 A linear system has the impulse response $h(t) = e^{-at}$, where $t \ge 0$ and $a > 0$. Find the power transfer function of the system.

11.10 Consider the system with the impulse response $h(t) = e^{-at}$, where $t \ge 0$ and $a > 0$. Assume that the input is white noise with power spectral density $N_0/2$. What is the power spectral density of the output process?

11.11 The power transfer function of a system is given by

$$|H(w)|^2 = \frac{64}{[16 + w^2]^2}$$

Use Table 10.1 to obtain the impulse function $h(t)$ of the system.

11.12 A wide-sense stationary process $X(t)$ has the autocorrelation function given by

$$R_{XX}(\tau) = \cos(w_0 \tau)$$

The process is input to a system with the power transfer function

$$|H(w)|^2 = \frac{64}{[16 + w^2]^2}$$

a. Find the power spectral density of the output process.
b. If $Y(t)$ is the output process, find the cross-power spectral density $S_{XY}(w)$.

11.13 A causal system is used to generate an output process $Y(t)$ with the power spectral density

$$S_{YY}(w) = \frac{2a}{a^2 + w^2}$$

Find the impulse response $h(t)$ of the system.

11.14 $X(t)$ is a wide-sense stationary process. It is the input to a linear system with impulse response $h(t)$, and $Y(t)$ is the output process. Consider another process $Z(t)$ that is obtained as follows: $Z(t) = X(t) - Y(t)$. The scheme is illustrated in Figure 11.8.
Determine the following in terms of the parameters of $X(t)$:
a. The autocorrelation function $R_{ZZ}(\tau)$
b. The power spectral density $S_{ZZ}(w)$
c. The crosscorrelation function $R_{XZ}(\tau)$
d. The cross-power spectral density $S_{XZ}(w)$

11.15 Consider the system shown in Figure 11.9 in which an output process $Y(t)$ is the sum of an input process $X(t)$ and a delayed version of $X(t)$ that is scaled (or multiplied) by a factor $a$.

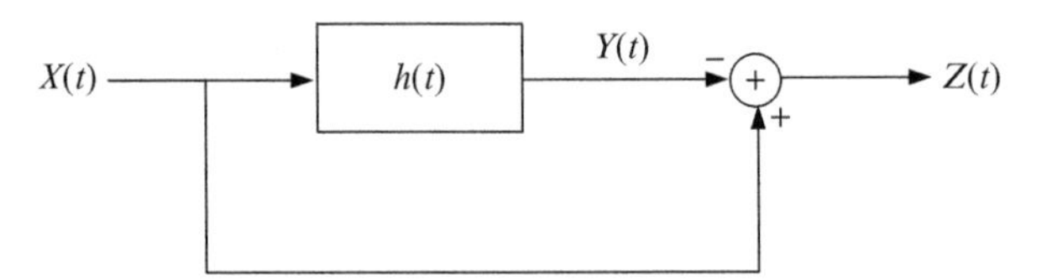


**FIGURE 11.8**
Figure for Problem 11.14

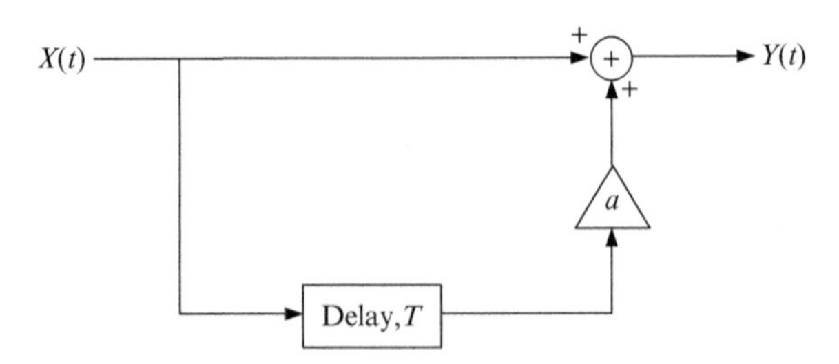


**FIGURE 11.9**
Figure for Problem 9.15

Determine the following:

a. The equation that governs the system (i.e., the equation that relates $Y(t)$ to $X(t)$)
b. The crosscorrelation function $R_{XY}(\tau)$
c. The cross-power spectral density $S_{XY}(w)$
d. The transfer function $H(w)$ of the system
e. The power spectral density of $Y(t)$

11.16 $X(t)$ and $Y(t)$ are two jointly wide-sense stationary processes. If $Z(t)=X(t)+Y(t)$ is the input to a linear system with impulse response $h(t)$, determine the following:

a. The autocorrelation function of $Z(t)$
b. The power spectral density of $Z(t)$
c. The cross-power spectral density $S_{ZV}(w)$ of the input process $Z(t)$ and the output process $V(t)$
d. The power spectral density of the output process $V(t)$.

11.17 $X(t)$ is a wide-sense stationary process. Assume that $Z(t)=X(t-d)$, where $d$ is a constant delay. If $Z(t)$ is the input to a linear system with impulse response $h(t)$, determine the following:

a. The autocorrelation function of $Z(t)$
b. The power spectral density $S_{ZZ}(w)$
c. The crosscorrelation function $R_{ZX}(\tau)$
d. The cross-power spectral density $S_{ZX}(w)$
e. The power spectral density $S_{YY}(w)$ of the output process $Y(t)$, which is obtained by passing $Z(t)$ through a linear system with the system response $H(w)$

11.18 $X(t)$ is a wide-sense stationary process that is the input to a linear system with the transfer function

$$H(w)=\frac{1}{a+jw}$$

where $a>0$. If $X(t)$ is a zero-mean white noise with power spectral density $N_0/2$, determine the following:

a. The impulse response $h(t)$ of the system
b. The cross-power spectral density $S_{XY}(w)$ of the input process and the output process $Y(t)$
c. The crosscorrelation function $R_{XY}(\tau)$ of $X(t)$ and $Y(t)$
d. The crosscorrelation function $R_{YX}(\tau)$ of $Y(t)$ and $X(t)$
e. The cross-power spectral density $S_{YX}(w)$ of $Y(t)$ and $X(t)$
f. The power spectral density $S_{YY}(w)$ of the output process

## Section 11.4 Linear Systems with Discrete Random Input

11.19 A linear system has an impulse response given by

$$h[n]=\begin{cases} e^{-an} & n\geq 0 \\ 0 & n<0 \end{cases}$$

where $a>0$ is a constant. Find the transfer function of the system.

11.20 A linear system has an impulse response given by

$$h[n]=\begin{cases} e^{-an} & n\geq 0 \\ 0 & n<0 \end{cases}$$

where $a>0$ is a constant. Assume that the autocorrelation function of the input sequence to this system is defined by

$$R_{XX}[n]=b^n \qquad 0<b<1,\ n\geq 0$$

Find the power spectral density of the output process.

11.21 The autocorrelation function of a discrete-time random sequence $X[n]$ is given by

$$R_{XX}[m]=e^{-b|m|}$$

where $b>0$ is a constant. Find the power spectral density of the sequence.

11.22 A linear system has an impulse response given by

$$h[n]=\begin{cases} e^{-an} & n\geq 0 \\ 0 & n<0 \end{cases}$$

where $a>0$ is a constant. Assume that the autocorrelation function of the input discrete-time random sequence $X[n]$ is given by

$$R_{XX}[m]=e^{-b|m|}$$

where $b>0$ is a constant. Find the power spectral density of the output process.

11.23 A wide-sense stationary continuous-time process has the autocorrelation function given by

$$R_{X_C X_C}(\tau)=e^{-4|\tau|}$$

If $X_C(t)$ is sampled with a sampling period 10 seconds to produce the discrete-time process $X[n]$, find the power spectral density of $X[n]$.

11.24 A wide-sense stationary continuous-time process has the autocorrelation function given by

$$R_{X_C X_C}(\tau)=e^{-4|\tau|}$$

$X_C(t)$ is sampled with a sampling period 10 seconds to produce the discrete-time sequence. The sequence is then input to a system with the impulse response

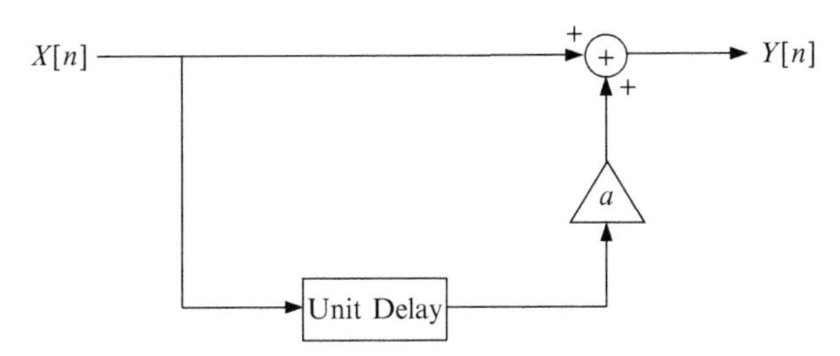


**FIGURE 11.10**
Figure for Problem 11.25

$$h[n] = \begin{cases} e^{-an} & n \geq 0 \\ 0 & n < 0 \end{cases}$$

where $a > 0$ is a constant. Find the power spectral density of the output process.

11.25 Consider the system shown in Figure 11.10 in which an output sequence $Y[n]$ is the sum of an input sequence $X[n]$ and a version of $X[n]$ that has been delayed by one unit and scaled (or multiplied) by a factor $a$. Determine the following:

a. The equation that governs the system
b. The crosscorrelation function $R_{XY}[m]$
c. The cross-power spectral density $S_{XY}(\Omega)$
d. The transfer function $H(\Omega)$ of the system

## Section 11.5 Autoregressive Moving Average Processes

11.26 A discrete-time feedback control system has the property that its output voltage $Y[n]$ at time $n$ is a linear combination of the output voltage at time $n-1$ scaled by a factor $a$ and a random error $W[n]$ at time $n$ that is independent of past outputs, as shown in Figure 11.11, where $|a| < 1$. The random process $W[n]$ is a sequence of independent and identically distributed random variables with zero mean and standard deviation $\beta$. Assume also that the random process $Y[n]$ has zero mean and $W[n] = 0$ for $n < 0$.

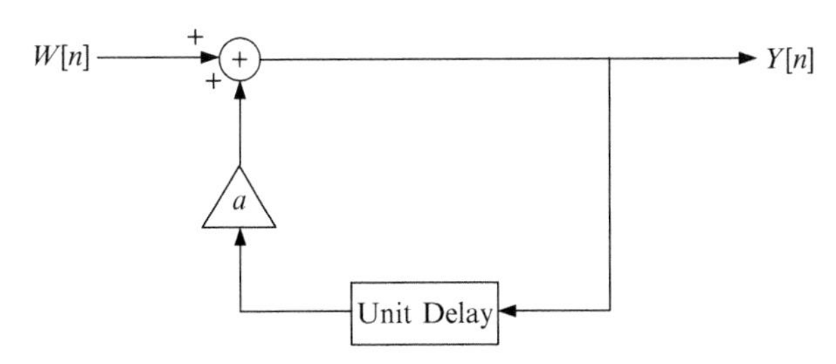


**FIGURE 11.11**
Figure for Problem 11.26

a. Determine the equation that governs the system
b. Is the output process $Y[n]$ a wide-sense stationary?
c. If $Y[n]$ is a wide-sense stationary process, find the power transfer function
d. Find the crosscorrelation function $R_{WY}[n, n+m]$
e. Find the autocorrelation function $R_{WW}[n, n+m]$

11.27 Find the mean, autocorrelation function and variance of the MA(2) process, assuming that $n > 2$ for the output process $Y[n]$.

11.28 Find the autocorrelation function of the following MA(2) process:

$$Y[n] = W[n] + 0.7W[n-1] - 0.2W[n-2]$$

11.29 Find the autocorrelation function of the following AR(2) process:

$$Y[n] = 0.7Y[n-1] - 0.2Y[n-2] + W[n]$$

11.30 Consider the following ARMA(1,1) process, where $|\alpha| < 1$, $|\beta| < 1$, and $Y[n] = 0$ for $n < 0$:

$$Y[n] = \alpha Y[n-1] + W[n] + \beta W[n-1]$$

Assume that $W[n]$ is a zero-mean random process with $W[n] = 0$ when $n < 0$, and the variance is $E[W[n]W[k]] = \sigma_W^2 \delta[n-k]$.

a. Carefully find a general expression for the $Y[n]$ in terms of only $W[n]$ and its delayed versions.
b. Using the above results, find the autocorrelation function of the ARMA(1, 1) process.

11.31 Write out the expression for the MA(5) process.

11.32 Write out the expression for the AR(5) process.

11.33 Write out the expression for the ARMA(4, 3) process.

CHAPTER 12

# Special Random Processes

## 12.1 INTRODUCTION

Chapter 10 deals with the definition and characterization of random processes. Chapter 11 deals with the response of linear systems when their inputs are random processes. In this chapter we consider some well-known random processes. These include the Bernoulli process, random walk process, Gaussian process, Poisson process, and Markov process.

## 12.2 THE BERNOULLI PROCESS

Consider a sequence of independent Bernoulli trials, such as coin tossing, where for each trial the probability of success is $p$ and the probability of failure is $1-p$. Let $X_i$ be the random variable that denotes the outcome of the $i$th trial and let it take values as follows: $X_i=1$ if a success occurs and $X_i=0$ if a failure occurs. Then the PMF of $X_i$ is given by

$$p_{X_i}(x) = \begin{cases} 1-p & x=0 \\ p & x=1 \end{cases}$$

The Bernoulli random variable is used when an experiment has only two outcomes: on/off, yes/no, success/failure, working/broken, hit/miss, early/late, heads/tails, and so on. The sequence of random variables $\{X_i, i=1,2,\ldots\}$ resulting from, say, tossing the same coin many times, is called a Bernoulli process. In such a process we may be interested in the number of successes in a given number of trials, the number of trials until the first success, or the number trials until the $k$th success.

Let the random variable $Y_n$ be defined as follows:

$$Y_n = \sum_{i=1}^{n} X_i \qquad n=1,2,\ldots$$

Then $Y_n$ denotes the number of successes in $n$ Bernoulli trials, which we know from Chapter 4 to be a *binomial random variable*. That is, the PMF of $Y_n$ is given by



Fundamentals of Applied Probability and Random Processes. http://dx.doi.org/10.1016/B978-0-12-800852-2.00012-2

$$p_{Y_n}(k) = \binom{n}{k} p^k (1-p)^{n-k} \qquad k = 0, 1, \ldots, n$$

Let $L_1$ be the random variable that denotes the arrival time of the first success; that is, $L_1$ is the number of times up to and including that trial in which the first success occurs. From Chapter 4 we know that $L_1$ is a geometrically distributed random variable with parameter $p$; that is, the PMF of $L_1$ is given by

$$p_{L_1}(l) = p(1-p)^{l-1} \qquad l = 1, 2, \ldots$$

Also as stated in Chapter 4, $L_1$ is a random variable that has no memory. That is, if we have observed a fixed number $n$ of Bernoulli trials and they are all failures, the number $K$ of additional trials until the first success has the PMF

$$\begin{aligned} p_{K|L_1>n}(k|L_1 > n) &= P[K = k|L_1 > n] = P[L_1 - n = k|L_1 > n] = p(1-p)^{k-1} \\ &= p_{L_1}(k) \end{aligned}$$

Finally, the number of trials up to and including that in which the $k$th success occurs is known to be a $k$th-order *Pascal random variable* $L_k$ whose PMF is given by

$$p_{L_k}(n) = \binom{n-1}{k-1} p^k (1-p)^{n-k} \qquad k = 1, 2, \ldots; \ n = k, k+1, \ldots$$

## EXAMPLE 12.1

Consider a sequence of independent tosses of a coin with probability $p$ of heads in any toss. Let $Y_n$ denote the number of heads in $n$ consecutive tosses of the coin. Evaluate the probability of the following event: $Y_5 = 3,\ Y_8 = 5, Y_{14} = 9$.

**Solution:**

We solve the problem by considering the following non-overlapping intervals and the associated numbers of successes:

$$Y_5 = 3,\ Y_8 - Y_5 = 2, Y_{14} - Y_8 = 4$$

Since these are non-overlapping intervals, $Y_5$, $Y_8 - Y_5$ and $Y_{14} - Y_8$ are independent binomial random variables. Thus, we have that

$$\begin{aligned} P[Y_5 = 3,\ Y_8 = 5, Y_{14} = 9] &= P[Y_5 = 3,\ Y_8 - Y_5 = 2, Y_{14} - Y_8 = 4] \\ &= P[Y_5 = 3]P[Y_8 - Y_5 = 2]P[Y_{14} - Y_8 = 4] \\ &= \binom{5}{3} p^3 (1-p)^2 \binom{3}{2} p^2 (1-p) \binom{6}{4} p^4 (1-p)^2 \\ &= 450 p^9 (1-p)^5 \end{aligned}$$

## 12.3 RANDOM WALK PROCESS

A random walk is derived from a sequence of Bernoulli trials as follows. Consider a Bernoulli trial in which the probability of success is $p$ and the probability of failure is $1-p$. Assume that the experiment is performed every $T$ time units, and let the random variable $X_k$ denote the outcome of the $k$th trial. Furthermore, assume that the PMF of $X_k$ is as follows:

$$p_{X_k}(x) = \begin{cases} 1-p & x=-1 \\ p & x=1 \end{cases}$$

Finally, let the random variable $Y_n$ be defined as follows:

$$Y_n = \sum_{k=1}^{n} X_k = Y_{n-1} + X_n \qquad n=1,2,\ldots \tag{12.1}$$

where $Y_0=0$. If we use $X_k$ to model a process where we take a step to the right if the outcome of the $k$th trial is a success and a step to the left if the outcome is a failure, then the random variable $Y_n$ represents the location of the process relative to the starting point (or origin) at the end of the $n$th trial in a one-dimensional space. The resulting trajectory of the process as it moves through the $xy$ plane, where the $x$ coordinate represents the time and the $y$ coordinate represents the location at a given time, is called a one-dimensional *random walk*.

From equation (12.1) we can give a more general definition of a random walk as a process where the current value of a random variable is the sum of the past values and a new observation. Specifically, let $\{X_1, X_2, X_3, \ldots\}$ be a sequence of independent and identically distributed random variables. For each integer $n>0$, let $Y_n = X_1 + X_2 + \cdots + X_n$. Then the sequence of partial sums $\{Y_1, Y_2, Y_3, \ldots\}$ is called a random walk.

If we define the random process

$$Y(t) = Y_n \qquad n \le t \le n+1$$

then Figure 12.1 shows an example of the sample path of $Y(t)$, where the length of each step is $s$. It is a staircase with discontinuities at $t=kT,\ k=1,2,\ldots$.

Suppose that at the end of the $n$th trial there are exactly $k$ successes. Then there are $k$ steps to the right and $n-k$ steps to the left. Thus,

$$Y(nT) = ks - (n-k)s = (2k-n)s \equiv rs$$

where $r=2k-n$. This implies that $Y(nT)$ is a random variable that assumes values $rs$, where $r=n, n-2, n-4, \cdots, -n$. Since the event $\{Y(nT)=rs\}$ is the event $\{k \text{ successes in } n \text{ trials}\}$, where $k=(n+r)/2$, we have that

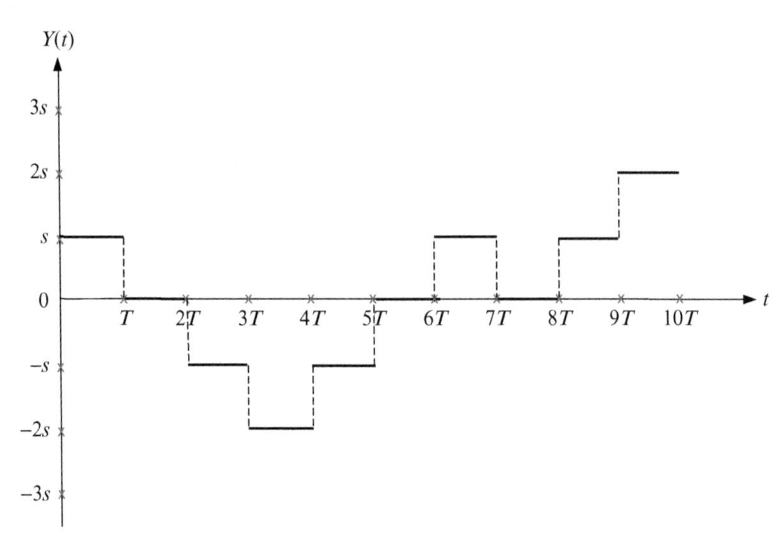


**FIGURE 12.1**
A Sample Path of the Random Walk

$$P[Y(nT)=rs]=P\left[\frac{n+r}{2}\text{ successes}\right]=\binom{n}{\frac{n+r}{2}}p^{\frac{n+r}{2}}(1-p)^{\frac{n-r}{2}} \tag{12.2}$$

Note that $(n+r)$ must be an even number. Also, since $Y(nT)$ is the sum of $n$ independent Bernoulli random variables, its mean and variance are given as follows:

$$E[Y(nT)]=nE[X_k]=n[ps-(1-p)s]=(2p-1)ns \tag{12.3a}$$

$$E\left[X_k^2\right]=ps^2+(1-p)s^2=s^2 \tag{12.3b}$$

$$\sigma^2_{Y(nT)}=n\sigma^2_{X_k}=n\left[s^2-s^2(2p-1)^2\right]=4p(1-p)ns^2 \tag{12.3c}$$

In the special case where $p=1/2, E[Y(nT)]=0$, and $\sigma^2_{Y(nT)}=ns^2$.

### 12.3.1 Symmetric Simple Random Walk

A simple random walk is a random walk where $X_i=1$ with probability $p$ and $X_i=-1$ with probability $1-p$ for $i=1,2,\ldots$. A symmetric random walk is a random walk in which $p=1/2$. Thus, a symmetric simple random walk is a random walk in which $X_i=1$ with probability $1/2$, and $X_i=-1$ with probability $1/2$. In a symmetric simple random walk,

$$P[Y_n=k]=\binom{n}{\frac{n+k}{2}}\left(\frac{1}{2}\right)^n \qquad k=-n,-n+2,\ldots,0,\ldots,n-2,n \tag{12.4a}$$

$$E[Y_n]=0 \tag{12.4b}$$

$$\sigma^2_{Y_n}=n \tag{12.4c}$$

### 12.3.2 Gambler's Ruin

The random walk described above assumes that the process can continue forever; in other words, it is unbounded. If the walk is bounded, then the ends of the walk are called *barriers*. These barriers can impose different characteristics on the process. For example, they can be *reflecting barriers*, which means that on hitting them the walk turns around and continues. They can also be *absorbing barriers*, which means that on hitting them the walk ends.

Consider the following random walk with absorbing barriers, which is generally referred to as the gambler's ruin. Suppose a gambler plays a sequence of independent games against an opponent. He starts out with $\$k$, and in each game he wins $\$1$ with probability $p$ and loses $\$1$ with probability $q=1-p$. When $p>q$, the game is advantageous to the gambler either because he is more skilled than his opponent or the rules of the game favor him. If $p=q$, the game is fair; and if $p<q$, the game is disadvantageous to the gambler.

Assume that the gambler stops when he has a total of $\$N$, which means he has additional $\$(N-k)$ over his initial $\$k$. (Another way to express this is that he plays against an opponent who starts out with $\$(N-k)$ and the game stops when either player has lost all his or her money.) We are interested in computing the probability $r_k$ that the player will be ruined (or he/she has lost all his or her money) after starting with $\$k$.

To solve the problem, we note that at the end of the first game, the player will have the sum of $\$(k+1)$ if he wins the game (with probability $p$) and the sum of $\$(k-1)$ if he loses the game (with probability $q$). Thus, if he wins the first game, the probability that he will eventually be ruined is $r_{k+1}$; and if he loses his first game, the probability that he will be ruined is $r_{k-1}$. There are two boundary conditions in this problem. First, $r_0=1$, since he cannot gamble when he has no money. Second, $r_N=0$, since he cannot be ruined. Thus, we obtain the following difference equation:

$$r_k = qr_{k-1} + pr_{k+1} \qquad 0<k<N$$

Since $p+q=1$, we obtain

$$(p+q)r_k = qr_{k-1} + pr_{k+1} \qquad 0<k<N$$

which we can write as

$$p(r_{k+1} - r_k) = q(r_k - r_{k-1})$$

From this we obtain the following:

$$r_{k+1} - r_k = (q/p)(r_k - r_{k-1}) \qquad 0<k<N$$

Thus,

$$\begin{aligned} r_2 - r_1 &= (q/p)(r_1 - r_0) = (q/p)(r_1 - 1) \\ r_3 - r_2 &= (q/p)(r_2 - r_1) = (q/p)^2(r_1 - 1) \\ r_4 - r_3 &= (q/p)(r_3 - r_2) = (q/p)^3(r_1 - 1) \\ &\vdots \\ r_{k+1} - r_k &= (q/p)^k(r_1 - 1) \end{aligned}$$

Now,

$$\begin{aligned} r_k - 1 = r_k - r_0 &= (r_k - r_{k-1}) + (r_{k-1} - r_{k-2}) + \cdots + (r_1 - 1) \\ &= \left[(q/p)^{k-1} + (q/p)^{k-2} + \cdots + 1\right](r_1 - 1) \\ &= \begin{cases} \dfrac{1-(q/p)^k}{1-(q/p)}(r_1 - 1) & p \neq q \\ k(r_1 - 1) & p = q \end{cases} \end{aligned}$$

The boundary condition $r_N = 0$ implies that

$$r_1 = \begin{cases} 1 - \dfrac{1-(q/p)}{1-(q/p)^N} & p \neq q \\ 1 - \dfrac{1}{N} & p = q \end{cases}$$

Thus,

$$r_k = \begin{cases} \dfrac{(q/p)^k - (q/p)^N}{1-(q/p)^N} & p \neq q \\ 1 - \dfrac{k}{N} & p = q \end{cases} \tag{12.5}$$

## EXAMPLE 12.2

A certain student wanted to travel during a break to visit his parents. The bus fare was \$20, but the student had only \$10. He figured out that there was a bar nearby where people play card games for money. The student signed up for one where he could bet \$1 per game. If he won the game, he would gain \$1; but if he lost the game, he would lose his \$1 bet. If the probability that he won a game is 0.6 independent of other games, what is the probability that he was not able to make the trip?

**Solution:**

In this example, $k = 10$ and $N = 20$. Define $a = q/p$, where $p = 0.6$ and $q = 1 - p = 0.4$. Thus, $a = 2/3$ and the probability that he was not able to make the trip is the probability that he was ruined given that he started with $k = 10$. This is $r_{10}$, which is given by

$$_{10} = \frac{(q/p)^{10} - (q/p)^{20}}{1 - (q/p)^{20}} = \frac{(2/3)^{10} - (2/3)^{20}}{1 - (2/3)^{20}} = 0.0170$$

hus, there is only a very small probability that he will not make the trip. In other words, he is very kely to make the trip.

## 12.4 THE GAUSSIAN PROCESS

Gaussian processes are important in many ways. First, many physical problems are the results of adding large numbers of independent random variables. According to the central limit theorem, such sums of random variables are essentially normal (or Gaussian) random variables. Also, the analysis of many systems is simplified if they are assumed to be Gaussian processes because of the properties of Gaussian processes. For example, noise in communication systems is usually modeled as a Gaussian process. Similarly, noise voltages in resistors are modeled as Gaussian processes.

A random process $\{X(t),\ t \in T\}$ is defined to be a Gaussian random process if and only if for any choice of $n$ real coefficients $a_1, a_2, \ldots, a_n$, and choice of $n$ time instants $t_1, t_2, \ldots, t_n$ in the index set $T$ the random variable $a_1X(t_1) + a_2X(t_2) + \ldots + a_nX(t_n)$ is a Gaussian (or normal) random variable. This definition implies that the random variables $X(t_1), X(t_2), \ldots, X(t_n)$ have a jointly normal PDF; that is,

$$f_{X(t_1)X(t_2)\cdots X(t_n)}(x_1, x_2, \ldots, x_n) = \frac{1}{(2\pi)^{n/2}|C_{XX}|^{1/2}} \exp\left[-\frac{(x-\mu_X)^T C_{XX}(x-\mu_X)}{2}\right] \quad (12.6)$$

where $\mu_X$ is the vector of the mean functions of the $X(t_k)$, $C_{XX}$ is the matrix of the autocovariance functions, $X$ is the vector of the $X(t_k)$, and $b^T$ denotes the transpose of $b$. That is,

$$\mu_X = \begin{bmatrix} \mu_X(t_1) \\ \mu_X(t_2) \\ \vdots \\ \mu_X(t_n) \end{bmatrix} \qquad X = \begin{bmatrix} X(t_1) \\ X(t_2) \\ \vdots \\ X(t_n) \end{bmatrix}$$

$$C_{XX} = \begin{bmatrix} C_{XX}(t_1, t_1) & C_{XX}(t_1, t_2) & \cdots & C_{XX}(t_1, t_n) \\ C_{XX}(t_2, t_1) & C_{XX}(t_2, t_2) & \cdots & C_{XX}(t_2, t_n) \\ \vdots & \vdots & \vdots & \vdots \\ C_{XX}(t_n, t_1) & C_{XX}(t_n, t_2) & \cdots & C_{XX}(t_n, t_n) \end{bmatrix}$$

If the $X(t_k)$ are mutually uncorrelated, then

$$C_{XX}(t_i, t_j) = \begin{cases} \sigma_X^2 & i=j \\ 0 & \text{otherwise} \end{cases}$$

In this case the autocovariance matrix and its inverse become

$$C_{XX} = \begin{bmatrix} \sigma_X^2 & 0 & \cdots & 0 \\ 0 & \sigma_X^2 & \cdots & 0 \\ \vdots & \vdots & \vdots & \vdots \\ 0 & 0 & \cdots & \sigma_X^2 \end{bmatrix} \qquad C_{XX}^{-1} = \begin{bmatrix} \frac{1}{\sigma_X^2} & 0 & \cdots & 0 \\ 0 & \frac{1}{\sigma_X^2} & \cdots & 0 \\ \vdots & \vdots & \vdots & \vdots \\ 0 & 0 & \cdots & \frac{1}{\sigma_X^2} \end{bmatrix}$$

Thus, we obtain

$$(x-\mu_X)^{\mathrm{T}} C_{XX}(x-\mu_X) = \sum_{k=1}^{n} \frac{[x_k - \mu_X(t_k)]^2}{\sigma_X^2}$$

$$f_{X(t_1)X(t_2)\cdots X(t_n)}(x_1, x_2, \ldots, x_n) = \frac{1}{(2\pi\sigma_X^2)^{n/2}} \exp\left[-\sum_{k=1}^{n} \frac{[x_k - \mu_X(t_k)]^2}{\sigma_X^2}\right]$$

If in addition to being mutually uncorrelated the random variables $X(t_1)$, $X(t_2), \ldots, X(t_n)$ have different variances such that $\mathrm{Var}(X(t_k)) = \sigma_k^2$, $1 \leq k \leq n$, then the covariance matrix and the joint PDF are given by

$$C_{XX} = \begin{bmatrix} \sigma_1^2 & 0 & \cdots & 0 \\ 0 & \sigma_2^2 & \cdots & 0 \\ \vdots & \vdots & \vdots & \vdots \\ 0 & 0 & \cdots & \sigma_n^2 \end{bmatrix}$$

$$f_{X(t_1)X(t_2)\cdots X(t_n)}(x_1, x_2, \ldots, x_n) = \frac{1}{(2\pi)^{n/2}\left(\prod_{k=1}^{n} \sigma_k\right)} \exp\left[-\sum_{k=1}^{n} \frac{[x_k - \mu_X(t_k)]^2}{\sigma_k^2}\right]$$

which implies that $X(t_1), X(t_2), \ldots, X(t_n)$ are also mutually independent. We list three important properties of Gaussian processes:

a. A Gaussian process that is wide-sense stationary is also strict-sense stationary.
b. If the input to a linear system is a Gaussian process, then the output is also a Gaussian process.
c. If the input $X(t)$ to a linear system is a zero-mean Gaussian process, the output process $Y(t)$ is also a zero-mean process. This property is the result of the following fact: As discussed in Chapter 11, if $h(t)$ is the impulse response of the system, the output process is given by

$$Y(t) = \int_{-\infty}^{\infty} h(u)X(t-u)du$$
$$E[Y(t)] = E\left[\int_{-\infty}^{\infty} h(u)X(t-u)du\right] = \int_{-\infty}^{\infty} h(u)E[X(t-u)]du = 0$$

## EXAMPLE 12.3

A wide-sense stationary Gaussian random process has an autocorrelation function

$$R_{XX}(\tau) = 6e^{-|\tau|/2}$$

Determine the covariance matrix of the random variables $X(t), X(t+1), X(t+2)$ and $X(t+3)$.

**Solution:**
First, note that

$$E[X(t)] = \mu_X(t) = \pm\sqrt{\lim_{|\tau|\to\infty}\{R_{XX}(\tau)\}} = 0$$

Let $X_1 = X(t), X_2 = X(t+1), X_3 = X(t+2), X_4 = X(t+3)$. Then the elements of the covariance matrix are given by

$$C_{ij} = \text{Cov}(X_i, X_j) = E\left[(X_i - \mu_{X_i})(X_j - \mu_{X_j})\right] = E[X_iX_j] = R_{ij}$$
$$= R_{XX}(i,j) = R_{XX}(j-i) = 6e^{-|j-i|/2}$$

where $R_{ij}$ is the $i$-$j$th element of the autocorrelation matrix. Thus,

$$C_{XX} = R_{XX} = \begin{bmatrix} 6 & 6e^{-1/2} & 6e^{-1} & 6e^{-3/2} \\ 6e^{-1/2} & 6 & 6e^{-1/2} & 6e^{-1} \\ 6e^{-1} & 6e^{-1/2} & 6 & 6e^{-1/2} \\ 6e^{-3/2} & 6e^{-1} & 6e^{-1/2} & 6 \end{bmatrix}$$

### 12.4.1 White Gaussian Noise Process

The Gaussian process is used to model noise voltages in resistors as well as receiver noise in communication systems. As we saw in Chapter 10, white noise $N(t)$ is characterized as follows:

a. The mean value is $\mu_{NN} = 0$.
b. The autocorrelation function is given by $R_{NN}(\tau) = \frac{N_0}{2}\delta(\tau)$.
c. The power spectral density is $S_{NN}(w) = \frac{N_0}{2}$.

The autocorrelation function is an impulse function because, according to the definition of a Gaussian process, for any collection of distinct time instants $t_1$, $t_2, \ldots, t_n$, the random variables $N(t_1), N(t_2), \ldots, N(t_n)$ are independent. Thus, except when $\tau = 0$,

$$R_{NN}(\tau) = E[N(t)N(t+\tau)] = E[N(t)]E[N(t+\tau)] = 0$$

While white Gaussian noise is popularly used as a mathematical modeling tool, it does not conform to any physically realizable signal because of the following observation:

$$E\left[N^2(t)\right] = R_{NN}(0) = \int_{-\infty}^{\infty} S_{NN}(w)dw = \int_{-\infty}^{\infty} \frac{N_0}{2}dw = \infty$$

That is, white noise has infinite average power, which is physically impossible. While such a process cannot exist physically, it is a convenient mathematical concept that greatly simplifies many computations in the analysis of linear systems that would otherwise be very difficult. Another frequently used concept in the analysis of linear systems is the *bandlimited white noise*, which has a nonzero and constant power spectral density over a finite frequency band, and zero elsewhere. That is, for a bandlimited white noise $N(t)$, the power spectral density is given by

$$S_{NN}(w) = \begin{cases} S_0 & -w_0 < w < w_0 \\ 0 & \text{otherwise} \end{cases}$$

## 12.5 POISSON PROCESS

Poisson processes are widely used to model arrivals (or occurrence of events) in a system. For example, they are used to model the arrival of telephone calls at a switchboard, the arrival of customers' orders at a service facility, and the random failures of equipment. Before we provide a formal definition of a Poisson process, we first consider some basic definitions.

### 12.5.1 Counting Processes

A random process $\{X(t), t \geq 0\}$ is called a counting process if $X(t)$ represents the total number of "events" that have occurred in the interval $[0, t)$. An example of a counting process is the number of customers that arrive at a bank from the time the bank opens its doors for business until some time $t$. A counting process satisfies the following conditions:

a. $X(t) \geq 0$, which means that it has nonnegative values.
b. $X(0) = 0$, which means that the counting of events begins at time 0.
c. $X(t)$ is integer-valued.
d. If $s < t$, then $X(s) \leq X(t)$. This means that it is a non-decreasing function of time.
e. $X(t) - X(s)$ represents the number of events that have occurred in the interval $[s, t]$.

Figure 12.2 represents a sample path of a counting process. The first event occurs at time $t_1$, and subsequent events occur at times $t_2$, $t_3$ and $t_4$. Thus, the number of events that occur in the interval $[0, t_4]$ is four.

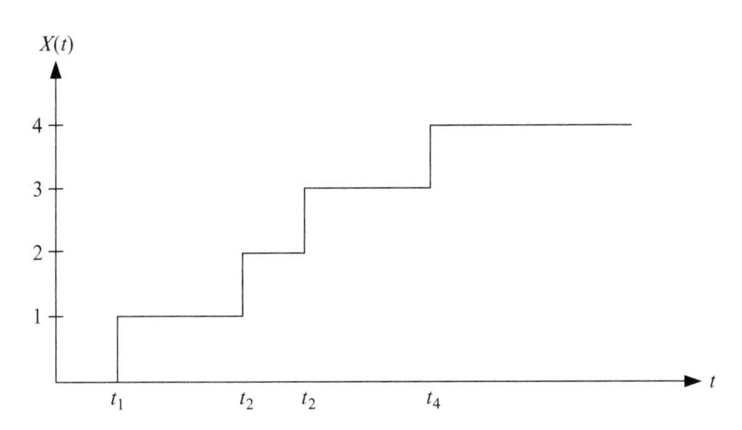


FIGURE 12.2
Sample Path of a Counting Process

### 12.5.2 Independent Increment Processes

A counting process is defined to be an independent increment process if the numbers of events that occur in disjoint time intervals are independent random variables. For example, in Figure 12.2, consider the two non-overlapping (i.e., disjoint) time intervals $[0,\ t_1]$ and $[t_2,\ t_4]$. If the number of events occurring in one interval is independent of the number of events that occur in the other, then the process is an independent increment process. Thus, $\{X(t), t \geq 0\}$ is an independent increment process if for every set of time instants $t_0 = 0 < t_1 < t_2 < \cdots < t_n$, the increments $X(t_1) - X(t_0)$, $X(t_2) - X(t_1)$, $\ldots$, $X(t_n) - X(t_{n-1})$ are mutually independent random variables.

### 12.5.3 Stationary Increments

A counting process $\{X(t), t \geq 0\}$ is defined to possess stationary increments if for every set of time instants $t_0 = 0 < t_1 < t_2 < \cdots < t_n$, the increments $X(t_1) - X(t_0)$, $X(t_2) - X(t_1)$, $\ldots, X(t_n) - X(t_{n-1})$ are identically distributed. In general, the mean of an independent increment process $\{X(t)\}$ with stationary increments has the form

$$E[X(t)] = mt \tag{12.7}$$

where the constant $m$ is the value of the mean at time $t = 1$. That is, $m = E[X(1)]$. Similarly, the variance of an independent increment process $\{X(t)\}$ with stationary increments has the form

$$\text{Var}[X(t)] = \sigma^2 t \tag{12.8}$$

where the constant $\sigma^2$ is the value of the variance at time $t = 1$; that is, $\sigma^2 = \text{Var}[X(1)]$.

### 12.5.4 Definitions of a Poisson Process

There are two ways to define a Poisson process. The first definition of the process is that it is a counting process $\{X(t)\}$ in which the number of events in any interval of length $t$ has a Poisson distribution with mean $\lambda t$. Thus, for all $s, t > 0$,

$$P[X(s+t) - X(s) = n] = \frac{(\lambda t)^n}{n!} e^{-\lambda t} \quad n = 0, 1, 2, \ldots \tag{12.9}$$

The second way to define the Poisson process $\{X(t)\}$ is that it is a counting process with stationary and independent increments such that for a rate $\lambda > 0$ the following conditions hold:

1. $P[X(t+\Delta t) - X(t) = 1] = \lambda \Delta t + o(\Delta t)$, which means that the probability of one event within a small time interval is approximately $\lambda \Delta t$, where $o(\Delta t)$ is a function of that goes to zero faster than $\Delta t$ does. That is,

   $$\lim_{\Delta t \to 0} \frac{o(\Delta t)}{\Delta t} = 0$$

2. $P[X(t+\Delta t) - X(t) \geq 2] = o(\Delta t)$, which means that the probability of two or more events within a small time interval $\Delta t$ is $o(\Delta t)$; that is, the probability is negligibly small.
3. $P[X(t+\Delta t) - X(t) = 0] = 1 - \lambda \Delta t + o(\Delta t)$.

These three properties enable us to derive the PMF of the number of events in a time interval of length $t$ as follows:

$$\begin{aligned}
P[X(t+\Delta t) = n] &= P[X(t) = n]P[X(\Delta t) = 0] \\
&\quad + P[X(t) = n-1]P[X(\Delta t) = 1] \\
&= P[X(t) = n](1 - \lambda \Delta t) + P[X(t) = n-1]\lambda \Delta t
\end{aligned}$$

$$P[X(t+\Delta t) = n] - P[X(t) = n] = -\lambda P[X(t) = n]\Delta t + \lambda P[X(t) = n-1]\Delta t$$

From this we obtain

$$\frac{P[X(t+\Delta t) = n] - P[X(t) = n]}{\Delta t} = -\lambda P[X(t) = n] + \lambda P[X(t) = n-1]$$

$$\begin{aligned}
\lim_{\Delta t \to 0} \left\{ \frac{P[X(t+\Delta t) = n] - P[X(t) = n]}{\Delta t} \right\} &= \frac{d}{dt} P[X(t) = n] \\
&= -\lambda P[X(t) = n] + \lambda P[X(t) = n-1]
\end{aligned}$$

Thus, we have that

$$\frac{d}{dt} P[X(t) = n] = -\lambda P[X(t) = n] + \lambda P[X(t) = n-1]$$

This equation may be solved iteratively for $n = 0, 1, 2, \ldots$, subject to the initial conditions

$$P[X(0) = n] = \begin{cases} 1 & n = 0 \\ 0 & n \neq 0 \end{cases}$$

This gives the PMF of the number of events (or "arrivals") in an interval of length $t$ as

$$p_{X(t)}(n,t) = \frac{(\lambda t)^n}{n!} e^{-\lambda t} \quad t \geq 0; \quad n = 0, 1, 2, \ldots$$

From the results obtained for Poisson random variables in Chapters 4 and 7, the z-transform, mean and variance are, respectively, as follows:

$$G_{X(t)}(z) = e^{-\lambda t(1-z)} \tag{12.10a}$$

$$E[X(t)] = \lambda t \tag{12.10b}$$

$$\sigma^2_{X(t)} = \lambda t \tag{12.10c}$$

The fact that the mean $E[X(t)] = \lambda t$ indicates that $\lambda$ is the expected number of arrivals per unit time in the Poisson process. Thus, the parameter $\lambda$ is called the *arrival rate* for the process. If $\lambda$ is independent of time, the Poisson process is called a *homogeneous Poisson process*. Sometimes the arrival rate is a function of time, and we represent it as $\lambda(t)$. Such processes are called *non-homogeneous Poisson processes*. In this chapter we are concerned mainly with homogeneous Poisson processes.

### 12.5.5 Interarrival Times for the Poisson Process

Let $L_r$ be a continuous random variable that is defined to be the interval between any event in a Poisson process and the $r$th event after it. Then $L_r$ is called the $r$th-order interarrival time. Let $f_{L_r}(l)$ be the PDF of $L_r$. To derive the expression for $f_{L_r}(l)$, we consider time of length $l$ over which we know that $r-1$ events have occurred. Assume that the next event (that is the $r$th event) occurs during the next time of length $\Delta l$, as shown in Figure 12.3.

Since the intervals $l$ and $\Delta l$ are non-overlapping, the number of events that occur within one interval is independent of the number of events that occur within the other interval. Thus, the PDF of $L_r$ can be obtained as follows. The probability that the $r$th arrival occurs between $l$ and $l + \Delta l$ is the area between $l$ and $l + \Delta l$ under the curve defined by $f_{L_r}(l)$; that is,

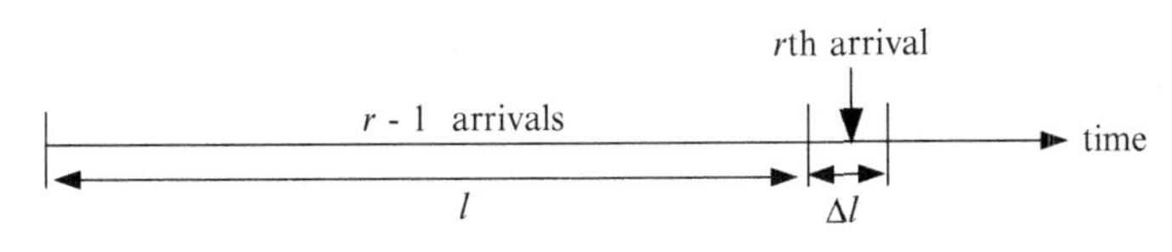


**FIGURE 12.3**
Definition of Event Intervals

$$\begin{aligned} f_{L_r}(l)\Delta l &= P[l < L_r \leq l + \Delta l] = P[\{X(l) = r-1\} \cap \{X(\Delta l) = 1] \\ &= P[X(l) = r-1]P[X(\Delta l) = 1] \\ &= \left\{\frac{(\lambda l)^{r-1}}{(r-1)!} e^{-\lambda l}\right\}\{\lambda \Delta l\} \end{aligned}$$

From this we obtain

$$f_{L_r}(l) = \left\{\frac{(\lambda l)^{r-1}}{(r-1)!} e^{-\lambda l}\right\}\{\lambda\} = \frac{\lambda^r l^{r-1}}{(r-1)!} e^{-\lambda l} \quad l \geq 0; r = 1, 2, \ldots \tag{12.11}$$

which is the Erlang-$r$ (or $r$th-order Erlang) distribution. The special case of $r=1$ is the exponential distribution. That is,

$$f_{L_1}(l) = \lambda e^{-\lambda l} \quad l \geq 0 \tag{12.12}$$

This result provides another definition of a Poisson process: It is a counting process with stationary and independent increments in which the intervals between consecutive events are exponentially distributed.

### 12.5.6 Conditional and Joint PMFs for Poisson Processes

Consider two times $t_1$ and $t_2$ such that $0 < t_1 < t_2$. Assume that $k_1$ Poisson events have occurred over the interval $(0, t_1)$. Let $\lambda$ be the arrival rate of the Poisson process $X(t)$. Then the PMF of $X(t_1)$ is given by

$$p_{X(t_1)}(k_1) = \frac{(\lambda t_1)^{k_1}}{k_1!} e^{-\lambda t_1} \quad k_1 = 0, 1, 2, \ldots$$

The conditional probability that $k_2$ events occur over the interval $(0, t_2)$ given that $k_1$ events have occurred over the interval $(0, t_1)$ is just the probability that $k_2 - k_1$ events occur over the interval $(t_1, t_2)$, which is given by

$$P[X(t_2) = k_2 | X(t_1) = k_1] = \frac{[\lambda(t_2 - t_1)]^{k_2 - k_1}}{(k_2 - k_1)!} e^{-\lambda(t_2 - t_1)} \tag{12.13}$$

where $k_2 \geq k_1$. Finally, the joint PMF of $k_2$ events occurring by time $t_2$ and $k_1$ events occurring by time $t_1$, where $t_1 < t_2$, is given by

$$\begin{aligned} p_{X(t_1)X(t_2)}(k_1, k_2) &= P[X(t_2) = k_2 | X(t_1) = k_1]P[X(t_1) = k_1] \\ &= \left\{\frac{[\lambda(t_2 - t_1)]^{k_2 - k_1}}{(k_2 - k_1)!} e^{-\lambda(t_2 - t_1)}\right\}\left\{\frac{(\lambda t_1)^{k_1}}{k_1!} e^{-\lambda t_1}\right\} \\ &= \frac{(\lambda t_1)^{k_1}[\lambda(t_2 - t_1)]^{k_2 - k_1}}{k_1!(k_2 - k_1)!} e^{-\lambda t_2} \quad k_2 \geq k_1 \end{aligned} \tag{12.14}$$

### 12.5.7 Compound Poisson Process

Let $\{N(t), t \geq 0\}$ be a Poisson process with arrival rate $\lambda$. Let $\{Y_i, i=1,2,\ldots\}$ be a family of independent and identically distributed random variables. Assume that the Poisson process $\{N(t), t \geq 0\}$ and the sequence $\{Y_i, i=1,2,\ldots\}$ are independent. We define a random process $\{X(t), t \geq 0\}$ to be a *compound Poisson process* if, for $t \geq 0$, it can be represented by

$$X(t) = \sum_{i=1}^{N(t)} Y_i \tag{12.15}$$

Thus, $X(t)$ is a Poisson sum of random variables. One example of the concept of compound Poisson process is the following. Assume students arrive at the university bookstore to buy books in a Poisson manner. If the number of books that each of these students buys is an independent and identically distributed random variable, then the number of books bought by time $t$ is a compound Poisson process.

Because the compound Poisson process has a rate that takes on a stochastic nature, it is also called a *doubly stochastic Poisson process*. This term is used to emphasize the fact that the process involves two kinds of randomness: There is a randomness that is associated with the main process that is sometimes called the *Poisson point process*, and there is another independent randomness that is associated with its rate.

Assume that the $Y_i$ are discrete random variables with the PMF $p_Y(y)$. The results can easily be modified to deal with the case when they are continuous random variables, where s-transforms rather than z-transforms are used. The value of $X(t)$, given that $N(t)=n$, is $X(t)=Y_1+Y_2+\cdots+Y_n$. Thus, the conditional z-transform of the PMF of $X(t)$, given that $N(t)=n$, is given by

$$G_{X(t)|N(t)}(z|n) = E\left[z^{Y_1+Y_2+\cdots+Y_n}\right] = \left(E\left[z^Y\right]\right)^n = [G_Y(z)]^n$$

where the last two equalities follow from the fact that the $Y_i$ are independent and identically distributed. Thus, the unconditional z-transform of the PMF of $X(t)$ is given by

$$G_{X(t)}(z) = \sum_{n=0}^{\infty} G_{X(t)|N(t)}(z|n) p_{N(t)}(n) = \sum_{n=0}^{\infty} [G_Y(z)]^n p_{N(t)}(n,t) = G_{N(t)}(G_Y(z))$$

Now,

$$G_{N(t)}(z) = \sum_{n=0}^{\infty} z^n \frac{(\lambda t)^n}{n!} e^{-\lambda t} = e^{-\lambda t} \sum_{n=0}^{\infty} \frac{(z\lambda t)^n}{n!} = e^{-\lambda t} e^{z\lambda t} = e^{-\lambda t(1-z)}$$

Thus,

$$G_{X(t)}(z) = G_{N(t)}(G_Y(z)) = e^{-\lambda t[1-G_Y(z)]} \tag{12.16}$$

The mean and variance of $X(t)$ can be obtained through differentiating the above function. These are given by

$$E[X(t)] = \frac{d}{dz}G_{X(t)}(z)\bigg|_{z=1} = \lambda t E[Y] \quad (12.17a)$$

$$E[X^2(t)] = \frac{d^2}{dz^2}G_{X(t)}(z)\bigg|_{z=1} + \frac{d}{dz}G_{X(t)}(z)\bigg|_{z=1} = \lambda t E[Y^2] + (\lambda t E[Y])^2 \quad (12.17b)$$

$$\sigma^2_{X(t)} = E[X^2(t)] - (E[X(t)])^2 = \lambda t E[Y^2] \quad (12.17c)$$

Note that this is a special case of the random sum of random variables discussed in Section 7.4 of Chapter 7. Using the current notation, it was shown in Section 7.4 that

$$\begin{aligned} E[X(t)] &= E[N(t)]E[Y] = \lambda t E[Y] \\ \sigma^2_{X(t)} &= E[N(t)]\sigma^2_Y + (E[Y])^2\sigma^2_{N(t)} = \lambda t \sigma^2_Y + \lambda t (E[Y])^2 = \lambda t\{\sigma^2_Y + (E[Y])^2\} \\ &= \lambda t E[Y^2] \end{aligned}$$

Note also that in the case when the $Y_i$ are continuous random variables, the result would be

$$M_{X(t)}(s) = G_{N(t)}(M_Y(s)) = e^{-\lambda t[1 - M_Y(s)]}$$

and the above results still hold.

## EXAMPLE 12.4

Customers arrive at a grocery store in a Poisson manner at an average rate of 10 customers per hour. The amount of money that each customer spends is uniformly distributed between $8.00 and $20.00. What is the average total amount of money that customers who arrive over a two-hour interval spend in the store? What is the variance of this total amount?

**Solution:**

This is a compound Poisson process with $\lambda = 10$ customers per hour. Let $Y$ be the random variable that represents the amount of money a customer spends in the store. Since $Y$ is uniformly distributed over the interval (8, 20), we have that

$$\begin{aligned} E[Y] &= \frac{20+8}{2} = 14 \\ \sigma^2_Y &= \frac{(20-8)^2}{12} = 12 \\ E[Y^2] &= \sigma^2_Y + (E[Y])^2 = 12 + 196 = 208 \end{aligned}$$

Therefore, the mean and the variance of the total amount of money that customers arriving over a two-hour time interval (i.e., $t = 2$) spend in the store are given by

$$\begin{aligned} E[X(2)] &= 2\lambda E[Y] = 2(10)(14) = 280 \\ \sigma^2_{X(2)} &= 2\lambda E[Y^2] = 2(10)(208) = 4160 \end{aligned}$$

### 12.5.8 Combinations of Independent Poisson Processes

Consider two independent Poisson processes $\{X(t), t \geq 0\}$ and $\{Y(t), t \geq 0\}$ with arrival rates $\lambda_X$ and $\lambda_Y$, respectively. Consider a random process $\{N(t), t \geq 0\}$ that is the sum of the two Poisson processes; that is, $N(t) = X(t) + Y(t)$. Thus, $\{N(t), t \geq 0\}$ is the process consisting of arrivals from the two Poisson processes. We want to show that $\{N(t)\}$ is also a Poisson process with arrival rate $\lambda = \lambda_X + \lambda_Y$.

To do this, we note that $\{X(t), t \geq 0\}$ and $\{Y(t), t \geq 0\}$ are specified as being independent. Therefore,

$$\begin{aligned} P[N(t+\Delta t) - N(t) = 0] &= P[X(t+\Delta t) - X(t) = 0]P[Y(t+\Delta t) - Y(t) = 0] \\ &= [1 - \lambda_X \Delta t + o(\Delta t)][1 - \lambda_Y \Delta t + o(\Delta t)] \\ &= 1 - (\lambda_X + \lambda_Y)\Delta t + o(\Delta t)] = 1 - \lambda \Delta t + o(\Delta t)] \end{aligned}$$

where $\lambda = \lambda_X + \lambda_Y$. Since the last equation is the probability that there is no arrival within an interval of length $\Delta t$, it follows that $\{N(t), t \geq 0\}$ is a Poisson process.

Another way to prove this is to note that when $t$ is fixed, $N(t)$ is a random variable that is a sum of two independent random variables. Thus, the z-transform of the PMF of $N(t)$ is the product of the z-transform of the PMF of $X(t)$ and the z-transform of the PMF of $Y(t)$. That is,

$$G_{N(t)}(z) = G_{X(t)}(z)G_{Y(t)}(z) = e^{-\lambda_X t(1-z)}e^{-\lambda_Y t(1-z)} = e^{-(\lambda_X + \lambda_Y)t(1-z)} = e^{-\lambda t(1-z)}$$

which is the z-transform of a Poisson random variable. Since for each $t$, $N(t)$ is a random variable, the collection of these random variables over time (that is, $\{N(t), t \geq 0\}$) constitutes a Poisson random process.

The third way to show this is via the interarrival time. Let $L$ denote the time until the first arrival in the process $\{N(t), t \geq 0\}$, let $L_X$ denote the time until the first arrival in the process $\{X(t), t \geq 0\}$, and let $L_Y$ denote the time until the first arrival in the process $\{Y(t), t \geq 0\}$. Then, since the two Poisson processes are independent,

$$P[L > t] = P[L_X > t]P[L_Y > t] = e^{-\lambda_X t}e^{-\lambda_Y t} = e^{-(\lambda_X + \lambda_Y)t} = e^{-\lambda t}$$

which shows that $\{N(t), t \geq 0\}$ exhibits the same memoryless property as $\{X(t), t \geq 0\}$ and $\{Y(t), t \geq 0\}$. Therefore, $\{N(t), t \geq 0\}$ must be a Poisson process.

## EXAMPLE 12.5

Two lightbulbs, labeled $A$ and $B$, have exponentially distributed lifetimes. If the two lifetimes of the two bulbs are independent and the mean lifetime of bulb $A$ is 500 hours, while the mean lifetime of bulb $B$ is 200 hours, what is the mean time to a bulb failure?

**Solution:**

Let $\lambda_A$ denote the burnout rate of bulb *A*, and let $\lambda_B$ be the burnout rate of bulb *B*. Since $1/\lambda_A = 500$ and $1/\lambda_B = 200$, the rates are $\lambda_A = 1/500$ and $\lambda_B = 1/200$. From the results obtained above, the two bulbs behave like a single system with exponentially distributed lifetime whose mean is $1/\lambda$, where $\lambda = \lambda_A + \lambda_B$. Thus, the mean time until a bulb fails in hours is

$$\frac{1}{\lambda} = \frac{1}{\lambda_A + \lambda_B} = \frac{1}{(1/500) + (1/200)} = \frac{1000}{7} = 142.86$$

## 12.5.9 Competing Independent Poisson Processes

In this section we extend the combination problem discussed in the previous section. Thus, we consider two independent Poisson processes $\{X(t), t \geq 0\}$ and $\{Y(t), t \geq 0\}$ with arrival rates $\lambda_X$ and $\lambda_Y$, respectively. The question we are interested in is this: What is the probability that an arrival from $\{X(t), t \geq 0\}$ occurs before an arrival from $\{Y(t), t \geq 0\}$? Since the interarrival times in a Poisson process are exponentially distributed, let $T_X$ be the random variable that denotes the interarrival time in the $\{X(t), t \geq 0\}$ process and let $T_Y$ be the random variable that denotes the interarrival time in the $\{Y(t), t \geq 0\}$ process. Thus, we are interested in computing $P[T_X < T_Y]$, where $f_{T_X}(x) = \lambda_X e^{-\lambda_X x}$, $x \geq 0$, and $f_{T_Y}(y) = \lambda_Y e^{-\lambda_Y y}$, $y \geq 0$. Because the two processes are independent, the joint PDF of $T_X$ and $T_Y$ is given by

$$f_{T_X T_Y}(x, y) = \lambda_X \lambda_Y e^{-\lambda_X x} e^{-\lambda_Y y} \quad x \geq 0, \ y \geq 0$$

In order to evaluate the probability, we consider the limits of integration by observing Figure 12.4. From the Figure we have that

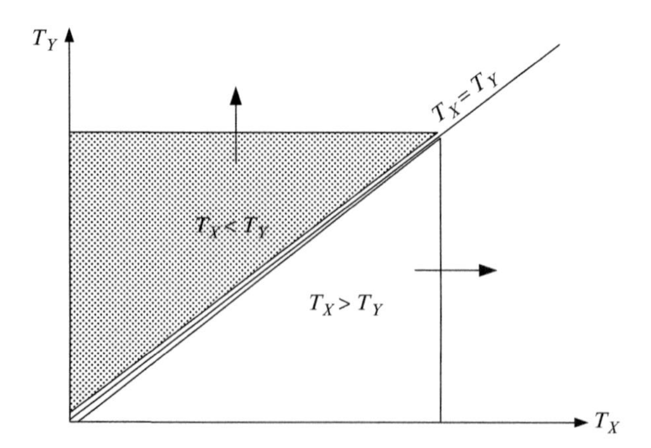


**FIGURE 12.4**
Partitioning the Regions Around the Line $T_X = T_Y$

$$P[T_X < T_Y] = \int_{x=0}^{\infty}\int_{y=x}^{\infty} f_{T_XT_Y}(x,y)dydx = \int_{x=0}^{\infty}\int_{y=x}^{\infty} \lambda_X\lambda_Y e^{-\lambda_X x}e^{-\lambda_Y y}dydx$$
$$= \int_{x=0}^{\infty} \lambda_X e^{-(\lambda_X+\lambda_Y)x}dx = \frac{\lambda_X}{\lambda_X+\lambda_Y} \quad (12.18)$$

Another way to derive this result is by considering events that occur within the small time interval $(t, t+\Delta t)$. Then, since the probability of an arrival from $X(t)$ within this interval is approximately $\lambda_X\Delta t$, and the probability of an arrival from either $X(t)$ or $Y(t)$) within this interval is approximately $(\lambda_X+\lambda_Y)\Delta t$, the probability that the $X(t)$ process occurs in the interval $[t, t+\Delta t]$, given an arrival in that interval is $\lambda_X\Delta t/(\lambda_X+\lambda_Y)\Delta t = \lambda_X/(\lambda_X+\lambda_Y)$.

The third way to solve the problem is to consider a time interval $T$. Within this interval, the total number of arrivals from the $\{X(t), t\geq 0\}$ process is $\lambda_X T$. Since the two processes form a combination of independent Poisson processes with rate $(\lambda_X+\lambda_Y)$, the total number of arrivals from both processes is $(\lambda_X+\lambda_Y)T$. Thus, the probability that an $\{X(t), t\geq 0\}$ process occurs is $\lambda_X T/(\lambda_X+\lambda_Y)T = \lambda_X/(\lambda_X+\lambda_Y)$.

### EXAMPLE 12.6

Two lightbulbs, labeled *A* and *B*, have exponentially distributed lifetimes. If the two lifetimes of the two lightbulbs are independent and the mean lifetime of lightbulb *A* is 500 hours, while the mean lifetime of lightbulb *B* is 200 hours, what is the probability that lightbulb *A* fails before lightbulb *B*?

**Solution:**

Let $\lambda_A$ denote the burnout rate of bulb *A*, and let $\lambda_B$ be the burnout rate of bulb *B*. Since $1/\lambda_A = 500$ and $1/\lambda_B = 200$, the rates are $\lambda_A = 1/500$ and $\lambda_B = 1/200$. Thus, the probability that bulb *A* fails before bulb *B* is

$$\frac{\lambda_A}{\lambda_A+\lambda_B} = \frac{(1/500)}{(1/500)+(1/200)} = \frac{2}{7}$$

### 12.5.10 Subdivision of a Poisson Process and the Filtered Poisson Process

Consider a Poisson process $\{X(t), t\geq 0\}$ with arrival rate $\lambda$. Assume that arrivals in $\{X(t), t\geq 0\}$ can be sent to one of two outputs, which we call output A and output B. Assume that the decision on which output an arrival is sent is made independently of other arrivals. Furthermore, assume that each arrival is sent to output A with probability $p$ and to output B with probability $1-p$, as shown in Figure 12.5.

The arrival rate at output A is $\lambda_A = p\lambda$, and the arrival rate at output B is $\lambda_B = (1-p)\lambda$. The two outputs are independent. Consider a small time interval $(t, t+\Delta t)$. The probability that there is an arrival in the original process over this

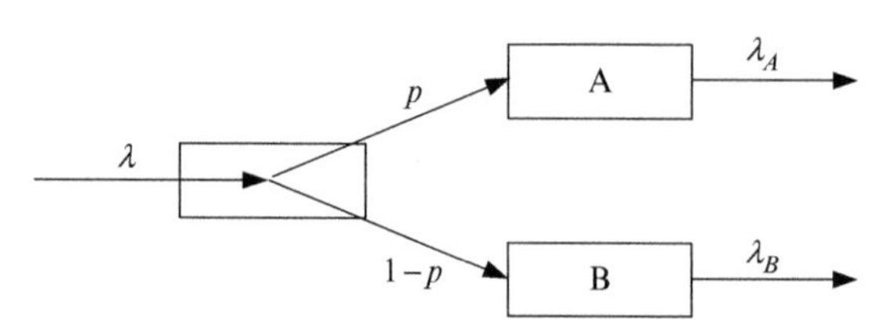


**FIGURE 12.5**
Subdivision of a Poisson Process

interval is approximately $\lambda\Delta t$, if we ignore higher-order terms of $\Delta t$. Thus, the probability that there is an arrival in output A over this interval is approximately $p\lambda\Delta t$, and the probability that there is an arrival in output B over this interval is $(1-p)\lambda\Delta t$. Since the original process is a stationary and independent increment process and the two outputs are independent, each output is a stationary and independent increment process. Thus, each output is a Poisson process. We can then refer to output A as the Poisson process $\{X_A(t), t\geq 0\}$ with arrival rate $\lambda_A = p\lambda$. Similarly, we can refer to output B as the Poisson process $\{X_B(t), t\geq 0\}$ with arrival rate $\lambda_B = (1-p)\lambda$.

A *filtered Poisson process* $Y(t)$ is a process in which events occur according to a Poisson process $X(t)$ with rate $\lambda$, but each event is independently recorded with a probability $p$. From the discussion above, we observe that $Y(t)$ is a Poisson process with rate $p\lambda$.

## EXAMPLE 12.7

A gas station is located next to a fast-food restaurant along a highway. Cars arrive at the restaurant according to a Poisson process at an average rate of 12 per hour. Independently of other cars, each car that stops at the restaurant will go to refuel at the gas station before going back to the highway with a probability of 0.25. What is the probability that exactly 10 cars have been refueled at the gas station within a particular two-hour period?

**Solution:**

The process that governs car arrivals at the gas station is a Poisson process with a rate of $\lambda_G = p\lambda = (0.25)(12) = 3$ cars per hour. Thus, if $K$ represents that number of cars that arrive at the gas within 2 hours, the probability that $K=10$ cars is given by

$$P[K=10] = \frac{(2\lambda_G)^{10}}{10!}e^{-2\lambda_G} = \frac{6^{10}}{10!}e^{-6} = 0.0413$$

### 12.5.11 Random Incidence

Consider a Poisson process $\{X(t), t\geq 0\}$ in which events (or arrivals) occur at times $T_0 = 0, T_1, T_2, \ldots$. Let the interarrival times $Y_k$ be defined as follows:

$$
\begin{aligned}
Y_1 &= T_1 - T_0 \\
Y_2 &= T_2 - T_1 \\
&\vdots \\
Y_k &= T_k - T_{k-1}
\end{aligned}
$$

These interarrival times are illustrated in Figure 12.6, where $A_k$ denotes the $k$th arrival.

The Poisson process belongs to a class of random processes called *renewal processes* that have the property that the $Y_k$ are mutually independent and identically distributed. For the Poisson process with mean arrival rate $\lambda$, the $Y_k$ are exponentially distributed with mean $1/\lambda$, as discussed earlier.

Consider the following problem in connection with the $Y_k$. Assume the $T_k$ are the points in time that buses arrive at a bus stop. A passenger arrives at the bus stop at a *random time* and wants to know how long he or she will wait until the next bus arrival. This problem is usually referred to as the *random incidence problem,* since the subject (or passenger in this example) is incident to the process at a random time. Let $R$ be the random variable that denotes the time from the moment the passenger arrived until the next bus arrival. $R$ is referred to as the *residual life* of the Poisson process. Also, let $W$ denote the length of the interarrival gap that the passenger entered by random incidence. Figure 12.7 illustrates the random incidence problem.

Let $f_Y(y)$ denote the PDF of the interarrival times; let $f_W(w)$ denote the PDF of $W$, the gap entered by random incidence; and let $f_R(r)$ denote the PDF of the

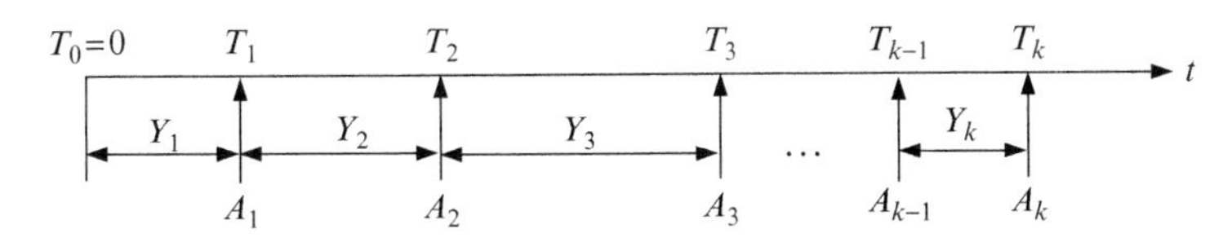


**FIGURE 12.6**
Interarrival Times of a Poisson Process

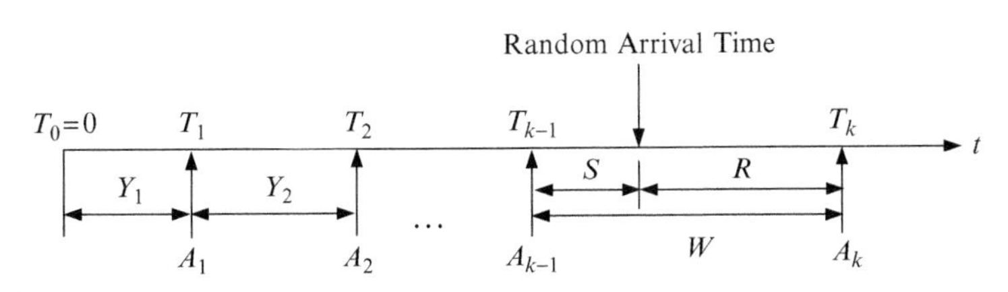


**FIGURE 12.7**
Random Incidence

residual life, $R$. The probability that the random arrival occurs in a gap of length between $w$ and $w+dw$ can be assumed to be directly proportional to the length $w$ of the gap and relative occurrence $f_Y(w)dw$ of such gaps. That is,

$$f_W(w)dw = \beta w f_Y(w)dw$$

where $\beta$ is a constant of proportionality. Thus, $f_W(w) = \beta w f_Y(w)$. Since $f_W(w)$ is a PDF, we have that

$$\int_{-\infty}^{\infty} f_W(w)dw = 1 = \beta \int_{-\infty}^{\infty} w f_Y(w)dw = \beta E[Y]$$

which means that $\beta = 1/E[Y]$, and we obtain

$$f_W(w) = \frac{w f_Y(w)}{E[Y]} \quad w \geq 0 \tag{12.19}$$

The expected value of $W$ is given by

$$E[W] = \frac{E[Y^2]}{E[Y]} \tag{12.20}$$

This result applies to all renewal processes. For a Poisson process, $Y$ is exponentially distributed with $E[Y] = 1/\lambda$ and $E[Y^2] = 2/\lambda^2$. Thus, for a Poisson process we obtain

$$f_W(w) = \lambda w f_Y(w) = \lambda^2 w e^{-\lambda w} \quad w \geq 0$$
$$E[W] = \frac{2}{\lambda}$$

This means that for a Poisson process the gap entered by random incidence has the second-order Erlang distribution; thus, the expected length of the gap is twice the expected length of an interarrival time. This is often referred to as the *random incidence paradox*. The reason for this fact is that the passenger is more likely to enter a large gap than a small gap.

Next, we consider the PDF of the residual life $R$ of the process. Given the passenger enters a gap of length $w$, he or she is equally like to be anywhere within the gap. That is, the gap has a uniform distribution and thus the conditional PDF of $R$ given that $W = w$ is given by

$$f_{R|W}(r|w) = \frac{1}{w} \quad 0 \leq r \leq w$$

When we combine this result with the previous one, we get the joint PDF of $R$ and $W$ as follows:

$$f_{RW}(r, w) = f_{R|W}(r|w) f_W(w) = \frac{1}{w}\left\{\frac{w f_Y(w)}{E[Y]}\right\} = \frac{f_Y(w)}{E[Y]} \quad 0 \leq r \leq w < \infty$$

he marginal PDF of $R$ becomes

$$f_R(r)=\int_r^{\infty} f_{RW}(r,w)dw=\int_r^{\infty}\frac{f_Y(w)}{E[Y]}dw=\frac{1-F_Y(r)}{E[Y]} \quad r\geq 0 \tag{12.21}$$

ince $Y$ is exponentially distributed, $1-F_Y(r)=e^{-\lambda r}$, which means that

$$f_R(r)=\lambda e^{-\lambda r} \quad r\geq 0 \tag{12.22}$$

hus, for a Poisson process, the residual life of the process has the same distribution as the interarrival time, which can be expected from the "forgetfulness" property of the exponential distribution. In Figure 12.7 the random variable $S$ denotes the time between the last bus arrival and the passenger's random arrival. Since $W=S+R$, the expected value of $S$ is $E[S]=E[W]-E[R]=1/\lambda$.

## EXAMPLE 12.8

City buses arrive at a particular bus stop according to a Poisson process with a rate of 5 buses per hour. Ross arrived at the bus stop and had to wait to catch the next bus.

a. What is the mean time between the instant Ross arrived at the bus stop until the next bus arrives?
b. What is the mean time between the last bus arrival and the arrival of the bus that Ross boarded?

**Solution:**

Since buses arrive at the bus stop according to a Poisson process, the time between bus arrivals is exponentially distributed with a mean of 1/5 hours (or 12 minutes). Thus, we obtain the following results:

a. According to the principles of random incidence, the time until the next bus arrives after Ross's arrival is exponentially distributed with a mean of 12 minutes. Thus, Ross waits an average of 12 minutes before the next bus arrives.
b. Also, the mean time between the last bus arrival and the arrival of the bus that Ross boarded is the mean length of the gap that Ross entered by random incidence. According to the principles of random incidence, the mean length of this gap is twice the mean length of time between bus arrivals. Thus, the mean time between the two bus arrivals is 24 minutes.

## 12.6 MARKOV PROCESSES

Markov processes are widely used in engineering, science, and business modeling. They are used to model systems that have a limited memory of their past. For example, in the gambler's ruin problem discussed earlier in this chapter, the amount of money the gambler will make after $n+1$ games is determined by the amount of money he has made after $n$ games. Any other information is irrelevant in making this prediction. In population growth studies, the population of the next generation depends mainly on the current population and possibly the last few generations.

A random process $\{X(t), t \in T\}$ is called a first-order Markov process if for any $t_0 < t_1 < \cdots < t_n$ the conditional CDF of $X(t_n)$ for given values of $X(t_0), X(t_1), \ldots, X(t_{n-1})$ depends only on $X(t_{n-1})$. That is,

$$\begin{aligned} P[X(t_n) \leq x_n | X(t_{n-1}) \leq x_{n-1}, X(t_{n-2}) \leq x_{n-2}, \ldots, X(t_0) \leq x_0] \\ = P[X(t_n) \leq x_n | X(t_{n-1}) \leq x_{n-1}] \end{aligned} \tag{12.23}$$

This means that, given the present state of the process, the future state is independent of the past. This property is usually referred to as the *Markov property*. In second-order Markov processes the future state depends on both the current state and the last immediate state, and so on for higher-order Markov processes. In this chapter we consider only first-order Markov processes.

Markov processes are classified according to the nature of the time parameter and the nature of the state space. With respect to state space, a Markov process can be either a discrete-state Markov process or continuous-state Markov process. A discrete-state Markov process is called a *Markov chain*. Similarly, with respect to time, a Markov process can be either a discrete-time Markov process or a continuous-time Markov process. Thus, there are four basic types of Markov processes:

1. Discrete-time Markov chain (or discrete-time discrete-state Markov process)
2. Continuous-time Markov chain (or continuous-time discrete-state Markov process)
3. Discrete-time Markov process (or discrete-time continuous-state Markov process)
4. Continuous-time Markov process (or continuous-time continuous-state Markov process)

This classification of Markov processes is illustrated in Figure 12.8.

The remainder of the discussion in this chapter deals with Markov chains (that is, discrete-state Markov processes).

| | | State Space | |
|---|---|---|---|
| | | Discrete | Continuous |
| Time | Discrete | Discrete-time Markov Chain | Discrete-time Markov Process |
| | Continuous | Continuous-time Markov Chain | Continuous-time Markov Process |

**FIGURE 12.8**
Classification of Markov Processes

## 12.7 DISCRETE-TIME MARKOV CHAINS

The discrete-time process $\{X_k, k=0, 1, 2, \ldots\}$ is called a Markov chain if for all $i, j, k, \ldots, m$, the following is true:

$$P[X_k = j|X_{k-1} = i, X_{k-2} = \alpha, \ldots, X_0 = \theta] = P[X_k = j|X_{k-1} = i] = p_{ijk}$$

The quantity $p_{ijk}$ is called the *state transition probability,* which is the conditional probability that the process will be in state $j$ at time $k$ immediately after the next transition, given that it is in state $i$ at time $k-1$. A Markov chain that obeys the preceding rule is called a *non-homogeneous Markov chain.* In this book we will consider only *homogeneous Markov chains,* which are Markov chains in which $p_{ijk} = p_{ij}$. This means that homogeneous Markov chains do not depend on the time unit, which implies that

$$P[X_k = j|X_{k-1} = i, X_{k-2} = \alpha, \ldots, X_0 = \theta] = P[X_k = j|X_{k-1} = i] = p_{ij} \tag{12.24}$$

The *homogeneous state transition probability* $p_{ij}$ satisfies the following conditions:

a. $0 \leq p_{ij} \leq 1$
b. $\sum_j p_{ij} = 1, \ i = 1, 2, \ldots,$ which follows from the fact that the states are mutually exclusive and collectively exhaustive

### 12.7.1 State Transition Probability Matrix

Consider a Markov chain with $N$ states. It is customary to display the state transition probabilities as elements of an $N \times N$ matrix $P$ such that $p_{ij}$ is the entry in the $i$th row and $j$th column. That is,

$$P = \begin{bmatrix} p_{11} & p_{12} & \cdots & p_{1N} \\ p_{21} & p_{22} & \cdots & p_{2N} \\ \cdots & \cdots & \cdots & \cdots \\ p_{N1} & p_{N2} & \cdots & p_{NN} \end{bmatrix}$$

$P$ is called the *transition probability matrix*. It is a *stochastic matrix* because for any row $i$, $\sum_{j=1}^{N} p_{ij} = 1$.

### 12.7.2 The *n*-Step State Transition Probability

Let $p_{ij}(n)$ denote the conditional probability that the system will be in state $j$ after exactly $n$ transitions, given that it is presently in state $i$. That is,

$$p_{ij}(n) = P[X_{m+n} = j|X_m = i]$$

$$p_{ij}(0) = \begin{cases} 1 & i = j \\ 0 & i \neq j \end{cases}$$

$$p_{ij}(1) = p_{ij}$$

Consider the two-step transition probability $p_{ij}(2)$, which is defined by

$$p_{ij}(2) = P[X_{m+2} = j | X_m = i]$$

Assume that $m = 0$, then

$$\begin{aligned} p_{ij}(2) &= P[X_2 = j | X_0 = i] = \sum_k P[X_2 = j, X_1 = k | X_0 = i] \\ &= \sum_k P[X_2 = j | X_1 = k, X_0 = i] P[X_1 = k | X_0 = i] \\ &= \sum_k P[X_2 = j | X_1 = k] P[X_1 = k | X_0 = i] = \sum_k p_{kj} p_{ik} \\ &= \sum_k p_{ik} p_{kj} \end{aligned}$$

where the third to the last equality is due to the Markov property. The final equation states that the probability of starting in state $i$ and being in state $j$ at the end of the second transition is the probability that we first go immediately from state $i$ to an intermediate state $k$ and then immediately from state $k$ to state $j$; the summation is taken over all possible intermediate states $k$.

The following proposition deals with a class of equations called the *Chapman-Kolmogorov equations,* which provide a generalization of the above results obtained for the two-step transition probability.

**Proposition 12.1:**
For all $0 < r < n$,

$$p_{ij}(n) = \sum_k p_{ik}(r) p_{kj}(n-r) \tag{12.25}$$

This proposition states that the probability that the process starts in state $i$ and finds itself in state $j$ at the end of the $n$th transition is the product of the probability that the process starts in state $i$ and finds itself in an intermediate state $k$ after $r$ transitions and the probability that it goes from state $k$ to state $j$ after additional $n-r$ transitions.

**Proof:**
The proof is a generalization of the proof for the case of $n = 2$ and is as follows.

$$\begin{aligned} p_{ij}(n) &= P[X_n = j | X_0 = i] = \sum_k P[X_n = j, X_r = k | X_0 = i] \\ &= \sum_k P[X_n = j | X_r = k, X_0 = i] P[X_r = k | X_0 = i] \\ &= \sum_k P[X_n = j | X_r = k] P[X_r = k | X_0 = i] = \sum_k p_{kj}(n-r) p_{ik}(r) \\ &= \sum_k p_{ik}(r) p_{kj}(n-r) \end{aligned}$$

This completes the proof.

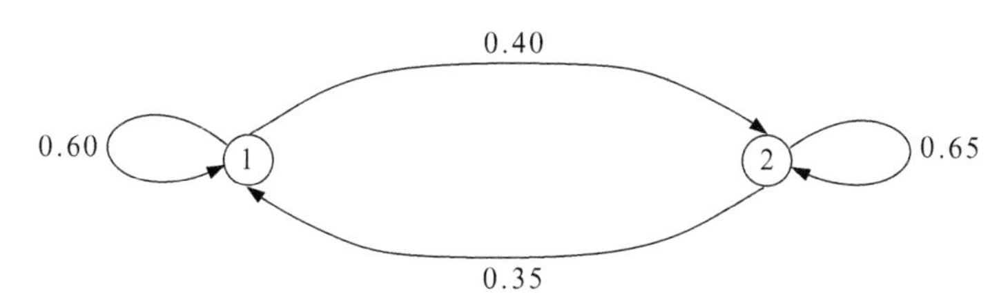


**FIGURE 12.9**
Example of State-Transition Diagram

### 12.7.3 State Transition Diagrams

Consider the following problem. It has been observed via a series of tosses of a particular biased coin that the outcome of the next toss depends on the outcome of the current toss. In particular, given that the current toss comes up heads, the next toss will come up heads with probability 0.6 and tails with probability 0.4. Similarly, given that the current toss comes up tails, the next toss will come up heads with probability 0.35 and tails with probability 0.65.

If we define state 1 to represent heads and state 2 to represent tails, then the transition probability matrix for this problem is the following:

$$P = \begin{bmatrix} 0.60 & 0.40 \\ 0.35 & 0.65 \end{bmatrix}$$

All the properties of the Markov process can be determined from this matrix. However, the analysis of the problem can be simplified by the use of the *state-transition diagram* in which the states are represented by circles and directed arcs represent transitions between states. The state transition probabilities are labeled on the appropriate arcs. Thus, with respect to the above problem, we obtain the state-transition diagram shown in Figure 12.9.

## EXAMPLE 12.9

Assume that people in a particular society can be classified as belonging to the upper class (U), middle class (M), and lower class (L). Membership in any class is inherited in the following probabilistic manner. Given that a person is raised in an upper-class family, he or she will have an upper-class family with probability 0.7, a middle-class family with probability 0.2, and a lower-class family with probability 0.1. Similarly, given that a person is raised in a middle-class family, he or she will have an upper-class family with probability 0.1, a middle-class family with probability 0.6, and a lower-class family with probability 0.3. Finally, given that a person is raised in a lower-class family, he or she will have a middle-class family with probability 0.3 and a lower-class family with probability 0.7. Determine (a) the transition probability matrix and (b) the state-transition diagram for this problem.

**Solution:**

a. Using state 1 to represent the upper class, state 2 to represent the middle class, and state 3 to represent the lower class, we obtain the following transition probability matrix:

$$P = \begin{bmatrix} 0.7 & 0.2 & 0.1 \\ 0.1 & 0.6 & 0.3 \\ 0.0 & 0.3 & 0.7 \end{bmatrix}$$

b. The state-transition diagram is as shown in Figure 12.10.

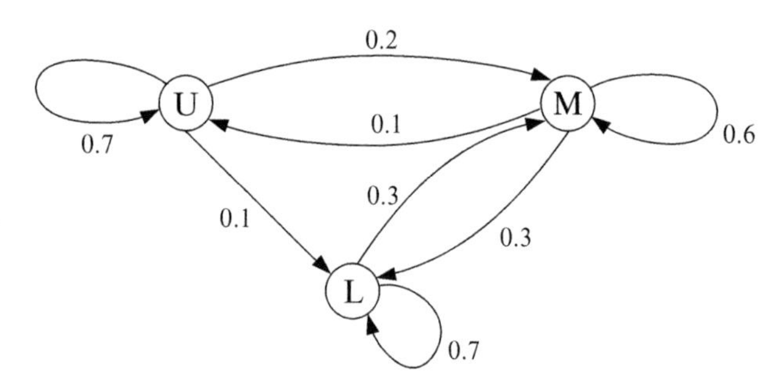


**FIGURE 12.10**
State-Transition Diagram for Example 12.9

## 12.7.4 Classification of States

A state $j$ is said to be *accessible* (or *can be reached*) from state $i$ if, starting from state $i$, it is possible that the process will ever enter state $j$. This implies that $p_{ij}(n) > 0$ for some $n > 0$. Thus, the $n$-step probability enables us to obtain reachability information between any two states of the process.

Two states that are accessible from each other are said to *communicate* with each other. The concept of communication divides the state space into different classes. Two states that communicate are said to be in the same *class*. All members of one class communicate with one another. If a class is not accessible from any state outside the class, we define the class to be a *closed communicating class*. A Markov chain in which all states communicate, which means that there is only one class, is called an *irreducible* Markov chain. For example, the Markov chains shown in Figures 12.9 and 12.10 are irreducible Markov chains.

The states of a Markov chain can be classified into two broad groups: those that the process enters infinitely often and those that it enters finitely often. In the long run, the process will be found to be in only those states that it enters infinitely often. Let $f_{ij}(n)$ denote the conditional probability that given that the process is presently in state $i$, the first time it will enter state $j$ occurs in exactly $n$ transitions (or steps). We call $f_{ij}(n)$ the probability of *first passage* from state $i$ to state $j$ in $n$ transitions. The parameter $f_{ij}$, which is defined by

$$f_{ij} = \sum_{n=1}^{\infty} f_{ij}(n) \tag{12.26}$$

is the probability of first passage from state $i$ to state $j$. It is the conditional probability that the process will ever enter state $j$, given that it was initially in state $i$. Obviously $f_{ij}(1) = p_{ij}$, and a recursive method of computing $f_{ij}(n)$ is

$$f_{ij}(n) = \sum_{l \neq j} p_{il} f_{lj}(n-1) \tag{12.27}$$

The quantity $f_{ii}$ denotes the probability that a process that starts at state $i$ will ever return to state $i$. Any state $i$ for which $f_{ii} = 1$ is called a *recurrent state*, and any state $i$ for which $f_{ii} < 1$ is called a *transient state*. More formally, we define these states as follows:

a. A state $j$ is called a *transient* (or *non-recurrent*) state if there is a positive probability that the process will never return to $j$ again after it leaves $j$.
b. A state $j$ is called a *recurrent* (or *persistent*) state if, with probability 1, the process will eventually return to $j$ after it leaves the state. A set of recurrent states forms a *single chain* if every member of the set communicates with all other members of the set.
c. A recurrent state $j$ is called a *periodic* state if there exists an integer $d$, $d > 1$, such that $p_{jj}(n)$ is zero for all values of $n$ other than $d, 2d, 3d, \ldots$; $d$ is called the period. If $d = 1$, the recurrent state $j$ is said to be *aperiodic*.
d. A recurrent state $j$ is called a *positive recurrent* state if, starting at state $j$ the expected time until the process returns to state $j$ is finite. Otherwise, the recurrent state is called a *null recurrent* state.
e. Positive recurrent, aperiodic states are called *ergodic* states.
f. A chain consisting of ergodic states is called an *ergodic chain*.
g. A state $j$ is called an *absorbing* (or *trapping*) state if $p_{jj} = 1$. Thus, once the process enters a trapping or absorbing state, it never leaves the state, which means that it is "trapped."

## EXAMPLE 12.10

Consider the Markov chain with the state-transition diagram shown in Figure 12.11. Identify the transient states, the recurrent states, and the periodic states with their periods. How many chains are there in the process?

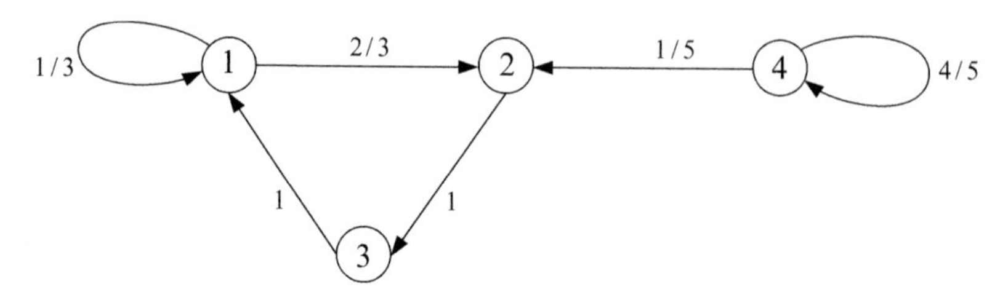


**FIGURE 12.11**
State-Transition Diagram for Example 12.10

**Solution:**

| | |
|---|---|
| Transient States: | 4 |
| Recurrent States: | 1, 2, 3 |
| Periodic States: | None |
| Single Chains: | 1 chain: {1, 2, 3} |

## EXAMPLE 12.11

Consider the state-transition diagram of Figure 12.12, which is a modified version of Figure 12.11. Here, the transition is now from state 2 to state 4 instead of from state 4 to state 2. For this case, states 1, 2, and 3 are now transient states because when the process enters state 2 and makes a transition to state 4, it does not return to these states again. Also, state 4 is a trapping (or absorbing) state because once the process enters the state, the process never leaves the state. As stated in the definition, we identify a trapping state from the fact that, as in this example, $p_{44}=1$ and $p_{4k}=0$ for $k$ not equal to 4.

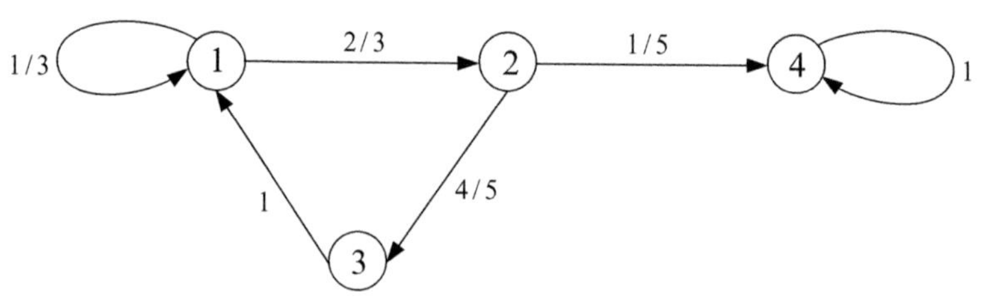


**FIGURE 12.12**
State-Transition Diagram for Example 12.11

## EXAMPLE 12.12

Identify the transient states, recurrent states, periodic states, and single chains in the Markov chain whose state-transition diagram is shown in Figure 12.13.

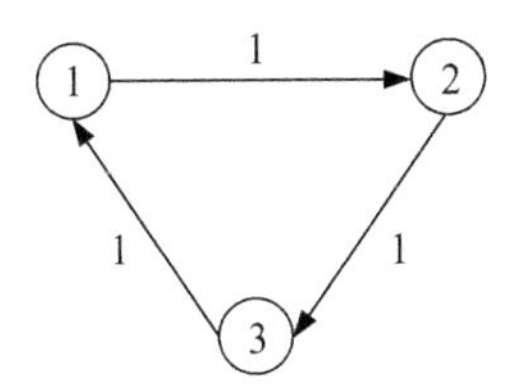


**FIGURE 12.13**
State-Transition Diagram for Example 12.12

**Solution:**

| | |
|---|---|
| Transient States: | None |
| Recurrent States: | 1, 2, 3 |
| Periodic States: | 1, 2, 3 |
| Period: | $d=3$ |
| Single Chains: | 1 chain: {1, 2, 3} |

## 12.7.5 Limiting-State Probabilities

Recall that the $n$-step state transition probability $p_{ij}(n)$ is the conditional probability that the system will be in state $j$ after exactly $n$ transitions, given that it is presently in state $i$. The $n$-step transition probabilities can be obtained by multiplying the transition probability matrix by itself $n$ times. For example, consider the following transition probability matrix:

$$P=\begin{bmatrix}0.4 & 0.5 & 0.1\\0.3 & 0.3 & 0.4\\0.3 & 0.2 & 0.5\end{bmatrix}$$

$$P^2=\begin{bmatrix}0.4 & 0.5 & 0.1\\0.3 & 0.3 & 0.4\\0.3 & 0.2 & 0.5\end{bmatrix}\times\begin{bmatrix}0.4 & 0.5 & 0.1\\0.3 & 0.3 & 0.4\\0.3 & 0.2 & 0.5\end{bmatrix}=\begin{bmatrix}0.34 & 0.37 & 0.29\\0.33 & 0.32 & 0.35\\0.33 & 0.31 & 0.36\end{bmatrix}$$

$$P^3=\begin{bmatrix}0.34 & 0.37 & 0.29\\0.33 & 0.32 & 0.35\\0.33 & 0.31 & 0.36\end{bmatrix}\times\begin{bmatrix}0.4 & 0.5 & 0.1\\0.3 & 0.3 & 0.4\\0.3 & 0.2 & 0.5\end{bmatrix}=\begin{bmatrix}0.334 & 0.339 & 0.327\\0.333 & 0.331 & 0.336\\0.333 & 0.330 & 0.337\end{bmatrix}$$

From the matrix $P^2$ we obtain the $p_{ij}(2)$. For example, $p_{23}(2)=0.35$, which is the entry in the second row and third column of the matrix $P^2$. Similarly, the entries of the matrix $P^3$ are the $p_{ij}(3)$.

For this particular matrix and matrices for a large number of Markov chains, we find that as we multiply the transition probability matrix by itself many times the entries remain constant. More importantly, all the members of one column will tend to converge to the same value.

If we define $P[X(0)=i]$ as the probability that the process is in state $i$ before it makes the first transition, then the set $\{P[X(0)=i]\}$ defines the initial condition for the process, and for an $N$-state process,

$$\sum_{i=1}^{N} P[X(0)=i]=1$$

Let $P[X(n)=j]$ denote the probability that it is in state $j$ at the end of the first $n$ transitions, then for the $N$-state process,

$$P[X(n)=j]=\sum_{i=1}^{N} P[X(0)=i]p_{ij}(n)$$

For the class of Markov chains referenced above, it can be shown that as $n\to\infty$ the $n$-step transition probability $p_{ij}(n)$ does not depend on $i$, which means that $P[X(n)=j]$ approaches a constant as $n\to\infty$ for this class of Markov chains. That is, the constant is independent of the initial conditions. Thus, for the class of Markov chains in which the limit exists, we define the *limiting-state probabilities* as follows:

$$\lim_{n\to\infty} P[X(n)=j]=\pi_j \quad j=1,2,\ldots,N$$

Recall that the $n$-step transition probability can be written in the form

$$p_{ij}(n)=\sum_{k} p_{ik}(n)p_{kj}$$

If the limiting-state probabilities exist and do not depend on the initial state, then

$$\lim_{n\to\infty} p_{ij}(n)=\pi_j=\lim_{n\to\infty}\sum_{k} p_{ik}(n)p_{kj}=\sum_{k}\pi_k p_{kj} \tag{12.28}$$

If we define the limiting-state probability vector, $\pi=[\pi_1,\pi_2,\ldots,\pi_N]$, then from (12.28) we have that $\pi=\pi P$, and

$$\pi_j=\sum_{k}\pi_k p_{kj} \tag{12.29a}$$

$$1=\sum_{j}\pi_j \tag{12.29b}$$

where the last equation is due to the law of total probability. These two equations give a system of linear equations that the $\pi_j$ must satisfy. The following propositions specify the conditions for the existence of the limiting-state probabilities:

Proposition 12.2:
In any irreducible, aperiodic Markov chain the limits $\pi_j = \lim_{n\to\infty} p_{ij}(n)$ exist and are independent of the initial distribution

Proposition 12.3:
In any irreducible, periodic Markov chain the limits $\pi_j = \lim_{n\to\infty} p_{ij}(n)$ exist and are independent of the initial distribution. However, they must be interpreted as the long-run probability that the process is in state $j$.

## EXAMPLE 12.13

Recall the biased coin problem whose state transition diagram is given in Figure 12.9 and reproduced in Figure 12.14. Find the limiting-state probabilities.

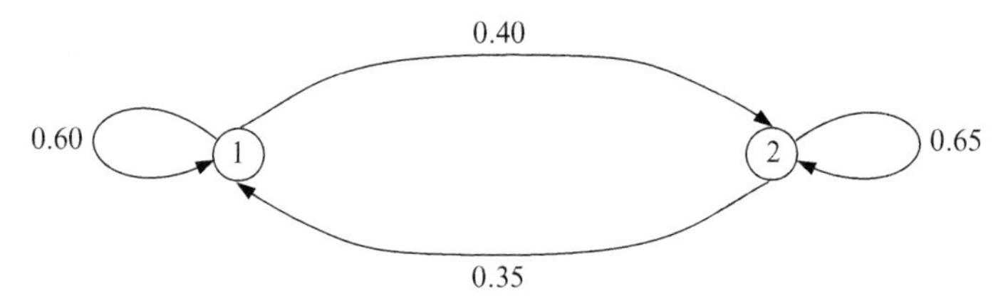


**FIGURE 12.14**
State-Transition Diagram for Example 12.13

**Solution:**
There are three equations associated with the above Markov chain, and they are

$$\pi_1 = 0.6\pi_1 + 0.35\pi_2$$
$$\pi_2 = 0.4\pi_1 + 0.65\pi_2$$
$$1 = \pi_1 + \pi_2$$

Since there are three equations and two unknowns, one of the equations is redundant. Thus, the rule-of-thumb is that for an $N$-state Markov chain, we use the first $N-1$ linear equations from the relation $\pi_j = \sum_k \pi_k p_{kj}$ and the total probability: $1 = \sum_k \pi_k$. For the given problem we have

$$\pi_1 = 0.6\pi_1 + 0.35\pi_2$$
$$1 = \pi_1 + \pi_2$$

From the first equation we obtain $\pi_1 = (0.35/0.4)\pi_2 = (7/8)\pi_2$. Substituting for $\pi_1$ and solving for $\pi_2$ in the second equation, we obtain the result

$$\pi = [\pi_1, \pi_2] = [7/15, 8/15].$$

Suppose we are also required to compute $p_{12}(3)$, which is the probability that the process will be in state 2 at the end of the third transition, given that it is presently in state 1. We can proceed in two ways: the direct method and the matrix method. We consider both methods.

a. **Direct Method**: Under this method we exhaustively enumerate all the possible ways of a state 1 to state 2 transition in 3 steps. If we use the notation $a \to b \to c$ to denote a transition from state $a$ to state $b$ and then from state $b$ to state $c$, the desired result is the following:

$$\begin{aligned} p_{12}(3) = P[&\{1 \to 1 \to 1 \to 2\} \cup \{1 \to 1 \to 2 \to 2\} \cup \{1 \to 2 \to 1 \to 2\} \\ &\cup \{1 \to 2 \to 2 \to 2\}] \end{aligned}$$

Since the different events are mutually exclusive, we obtain

$$\begin{aligned} p_{12}(3) &= P[1 \to 1 \to 1 \to 2] + P[1 \to 1 \to 2 \to 2] + P[1 \to 2 \to 1 \to 2] \\ &\quad + P[1 \to 2 \to 2 \to 2] \\ &= p_{11}p_{11}p_{12} + p_{11}p_{12}p_{22} + p_{12}p_{21}p_{12} + p_{12}p_{22}p_{22} \\ &= (0.6)(0.6)(0.4) + (0.6)(0.4)(0.65) + (0.4)(0.35)(0.4) + (0.4)(0.65)(0.65) \\ &= 0.525 \end{aligned}$$

b. **Matrix Method**: One of the limitations of the direct method is that it is difficult to exhaustively enumerate the different ways of going from state 1 to state 2 in $n$ steps, especially when $n$ is large. This is where the matrix method becomes very useful. As discussed earlier, $p_{ij}(n)$ is the $ij$th ($i$th row, $j$th column) entry in the matrix $P^n$. Thus, for the current problem, we are looking for the entry in the first row and second column of the matrix $P^3$. Therefore, we have

$$P = \begin{bmatrix} 0.60 & 0.40 \\ 0.35 & 0.65 \end{bmatrix}$$

$$P^2 = P \times P = \begin{bmatrix} 0.60 & 0.40 \\ 0.35 & 0.65 \end{bmatrix} \times \begin{bmatrix} 0.60 & 0.40 \\ 0.35 & 0.65 \end{bmatrix} = \begin{bmatrix} 0.5000 & 0.5000 \\ 0.4375 & 0.5625 \end{bmatrix}$$

$$P^3 = P \times P^2 = \begin{bmatrix} 0.60 & 0.40 \\ 0.35 & 0.65 \end{bmatrix} \times \begin{bmatrix} 0.5000 & 0.5000 \\ 0.4375 & 0.5625 \end{bmatrix} = \begin{bmatrix} 0.475000 & 0.525000 \\ 0.459375 & 0.540625 \end{bmatrix}$$

The required result (first row, second column) is 0.525, which is the result obtained via the direct method.

### 12.7.6 Doubly Stochastic Matrix

A transition probability matrix $P$ is defined to be a doubly stochastic matrix if each of its columns sums to 1. That is, not only does each row sum to 1 because $P$ is a stochastic matrix, each column also sums to 1. Thus, for every column $j$ of a doubly stochastic matrix, we have that $\sum_i p_{ij} = 1$.

Doubly stochastic matrices have interesting limiting-state probabilities, as the following theorem shows.

---

**Theorem:**
If $P$ is a doubly stochastic matrix associated with the transition probabilities of a Markov chain with $N$ states, then the limiting-state probabilities are given by $\pi_i = 1/N,\ i = 1, 2, \ldots, N$.

**Proof:**
We know that the limiting-state probabilities satisfy the condition

$$\pi_j = \sum_k \pi_k p_{kj}$$

To check the validity of the theorem, we observe that when we substitute $\pi_i = 1/N,\ i = 1, 2, \ldots, N$, in the above equation we obtain

$$\frac{1}{N} = \frac{1}{N} \sum_k p_{kj}$$

This shows that $\pi_i = 1/N$ satisfies the condition $\pi = \pi P$, which the limiting-state probabilities are required to satisfy. Conversely, from the above equation, we see that if the limiting-state probabilities are given by $\pi_i = 1/N$, then each column $j$ of $P$ sums to 1; that is, $P$ is doubly stochastic. This completes the proof.

---

## EXAMPLE 12.14

Find the transition probability matrix and the limiting-state probabilities of the process represented by the state-transition diagram shown in Figure 12.15.

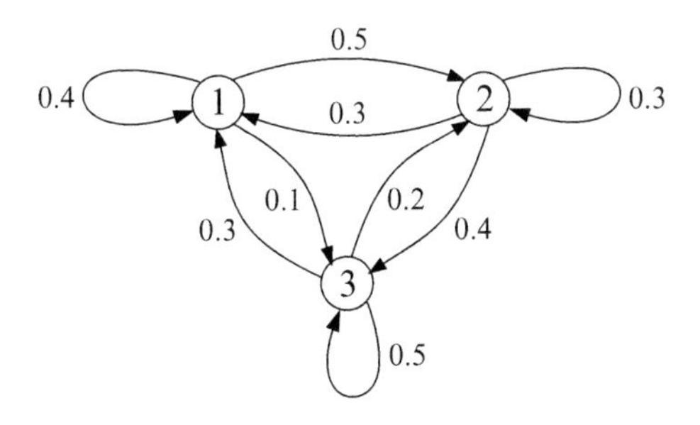


**FIGURE 12.15**
State-Transition Diagram for Example 12.14

**Solution:**
The transition probability matrix is given by

$$P = \begin{bmatrix} 0.4 & 0.5 & 0.1 \\ 0.3 & 0.3 & 0.4 \\ 0.3 & 0.2 & 0.5 \end{bmatrix}$$

It can be seen that each row of the matrix sums to 1 and each column also sums to 1; that is, $P$ is a doubly stochastic matrix. Since the process is an irreducible, aperiodic Markov chain, the limiting state probabilities exist and are given by $\pi_1 = \pi_2 = \pi_3 = 1/3$.

## 12.8 CONTINUOUS-TIME MARKOV CHAINS

A random process $\{X(t), t \geq 0\}$ is a continuous-time Markov chain if, for all $s$, $t \geq 0$ and nonnegative integers $i, j, k$,

$$P[X(t+s) = j | X(s) = i, X(u) = k, 0 \leq u \leq s] = P[X(t+s) = j | X(s) = i]$$

This means that in a continuous-time Markov chain the conditional probability of the future state *at* time $t+s$, given the present state at $s$ and all past states depends only on the present state and is independent of the past. If, in addition $P[X(t+s) = j | X(s) = i]$ is independent of $s$, then the process $\{X(t), t \geq 0\}$ is said to be *time homogeneous* or have the *time homogeneity property*. Time homogeneous Markov chains have stationary (or homogeneous) transition probabilities. Let

$$p_{ij}(t) = P[X(t+s) = j | X(s) = i]$$
$$p_j(t) = P[X(t) = j]$$

That is, $p_{ij}(t)$ is the probability that a Markov chain that is presently in state $i$ will be in state $j$ after an additional time $t$, and $p_j(t)$ is the probability that a Markov chain is in state $j$ at time $t$. Thus, the $p_{ij}(t)$ are the transition probabilities that satisfy the following condition $0 \leq p_{ij}(t) \leq 1$. Also,

$$\sum_j p_{ij}(t) = 1$$
$$\sum_j p_j(t) = 1$$

The last equation follows from the fact that at any given time the process must be in some state. Also,

$$\begin{aligned}
p_{ij}(t+s) &= \sum_k P[X(t+s) = j, X(t) = k | X(0) = i] \\
&= \sum_k \left\{ \frac{P[X(0) = i, X(t) = k, X(t+s) = j]}{P[X(0) = i]} \right\} \\
&= \sum_k \left\{ \frac{P[X(0) = i, X(t) = k]}{P[X(0) = i]} \right\} \left\{ \frac{P[X(0) = i, X(t) = k, X(t+s) = j]}{P[X(0) = i, X(t) = k]} \right\} \\
&= \sum_k P[X(t) = k | X(0) = i] P[X(t+s) = j | X(t) = k, X(0) = i] \\
&= \sum_k P[X(t) = k | X(0) = i] P[X(t+s) = j | X(t) = k] \\
&= \sum_k p_{ik}(t) p_{kj}(s)
\end{aligned} \tag{12.30}$$

This equation is called the Chapman-Kolmogorov equation for the continuous-time Markov chain. Note that the second to last equation is due to the Markov property.

Whenever a continuous-time Markov chain enters a state $i$, it spends an amount of time called the *dwell time* (or *holding time*) in that state. The holding time in state $i$ is exponentially distributed with mean $1/v_i$. At the expiration of the holding time the process makes a transition to another state $j$ with probability $p_{ij}$, where $\sum_j p_{ij} = 1$.

Because the mean holding time in state $i$ is $1/v_i$, $v_i$ represents the rate at which the process leaves state $i$ and $v_i p_{ij}$ represents the rate when in state $i$ that the process makes a transition to state $j$. Also, since the holding times are exponentially distributed, the probability that when the process is in state $i$ a transition to state $j \neq i$ will take place in the next small time $\Delta t$ is $p_{ij} v_i \Delta t$. The probability that no transition out of state $i$ will take place in $\Delta t$, given that the process is presently in state $i$ is $1 - \sum_{j \neq i} p_{ij} v_i \Delta t$; and $\sum_{j \neq i} p_{ij} v_i \Delta t$ is the probability that it leaves state $i$ in $\Delta t$.

With these definitions we consider the state-transition diagram for the process, which is shown in Figure 12.16 for state $i$. We consider the transition equations for state $i$ for the small time interval $\Delta t$.

From Figure 12.16, we obtain the following equation:

$$p_i(t + \Delta t) = p_i(t)\left\{1 - \sum_{j \neq i} p_{ij} v_i \Delta t\right\} + \sum_{j \neq i} p_j(t) p_{ji} v_j \Delta t$$

From this we have that

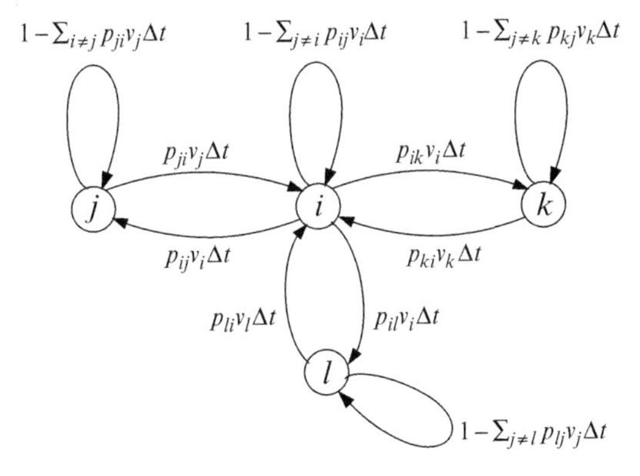


**FIGURE 12.16**
State-Transition Diagram for State *i* over Small Time

$$p_i(t+\Delta t) - p_i(t) = -v_i p_i(t) \sum_{j \neq i} p_{ij} \Delta t + \sum_{j \neq i} p_j(t) p_{ji} v_j \Delta t$$

$$\frac{p_i(t+\Delta t) - p_i(t)}{\Delta t} = -v_i p_i(t) \sum_{j \neq i} p_{ij} + \sum_{j \neq i} p_j(t) p_{ji} v_j$$

$$\lim_{\Delta t \to 0} \left\{ \frac{p_i(t+\Delta t) - p_i(t)}{\Delta t} \right\} = \frac{dp_i(t)^i}{dt} = -v_i p_i(t) \sum_{j \neq i} p_{ij} + \sum_{j \neq i} p_j(t) p_{ji} v_j$$

In the steady state, $p_j(t) \to p_j$ and

$$\lim_{t \to \infty} \left\{ \frac{dp_i(t)^i}{dt} \right\} = 0$$

Thus, we obtain

$$0 = -v_i p_i \sum_{j \neq i} p_{ij} + \sum_{j \neq i} p_j p_{ji} v_j$$

$$1 = \sum_i p_i$$

Alternatively, we may write

$$v_i p_i \sum_{j \neq i} p_{ij} = \sum_{j \neq i} p_j p_{ji} v_j \tag{12.31a}$$

$$1 = \sum_i p_i \tag{12.31b}$$

The left-hand side of the first equation is the rate of transition out of state $i$, while the right-hand side is the rate of transition into state $i$. This "balance" equation states that in the steady state the two rates are equal for any state in the Markov chain.

### 12.8.1 Birth and Death Processes

Birth and death processes are a special type of continuous-time Markov chains. Consider a continuous-time Markov chain with states $0, 1, 2, \ldots$. If $p_{ij} = 0$ whenever $j \neq i-1$ or $j \neq i+1$, then the Markov chain is called a birth and death process. Thus, a birth and death process is a continuous-time Markov chain with states $0, 1, 2, \ldots$, in which transitions from state $i$ can only go to either state $i+1$ or state $i-1$. That is, a transition either causes an increase in state by one or a decrease in state by one. A birth is said to occur when the state increases by one, and a death is said to occur when the state decreases by one. For a birth and death process, we define the following *transition rates* from state $i$:

$$\lambda_i = v_i p_{i(i+1)}$$

$$\mu_i = v_i p_{i(i-1)}$$

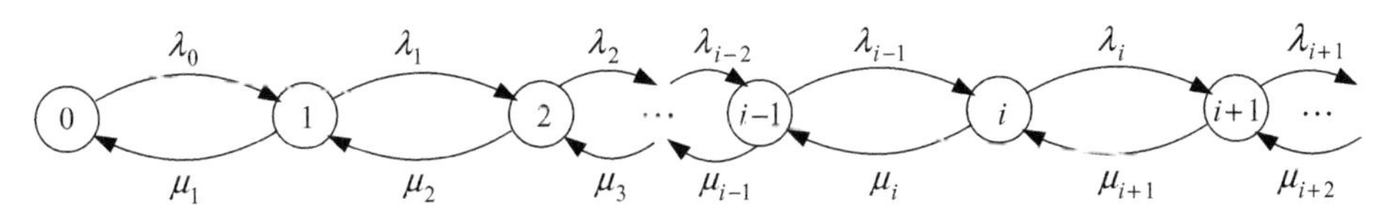


**FIGURE 12.17**
State-Transition-Rate Diagram for Birth and Death Process

Thus, $\lambda_i$ is the rate at which a birth occurs when the process is in state $i$ and $\mu_i$ is the rate at which a death occurs when the process is in state $i$. The sum of these two rates is $v_i = \lambda_i + \mu_i$, which is the rate of transition out of state $i$. The *state-transition-rate diagram* of a birth and death process is shown in Figure 12.17. It is called a state-transition-rate diagram as opposed to a state-transition diagram because it shows the rate at which the process moves from state to state and not the probability of moving from one state to another. Note that $\mu_0 = 0$, since there can be no death when the process is in empty state.

The actual state-transition probabilities when the process is in state $i$ are $p_{i(i+1)}$ and $p_{i(i-1)}$. By definition, $p_{i(i+1)} = \lambda_i/(\lambda_i + \mu_i)$ is the probability that a birth occurs before a death when the process is in state $i$. Similarly, $p_{i(i-1)} = \mu_i/(\lambda_i + \mu_i)$ is the probability that a death occurs before a birth when the process is in state $i$.

Recall that the rate at which the probability of the process being in state $i$ changes with time is given by

$$\frac{dp_i(t)}{dt} = -v_i p_i(t) \sum_{j \neq i} p_{ij} + \sum_{j \neq i} p_j(t) p_{ji} v_j = -(\lambda_i + \mu_i) p_i(t) + \mu_{i+1} p_{i+1}(t) + \lambda_{i-1} p_{i-1}(t)$$

In the steady state,

$$\lim_{t \to \infty} \left\{ \frac{dp_i(t)}{dt} \right\} = 0$$

which gives

$$(\lambda_i + \mu_i) p_i(t) = \mu_{i+1} p_{i+1}(t) + \lambda_{i-1} p_{i-1}(t) \tag{12.32}$$

The equation states that the rate at which the process leaves state $i$ either through a birth or a death is equal to the rate at which it enters the state through a birth when the process is in state $i-1$ or through a death when the process is in state $i+1$.

If we assume that the limiting probabilities $\lim_{t \to \infty} p_{ij}(t) = p_j$ exist, then from the above equation we obtain the following:

$$(\lambda_i + \mu_i) p_i = \mu_{i+1} p_{i+1} + \lambda_{i-1} p_{i-1} \tag{12.33a}$$

$$\sum_i p_i = 1 \tag{12.33b}$$

This is called the *balance equation* because it balances (or equates) the rate at which the process enters state $i$ with the rate at which it leaves state $i$.

## EXAMPLE 12.15

A machine is operational for an exponentially distributed time with mean $1/\lambda$ before breaking down. When it breaks down, it takes a time that is exponentially distributed with mean $1/\mu$ to repair it. What is the fraction of time that the machine is operational (or available)?

**Solution:**

This is a two-state birth and death process. Let U denote the up state and D the down state. Then, the state-transition-rate diagram is shown in Figure 12.18.

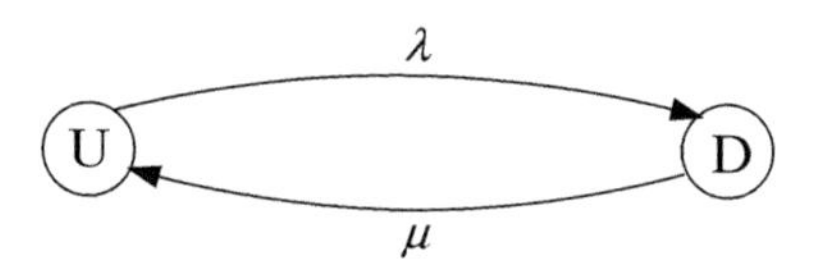


**FIGURE 12.18**
State-Transition-Rate Diagram for Example 12.15

Let $p_U$ denote the steady-state probability that the process is in the operational state, and let $p_D$ denote the steady-state probability that the process is in the down state. Then the balance equations become

$$\lambda p_U = \mu p_D$$
$$p_U + p_D = 1 \Rightarrow p_D = 1 - p_U$$

Substituting for $p_D$ in the first equation gives $p_U = \mu/(\lambda + \mu)$.

## EXAMPLE 12.16

Customers arrive at a bank according to a Poisson process with rate $\lambda$. The time to serve each customer is exponentially distributed with mean $1/\mu$. There is only one teller at the bank, and an arriving customer who finds the teller busy when she arrives will join a single queue that operates on a first-come, first-served basis. Determine the limiting-state probabilities given that $\mu > \lambda$.

**Solution:**

This is a continuous-time Markov chain in which arrivals constitute births and service completions constitute deaths. Also, for all $i$, $\lambda_i = \lambda$ and $\mu_i = \mu$. Thus if $p_k$ denotes the steady-state probability that there are $k$ customers in the system, the balance equations are as follows:

$$\lambda p_0 = \mu p_1 \Rightarrow p_1 = \left(\frac{\lambda}{\mu}\right) p_0$$

$$(\lambda+\mu)p_1 = \lambda p_0 + \mu p_2 \Rightarrow p_2 = \left(\frac{\lambda}{\mu}\right) p_1 = \left(\frac{\lambda}{\mu}\right)^2 p_0$$

$$(\lambda+\mu)p_2 = \lambda p_1 + \mu p_3 \Rightarrow p_3 = \left(\frac{\lambda}{\mu}\right) p_2 = \left(\frac{\lambda}{\mu}\right)^3 p_0$$

In general it can be shown that

$$p_k = \left(\frac{\lambda}{\mu}\right)^k p_0 \qquad k = 0, 1, 2, \ldots$$

Now,

$$\sum_{k=0}^{\infty} p_k = 1 = p_0 \sum_{k=0}^{\infty} \left(\frac{\lambda}{\mu}\right)^k = \frac{p_0}{1 - \frac{\lambda}{\mu}}$$

where the equality follows from the fact that $\mu > \lambda$. Thus,

$$p_0 = 1 - \frac{\lambda}{\mu}$$

$$p_k = \left(1 - \frac{\lambda}{\mu}\right)\left(\frac{\lambda}{\mu}\right)^k \qquad k = 0, 1, 2, \ldots$$

## 12.9 GAMBLER'S RUIN AS A MARKOV CHAIN

Recall the gambler's ruin problem discussed in section 12.3.2. Two players A and B play a series of games with A starting with \$$a$ and B starting with \$$b$, where $a+b=N$. With probability $p$ player A wins each game from player B and thus gains an additional \$1, and with probability $q=1-p$ player A loses each game to B, and thus loses \$1. If A reaches \$0 he is ruined, and the game ends. Similarly if B reaches \$0, he is ruined and the game is over. Let the state of the game be $k$, which denotes the total amount that A currently has. Thus, the game ends when $k=0$ and A is ruined; it also ends when $k=N$ and B is ruined. This means that states 0 and $N$ are absorbing states. Let $p_{ik}$ denote the conditional probability that the game will move to state $k$ next, given that it is currently in state $i$. Then $p_{ik}$ is the state transition probability, which is given by

$$p_{ik} = \begin{cases} p & k = i+1, i \neq 0 \\ 1-p & k = i-i, i \neq N \\ 1 & k = i = 0 \\ 1 & k = i = N \\ 0 & \text{otherwise} \end{cases}$$

Thus, the state transition probability matrix is given by

$$P=\begin{bmatrix}
1 & 0 & 0 & 0 & 0 & \cdots & 0 & 0 & 0 & 0 & 0 \\
1-p & 0 & p & 0 & 0 & \cdots & 0 & 0 & 0 & 0 & 0 \\
0 & 1-p & 0 & p & 0 & \cdots & 0 & 0 & 0 & 0 & 0 \\
0 & 0 & 1-p & 0 & p & \cdots & 0 & 0 & 0 & 0 & 0 \\
\cdots & \cdots & \cdots & \cdots & \cdots & \cdots & \cdots & \cdots & \cdots & \cdots & \cdots \\
\cdots & \cdots & \cdots & \cdots & \cdots & \cdots & \cdots & \cdots & \cdots & \cdots & \cdots \\
0 & 0 & 0 & 0 & 0 & \cdots & p & 0 & 0 & 0 & 0 \\
0 & 0 & 0 & 0 & 0 & \cdots & 0 & p & 0 & 0 & 0 \\
0 & 0 & 0 & 0 & 0 & \cdots & 1-p & 0 & p & 0 & 0 \\
0 & 0 & 0 & 0 & 0 & \cdots & 0 & 0 & 1-p & 0 & p \\
0 & 0 & 0 & 0 & 0 & \cdots & 0 & 0 & 0 & 0 & 1
\end{bmatrix}$$

Similarly, the state transition diagram for the game is shown in Figure 12.19.

The preceding process assumes that both the gambler and his adversary are playing to obtain each other's fortune and will stop when either one is out of money (i.e., either one is ruined). Sometimes they can play for sports, where when one is ruined, the other gives him \$1 (or whatever is the cost of each play) to continue. That is, when the process enters state 0, then with probability $1-p_0$ it stays in state 0, and with probability $p_0$ it makes a transition to state 1. Similarly, when it enters state $N$, it stays in state $N$ with probability $p_N$ and makes a transition to state $N-1$ with probability $1-p_N$. Note that $p_0$ and $p_N$ need not be equal to $p$. This scheme is a type of random walk with reflecting barriers and is illustrated in Figure 12.20. Note that when $p_0=0$ and $p_N=1$ we obtain the classical gambler's ruin shown in Figure 12.19.

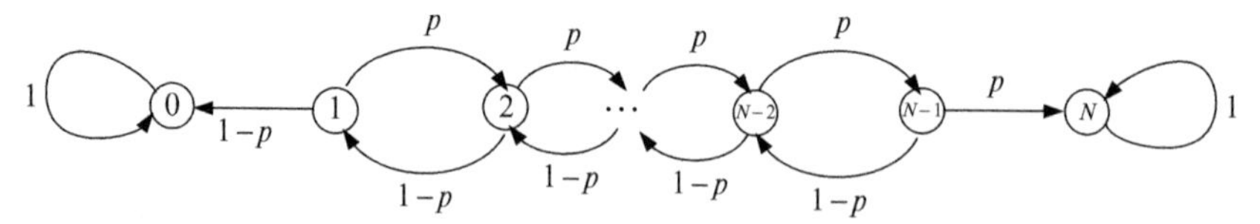


**FIGURE 12.19**

State Transition Diagram for the Gambler's Ruin

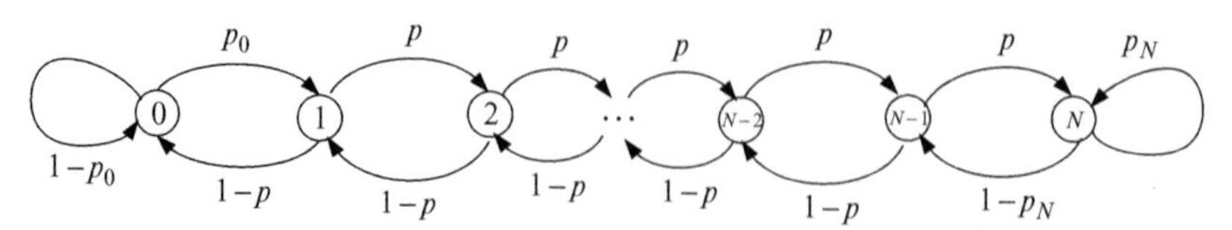


**FIGURE 12.20**

State Transition Diagram for Gambler's Ruin Game for Sports

## 12.10 CHAPTER SUMMARY

This chapter has considered some of the frequently used random processes in system modeling. These are the Bernoulli process, the Gaussian process, the random walk process, the Poisson process, and the Markov process. The Bernoulli process is used to model a sequence of trials, each of which results in one of two outcomes that are generally described as success or failure. For this process the number of trials between two successes is geometrically distributed, and the number of successes in a given number of trials has the binomial distribution. Finally, the number of trials up to and including the trial that results in the $k$th success is the $k$th-order Pascal random variable.

The random walk process is an extension of the Bernoulli process. Here the process takes a step in the positive direction in the $k$th trial if the outcome of the trial is a success and a step in the negative direction if the outcome is a failure. The resulting trajectory of the process as it moves through the $xy$ plane, where the $x$ coordinate represents time and the $y$ coordinate represents the location at a given time, is the so-called one-dimensional random walk. It is commonly used in a class of problems called the gambler's ruin.

A random process $\{X(t), t \in T\}$ is defined to be a Gaussian random process if and only if for any choice of $n$ time instants $t_1, t_2, \ldots, t_n$ in the index set $T$ the random variables $X(t_1), X(t_2), \ldots, X(t_n)$ have a jointly normal PDF. Gaussian processes are important because many physical problems are the results of adding large numbers of independent random variables. According to the central limit theorem such sums of independent random variables are essentially normal (or Gaussian) random variables.

The Poisson process is a counting process that is used to model systems where the interarrival times are exponentially distributed. The time until the $k$th arrival has the Erlang-$k$ distribution. Because of the forgetfulness property of the exponential distribution, Poisson processes are popularly used to model many arrival processes like customers at a restaurant or library, and message arrivals at switchboards.

Finally, the Markov process is used to model systems with limited memory of the past. For the so-called first-order Markov process, given the present state of the process the future state is independent of the past. This property is usually referred to as the Markov property.

## 12.11 PROBLEMS

### Section 12.2 Bernoulli Process

12.1 A random process $Y[n]$ is defined by $Y[n] = 3X[n] + 1$, where $X[n]$ is a Bernoulli process with a success probability $p$. Find the mean and variance of $Y[n]$.

12.2 A sequence of Bernoulli trials consists of choosing seven components at random from a batch of components. A selected component is classified as either defective or non-defective. A non-defective component is considered a success, while a defective component is considered a failure. If the probability that a selected component is non-defective is 0.8, what is the probability of three successes?

12.3 The probability that a patient recovers from a rare blood disease is 0.3. If 15 people are known to have contracted this disease, find the following probabilities:

a. At least 10 survive.
b. From 3 to 8 survive.
c. Exactly 6 survive.

12.4 A sequence of Bernoulli trials consists of choosing components at random from a batch of components. A selected component is classified as either defective or non-defective. A non-defective component is considered a success, while a defective component is considered a failure. If the probability that a selected component is non-defective is 0.8, determine the probabilities of the following events:

a. The first success occurs on the fifth trial.
b. The third success occurs on the eighth trial.
c. There are 2 successes by the fourth trial, there are 4 successes by the tenth trial, and there are 10 successes by the eighteenth trial.

12.5 A lady invites 12 people for dinner at her house. Unfortunately the dining table can only seat 6 people. Her plan is that if 6 or fewer guests come, then they will be seated at the table (i.e., they will have a sit-down dinner); otherwise she will set up a buffet-style meal. The probability that each invited guest will come to dinner is 0.4, and each guest's decision is independent of other guests' decisions. Determine the following:

a. The probability that she has a sit-down dinner
b. The probability that she has a buffet-style dinner
c. The probability that there are at most three guests

12.6 A Girl Scout troop sells cookies from house to house. One of the parents of the girls figured out that the probability that they sell a set of packs of cookies at any house they visit is 0.4, where it is assumed that they sell exactly one set to each house that buys their cookies.

a. What is the probability that the house where they make their first sale is the fifth house they visit?
b. Given that they visited 10 houses on a particular day, what is the probability that they sold exactly 6 sets of cookie packs?
c. What is the probability that on a particular day the third set of cookie packs is sold at the seventh house that the girls visit?

## Section 12.3 Random Walk

12.7 A bag contains 3 red balls, 6 green balls, and 2 blue balls. Jack plays a game in which he bets $1 to draw a ball from the bag. If he draws a green ball, he wins $1 dollar; otherwise he loses a dollar. Assume that the balls are drawn with replacement and that Jack starts the game with $50 with the hope of reaching $100 before going bankrupt. What is the probability that he will succeed?

12.8 Consider a gambler's ruin game in which $p=q=1/2$. Let $d_i, 0<i<N$, denote the expected duration of a game in which a gambler starts from state $i$, such as starting with $\$i$. He gets $1 when he wins a game and loses $1 when he loses a game. The boundary conditions for the series of games are $d_0=0$ and $d_N=0$; that is, the gambler is ruined when he enters state 0, and the series ends when he enters state $N$.

a. Show that $d_i$ is given by

$$d_i = \begin{cases} 0 & i=0,N \\ 1+\dfrac{d_{i+1}+d_{i-1}}{2} & i=1,2,\ldots,N-1 \end{cases}$$

b. Obtain the general expression for $d_i$ from the above equation terms of $N$.

12.9 Consider a variation of the gambler's ruin problem with parameters $p$ (i.e., the probability that player A wins a game) and $N$ modeled in terms of a random walk with *reflecting barrier* at zero. Specifically, when state 0 is reached, the process moves to state 1 with probability $p_0$ or stays at state 0 with probability $1-p_0$. Thus, the only trapping state is $N$. That is, only player B can be ruined.

a. Give the state transition diagram of the process.

b. If $r_i$ is the probability of player B being ruined when the process is currently in state $i$, obtain the expression for $r_i$ to show what happens on the first game when the process is in state $i$.

12.10 Ben and Jerry play a series of games of checkers. During each game each player bets a $1, and whoever wins the game gets the $2. Ben is a better player than Jerry and has a probability 0.6 of winning each game. Initially Ben had $9, while Jerry had $6, and the game is over when either player is wiped out.

a. What is the probability that Ben is ruined; that is, that Jerry will wipe him out?

b. What is the probability that Jerry is ruined?

12.11 Ben and Jerry play a series of games of cards. During each game each player bets a $1, and whoever wins the game gets the $2. Sometimes a game can end in a tie, in which case neither player loses his money. Ben is a better player than Jerry and has a probability 0.5 of winning each game, a probability 0.3 of losing each game, and probability 0.2 of tying with Jerry. Initially Ben had \$9, while Jerry had \$6, and the game is over when either player is wiped out.

a. Give the state transition diagram of the process.
b. If $r_k$ denotes the probability that Ben is ruined, given that the process is currently in state $k$, obtain an expression for $r_k$ in the first game when the process is in state $k$.

## Section 12.4 Gaussian Process

12.12 Suppose that $X(t)$ is a wide-sense stationary Gaussian process with the autocorrelation function

$$R_{XX}(\tau) = 4 + e^{-|\tau|}$$

Determine the covariance matrix for the random variables $X(0), X(1), X(3)$, and $X(6)$.

12.13 A Gaussian process $X(t)$ has an autocorrelation function

$$R_{XX}(\tau) = \frac{4\sin(\pi\tau)}{\pi\tau}$$

Determine the covariance matrix for the random variables $X(t), X(t+1), X(t+2)$ and $X(t+3)$.

12.14 Suppose $X(t)$ is a Gaussian random process with a mean $E[X(t)] = 0$ and autocorrelation function $R_{XX}(\tau) = e^{-|\tau|}$. Assume that the random variable $A$ is defined as follows:

$$A = \int_0^1 X(t)dt$$

Determine the following:

a. $E[A]$
b. $\sigma_A^2$

12.15 Suppose $X(t)$ is a Gaussian random process with a mean $E[X(t)] = 0$ and autocorrelation function $R_{XX}(\tau) = e^{-|\tau|}$. Assume that the random variable $A$ is defined as follows:

$$A = \int_0^B X(t)dt$$

where $B$ is a uniformly distributed random variable with values between 1 and 5 and is independent of the random process $X(t)$. Determine the following:

a. $E[A]$
b. $\sigma_A^2$

## Section 12.5 Poisson Process

12.16 University buses arrive at the Students Center to take students to their classes according to a Poisson process with an average rate of 5 buses per hour. Chris just missed the last bus. What is the probability that he waits more than 20 minutes before boarding a bus?

12.17 Cars arrive at a gas station according to a Poisson process at an average rate of 12 cars per hour. The station has only one attendant. If the attendant decides to take a 2-minute coffee break when there are no cars at the station, what is the probability that one or more cars will be waiting when he comes back from the break, given that any car that arrives when he is on coffee break waits for him to get back?

12.18 Cars arrive at a gas station according to a Poisson process at an average rate of 50 cars per hour. There is only one pump at the station, and the attendant takes 1 minute to fill up each car. What is the probability that a waiting line will form at the station? (Note: A waiting line occurs if two or more cars arrive in any 1-minute period.)

12.19 Studies indicate that the probability that three cars will arrive at a parking lot in a 5-minute interval is 0.14. If cars arrive according to a Poisson process, determine the following:

a. The average arrival rate of cars
b. The probability that no more than 2 cars arrive in a 10-minute interval

12.20 Telephone calls arrive at a switching center according to a Poisson process at an average rate of 75 calls per minute. What is the probability that more than three calls arrive within a 5-second period.

12.21 An insurance company pays out claims on its life insurance policies in accordance with a Poisson process with an average rate of 5 claims per week. If the amount of money paid on each policy is uniformly distributed between \$2,000 and \$10,000, what is the mean of the total amount of money that the company pays out in a four-week period?

12.22 Customers arrive at the neighborhood bookstore according to a Poisson process with an average rate of 10 customers per hour. Independent of other customers, each arriving customer buys a book with probability 1/8.

a. What is the probability that the bookstore sells no book during a particular hour?
b. What is the PDF of the time until the first book is sold?

12.23 Joe is a student who is conducting experiments with a series of lightbulbs. He started with 10 identical lightbulbs, each of which has an exponentially distributed lifetime with a mean of 200 hours. Joe wants to know how long it will take until the last bulb burns out (or fails). At noontime Joe stepped out to get some lunch with 6 bulbs still on. Assume that Joe came back and found that none of the 6 bulbs has failed.

a. After Joe came back, what is the expected time until the next bulb failure?
b. What is the expected length of time between the fourth bulb failure and the fifth bulb failure?

12.24 Three customers $A$, $B$, and $C$ simultaneously arrive at a bank with two tellers on duty. The two tellers were idle when the three customers arrived, and $A$ goes directly to one teller, $B$ goes to the other teller, and $C$ waits until either $A$ or $B$ leaves before she can begin receiving service. If the service times provided by the tellers are exponentially distributed with a mean of 4 minutes, what is the probability that customer $A$ is still in the bank after the other two customers leave?

12.25 The times between component failures in a certain system are exponentially distributed with a mean of 4 hours. What is the probability that at least one component failure occurs within a 30-minute period?

12.26 Students arrive at the professor's office for extra help according to a Poisson process with an average rate of 4 students per hour. The professor does not start the tutorial until at least 3 students are available. Students who arrive while the tutorial is going on will have to wait for the next session.
a. Given that a tutorial has just ended and there are no students currently waiting for the professor, what is the mean time until another tutorial can start?
b. Given that one student was waiting when the tutorial ended, what is the probability that the next tutorial does not start within the first 2 hours?

12.27 Customers arrive at a bank according to a Poisson process with an average rate of 6 customers per hour. Each arriving customer is either a man with probability $p$ or a woman with probability $1-p$. It was found that in the first 2 hours the average number of men who arrived at the bank was 8. What is the average number of women who arrived over the same period?

12.28 Alan is conducting an experiment to test the mean lifetimes of two sets of electric bulbs labeled $A$ and $B$. The manufacturer claims that the mean lifetime of bulbs in set $A$ is 200 hours, while the mean lifetime of the bulbs in set $B$ is 400 hours. The lifetimes for both sets are exponentially distributed. Alan's experimental procedure is as follows. He started with one bulb from each set. As soon as a bulb from a given set fails (or burns out), he immediately replaces it with a new bulb from the same set and writes down the lifetime of the burnt-out bulb. Thus, at any point in time he has two bulbs on, one from each set. If at the end of the week Alan tells you that 8 bulbs have failed, determine the following:
a. The probability that exactly 5 of those 8 bulbs are from set $B$

b. The probability that no bulb will fail in the first 100 hours
c. The mean time between two consecutive bulb failures

12.29 The Merrimack Airlines company runs a commuter air service between Manchester, New Hampshire, and Cape Cod, Massachusetts. Since the company is a small one, there is no set schedule for their flights, and no reservation is needed for the flights. However, it has been determined that their planes arrive at the Manchester airport according to a Poisson process with an average rate of 2 planes per hour. Vanessa arrived at the Manchester airport and had to wait to catch the next flight.
a. What is the mean time between the instant Vanessa arrived at the airport until the time the next plane arrived?
b. What is the mean time between the arrival time of the last plane that took off from the Manchester airport before Vanessa arrived and the arrival time of the plane that she boarded?

12.30 Bob has a pet that requires the light in his apartment to always be on. To achieve this, Bob keeps three lightbulbs on with the hope that at least one bulb will be operational when he is not at the apartment. The lightbulbs have independent and identically distributed lifetimes $T$ with PDF $f_T(t)=\lambda e^{-\lambda t}, t\geq 0$.
a. Probabilistically speaking, given that Bob is about to leave the apartment and all three bulbs are working fine, what does he gain by replacing all three bulbs with new ones before he leaves?
b. Suppose $X$ is the random variable that denotes the time until the first bulb fails. What is the PDF of $X$?
c. Given that Bob is going away for an indefinite period of time and all three bulbs are working fine before he leaves, what is the PDF of $Y$, the time until the third bulb failure after he leaves?
d. What is the expected value of $Y$?

12.31 Joe just replaced two lightbulbs one of which is rated 60 watts with an exponentially distributed lifetime whose mean is 200 hours, and the other is rated 100 watts with an exponentially distributed lifetime whose mean is 100 hours.
a. What is the probability that the 60-watt bulb fails before the 100-watt bulb?
b. What is the mean time until the first of the two bulbs fails?
c. Given that the 60-watt bulb has not failed after 300 hours, what is the probability that it will last at least another 100 hours?

12.32 A 5-motor machine can operate properly if at least 3 of the 5 motors are functioning. If the lifetime $X$ of each motor has the PDF $f_X(x)=\lambda e^{-\lambda x}$, $x\geq 0, \lambda>0$, and if the lifetimes of the motors are independent, what is the mean of the random variable $Y$, the time until the machine fails?

12.33 Alice has two identical personal computers, which she never uses at the same time. She uses one PC at a time and the other is a backup. If the one she is currently using fails, she turns it off, calls the PC repairman and turns on the backup PC. The time until either PC fails when it is in use is exponentially distributed with a mean of 50 hours. The time between the moment a PC fails until the repairman comes and finishes repairing it is also exponentially distributed with a mean of 3 hours. What is the probability that Alice is idle because neither PC is operational?

12.34 Cars arrive from the northbound section of an intersection according to a Poisson process at the rate of $\lambda_N$ cars per minute and from the eastbound section according to a Poisson process at the rate of $\lambda_E$ cars per minute.

a. Given that there is currently no car at the intersection, what is the probability that a northbound car arrives before an eastbound car?
b. Given that there is currently no car at the intersection, what is the probability that the fourth northbound car arrives before the second eastbound car?

12.35 A one-way street has a fork in it, and cars arriving at the fork can either bear right or left. A car arriving at the fork will bear right with probability 0.6 and will bear left with probability 0.4. Cars arrive at the fork according to a Poisson process with a rate of 8 cars per minute.

a. What is the probability that at least four cars bear right at the fork in three minutes?
b. Given that three cars bear right at the fork in three minutes, what is the probability that two cars bear left at the fork in three minutes?
c. Given that 10 cars arrive at the fork in three minutes, what is the probability that four of the cars bear right at the fork?

## Section 12.7 Discrete-Time Markov Chains

12.36 Determine the missing elements denoted by $x$ in the following transition probability matrix:

$$P = \begin{bmatrix} x & 1/3 & 1/3 & 1/3 \\ 1/10 & x & 1/5 & 2/5 \\ x & x & x & 1 \\ 3/5 & 2/5 & x & x \end{bmatrix}$$

12.37 Draw the state-transition diagram for the Markov chain with the following transition probability matrix.

$$P = \begin{bmatrix} 1/2 & 0 & 0 & 1/2 \\ 1/2 & 1/2 & 0 & 0 \\ 1/4 & 0 & 1/2 & 1/4 \\ 0 & 1/2 & 1/4 & 1/4 \end{bmatrix}$$

12.38 Consider a Markov chain with the state-transition diagram shown in Figure 12.21.

a. Give the transition probability matrix.
b. Identify the recurrent states.
c. Identify the transient states.

12.39 Consider the Markov chain with the state-transition diagram shown in Figure 12.22.

a. List the transient states, the recurrent states, and the periodic states.
b. Identify the members of each chain of recurrent states.
c. Give the transition probability matrix of the process.
d. Given that the process starts in state 1, either determine the numerical value of the probability that the process is in state 8 after an infinitely large number of transitions or explain why this quantity does not exist.

12.40 Consider the three-state Markov chain shown in Figure 12.23:

a. Identify the transient states, the recurrent states, the periodic states, and the members of each chain of recurrent states.
b. Either determine the limiting-state probabilities or explain why they do not exist.
c. Given that the process is currently in state 1, determine $P[A]$, the probability that it will be in state 3 at least once during the next two transitions.

12.41 Consider the Markov chain shown in Figure 12.24:

a. Which states are transient?
b. Which states are periodic?
c. Does state 3 have a limiting-state probability? If so, determine this probability.

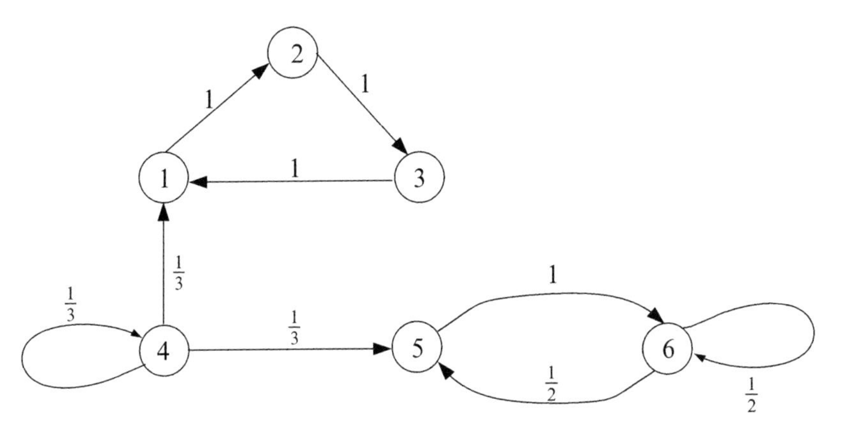


**FIGURE 12.21**
Figure for Problem 12.38

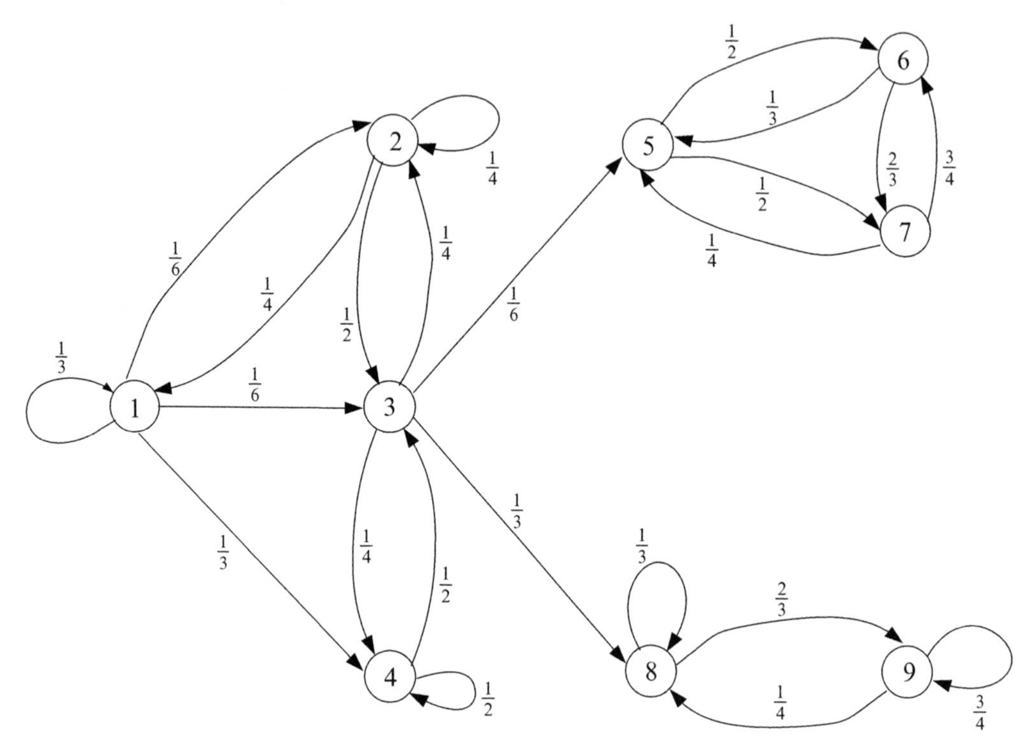


**FIGURE 12.22**
Figure for Problem 12.39

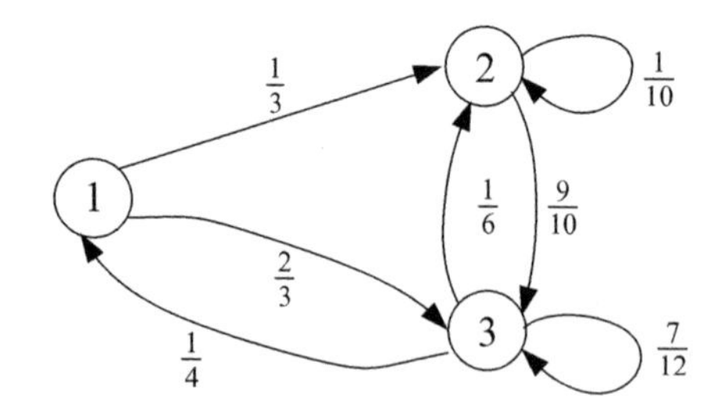


**FIGURE 12.23**
Figure for Problem 12.40

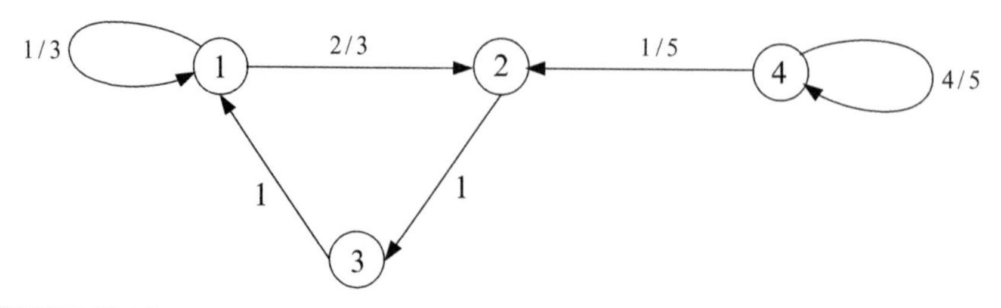


**FIGURE 12.24**
Figure for Problem 12.41

d. Assuming that the process begins in state 4, determine the z-transform of the PMF of $K$, where $K$ is the number of trials up to and including the trial in which the process enters state 2 for the second time.

12.42 Find the limiting-state probabilities associated with the following transition probability matrix:

$$P = \begin{bmatrix} 0.4 & 0.3 & 0.3 \\ 0.3 & 0.4 & 0.3 \\ 0.3 & 0.3 & 0.4 \end{bmatrix}$$

12.43 Consider the following transition probability matrix:

$$P = \begin{bmatrix} 0.6 & 0.2 & 0.2 \\ 0.3 & 0.4 & 0.3 \\ 0.0 & 0.3 & 0.7 \end{bmatrix}$$

a. Give the state-transition diagram.
b. Given that the process is currently in state 1, what is the probability that it will be in state 2 at the end of the third transition?
c. Given that the process is currently in state 1, what is the probability that the first time it enters state 3 is the fourth transition?

12.44 Consider the following social mobility problem. Studies indicate that people in a society can be classified as belonging to the upper class (state 1), middle class (state 2), and lower class (state 3). Membership in any class is inherited in the following probabilistic manner. Given that a person is raised in an upper-class family, he will have an upper-class family with probability 0.45, a middle-class family with probability 0.48, and a lower-class family with probability 0.07. Given that a person is raised in a middle-class family, he will have an upper-class family with probability 0.05, a middle-class family with probability 0.70, and a lower-class family with probability 0.25. Finally, given that a person is raised in a lower-class family, he will have an upper-class family with probability 0.01, a middle-class family with probability 0.50, and a lower-class family with probability 0.49. Determine the following:

a. The state-transition diagram of the process
b. The transition probability matrix of the process
c. The limiting-state probabilities. Interpret what they mean to the layperson.

12.45 A taxi driver conducts his business in three different towns 1, 2, and 3. On any given day, when he is in town 1, the probability that the next passenger he picks up is going to a place in town 1 is 0.3, the probability that the next passenger he picks up is going to town 2 is 0.2, and the probability that the next passenger he picks up is going to town 3 is 0.5. When he is in town 2, the probability that the next passenger he picks up is going to town 1 is 0.1, the probability that the next passenger he picks up is going to town 2 is 0.8, and the probability that the next passenger he picks up is going to town 3 is 0.1. When he is in town 3, the probability that the next passenger he picks up is going to town 1 is 0.4, the probability that the next passenger he picks up is going to town 2 is 0.4, and the probability that the next passenger he picks up is going to town 3 is 0.2.

a. Determine the state-transition diagram for the process.
b. Give the transition probability matrix for the process.
c. What are the limiting-state probabilities?
d. Given that the taxi driver is currently in town 2 and is waiting to pick up his first customer for the day, what is the probability that the first time he picks up a passenger to town 2 is when he picks up his third passenger for the day?
e. Given that he is currently in town 2, what is the probability that his third passenger from now will be going to town 1?

12.46 New England fall weather can be classified as sunny, cloudy, or rainy. A student conducted a detailed study of the weather pattern and came up with the following conclusion: Given that it is sunny on any given day, then on the following day it will be sunny again with probability 0.5, cloudy with probability 0.3 and rainy with probability 0.2. Given that it is cloudy on any given day, then on the following day it will be sunny with probability 0.4, cloudy again with probability 0.3 and rainy with probability 0.3. Finally, given that it is rainy on any given day, then on the following day it will be sunny with probability 0.2, cloudy with probability 0.5 and rainy again with probability 0.3.

a. Give the state-transition diagram of New England fall weather with the state "sunny" as state 1, the state "cloudy" as state 2, and the state "rainy" as state 3.
b. Using the same convention as in part (a), give the transition probability matrix of New England fall weather.
c. Given that it is sunny today, what is the probability that it will be sunny four days from now?
d. Determine the limiting-state probabilities of the weather.

12.47 A student went to a gambling casino with \$3. He wins \$1 at each round with a probability $p$ and loses \$1 with a probability $1-p$. Being a very cautious player, the student has decided to stop playing when he doubles his original \$3 (i.e., when he has a total of \$6) or when he loses all his money.

a. Give the state-transition diagram of the process.
b. What is the probability that he stops after being ruined (i.e., he lost all his money)?
c. What is the probability that he stops after he has doubled his original amount?

## Section 12.8 Continuous-Time Markov Chains

12.48 A small company has two identical servers that are running at the same time. The time until either server fails is exponentially distributed with a mean of $1/\lambda$. When a server fails, a technician starts repairing it immediately. The two servers fail independently of each other. The time to repair a failed server is exponentially distributed with a mean of $1/\mu$. As soon as the repair is completed the server is brought back on line and is assumed to be as good as new.

a. Give the state-transition-rate diagram of the process.
b. What is the fraction of time that both servers are down?

12.49 Customers arrive at Mike's barber shop according to a Poisson process with rate $\lambda$ customers per hour. Unfortunately Mike, the barber, has only five chairs in his shop for customers to wait when there is already a customer receiving a haircut. Customers who arrive when Mike is busy and all the chairs are occupied leave without waiting for a haircut. Mike is the only barber in the shop, and the time to complete a haircut is exponentially distributed with a mean of $1/\mu$ hours.

a. Give the state-transition-rate diagram of the process.
b. What is the probability that there are three customers waiting in the shop?
c. What is the probability that an arriving customer leaves without receiving a haircut?
d. What is the probability that an arriving customer does not have to wait?

12.50 A small company has two servers: A and B. The time to failure for server A is exponentially distributed with a mean of $1/\lambda_A$ hours, and the time to failure for server B is exponentially distributed with a mean of $1/\lambda_B$ hours. The servers also have different repair times. The time to repair server A when it fails is exponentially distributed with a mean if $1/\mu_A$ hours and the time to repair server B when it fails is exponentially distributed with a mean of $1/\mu_B$ hours. There is only one repair person available to work on both servers when failure occurs, and each server is considered as good as new after it has been repaired.

a. Give the state-transition-rate diagram of the process.
b. What is the probability that both servers are down?
c. What is the probability that server A is the first to fail given that both servers have failed?
d. What is the probability that both servers are up?

12.51 Lazy Lou has three identical lightbulbs in his living room that he keeps on all the time. But because of his laziness Lou does not replace a lightbulb when it fails. (Maybe Lou does not even notice that the bulb has failed!) However, when all three bulbs have failed, Lou replaces them at the same time. The lifetime of each bulb is exponentially distributed with a mean of $1/\lambda$, and the time to replace all three bulbs is exponentially distributed with a mean of $1/\mu$.

a. Give the state-transition-rate diagram of the process.
b. What is the probability that only one lightbulb is working?
c. What is the probability that all three lightbulbs are working?

12.52 A switchboard has two outgoing lines serving four customers who never call each other. When a customer is not talking on the phone, he or she generates calls according to a Poisson process with rate $\lambda$ calls/minute. Call lengths are exponentially distributed with a mean of $1/\mu$ minutes. If a customer finds the switchboard blocked (i.e., both lines are busy) when attempting to make a call, he or she never tries to make that particular call again; that is, the call is lost.

a. Give the state-transition-rate diagram of the process.
b. What is the fraction of time that the switchboard is blocked?

12.53 A service facility can hold up to six customers who arrive according to a Poisson process with a rate of $\lambda$ customers per hour. Customers who arrive when the facility is full are lost and never make an attempt to return to the facility. Whenever there are two or fewer customers in the facility, there is only one attendant serving them. The time to service each customer is exponentially distributed with a mean of $1/\mu$ hours. Whenever there are three or more customers, the attendant is joined by a colleague, and the service time is still the same for each customer. When the number of customers in the facility goes down to 2, the last attendant to complete service will stop serving. Thus, whenever there are 2 or less customers in the facility, only one attendant can serve.

a. Give the state-transition-rate diagram of the process.
b. What is the probability that both attendants are busy attending to customers?
c. What is the probability that neither attendant is busy?

12.54 A taxicab company has a small fleet of three taxis that operate from the company's station. The time it takes a taxi to take a customer to his or her destination and return to the station is exponentially distributed with a mean of $1/\mu$ hours. Customers arrive according to a Poisson process with average rate of $\lambda$ customers per hour. If a potential customer arrives at the station and finds that no taxi is available, he or she goes to another taxicab company. The taxis always return to the station after dropping off a customer without picking up any new customers on their way back.

a. Give the state-transition-rate diagram of the process.
b. What is the probability that an arriving customer sees exactly one taxi at the station?
c. What is the probability that an arriving customer goes to another taxicab company?

12.55 Consider a collection of particles that act independently in giving rise to succeeding generations of particles. Suppose that each particle, from the time it appears, waits a length of time that is exponentially distributed with a mean of $1/\lambda$ and then either splits into two identical particles with probability $p$ or disappears with probability $1-p$. Let $X(t), 0 \leq t < \infty$, denote the number of particles that are present at time $t$.

a. Find the birth and death rates of the process.

b. Give the state-transition-rate diagram of the process.

# Appendix: Table of CDF of the Standard Normal Random Variable

**Table 1** Area $\Phi(x)$ under the standard normal curve to the left of $x$

| $x$ | 0.00 | 0.01 | 0.02 | 0.03 | 0.04 | 0.05 | 0.06 | 0.07 | 0.08 | 0.09 |
|---|---|---|---|---|---|---|---|---|---|---|
| 0.0 | 0.5000 | 0.5040 | 0.5080 | 0.5120 | 0.5160 | 0.5199 | 0.5239 | 0.5279 | 0.5319 | 0.5359 |
| 0.1 | 0.5398 | 0.5438 | 0.5478 | 0.5517 | 0.5557 | 0.5596 | 0.5636 | 0.5675 | 0.5714 | 0.5753 |
| 0.2 | 0.5793 | 0.5832 | 0.5871 | 0.5910 | 0.5948 | 0.5987 | 0.6026 | 0.6064 | 0.6103 | 0.6141 |
| 0.3 | 0.6179 | 0.6217 | 0.6255 | 0.6293 | 0.6331 | 0.6368 | 0.6406 | 0.6443 | 0.6480 | 0.6517 |
| 0.4 | 0.6554 | 0.6591 | 0.6628 | 0.6664 | 0.6700 | 0.6736 | 0.6772 | 0.6808 | 0.6844 | 0.6879 |
| 0.5 | 0.6915 | 0.6950 | 0.6985 | 0.7019 | 0.7054 | 0.7088 | 0.7123 | 0.7157 | 0.7190 | 0.7224 |
| 0.6 | 0.7257 | 0.7291 | 0.7324 | 0.7357 | 0.7389 | 0.7422 | 0.7454 | 0.7486 | 0.7517 | 0.7549 |
| 0.7 | 0.7580 | 0.7611 | 0.7642 | 0.7673 | 0.7704 | 0.7734 | 0.7764 | 0.7794 | 0.7823 | 0.7852 |
| 0.8 | 0.7881 | 0.7910 | 0.7939 | 0.7967 | 0.7995 | 0.8023 | 0.8051 | 0.8078 | 0.8106 | 0.8133 |
| 0.9 | 0.8159 | 0.8186 | 0.8212 | 0.8238 | 0.8264 | 0.8289 | 0.8315 | 0.8340 | 0.8365 | 0.8389 |
| 1.0 | 0.8413 | 0.8438 | 0.8461 | 0.8485 | 0.8508 | 0.8531 | 0.8554 | 0.8577 | 0.8599 | 0.8621 |
| 1.1 | 0.8643 | 0.8665 | 0.8686 | 0.8708 | 0.8729 | 0.8749 | 0.8770 | 0.8790 | 0.8810 | 0.8830 |
| 1.2 | 0.8849 | 0.8869 | 0.8888 | 0.8907 | 0.8925 | 0.8944 | 0.8962 | 0.8980 | 0.8997 | 0.9015 |
| 1.3 | 0.9032 | 0.9049 | 0.9066 | 0.9082 | 0.9099 | 0.9115 | 0.9131 | 0.9147 | 0.9162 | 0.9177 |
| 1.4 | 0.9192 | 0.9207 | 0.9222 | 0.9236 | 0.9251 | 0.9265 | 0.9279 | 0.9292 | 0.9306 | 0.9319 |
| 1.5 | 0.9332 | 0.9345 | 0.9357 | 0.9370 | 0.9382 | 0.9394 | 0.9406 | 0.9418 | 0.9429 | 0.9441 |
| 1.6 | 0.9452 | 0.9463 | 0.9474 | 0.9484 | 0.9495 | 0.9505 | 0.9515 | 0.9525 | 0.9535 | 0.9545 |
| 1.7 | 0.9554 | 0.9564 | 0.9573 | 0.9582 | 0.9591 | 0.9599 | 0.9608 | 0.9616 | 0.9625 | 0.9633 |
| 1.8 | 0.9641 | 0.9649 | 0.9656 | 0.9664 | 0.9671 | 0.9678 | 0.9686 | 0.9693 | 0.9699 | 0.9706 |
| 1.9 | 0.9713 | 0.9719 | 0.9726 | 0.9732 | 0.9738 | 0.9744 | 0.9750 | 0.9756 | 0.9761 | 0.9767 |
| 2.0 | 0.9772 | 0.9778 | 0.9783 | 0.9788 | 0.9793 | 0.9798 | 0.9803 | 0.9808 | 0.9812 | 0.9817 |
| 2.1 | 0.9821 | 0.9826 | 0.9830 | 0.9834 | 0.9838 | 0.9842 | 0.9846 | 0.9850 | 0.9854 | 0.9857 |
| 2.2 | 0.9861 | 0.9864 | 0.9868 | 0.9871 | 0.9875 | 0.9878 | 0.9881 | 0.9884 | 0.9887 | 0.9890 |
| 2.3 | 0.9893 | 0.9896 | 0.9898 | 0.9901 | 0.9904 | 0.9906 | 0.9909 | 0.9911 | 0.9913 | 0.9916 |
| 2.4 | 0.9918 | 0.9920 | 0.9922 | 0.9925 | 0.9927 | 0.9929 | 0.9931 | 0.9932 | 0.9934 | 0.9936 |
| 2.5 | 0.9938 | 0.9940 | 0.9941 | 0.9943 | 0.9945 | 0.9946 | 0.9948 | 0.9949 | 0.9951 | 0.9952 |
| 2.6 | 0.9953 | 0.9955 | 0.9956 | 0.9957 | 0.9959 | 0.9960 | 0.9961 | 0.9962 | 0.9963 | 0.9964 |
| 2.7 | 0.9965 | 0.9966 | 0.9967 | 0.9968 | 0.9969 | 0.9970 | 0.9971 | 0.9972 | 0.9973 | 0.9974 |

*Continued*

**Table 1** Area $\Phi(x)$ under the standard normal curve to the left of $x$—cont'd

| | | | | | | | | | | |
|---|---|---|---|---|---|---|---|---|---|---|
| 2.8 | 0.9974 | 0.9975 | 0.9976 | 0.9977 | 0.9977 | 0.9978 | 0.9979 | 0.9979 | 0.9980 | 0.9981 |
| 2.9 | 0.9981 | 0.9982 | 0.9982 | 0.9983 | 0.9984 | 0.9984 | 0.9985 | 0.9985 | 0.9986 | 0.9986 |
| 3.0 | 0.9987 | 0.9987 | 0.9987 | 0.9988 | 0.9988 | 0.9989 | 0.9989 | 0.9989 | 0.9990 | 0.9990 |
| 3.1 | 0.9990 | 0.9991 | 0.9991 | 0.9991 | 0.9992 | 0.9992 | 0. 9992 | 0.9992 | 0.9993 | 0.9993 |
| 3.2 | 0.9993 | 0.9993 | 0.9994 | 0.9994 | 0.9994 | 0.9994 | 0.9994 | 0.9995 | 0.9995 | 0.9995 |
| 3.3 | 0.9995 | 0.9995 | 0.9995 | 0.9996 | 0.9996 | 0.9996 | 0.9996 | 0.9996 | 0.9996 | 0.9997 |
| 3.4 | 0.9997 | 0.9997 | 0.9997 | 0.9997 | 0.9997 | 0.9997 | 0.9997 | 0.9997 | 0.9998 | 0.9998 |
| 3.5 | 0.9998 | 0.9998 | 0.9998 | 0.9998 | 0.9998 | 0.9998 | 0.9998 | 0.9998 | 0.9998 | 0.9998 |
| 3.6 | 0.9998 | 0.9999 | 0.9999 | 0.9999 | 0.9999 | 0.9999 | 0.9999 | 0.9999 | 0.9999 | 0.9999 |
| 3.7 | 0.9999 | 0.9999 | 0.9999 | 0.9999 | 0.9999 | 0.9999 | 0.9999 | 0.9999 | 0.9999 | 0.9999 |
| 3.8 | 0.9999 | 0.9999 | 0.9999 | 0.9999 | 0.9999 | 0.9999 | 0.9999 | 1.0000 | 1.0000 | 1.0000 |

# Bibliography

## Bibliography

Some chapters in the following books provide information on probability, random processes, and statistics at a level that is close to that presented in this book:


Allen, A. O. (1978). *Probability, statistics, and queueing theory with computer science applications.* New York: Academic Press.

Ash, C. (1993). *The probability tutoring book.* New York: IEEE Press.

Bertsekas, D. P., & Tsitsiklis (2002). *Introduction to probability.* Belmont, Massachusetts: Athena Scientific.

Chatfield, C. (2004). *The analysis of time series: An introduction* (6th ed.). Boca Raton, Florida: Chapman & Hall/CRC.

Chung, K. L. (1979). *Elementary probability theory with stochastic processes* (3rd ed.). New York: Springer-Verlag.

Clarke, A. B., & Disney, R. L. (1985). *Probability and random processes: A first course with applications.* New York: John Wiley.

Cogdell, J. R. (2004). *Modeling random systems.* Upper Saddle River, New Jersey: Prentice-Hall.

Cooper, R. G., & McGillem, C. D. (2001). *Probabilistic methods of signal and system analysis.* New York: Oxford University Press.

Davenport, W. B., Jr., & Root, W. L. (1958). *An introduction to the theory of random signals and noise.* New York: McGraw-Hill Book Company.

Davenport, W. B., Jr. (1970). *Probability and random processes: An introduction for applied scientists and engineers.* New York: McGraw-Hill Book Company.

Drake, A. W. (1967). *Fundamentals of applied probability theory.* New York: McGraw-Hill Book Company.

Durrett, R. (1999). *Essentials of stochastic processes.* New York: Springer-Verlag.

Falmagne, J. C. (2003). *Lectures in elementary probability theory and stochastic processes.* New York: McGraw-Hill Book Company.

Freund, J. E. (1973). *Introduction to probability.* Encino, California: Dickenson Publishing Company, Reprinted by Dover Publications, Inc., New York, in 1993.

Gallager, R. G. (1996). *Discrete stochastic processes.* Boston, Massachusetts: Kluwer Academic Publishers.

Goldberg, S. (1960). *Probability: An introduction.* Englewood Cliffs, New Jersey: Prentice-Hall, Inc, Reprinted by Dover Publications, Inc. New York.

Grimmett, G., & Stirzaker, D. (2001). *Probability and random processes* (3rd ed.). Oxford, England: Oxford University Press.

Haigh, J. (2002). *Probability models.* London: Springer-Verlag.

Hsu, H. (1996). *Probability, random variables, & random processes.* Schaum's Outline Series, New York: McGraw-Hill Book Company.

Isaacson, D. L., & Madsen, R. W. (1976). *Markov chains.* New York: John Wiley & Sons.

Jones, P. W., & Smith, P. (2001). *Stochastic Processes: An introduction.* London: Arnold Publishers.

Leon-Garcia, A. (1994). *Probability and random processes for electrical engineering* (2nd ed.). Reading, Massachusetts: Addison Wesley Longman.

Kemeny, J. G., & Snell, J. L. (1975). *Finite Markov chains.* New York: Springer-Verlag.
Ludeman, L. C. (2003). *Random processes: Filtering, estimation, and detection.* New York: John Wiley & Sons.
Maisel, L. (1971). *Probability, statistics and random processes.* Tech Outline Series, New York: Simon and Schuster.
Mood, A. M., Graybill, F. A., & Boes, D. C. (1974). *Introduction to the theory of statistics* (3rd ed.). New York: McGraw-Hill Book Company.
Papoulis, A., & Pillai, S. U. (2002). *Probability, random variables and stochastic processes* (4th ed.). New York: McGraw-Hill Book Company.
Parzen, E. (1960). *Modern probability theory and its applications.* New York: John Wiley & Sons.
Parzen, E. (1999). *Stochastic processes,* Society of Industrial and Applied Mathematics. Philadelphia: Pennsylvania.
Peebles, P. Z., Jr. (2001). *Probability, random variables and random signal principles.* New York: McGraw-Hill Book Company.
Pfeiffer, P. E. (1965). *Concepts of probability theory.* New York: McGraw-Hill Book Company, Reprinted by Dover Publications, Inc. New York in 1978.
Pursley, M. B. (2002). *Random processes in linear systems.* Upper Saddle River, New Jersey: Prentice-Hall.
Ross, S. (2002). *A first course in probability* (6th ed). Upper Saddle River, New Jersey: Prentice-Hall.
Ross, S. (2003). *Introduction to probability models* (8th ed.). San Diego, California: Academic Press.
Rozanov, Y. A. (1977). *Probability theory: a concise course.* New York: Dover Publications.
Spiegel, M. R. (1961). *Theory and problems of statistics.* Schaum's Outline Series, New York: McGraw-Hill Book Company.
Spiegel, M. R., Schiller, J., & Srinivasan, R. A. (2000). *Probability and statistics* (2nd ed). Schaum's Outline Series, New York: McGraw-Hill Book Company.
Stirzaker, D. (1999). *Probability and random Variables: A beginner's guide.* Cambridge, England: Cambridge University Press.
Taylor, H. M., & Karlin, S. (1998). *An introduction to stochastic modeling.* San Diego, California: Academic Press.
Thomas, J. B. (1981). *An introduction to applied probability and random processes.* Malabar, Florida: Robert E. Krieger Publishing Company.
Trivedi, K. S. (2002). *Probability and statistics with reliability, queueing and computer science applications* (2nd ed.). New York: John Wiley & Sons.
Tuckwell, H. C. (1995). *Elementary applications of probability theory* (2nd ed.). London: Chapman and Hall.
Yates, R. D., & Goodman, D. J. (1999). *Probability and stochastic processes: A friendly introduction for electrical and computer engineers.* New York: John Wiley & Sons.

# Index

Note: Page numbers followed by *b* indicate boxes, *f* indicate figures and *t* indicate tables.

Printed and bound by CPI Group (UK) Ltd, Croydon, CR0 4YY

08/05/2025

01864914-0002